Sustainable Textile Chemical Processing

Sustainable Textile Chemical Processing

Edited by
Dr. Javed N. Sheikh
Prof. (Dr.) M. D. Teli

WOODHEAD PUBLISHING INDIA PVT LTD

New Delhi

Published by Woodhead Publishing India Pvt. Ltd.
Woodhead Publishing India Pvt. Ltd.,
303, Vardaan House, 7/28, Ansari Road,
Daryaganj, New Delhi - 110002, India
www.woodheadpublishingindia.com

First published 2022, Woodhead Publishing India Pvt. Ltd.

Woodhead Publishing India Pvt. Ltd. ISBN : 978-81-954048-5-8
Woodhead Publishing India Pvt. Ltd. e-ISBN : 978-81-957618-7-6

Typeset by Bhumi Graphics, New Delhi
Printed and bound in India by Replika Press Pvt. Ltd

Contents

Preface

We are extremely pleased to release our book titled "Sustainable Textile Chemical Processing". The books are the best friends of humankind, and we sincerely feel this book will be an excellent resource for the textile industry in general and textile chemical processing in particular.

We always believed in this Universal Truth that the "*Earth is but one country and Mankind its Citizens*". However, to understand it, in reality, several people required a different kind of lessons. Today at the backdrop of COVID 19, majority of humans has come to this understanding that we are all interconnected, and the only way to live in this world is "cooperation and respect for coexistence".

It is also well known that every good cause and welfare scheme prove to be effective as much as the honesty of those who implement them. So it all depends on the honesty and integrity of the people involved in fund disbursing system. In fact, no efforts towards sustainability can achieve real success unless an ethical commitment is there.

SUSTAINABILITY-in this context makes us think proactively and introspect on our *business as usua*l practices for higher productivity, lower costs and more profits. In the race characterised by a competitive atmosphere, an environment is ignored, social accountability is conveniently bypassed, and profitability is pursued. These acts of ours will eventually bring us again on the verge of extinction if we do not become cognizant of these dangers. The very future of our children for whom we are ready to sacrifice everything, is being brought in jeopardy and hence it is time-yes, Covid19 gives us that time to think about SUSTAINABILITY in our way of living the life in the society.

Friends, this is not the time to criticise any strata of society, including governing bodies; We all have to be blamed for getting into this situation because of following or ignoring unsustainable means to earn fast profits or fix the problems. Now is the time, we must take a lesson from this and see that anyone carrying operations which are dangerous to sustainability are to be nixed in the bud. No one should be in fool's paradise that he or she is safe because they are far-off geographically from the place where they live. Nature will sweep the effects of those unsustainable activities from one corner of the

Globe and bring the adverse impact of the same on the other side of the Globe, and technically no one is safe when nature's fury is at work. Then why in the first place should we be the cause of it?

Being Textile Chemists, we undertook this venture of editing the Book on Sustainability in Textile Chemical Processing. Why sustainability has become so important is for all to see the ill effects of how various countries are following different practices which are not environment friendly and thus are depleting precious water reserves, endangering aquatic life, polluting the environment and causing the contamination of our very food chain. The very existence of human life is thus getting into danger if we continue to follow the textile processing the way we have been traditionally doing it.

Every step of ours in every sphere of activities in textile processing chain has to be considered in terms of sustainability. In other words, while we meet our needs, we do not use unsustainable means. Otherwise, we solve one problem but create many more.

As far as this book is concerned, the various chapters in it, cover different aspects of efforts being put in making the textile chemical processing sustainable. Hence right from understanding the importance of sustainability, it covers various different approaches towards sustainable textile processing; e.g., use of microbial colourants, application of biotechnology and enzymes in pretreatment and finishing, natural dyes and their scope in making operation sustainable, role of right first time approach towards sustainability, various developments which led to minimisation of use of auxiliaries and specialty chemicals and colourants to decrease the effluent load, use of functional colourants, technology of effluent treatment, replacement of harmful chemicals and reuse as well as recycle of specialty chemicals to make the operation more sustainable, technology of conservation of energy, chemicals as well as water, assessment standards to gauge the extent of sustainability achieved, compliance requirement in textile processing and various waterless technologies, herbal finishing, simultaneous dyeing and finishing using functional colourants, digital printing, etc.

The very first chapter of this book precisely discusses the importance of series of various chapters, and thus we chose to restrict this preface for understanding sustainability at a much deeper level of our conscience. Indeed sustainability is the subject wherein while it depends on technology, but more so it depends on the conscious commitment of all the stakeholders, and thus this subject has to have HEARTS of the friends in the right place; once that is guaranteed, rest of the knowhow can be obtained from various sources including this book.

We thank all the esteemed authors who stuck to the timelines and took pains to write their respective Chapters, enabling us to adhere to the final deadline.

Our special thanks are also due to the Woodhead Publishing in India (WPI), because of whom this book is seeing the light of the day.

Dr Javed Sheikh
(C.Col., FSDC), Assistant Professor,
Dept. of Textile and Fibre Engineering,
Indian Institute of Technology, Delhi.
Dr. Mangesh D. Teli
Former Dean, Professor and Head,
Department of Fibres and Textile
Processing Technology,
Institute of Chemical Technology, Mumbai

1

Introduction to sustainable textile chemical processing

Annu[1], Javed N. Sheikh[2*]

[1]Bio/Polymer Research Laboratory, Department of Chemistry, Jamia Millia Islamia, New Delhi-110025, India

[2]Department of Textile and Fibre Engineering, Indian Institute of Technology Delhi, New Delhi-110016, India

*Corresponding Author, Email: jnsheikh@textile.iitd.ac.in

Abstract: Sustainability is a buzzword in today's world. The air, the water, and the soil need to be preserved for future generations, which necessitates a careful look into the sustainable aspects in each engineering discipline, including textiles. Textile chemical processing utilises plenty of water and energy, discharges a large quantity of heavy-load effluents, and utilises some toxic chemicals. Moreover, the working environments might be unhealthy, and all ethical guidelines may not be followed. This necessitates the application of best available practices leading to the ultimate sustainability in textile chemical processing. The chapter summarises the various aspects involved in the sustainable processing of textiles.

1.1 Sustainability and textile chemical processing

Sustainability is mainly a long-term practice for developing stability in terms of integration of three main elements: economic, environmental, and social sustainability. It is a subset of social, economic, and environmental sustainability. United Nations defined sustainability as the development that meets the needs of the present without compromising the ability of future generations to meet their own needs [1-3]. The most obvious, **environmental sustainability** includes the practices that preserve the natural resources to the maximum extent and will not display long-term adverse effects on the environment. **Economic sustainability** is designing and creating the products that are cost-efficient and, look attractive in addition to the usage of environment-caring materials. **Social sustainability** encourages social equity to avoid innate exploitation or to avoid detriment to others by providing benefit to some specific communities [4]. In the quest for rapid industrialisation, nature was neglected by human beings. This led to the development of various harmful effects on the various forms of life. Apart from the living organisms, the environment is also affected which can lead to a danger to the existence of mother earth. This is well understood now and the necessary measures are suggested in the various fields. Even though the economic aspects of successful business remained the prime focus of the various industries, the ecological and social aspects which were neglected initially, are now becoming the major area of concern. The industries are facing strict norms regarding safer work practices, health and safety, and effluent discharges. The user ecology is coming into focus with the increasing awareness of the consumer regarding safety and a need for protection.

Textile chemical processing, which is widely known as "wet processing" always remains one of the most water-intensive industries. The textile industry, one of the highest contributors of pollution throughout the globe for decades, is known for the complexity of hazardous chemicals utilised and the effluent discharged. The toxicity of the chemicals used during the processing of the textile, from cultivation to the production of garments to the market, is not only perishable to the environment; besides, it adversely affects human health and hence generations too [1]. Additionally, textile processing consumes a huge quantity of water and discharges a huge amount of wastewater from the industries. Poor wastewater management causes water pollution, which leads to a scarcity of natural water resources. Currently, the fast conversion in fashion trends has amplified the demand on one hand and reduced the apparel life cycle on the other hand. Looking at the seriousness of the consequences for a long time, scientists and researchers nowadays have turned towards natural or environmentally benign approaches. In this context, biological and biotechnological approaches have gained momentum to minimise the harmful effects and make textiles sustainable [1]. Also, the subsequent demand of consumers for sustainable products has emerged the sustainable practices in textile and its chain supplies [5].

In recent years, textile industries have been facing ecological and economic challenges, therefore leading to the development of advanced and sustainable modern strategies. The development of innovative processes, modification of existing processes, innovative machine designs, and recycling and reuse of chemicals are thus coming into the picture. The requirement for the use of safer dyes and chemicals led to a revival of the interest in natural dyes, research in the field of microbial dyes, and the exploration of eco-friendly alternatives to existing chemicals and auxiliaries. The new methods with the need of lower temperatures and shorter processing time led to energy conservation. Various batch machines with lower liquor ratios, continuous processing machines, and novel concepts like recycling and reuse of water are highly appreciated. The standard methods of evaluation and the labels indicating the eco-friendliness of the product also came in to demand, leading to the development of various standards and eco-labels. Different dry processes of textile chemical processing were also invented and established to suppress the release of hazardous and toxic chemicals.

Some of the consumer brands in garments led the battle, which pressed to supply chain at various levels leading to the implementation of sustainability concepts in the entire supply chain of textiles. The book mainly deals with the sustainable aspects of textile chemical processing, and the following subheading represents the concepts that, in combination, can lead to the achievement of the same. It must be understood that the combined inputs from various fields, including machines, chemicals, dyes, fabric manufacturers, management, and brands, will be required. Standalone development may discourage the concept of sustainability.

1.2 Various techniques to achieve sustainability in Textiles

To describe this, we need to define the factors which are hindering the sustainability in textile chemical processing. The details are provided in individual chapters. The following subsections describe the various technologies and concepts for achieving sustainable textile chemical processing.

1.2.1 Concepts in selecting dyes and chemicals

This involves dyes and chemicals which are toxic either for the user or the environment, water-intensiveness, and energy-intensiveness. To tackle this, the techniques like a replacement of harmful chemicals and dyes with safer ones like enzymes, natural dyes, microbial dyes, dyes which are not releasing harmful carcinogenic amines, non-formaldehyde based finishes, heavy metal-free chemicals, and non-phthalate chemicals can be used. The toxic and hazardous chemicals used during textile processing and their disadvantages are listed in Table 1.1.

Even though some chemicals and auxiliaries show superior performance and cannot be replaced by the available substitutes, the continuous research and developments in the field of textile chemicals will lead to chemistries that can be effective and sustainable. The major focus is on the development of chemicals with lower TDS, COD, BOD, and negligible human and aquatic toxicity. The concepts like carbon footprints during preparation, application and disposal are also crucial while discussing sustainability. Even though sustainability in textile chemical processing is discussed, interdisciplinary and collaborative research is essential to solve many issues.

Table 1.1 Examples of toxic chemicals used during textile processing and their disadvantages

Textile processing	Toxic chemicals	Disadvantages
Sizing	Chlorinated and phenolic compounds, starch paste, PCP as a preservative	Carcinogenic, cause algal blooming in water bodies and skin problems
Spinning and Desizing	Floating fibres and Starches, PCP as a preservative, pyridine based softeners	Cause byssinosis, air pollution, carcinogenic, increase biological and chemical oxygen demand (COD) and biochemical oxygen demand (BOD) and due to algal blooming in water bodies
Bleaching and scouring	Chlorine	Absorbable organic halides causing cancer mutants
Dye fixing and printing	Fixing agents having formaldehyde and metals such as Zn, Ni, Hg, Pb, Cd, Co, Cu and As	Skin irritation, cardiac attack, vomiting, nervous breakdown, anaemia, anorexia as metals can bind with Fe in blood causing reduction in O_2 in blood and may lead to death
Carrier dyeing and pigment printing	Phenol-based carriers and kerosene	Produce non-biodegradable effluent and cause air pollution
Fabric finishing	Formaldehyde	Skin irritation or allergies
Garments to packaging	Stain removers having carbon tetrachloride, CCl_4 and chlorofluorocarbon, CFC	Ozone depletion, therefore, cause skin cancer due to direct exposure of UV rays

Azo dyes remained dominant in the field of textile colouration. However, the dyes which can liberate banned carcinogenic amines are banned. Various dyes based on such banned amines are listed. The screening of azo dyes and intermediates is further required for allergenicity and carcinogenicity. The list of banned amines and the azo dyes synthesised from them will be expanded in the near future. Although some of the azo dyes are banned, a wide range of colours is available based on other intermediates, which can serve the purpose of the textile industry. Natural dyes are also gaining momentum due to increased awareness regarding the possible toxicity associated with some of the synthetic dyes. However, most natural dyes have limitations in terms of availability of limited shades, non-reproducibility of shades, poor affinity for textile substrates, and poor fastness properties. Apart from this, the mordants are required the during application of the majority of natural dyes. The use of metallic mordants generally lowers the sustainability advantage of natural dyes, and optimisation is generally required regarding safer limits in which such mordants can be used. Natural mordants like tannins, chitosan, and hydrolysed proteins can also serve the purpose; however, these mordants have limitations in terms of poor fastness properties of dyed substrates. Continuous research in the area of natural dyeing to solve their current limitations is required. The added advantage of natural dyeing is the impartment of functional properties to textiles by using most of the natural dyes. This can be explored as functional colouration, thus combining dyeing and finishing to lower the environmental impact in separate dyeing and finishing processes.

Microbial dyes are also emerging as new sustainable alternatives. As a secretion product of living beings, microbial dyes and pigments have a distinct advantage in terms of biodegradability and safety. Apart from this, most of the microbial dyes and pigments also possess functional properties that could be utilised for functional colouration. However, the production process is still confined to laboratory scales. Production should be scaled-up to verify the probability of replacement of some parts of synthetic dyes and pigments used today for the colouration of textiles. Combined efforts from biotechnologists and textile chemists are desirable to open a new area of microbial dyes and pigments for the sustainable colouration of textiles.

1.2.2 Enzymes: The sustainable alternatives

Being biological catalysts, enzymes possess significant properties essential for sustainable textile processing. The efficient properties include non-toxicity, a high degree of selectivity, biodegradability, sustainability, environment-friendliness, requirement of mild conditions, catalysis of a broad spectrum of reactions as an active catalyst, harmless by-products, reusability, minimum energy consumption, production in large quantities, etc [6]. Generally, enzymatic processes replace the conventional methods; for instance, in bio-washing, the disposal of pumice stone is a major problem that can be replaced by enzymes. In textile, hydrolases type of enzymes is of main interest. Some of the enzymes used in textile processing and their effects on various fabrics are listed in Table 1.2. Besides, surface modification has also been observed in the case of synthetic fibres such as polyester (polyethylene terephthalate, PET), polyamide, and polyacrylonitrile (PAN). As the enzymes are site-specific and large enough not to penetrate inside the fabric material, no alteration has been found in the inherent properties of the synthetic fibre (PET) when modified with the cutinase enzyme [7].

Table 1.2 Enzymes used in textile processing and their effects on fabric

Enzyme	Textile processing and effects	References
Amylases	Desizing or removal of starch	[8]
Pectinases	Scouring of cotton by hydrolysis of pectin	[9]
Cellulases	Anti-pilling finishing, bio-finishing, biopolishing of cotton fabric, effects on denim, smoothness, softness and lustre	[10–12]
Laccase	Dyeing of wool and nylon fabric	[13]
Proteases	Degumming of silk, removal of oil or stains, anti-felting and shrink resistant of wool,	[14]
Catalases and Peroxidases	Neutralise harmful traces of hydrogen peroxide after bleaching, especially in silk	[15, 16]
Transglutaminase	Modification in wool and leather	[17]
Lipases	Surface modification by moisture regaining and hydrolysis of polyester	[18, 19]
Polygalacturonase	Retting of flax	[20]
Pectinase-rich mixture	Degumming of bast fibres, retting of flax	[21]
Esterase	Hydrolysis and modification in polyester	[22]
Nitrilase	For better colouration of Polyacrylonitrile fabrics	[23]

1.2.3 Biopolymers for sustainable textile processing

As biopolymers have been extracted from natural resources, they provide a sustainable approach in textiles. A list of various biopolymers/biopolymer-derived products utilised in textile processing is depicted in Table 1.3.

Table 1.3 Examples of biopolymers/biopolymer-derived products for textile processing

Biopolymers/biopolymer-derived products used in formulation	Application in textile processing	References
Chitosan	Functionalisation of linen fabric and flame retardant treatment of the wool fabric	[24, 25]
Sericin	Improved moisture content, enhanced dyeability, higher UV-protection in NaOH-pretreated polyester. Radical scavenging and antistatic property on UV excimer irradiated polyester.	[26, 27]
Alginate	Antibacterial finishing of textile	[28]
Cyclodextrin	Coated on denim fabric for transferring active ingredients to the skin and for controlled drug release	[29, 30]

1.2.4 Recycling and reuse of dyes and chemicals

In the cases where total replacement is not possible, techniques like recycling and reuse can be used. Recycling and reuse of chemicals can be beneficial for all chemicals used; however, standardisation of methods for evaluation of accurate concentration of chemicals should be established on a commercial scale. Standing bath techniques can be used to maintain the accurate concentrations of dyes and chemicals in the subsequent processes. Trial and error can also be made to establish the exhaustion of dye from dyebaths and subsequent replenishment required. Although the concept sounds attractive, challenges do exist in the case of a mixture of a variety of chemicals in a bath, and applicability is not available for all chemicals.

1.2.5 Water and energy

The issue of water-intensiveness can be solved using various technologies requiring less quantity of water. The simplest is to optimise the individual processes toward water requirements. The newer processes and machines with low liquor processing can serve the purpose. Obviously, for large large-scale production, the use of continuous processes is beneficial in terms of water and energy consumption; however, this should also be optimised based on the batch size, as for smaller batch sizes, the processes might require a higher quantity of water and energy. Concepts like Right-First-Time (RFT) chemical processing are also advantageous in terms of savings in water, energy, manpower cost and chemicals, as it saves the re-processing cost. The water management techniques like counter current flows in machines, reuse of water (fresh as well as less contaminated), etc., can be utilised. The energy is mainly required to heat the liquors used for processing, boiler operations, and the maintenance of temperatures for drying/curing of textiles. The replacement of processes with the low-temperatures ones can save a lot of energy. The efficient removal of water by centrifuging or mangle squeezing before drying can help in saving the energy required for drying of textiles. The use of renewable sources of energy is also rising in textile industries.

1.2.6 Effluent management

As textile chemical processing is water-intensive, a huge quantity of effluent with a heavy load of contaminants is produced. The various unit operations like desizing, scouring, bleaching, mercerization, dyeing, printing and finishing produce a variety of effluent, and a composite effluent comes in the effluent treatment plant. The effluent is required to be treated with various processes like preliminary, primary, secondary and tertiary processes to meet the discharge characteristics of water as provided by various pollution control norms. Sometimes, the cost of effluent treatment is higher than the cost of textile chemical processing, which ultimately results in uneconomical chemical processing. The effluent treatment plants might not be available with small size textile units. Thus, it is essential to reduce the volume of effluent along with the effluent load. The use of various low-liquor ratio processes and machines can help in this respect. For large batch sizes, the shifting to continuous processes is desirable. The difficult-to-treat chemicals with high COD/BOD, high TDS, and higher toxicity can be replaced with suitable alternatives. The chemical with higher aquatic toxicity can be avoided. The concepts like zero liquid discharge (ZLD) are coming up, which need to be followed by various textile processors. Water is a vital resource that needs to be preserved. The use of a lower quantity of water and treatment of resultant effluent to meet the discharge standards are absolutely essential to achieve ultimate sustainability.

1.2.7 Ethical commitments

As the concept of sustainability requires ethical commitments apart from technological innovations, in the end, ethics plays a major role. The major drivers for sustainable textile chemical processing can be the brands, as they can press the entire supply chain for sustainable practices. The added advantages can be offered by the buyers to those who indulge in sustainable chemical processes. However, to achieve overall sustainability, ethical commitments from all the parts of the supply chain are required.

1.2.8 Innovative technologies to achieve sustainable textile chemical processing

1.2.8.1 Foam technology

In recent years, foam dyeing and finishing have been given major attention as they can save a considerable quantity of water. Foam is simply an air dispersion in a liquid, generally water, a colloidal system made by vigorous mechanical agitation [31]. It is a well-known phenomenon in screen printing, pretreatments, finishing and dyeing. In the foam finishing technique, foam is the medium used to deliver dyes and chemicals for treating porous material with minimal thermal and energy consumption. The processing involves pressure-driven impregnation of the foam inside the substrate and an applicator system for generating a single step, high-speed collapse of the foam. However, the stable foam generation process requires an additional step to break and distribute it upon the fabric material. In this regard, Shen et al. have utilised foam single-face pretreatment to modify silk fabric and improve inkjet printing performance. For that, ethylene base-double octadecyl dimethyl ammonium chloride (EBODAC) agent has been used as a cationic modifier revealing high compatibility with Tween 80 and interaction with silk fabric, leading to the improvisation of inkjet printing performance [32]. Similarly, Song et al. revealed the low take-up and less pollution in water repellent and crease-resistant foam finishing of cotton fabric [33]. Some beneficial foam application techniques have been developed for sustainable processing of the textile, such as kiss roll, horizontal pad and vacuum-suction technique. Additionally, the foam technology provides better colour yield, negligible pollution, and minimal effect on fabric material [31].

1.2.8.2 Supercritical CO_2 ($scCO_2$) dyeing

Generally, CO_2 above its critical temperature and pressure behaves midway between gas and liquid, known as supercritical CO_2, possesses density like liquid and expansion like gas. $scCO_2$ is commercially becoming a vital solvent due to its non-toxicity, chemical stability, low cost, easy availability, non-flammability, recyclability and low environmental impact [34]. Eco-friendly $scCO_2$ illustrates the remarkable potential to develop alternative progressions that can eliminate the use of water and organic solvents. In the textile industry, $scCO_2$ dyeing considerably resolves the problem of wastewater treatment. Its high rate of diffusion and low viscosity allows the dye to penetrate the fabric [35]. Luo et al. revealed high colour strength and fastness properties under moderate conditions for dyeing wool and cotton using $scCO_2$ and reported no effluent release during dyeing [36]. Gao et al. worked for PET and cotton fabrics by circulating $scCO_2$ fluid to facilitate the quick uptake and adsorption of dye molecules on the fabric surface. For this study, the fabric was kept under mild conditions of 353.2 K and 18.0 MPa for 60 min in $scCO_2$ and obtained reasonably good results in colour uniformity, colour strength, and colour fastness [37].

1.2.8.3 Plasma technology

Plasma consists of partially ionized gas having neutral particles, electrons, and ions. The plasma technology allows the fibres to expose to gaseous plasma either by depositing or non-depositing plasma procedures. In respect of textile, plasma technology has opened green and sustainable attainment for heat shielding nano-coating on fabric surfaces [38]. Being a dry, compatible and environmentally benign technology, plasma technology does not use water and is an alternative to a chemical solvent-based conventional method for textile surface treatment. It generally improves capillary action, wettability, surface area, and dyeing ability. The plasma coating used in blood filtration membranes helps the fabric to be antimicrobial, and surface activation provides the hydrophilic properties with enhanced biocompatibility in the case of medical textiles for woven and non-woven fabric [39, 40]. Plasma-enhanced chemical vapours, or depositing plasma involves the application of saturated and unsaturated gases such as ethylene. The physical sputtering and chemical vapour deposition permit straight metallic coating of nanoparticles onto the fabrics due to their low-pressure process. This technology has also been used in finishing to introduce the functionality in textile. Although it is an expensive technology, at the same time, it offers a better production rate, low-cost production, improved results, and, most importantly, environment-friendliness[38–40].

1.2.9 Standards for guideline

It is essential to analyse sustainable practices in various production stages and finished products at various stages of the supply chain. This necessitates the standards dealing with the measurement of sustainability, which can also summarise the criteria to be followed by the industry. Such standards can be obligatory standards, quality standards, voluntary standards, eco-labelling programs, and ethical sourcing or production standards. These standards can help both producers and buyers to select the right sustainable textile products.

Concluding remarks and future perspectives

With fast pacing technologies, the textile is also gradually facing the sustainability procedures in order to prevent the adverse effects on the environment caused by the discharge from the textile industry. Different procedures have been developed and adopted for the sake of environmental, economic, and social sustainability in textiles. Organic fibres and natural fibres extracted from the natural resources have nowadays an attractive tools to resolve the issues associated with textile industries. Besides, the recent awareness among the people has a growing alteration in the market production of the textile industries, and it also brings new income streams to the subsistence of farmers and allows them to utilise their crops fully. Besides, some manufacturers nowadays illustrate their sustainability efforts in their corporate social responsibility (CSR) reports. Hence, an overall effort in attaining sustainability in textile will attract a huge textile market.

References

1. Kumar PS, Suganya S (2017). Introduction to sustainable fibres and textiles. In: Muthu SS (ed) *Sustainable Fibres and Textiles*, Woodhead Publishing, pp 1–18.
2. Kumar V, Agrawal TK, Wang L, Chen Y (2017) Contribution of traceability towards attaining sustainability in the textile sector. *Text. Cloth. Sustain.* 3, 5. https://doi.org/10.1186/s40689-017-0027-8.
3. Assembly UNG (1987) Report of the World Commission on Environment and Development: Our Common Future. https://digitallibrary.un.org/record/139811?ln=en. (Accessed 25 Feb 2020).
4. Sustainability in Textiles: Definition & Design Chapter 2 / Lesson 11. https://study.com/academy/lesson/sustainability-in-textiles-definition-design.html. (Accessed 25 Feb 2020).
5. Karan K, Marco R (2016). Two decades of sustainable supply chain management in the fashion business, an appraisal. *J. Fash. Mark. Manag.* 20, 89–104. https://doi.org/10.1108/JFMM-05-2015-0040.
6. Paul R, Genescà E (2013) The use of enzymatic techniques in the finishing of technical textiles. In: Gulrajani ML (ed) *Advances in Dyeing and Finishing of Technical Textiles*, Woodhead Publishing, pp 177–198.
7. Kanelli M, Vasilakos S, Nikolaivits E, Ladas S, Christakopoulos P, Topakas E (2015). Surface modification of poly(ethylene terephthalate) (PET) fibers by a cutinase from Fusarium oxysporum. *Process Biochem.* 50, 1885–1892. https://doi.org/10.1016/j.procbio.2015.08.013.
8. Chand N, Sajedi RH, Nateri AS, Khajeh K, Rassa M (2014). Fermentative desizing of cotton fabric using an α-amylase-producing Bacillus strain: Optimization of simultaneous enzyme production and desizing. *Process Biochem.* 49, 1884–1888. https://doi.org/10.1016/j.procbio.2014.07.007.
9. Joshi M, Nerurkar M, Badhe P, Adivarekar R (2013). Scouring of cotton using marine pectinase. *J. Mol. Catal. B Enzym.* 98, 106–113. https://doi.org/10.1016/j.molcatb.2013.10.010.
10. Wang HE (2011). Anti-Pilling Finishing of Bamboo Pulp Knitted Fabrics by Cellulase. *Adv. Mater. Res.* 233–235, 1292–1295.
11. Ibrahim NA, EL-Badry K, Eid BM, Hassan TM (2011). A new approach for biofinishing of cellulose-containing fabrics using acid cellulases. *Carbohydr. Polym.* 83, 116–121. https://doi.org/10.1016/j.carbpol.2010.07.025.
12. Uddin MG (2015). Effects of biopolishing on the quality of cotton fabrics using acid and neutral cellulases. *Text. Cloth. Sustain.* 1, 9. https://doi.org/10.1186/s40689-015-0009-7.
13. Pezzella C, Giacobbe S, Giacobelli VG, Guarino L, Kylic S, Sener M, Sannia G, Piscitelli A (2016). Green routes towards industrial textile dyeing: A laccase based approach. *J. Mol. Catal. B Enzym.* 134, 274–279. https://doi.org/10.1016/j.molcatb.2016.11.016.

14. Shen J (2010). Enzymatic treatment of wool and silk fibres. In: Nierstrasz VA, Cavaco-Paulo A (eds) *Advances in Textile Biotechnology*, Woodhead Publishing, pp 171–192.

15. Tzanov T, Costa S, Guebitz GM, Cavaco-Paulo A (2001). Dyeing in catalase-treated bleaching baths. *Color. Technol.* 117, 1–5. https://doi.org/10.1111/j.1478-4408.2001.tb00327.x.

16. Paar A, Raninger A, Costa de Sousa MF, Beurer I, Cavaco-Paulo A, Gübitz G (2003). Production of Catalase-Peroxidase and Continuous Degradation of Hydrogen Peroxide by an Immobilised Alkalothermophilic Bacillus sp. *Food Technol. Biotechnol.* 41, 101–104.

17. Tesfaw A, Assefa F (2014). Applications of Transglutaminase in Textile, Wool, and Leather Processing. *Int. J. Text. Sci.* 3, 64–69.

18. Kim HR, Song WS (2006). Lipase treatment of polyester fabrics. *Fibers Polym.* 7, 339–343. https://doi.org/10.1007/BF02875764

19. Rehman A, Raza ZA, Masood R (2019). Optimization of lipase activity under various chemo-physical conditions for hydrolysis of polyester fabric using multiple statistical approaches. *J. Text. Inst.* 111, 826–834. https://doi.org/10.1080/00405000.2019.1663631.

20. Zhang J, Henriksson G, Johansson G (2000). Polygalacturonase is the key component in enzymatic retting of flax. *J. Biotechnol.* 81, 85–89. https://doi.org/10.1016/S0168-1656(00)00286-8.

21. De Prez, J., Van Vuure AW, Ivens J, Aerts G, Van de Voorde I (2018). Enzymatic treatment of flax for use in composites. *Biotechnol. Rep.* 20, e00294. https://doi.org/10.1016/j.btre.2018.e00294.

22. Biundo A, Ribitsch D, Steinkellner G, Gruber K, Guebitz GM (2017). Polyester hydrolysis is enhanced by a truncated esterase: Less is more. *Biotechnol. J.* 12. https://doi.org/10.1002/biot.201600450

23. Matamá T, Cavaco-Paulo A (2010). Enzymatic modification of polyacrylonitrile and cellulose acetate fibres for textile and other applications. In: Nierstrasz VA, Cavaco-Paulo A (eds) *Advances in Textile Biotechnology,* Woodhead Publishing, pp 98–131.

24. Saini S, Gupta A, Singh N, Sheikh J (2020) Functionalization of linen fabric using layer by layer treatment with chitosan and green tea extract. *J. Ind. Eng. Chem.* 82, 138–143. https://doi.org/10.1016/j.jiec.2019.10.005

25. Cheng X-W, Guan J-P, Yang X-H, Tang R-C, Yao F (2019). A bio-resourced phytic acid/chitosan polyelectrolyte complex for the flame retardant treatment of wool fabric. *J. Clean. Prod.* 223, 342–349. https://doi.org/10.1016/j.jclepro.2019.03.157

26. Gulrajani ML, Brahma KP, Kumar PS, Purwar R (2008). Application of silk sericin to polyester fabric. *J. Appl. Polym. Sci.* 109, 314–321. https://doi.org/10.1002/app.28061.

27. Gupta D, Chaudhary H, Gupta C (2015). Sericin-based polyester textile for medical applications. *J. Text. Inst.* 106, 366–376. https://doi.org/10.1080/00405000.2014.922244.

28. Li J, He J, Huang Y (2017). Role of alginate in antibacterial finishing of textiles. *Int. J. Biol. Macromol.* 94, 466–473. https://doi.org/10.1016/j.ijbiomac.2016.10.054.

29. Issazadeh- Baltorki H, Khoddami A (2014). Cyclodextrin-coated denim fabrics as novel carriers for ingredient deliveries to the skin. *Carbohydr. Polym.* 110, 513–517. https://doi.org/10.1016/j.carbpol.2014.03.008

30. Martin A, Tabary N, Leclercq L, Junthip J, Degoutin S, Aubert-Viard F, Cazaux F, Lyskawa J, Janus L, Bria M, Martel B (2013). Multilayered textile coating based on a β-cyclodextrin polyelectrolyte for the controlled release of drugs. *Carbohydr. Polym.* 93, 718–730. https://doi.org/10.1016/j.carbpol.2012.12.055

31. Capponi M, Flister A, Hasler R, Oschatz C, Robert G, Robinson T, Stakelbeck HP, Tschudin P and Vierlina JP (1982). Foam Technology in Textile Processing, *Rev. Prog. Color. Relat. Top.* 12, 48-57. https://doi.org/10.1111/j.1478-4408.1982.tb00225.x.

32. Shen Q, Chen S, Wang C, Liu C, Tian A (2014). A foam single-face pretreatment to modify silk fabric using EBODAC to improve inkjet printing performance. *J. Text. Inst.* 105, 799–805. doi: https://doi.org/10.1080/00405000.2013.852735.

33. Song MS, Hou JB, Lu YH, Lin J, Cheng DH (2013). Performance of Foam and Application in Foam Finishing of Textile. *Adv. Mater. Res.* 821–822, 661–664. https://doi.org/10.4028/www.scientific.net/amr.821-822.661.

34. Elmaaty TA, Abd El-Aziz, E (2018). Supercritical carbon dioxide as a green media in textile dyeing: A review. *Text. Res. J.* 88, 1184-1212. https://doi.org/10.1177/0040517517697639.

35. Ramsey E, Sun Q, Zhang Z, Zhang C, Gou W (2009). Mini-Review: Green sustainable processes using supercritical fluid carbon dioxide. *J. Environ. Sci.* 21, 720–726. https://doi.org/10.1016/S1001-0742(08)62330-X.

36. Luo X, White J, Thompson R, Rayner C, Kulik B, Kazlauciunas A, He W, Lin L (2018). Novel sustainable synthesis of dyes for clean dyeing of wool and cotton fibres in supercritical carbon dioxide. *J. Clean. Prod.* 199, 1–10. https://doi.org/10.1016/j.jclepro.2018.07.158.

37. Gao D, Yang D, Cui H, Huang T, Lin J (2015). Supercritical Carbon Dioxide Dyeing for PET and Cotton Fabric with Synthesized Dyes by a Modified Apparatus. *ACS Sustain. Chem. Eng.* 3, 668–674. https://doi.org/10.1021/sc500844d.

38. Neisius M, Stelzig T, Liang S, Gaan S (2015). Flame retardant finishes for textiles. In: Paul R (ed) *Functional Finishes for Textiles: Improving Comfort, Performance and Protection*, Woodhead Publishing, pp 429–461.

39. Tessier D (2013). Surface modification of biotextiles for medical applications. In: King MW, Gupta BS, Guidoin R (eds) *Biotextiles as Medical Implants*, Woodhead Publishing, pp 137–156.

40. Lakshmanan SO, Raghavendran G (2017). Low water-consumption technologies for textile production. In: Muthu SS (ed) *Sustainable Fibres and Textiles,* Woodhead Publishing, pp 243–265.

2

Application of enzymes for sustainable textile chemical processing

Edward Menezes[a*], Nagender Singh[b], Ashok Athalye[c]
[a]Rossari Biotech Limited, Kanjurmarg (W) Mumbai 400 078.
[b]Department of Textile and Fibre Engineering, IIT Delhi, New Delhi, India
[c]Department of Fibres and Textile Processing Technology, Institute of Chemical Technology, Mumbai, India
*Corresponding Author, Email: edward@rossarimail.com

Abstract: Textile wet processing is considered to be one of the largest sources of industrial wastewater pollution. Textile manufacturers use several strong and difficult-handle chemicals, many of which are hazardous and have been associated with health-related problems to humans and detrimental ecological impact on nature. Replacement of such polluting chemicals with non-hazardous, easy-handle, and eco-friendly biocatalyst is emerging as a major solution to overcome the environmental threat. Enzymes are such eco-saviours and are rapidly gaining global recognition owing to their non-toxic, non-polluting, biodegradable characteristics and practical application under the reduced utility of water, energy, and time. The combined application of different enzymes is a new trend in textile processing, which further helps in the reduction of the number of processing steps and increases productivity. Immobilization of enzymes is an emerging area of research that helps to enhance the activity and reusability of enzymes for their long-lasting and improved applications.

2.1 Introduction

Enzymes are biocatalysts, which have been broadly accepted by various industrial applications because of their green chemistry and ecological benefits. Currently, enzymes are becoming an important element in the textile industry, specifically in wet processing, because of their unique characteristics to replace conventionally used harsh chemicals. Different types of enzymes are being used on a variety of textile substrates and at different processing stages. However, the lack of long-term storage stability, typical process conditions, and high cost make their application difficult. Nevertheless, enzymes are environment-friendly, save precious water and energy and work under mild conditions. They were brought in an era of white biotechnology for practicing eco-friendly usage through renewable resources [1–3].

The enzyme term originated from the Greek word *enzume,* meaning "in (en) yeast (zume)," and the term *enzyme* was introduced by Kirchoff in 1811 [4]. It is considered that the first purified and crystallised enzyme was prepared in 1926 by James Summer of Cornell University. After that, various researchers have studied and highlighted the properties, classification, and structure of enzymes. Nowadays, enzymes are commercially used in many applications such as textiles, detergents, leather, and paper industries. Moreover, in advanced applications, they are used for dairy, brewing, protein hydrolysis, distillery, and animal feed. Enzymes comprise of amino acids, which are obtained from living organisms. The large protein molecules are made up of long chains of about 200 to 250 amino acid monomers and are held by peptide bonds. It loops and folds by itself into secondary, tertiary, and often quaternary structures, exposing its complementary surface to the substrate with which it is expected to react. Enzymes act as highly efficient catalysts, which help in accelerating the chemical reaction [5,6].

Conventionally, the enzymes were used by simply adding them in a process and later drained in an effluent once they were inactive. Such conventionally used enzymes were neither efficient nor sustainable. To make the process efficient and sustainable, various new-age and tailor-made enzymes were developed, which could work under customized process conditions (pH, temperature, and auxiliaries) [7].

Though the use of enzymes in cotton desizing was established decades ago, the applications have widened with the introduction of new enzymes. Now, enzymatic processes are used in almost all steps of textile production, starting from desizing to bio-finishing [8,9,10,11]. Therefore, with the increased awareness and regulation about environmental concerns regarding hazardous and polluting chemicals, the use of enzymes has become an obvious choice. Enzymes, being biocatalysts, are particular in their specific application used in minimum quantities and have a direct impact on reduced storage space and lower transportation costs. Thus, the enzymes are considered a boon to textile wet processing [10,11].

The objective of this chapter is to discuss the critical aspects of enzymes. This chapter also reviews the latest development in the application of enzymes in the textile industry and highlights the innovative approaches used to eliminate harsh chemicals from the textile industry.

2.2 Role of enzymes in the textile industry

The introduction of enzymes in textile processing has great potential to reduce environmental issues associated with chemicals. The enzymes in the textile industry were used for a long time; however, the researchers are focusing on the modified and new enzymes to enhance the process efficiency [12].

Bio-desizing of cotton is the classic example of an application of enzyme in textile processing. In the early 1950s, amylase was the first microbial enzyme introduced by Novo Nordisk [13]. Since then, various enzymes based systems have been well-established in the textile industry, like amylases for the desizing of cotton, cellulases for the bio-finishing of cellulosic textile, catalases for the peroxide neutralisation, and proteases for the degumming of silk as well as anti-pilling and softening of wool [12,14]. The enzymes like xylanase, pectinase or hemicellulases are extensively used for the retting of bast fibres (jute, flax, ramie) [15,17,18]. In addition to that, hydrolases class of enzymes are reported on modification of synthetics textile material to impart antistatic properties and hydrophilicity [16]. Moreover, enzymes are used with detergents in garment laundering, which can remove a variety of stains from the garment [17].

Recently, researchers have been investigating the utilisation of enzymes in a combined desizing, scouring and bleaching of textile, which can eliminate the separate processes and also save precious water and energy [18,19]. Furthermore, the textile chemical processing unit is a highly chemical-intensive unit in the textile industry. The effluent discharged from the unit is highly toxic, non-biodegradable, coloured, and contains heavy metals, inorganic salts, acids, alkalis and other components. Therefore, researchers are investigating the novel enzymes which can be used for the textile effluent treatment [20,21].

2.3 Enzyme classification

The enzyme classification by the International Biochemical Union, provides every enzyme an "EC number" with 4 digits [22,23]. The enzymes are categorised into 6 classes, as shown in Table 2.1, based on the degree of specificity to catalyse a particular reaction and the reaction methodology. The enzymes are then numbered with 4 digits, as shown in Table 2.2.

Table 2.1 Classification of enzymes

EC number	Reaction methodology
EC 1	Oxidoreductase (oxidation or reduction of substrates)
EC 2	Transferase (group transfer reaction)
EC 3	Hydrolase (bond breakage, splitting of the substrate)
EC 4	Lyase (removes a group from substrate)
EC 5	Isomerase (bring about intra-molecular rearrangement)
EC 6	Ligase (joining of 2 molecules)

Table 2.2 The enzymes numbers with 4 digits as a class and sub-class

Digits	Enzyme Class
1st	Class
2nd and 3rd	Sub-class
4th	Order number in the sub-class

2.4 Mode of enzyme action

Enzymes, being a protein, act as highly efficient catalysts to increase the rate of a biochemical reaction. The enzyme-catalysed biochemical reaction, starts with a binding of the active site of the enzyme to the substrate, followed by a change in the electron distribution in the chemical bonds of the substrate, which ultimately results in the reaction that leads to the development of products. Then the developed products are liberated from the enzyme to regenerate the enzyme for another biochemical reaction cycle.

Enzymes show the maximum performance at an optimised reaction condition. The reaction rate increases with increasing temperature and the activity decrease sharply above the optimum value until a point where they become permanently deactivated [24].

2.4.1 Lock and key theory

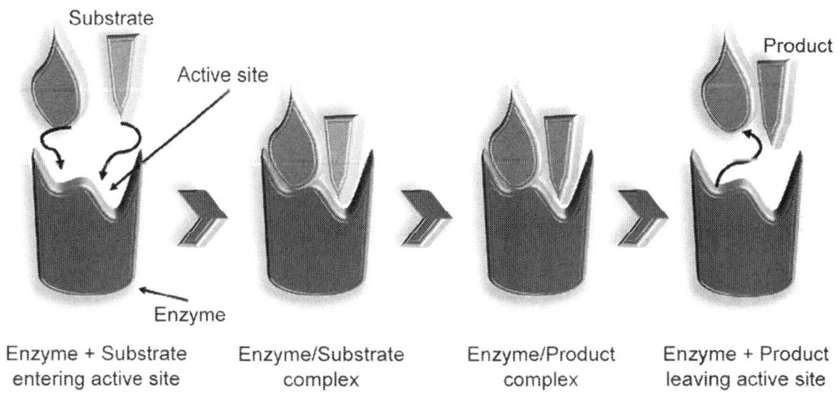

Figure 2.1 Hypothesis of lock and key theory

The active site of the enzyme has a unique geometrical shape, which is similar to the geometric shape of substrate molecules. Therefore, the enzymes exactly react with only those compounds which are similar to substrate molecules. In the lock and key theory, the enzyme acts as a lock and the substrate as a key. The adequately sized key fits into the lock's keyhole. The incorrect size of keys cannot open the lock. Thus, only the proper size and shape of keys opens up the lock's keyhole [25]. The hypothesis of the lock and key theory is shown in Fig. 2.1.

2.4.2 Induced fit theory

The substrate determines the final shape of enzymes because it is contemplated that the enzymes are partially flexible, which can take a similar shape as that of the substrate. However, it is considered that some compounds can be attached to the enzyme but still do not react when the enzymes have different shapes. Therefore, according to induced fit theory, only a correctly shaped substrate can persuade the proper shape of the active site of the enzyme, as shown in Fig. 2.2 [25].

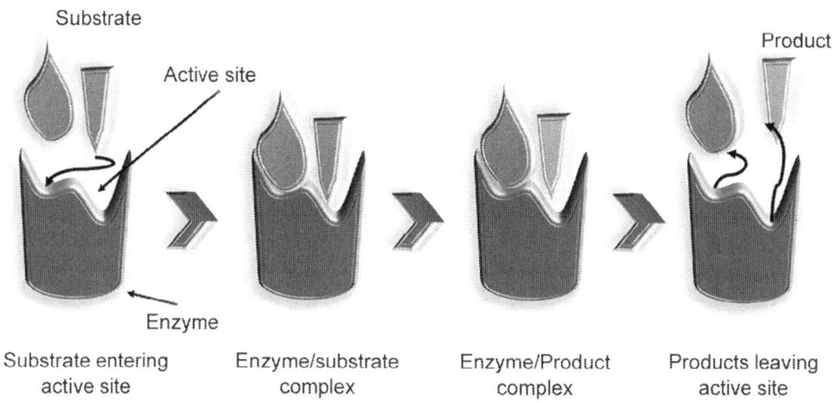

Figure 2.2 Hypothesis of induced fit theory

2.5 Application of enzymes in textile industry

Enzymes are manufactured by a fermentation process that is carried out under controlled conditions such as CO_2, pH, nutrients, feed additives, and oxygen. Generally, for a peak production level of such fermentation, it requires around 5-9 days, followed by filtration and transformation into a storage stage for commercial application. In the textile industry, a specific number of enzymes are used, and their benefits are enormous, as shown in Table 2.3.

Table 2.3 Summary of enzymes and their usage in textile [6,12,23,26,27]

Enzymes	Application	Benefit
Amylase	Desizing of cotton	Improves speed, economics and consistency of the process. Use of thermostable enzyme, characterisation of the process, etc., are recent developments in the amylase applications.
Cellulase	Bio-polishing and Bio-fading	Removes surface fuzz (protruding fibres), surface pills (balls of entangled fibres) from cotton and viscose fabrics. Imparts cleaner, softer, smoother look and feel. Provides effective colour fading of denim. Replaces pumice stone used to get characteristic abraded, faded appearance.
Catalase	Neutralizes hydrogen peroxide	Quenches residual peroxide. Reactive dyes are especially sensitive to peroxides and require extended rinsing and/or use of chemical scavengers.
Pectinase	Carbonization of wool, Scouring of cotton and Retting of flax	Degrades vegetable matter in wool. Replaces conventional treatment using a strong acid, followed by mechanical crushing. Removes pectineous impurities from cotton and enhances absorbency. Achieves rapid and uniform retting and avoids the risk of bacterial or fungal contamination which tends to occur in case of dew and water retting.
Laccase	Denim finishing	Decolourises indigo dyestuff and enhances the apparent abrasion effect with little or no impact on cellulosic fibre strength.
Lipase	Desizing	Removes triglyceride-based size lubricants from fabrics.
Protease	Wool finishing and Silk degumming	Induces sand-washed effect on silk garments. Enhances comfort (reduced prickle, greater softness), improves the surface appearance and pilling performance.
Xylanase	Scouring and bleaching	Removes wax, colour, residual seed coatings, etc., which otherwise inhibits the natural absorbency of the fibre and prevents subsequent dyeing, printing or finishing of cotton yarns and fabrics.

Advantages of enzymatic textile processing:

- Enhance the rate of chemical reactions
- Work under specific process conditions such as pH, pressure, and temperature
- Reduce impact on polluting solid waste, liquid effluent and gaseous emission

- Ensure bio-degradable characteristics and save the environment
- Save water, energy and time
- Minimise effluent load, in terms of BOD, COD and TDS

However, there are factors which need to be considered, monitored and controlled in order to achieve optimum performance efficiency of enzymes.

- Active over a narrow range of temperature and pH
- Work until get deactivated
- Once destroyed, cannot be reactivated
- Certain antiseptics, acids and alkalis tend to hamper the enzyme activity
- Certain enzymes may cause respiratory problems and skin and eyes irritation
- Many enzymes get affected by ionic nature (cationic or anionic) of other chemicals
- Heavy metal ions tend to deactivate particular type of enzymes

Considering the above factors the storage, usage and industrial application of enzymes need to be carried out under precise and specific pH, temperature, time and conditions taking into account the machinery and the nature of other chemicals used in combination.

2.5.1 Enzymes for cotton processing

As a natural cellulosic substrate, cotton is one of the most widely used fibre in textile and is processed in various forms depending on the end-use requirement. The wet processing of cotton involves various steps, from desizing to the finishing that needs a different type of biocatalyst. Given below are some major enzyme categories and their application usage.

2.5.1.1 Bio-desizing

In the case of woven fabric, terry towel and garment manufacturing, sizing of warp yarn is carried out to prevent their breakage during subsequent processes. During further processing, it becomes essential to remove such applied size from the surface of cotton, which otherwise hinders the effectiveness of wet processing steps. Desizing means the removal of size material from grey cotton fabric and it allows further wet processing steps to occur at optimum efficiency, as shown in Fig. 2.3. Therefore, desizing of cotton is considered to be an important step, which may decide the performance of subsequent steps such as scouring, bleaching, dyeing, printing and finishing [9,28].

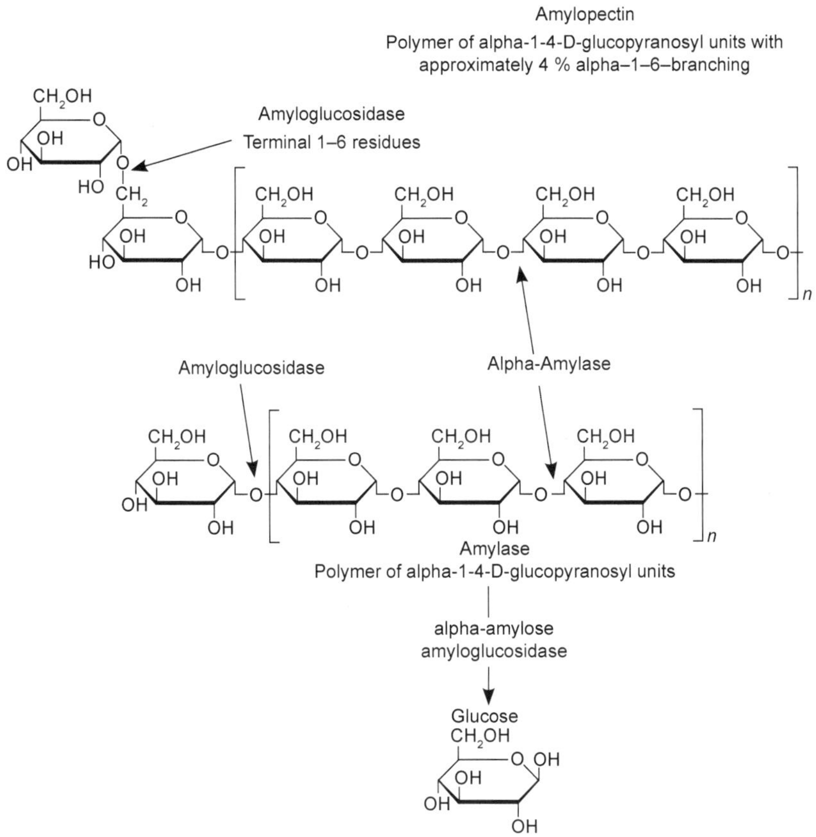

Figure 2.3 The hydrolysis of starch to glucose.

Conventionally, inorganic acids and oxidizing agents are used to desize the size material from the yarns. However, the application of enzymes selectively removes the size material explicitly. For example, amylases have been classified based on their working mechanism. α-amylases attack randomly on the cleavage of 1,4-glucosidic linkage, while β-amylases release maltose by stepwise hydrolysis, and amyl glucosidases split both amylopectin and amylase through stepwise hydrolysis which removes D-glucose from the non-reducing end [2,29].

Various researches and literature reviews are available, which discuss the conventional and advanced applications of enzymes for the desizing of textile [1,2,23]. Sreelakshmi et al. prepared the low-temperature acidic amylases from *Aspergillus niger* and *Aspergillus flavus* for desizing of cotton. Fig. 2.4 shows the effect of starch materials on amylase production during incubation.

The results showed a higher desizing efficiency with significant improvement in absorbency [30]. Rehman et al. optimised the enzymatic desizing of cotton using the Taguchi design. The results showed that the cotton fabric exhibited maximum TEGEWA rating; however, minimum bending length, weight loss, and absorbency [31]. Chand et al. investigated the simultaneous enzyme production and optimised desizing of cotton fabric using α-amylase [32].

Efforts have been made to combine two or more processes in a single step; therefore, researchers have explored the combination of pre-treatments and dyeing process in a single step to develop a sustainable system. Lenting et al. used a mixture of enzymes containing an α-amylase and a pectate lyase to obtain the combined desizing and scouring [33]. Aly et al. used a mixture of α-amylase and polygalacturonase enzymes to investigate a one-step desizing and scouring of cotton fabric [18]. Ali et al. reported an integrated three process (desizing-scouring-reactive dyeing) in one step for cotton fabric using glucose oxidase enzymes [19].

Some researchers have tried to improve the effectiveness of enzymes using the assistance of salt or ultrasonic waves, which helps in reducing dosage and optimised process conditions [34–37, 42]. There are many studies related to the application of immobilized amylases; however, the uses of immobilized amylases in desizing are limited because they are not able to work on the macromolecular substrates [38,39].

2.5.1.2 Bio-scouring

Generally, cotton contains about 8-10 % of natural impurities in the form of oil, fat, wax, pectin, and protein, and the process which involves the removal of these impurities is called scouring. Conventionally, scouring is done with the help of an active chemical like caustic soda under high-temperature conditions. Such alkaline scouring tends to generate high COD, BOD and TDS load on effluent. It consumes much water for washing, energy for heating, and hampers the hand-feel of cotton. The safe and effective alternative is the enzymatic process which is called as bio-scouring. As shown in Fig. 2.4, the pectinase enzyme catalyzes 1,4-alpha-D-galactosiduronic linkages by random hydrolysis in pectin. Protopectinase, pectinesterases, polygalacturonases, and pectin lyases are the four types of enzymes used to hydrolyse the pectin substances [40,41].

The bio-scouring process removes pectin from cotton and helps loosen out other impurities like wax, which subsequently gets removed during washing. The ultimate benefit is in terms of achieving the desired removal of impurities from cotton, attaining adequate water absorbency, controlling weight loss, minimising the impact on tensile strength and optimising the hand feel of the fabric surface [42,43]. There are various studies on the application of pectinase for the souring of cotton fabric [42,44]. Agrawal et al. designed an enzymatic scouring process for the removal of wax from cotton fabric using alkaline pectinase. The results showed that alkaline pectinase displayed a 75% higher effect as compared to acidic pectinase [40]. Rajendran et al. explored and optimised the bio-scouring of cotton using pectinase and compared it with conventional scouring [45]. The acidic pectinase was used for bio-scouring of cotton with varied process parameters (pH, temperature, and surfactant) was also investigated [46].

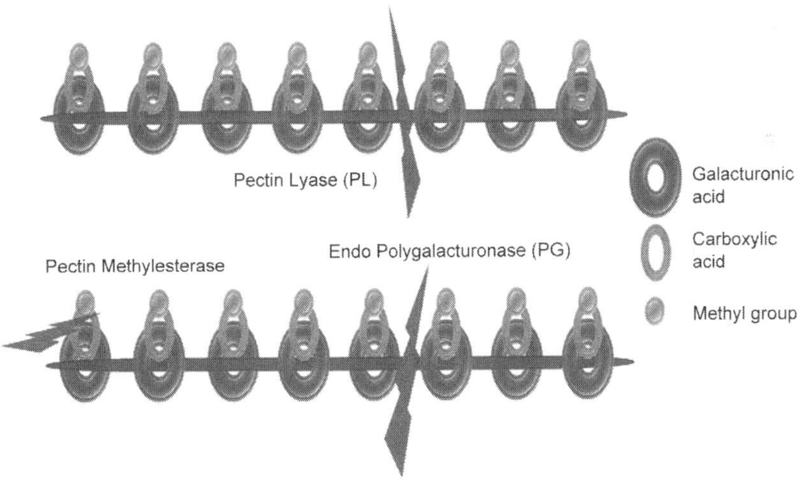

Figure 2.4 Mode of action of the main pectolytic enzymes

Protease-based enzymes catalyse the hydrolysis of proteins and their removal from the fibre. Depending on the type of protease, it effectively removes the proteinaceous materials from the lumen, which is the innermost part of cotton, as well as the proteins embedded beneath the surface layers. Efforts have been made on the application of protease for bio-scouring of cotton; however, it has only a slight effect on improving wettability, and there is no substantial change in surface friction and the tensile strength of cotton [41,47,48]

The cotton bio-scouring was carried out using a mixture of pectinase with other enzymes, which displayed a notable improvement in the overall performance of bio-scouring [49,50]. Lipase hydrolyses the fat present in cotton fibre, where the common fats are fatty acids and glycerol. During the bio-scouring lipase hydrolysed the ester bonds, which resulted in the regeneration of the water-insoluble fatty acid and water-soluble glycerol.

However, bio-scouring using lipase does not show significant improvement in water retention and absorbency property of cotton but helps in reducing the friction coefficient without affecting tensile strength [51].

The one-step bio-scouring and bleaching process for pre-treatment of cotton has also been explored by researchers [52,53]. Preša et al. investigated a single-bath bio-scouring and bleaching process using pectinase enzyme in the presence of peracetic acid. A significant amount of wax and pectin was removed from the fibres without damaging the fibre surface [54].

2.5.1.3 *Bio-bleaching*

Bleaching is a very important process in cotton pre-treatment; it involves the removal of natural colouring components from fibre and enhancing the whiteness of the material. Bleaching is carried out by oxidative or reductive methods, and H_2O_2 is the most commonly used oxidative bleaching agent. Moreover, conventional bleaching requires the use of high amounts of alkaline chemicals and generates massive effluent due to repetitive water rinsing to get rid of residual alkali and peroxide. The enzymatic bleaching replaces hydrogen peroxide and helps reduce the impact on effluent in terms of less alkali and water wastage [55,56].

Further, this can help carry out dyeing in the same bath resulting in reduced water consumption which in turn requires less power to dye fabric which eventually reduces the amount of effluent produced [57]. Therefore, the ultimate benefit is in terms of conservation of water, energy and chemicals [58]. The enzymatic bleaching is commonly carried out through glucose-oxidase, laccase/mediator systems and peroxidase enzymes, as shown in Fig. 2.5. The bleaching of cotton carried out using laccase/mediator systems specifically targets the colouring components on the fabric[1,58].

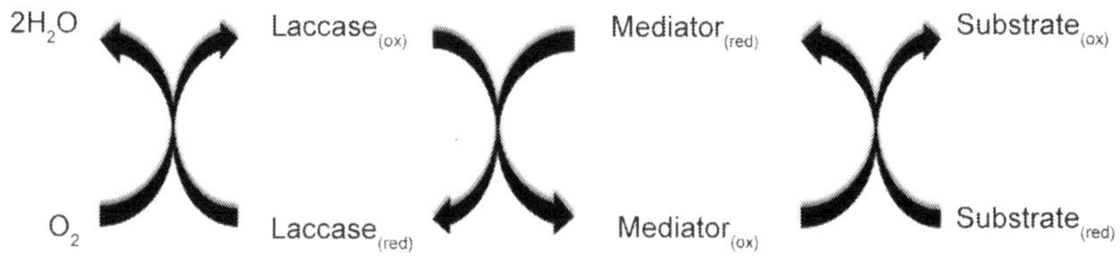

Figure 2.5 Schematic presentation of the laccase–mediator system [58]

The glucose oxidase enzymes belong to the oxidoreductase category, which helps reduce a substrate by transfer of hydrogen(s) and/or electron(s). These enzymes tend to form H_2O_2 and gluconic acid from oxygen and glucose [19,59]. Thus, glucose is oxidised by glucose oxidase to H_2O_2 and gluconic acid in the presence of oxygen [59,60]. Shin used glucose oxidase for enzymatic bleaching of scoured cotton by producing H_2O_2. It has been found that the obtained whiteness index was comparable with commercially available hydrogen peroxide (H_2O_2) [61]. Tzanov et al. introduced a bio-process for single-step scouring and bleaching of cotton fabric with the help of enzymatically generated peroxide [62]. In another study, Tzanov et al. used immobilized glucose oxidase enzymes for the generation of H_2O_2 for bleaching of cellulosic textile [63]. Eren and colleagues have developed a novel process of one bath desizing-bleaching and dyeing. The desizing was performed by amyloglucosidase/pullanase to hydrolyze the size material into glucose, followed by the generation of H_2O_2 with the help of glucose oxidase; after that, hydrogen peroxide present in the bath was decomposed by using catalase enzyme. Finally, the same bath was utilised for dyeing of the fabric using monochlortriazine reactive dyes [64].

Peroxide quenching is essential after bleaching. If the residual peroxide on the fabric is not neutralised, it results in fabric tendering and uneven dyeing. Conventional reducing agents used for peroxide neutralization are non-eco-friendly and tend to increase the TDS of the effluent. Catalase-based enzymes offer an effective solution for bleach clean-up [64]. The catalase enzyme is a hydro-peroxidase enzyme that decomposes the H_2O_2 into water and oxygen. Catalases are obtained from a variety of microorganisms and have optimum activity at neutral pH and moderate temperatures of 50 ^{o}C. Each molecule of catalase is capable of processing around five million molecules of hydrogen peroxide per second. Apart from their specificity, high catalytic power and eco-friendliness, they do not interfere in dyeing, and hence dyeing can be continued in the same bath [20,21]. Therefore, Amorim et al. explored the elimination of H_2O_2 residue after bleaching of cotton by using catalase enzymes. The results showed that the catalase treated system significantly reduced the water consumption during the dyeing of cotton [21]. Tzanov studied the enzymatic cleaning of hydrogen peroxide from a bleaching bath and used the same bath for dyeing of fabric [65].

2.5.1.4 Bio-Polishing

The surface appearance and feel of cotton fabric depend on the extent of surface fuzz and loosely held fibres. The removal of such surface hairiness is essential to enhance the aesthetic appeal of the textile material. This is achieved by the enzymatic bio-polishing treatment using cellulase enzyme.

Bio-polishing is mostly carried out after bleaching because at this stage, the fabric is clean, hydrophilic and more accessible to dyes. However, if the bio-polishing is performed after dyeing, there could be a risk of shade change and strength loss of dyes; also, some colourants tend to reduce the performance of enzyme, which necessitates a higher concentration of enzyme. The cellulase enzymes are widely used for bio-polishing and modifying the surface properties of cellulosic fibres, as shown in Fig. 2.6 [66,67].

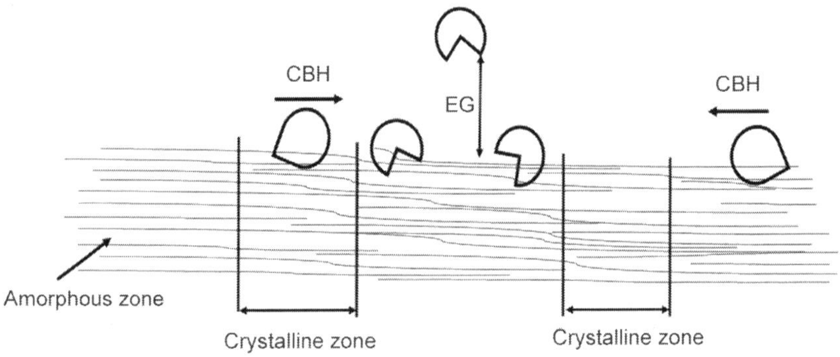

Figure 2.6 Action of cellulases on cotton

Mechanism of cellulase action

Commercially cellulase is the mixture of cellobiases, endoglucanses and cellobiohydrolases. Endoglucanases attack cellulose randomly and hydrolyse internal glycosidic bonds, and the cellobiohydrolases remove terminal cellobiose residues from both cellulosic chain ends, while cellobiases hydrolyse small cellobiose to glucose. Various studies have been conducted to explain the mechanism of cellulase for the bio-polishing of cellulosic textiles. Generally, it is considered that there is an adsorption of cellobiohydrolase, β-glucanase or endo-glucanase onto the fibre surface, which results in the formation of a complex with the water and cellulose chain. After hydrolysis, the enzyme is desorbed and is available for subsequent adsorption and reaction. The cellulase mixture works synergistically, where endoglucanase successfully opens up fibre structure to attack by cellobiohydrolase and β-glucanase, which results in a reduction of fibre strength, as shown in Fig. 2.7 [66,68].

There are several studies that reported the bio-polishing of cellulosic material to enhance the comfort and aesthetic of fabric [69,70]. Ibrahim et al. developed a new approach for bio-finishing of cellulosic textiles using acidic cellulase enzyme. The obtained results showed a smooth and soft cotton fabric surface [69]. Esfandiari and co-workers studied the effect of mechanical treatment after combined enzymatic desizing and bio-polishing using rocolase and amylase enzymes. It was observed that the combined enzymatic and mechanical treatment displayed more fabric weight loss and more water absorption as compared to fabric treated without mechanical treatment [71]. Tavcer bio-scoured the terry-towel, followed by bleaching and bio-polishing with cellulase mixture and endoglucanase. The results showed that the towel bio-polished with cellulase deteriorated the properties of terry-towel [72]. Pazarlioğlu and colleagues used immobilized cellulase with chemically modified pumice particles for denim washing. The indigo-dyed denim was efficiently abraded by immobilized acid cellulase [73].

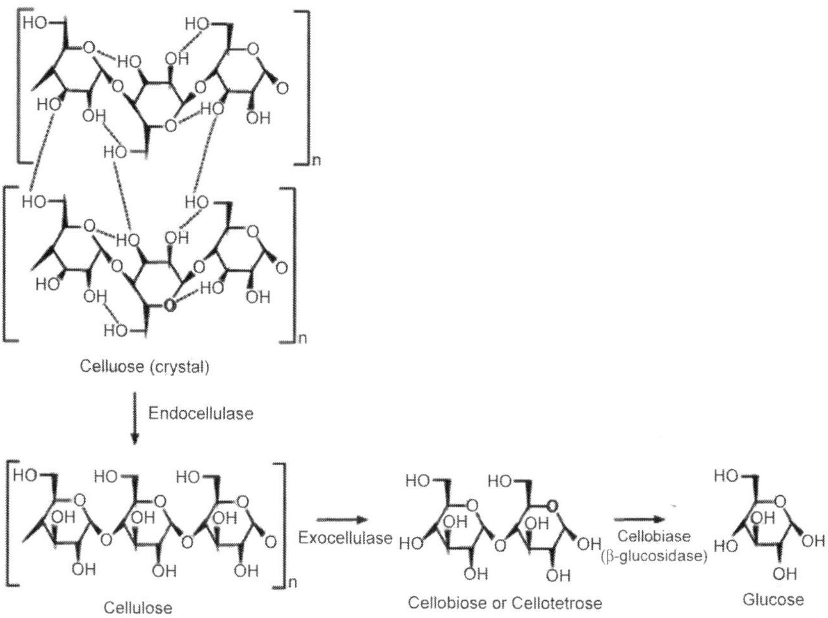

Figure 2.7 Mechanism of cellulase on cellulose

Similarly, Yu et al. studied the effect of bio-polishing on denim using immobilized cellulase with methacrylate copolymer and native cellulase. The results suggest that the denim finished with immobilized cellulase displayed lower tensile strength and weight loss as compared to denim finished with native cellulase [74]. Hao et al. cationized the cotton fabric using a cationic agent and bio-polished it with cellulase to enhance the effect of cellulase on cotton [75].

Benefits of enzymatic Bio-Polishing

Enzymatic removal of the surface fibre fuzz is wash-durable, and it enhances the dyeability, drapeability and hand-feel of cotton. The significant benefits of cellulose-based enzymatic bio-polishing are:

- Prevents pills formation
- Increases smoothness and softness
- Increases lustre and colour brightness
- Improves handle and drapability

2.5.2 Other important textile applications of enzymes

2.5.2.1 *Bio-washing off*

Generally, the post dyeing or printing washing-off is a lengthy, water, time and energy-consuming process, and it generates a large amount of effluent. The use of laccase-based enzyme in soaping removes unfixed reactive dye, and the soaping time can be shortened to reduce the amount of discharge, which contributes to environmental protection and reduce the cost of effluent treatment. The wet fastness properties are found to be comparable to the conventional washing-off chemicals [76].

In the textile industry, the application of laccases enzymes for the degradation of dyes from effluent has been extensively explored by a number of researchers [77–79]. Tavares et al. studied the utilisation of laccases enzymes for the degradation of reactive dye from the effluent [80]. Cristóvão et al. used kinetic modelling based on Michaelis–Menten equation for simulation of laccases behaviour on decolourisation of reactive dyes [81].

2.5.2.2 *Seed coat removal*

Seed coat fragments are the impurities present in cotton, which are difficult to remove entirely even after a harsh chemical wash, and the fragments are bleached in the course of fabric preparation. The enzymatic treatment of cotton with xylanase helps in the hydrolysis of tiny fibres of seed-coat fragments, which make the seed-coat fragments easily accessible to chemicals. The xylanase treatment of cotton reduces the overall consumption of hydrogen peroxide in the chemical bleaching process [49,50]. Dhiman produced xylanase from *Bacillus stearothermophilus* and used it for the bioprocessing of cotton fabric [82]. Losonczi and co-workers studied the role of EDTA chelating agent in the performance improvement of the enzymatic process. The result indicates the synergetic effect of EDTA and enzymes [83].

2.5.2.3 *Wax removal*

It is important to remove the cuticle waxes from the cotton fibre to improve enzymatic scouring. The cuticle waxes are cross-linked to the primary cell wall through an esterified pectic component, which inhibits pectinase enzymatic action on the pectin present on cotton fibre. The the cutinase enzymes from bacteria degrade wax during cotton scouring, and the combined application of cutinase and pectinase synergistically enhanced scouring process [84,85]. Agrawal et al. performed an ultrasound-assisted enzymatic bio-scouring of cotton using cutinase and pectate lyase enzymes. The result suggests that the ultrasound-assisted enzymatic process significantly reduced the reaction time [86]. In another research, Agrawal et al. investigated the role of mechanical action in improving the bio-scouring process. The use of mechanical action with cutinase has effectively degraded the wax from the cotton, which results in the reduction of pectinase incubation time [87].

2.5.2.4 *Linen fibre processing*

Enzymatic retting for linen fibre processing is an important process that effectively and precisely separates fibres from the stem. Enzymatic retting has several advantages as compared to water-retting, such as a reduction in water consumption, and the process is cost-effective, odourless and eco-friendly. Various researchers have studied the application of pectinase-containing enzyme mixtures for the retting of linen [88,89]. Zhang et al. studied the enzymatic retting of flax and have identified that the polygalacturonase is a crucial element for the retting of linen [90].

2.5.2.5 *Silk degumming*

The enzymatic degumming of silk is used for the removal of sericin from the silk without damaging its properties. The protease is extensively used for degumming of silk, which includes proteolytic degradation of sericin without attacking fibroin. Protease is a class of enzymes that catalyse the hydrolysis of peptide bonds [91]. There are various researches available that report the enzymatic degumming of silk [92,93]. Mahmoodi et al. studied the feasibility of Persian silk degumming with ultrasonic, ultrasonic with soap and ultrasonic with an enzyme. The ultrasonic-enzymatic treatment improves the silk yarn strength and elongation as compared to other employed methods [94]. Gulrajani and colleagues have reported the combined use of lipase and protease for dewaxing and degumming of silk. The silk obtained showed a cleaner longitudinal surface and better wettability [95].

2.5.2.6 *Wool processing*

Conventional wool processing consumes a lot of energy and time, while the enzymatic treatment reduces the treatment energy and time. Different types of enzymes are used in wool processing based on the required outcome; enzyme for pre-treatment, enzyme assisted processing and enzyme for enhancement of performance properties of the wool [24].

Carbonization

This process removes the cellulosic portion from greasy wool by strong acidic treatment. Instead of such harsh and difficult to handle acid, the enzymatic process is used to separate the sticky vegetable matter. A mixture of cellulase, ligninase and pectinase enzymes is utilised to remove vegetable matters from a range of wool grease, and oxidoreductase is used to remove plant impurities from wool. After incubation of wool with cellulase enzyme, the removal of burr becomes easy because of the cohesion weakening between wool and burr [96].

Scouring and bleaching

Scouring of wool is used to remove the natural impurities such as fat, wax, and other impurities. Conventionally, it requires high energy, chemicals, and consumption of a huge amount of water. An enzymatic scouring containing the alkali-stable protease enzyme is preferred to reduce the amount of water, chemicals and achieve an enhancement in softness, dyeability and whiteness index [97].

Bleaching of wool is widely carried out by hydrogen peroxide, whereas several research works revealed that peroxide bleaching significantly improved in the presence of protease enzyme [98]. It has been reported that a single bath peroxide bleaching of wool using protease can improve the whiteness [99].

Dyeing

The conventional dyeing of wool requires high temperature and time to attain the desired results. However, a high dyeing temperature tends to degrade the quality of wool. An enzyme helps to digest the amorphous region, induce adequate dyeability and decrease affinity for small-size dye molecules. Thus, enzymatic hydrolysis of wool fibre increases dyeing rate and dye affinity, specifically for large-size dye molecules [100]. Lipase considerably improves the dyeability of wool fibre because of the degradation of the hydrophobic layer and the increase in accessibility of fibre to the aqueous dye liquor [101].

Shrink proofing

This treatment is given to avoid a natural tendency of wool to shrink during washing. The traditional process involves the use of chlorination, which causes pollution. To avoid this problem, shrink proofing is done using enzymes. The protease enzyme is used for shrink-resistance, which improves the whiteness, dyeability and handle of wool [102]. However, excessive enzymatic treatment may damage the wool fibre and also cause strength and weight losses [103].

Researchers have developed a novel trifunctional protease to obtain machine washable wool fabric. The modified protease offers better dimensional stability to felting, dyeability, wettability and alkali solubility [104]. Chen et al. modified protease by covalently attaching chitosan for anti-felting of wool fabric. The modified protease showed much higher stability compared to the native protease [105].

2.2.2.7 *Synthetic fibre processing*

Polyester is the most significantly consumed synthetic fibre in the textile industry. Polyester possesses some key features such as soft hand, high strength, stain resistance, abrasion resistance, wrinkle resistance and machine washability. The alkaline hydrolysis of polyester is carried out to improve fabric handle and lustre; and however, sometimes, this process deteriorates the properties of the fabric. Thus, the enzymatic hydrolyses are capable of hydrolysing carboxylic ester or fatty acid, and they can hydrolyse the ester linkage in polyester. Polyesterase enzyme helps in surface modification, and it can impart hydrophilicity in polyester fabric, which results in the improvement of fabric characteristics like wettability, dyeability and stain resistance [16]. Various research articles reported the use of lipases and esterases and cutinases enzymes for the modification of polyester [106–108].

Similarly, enzymatic treatment of polyacrylonitrile fabric enhanced hydrophilicity and improved the adsorption of dye. The nitrile hydratases treatment increases the amide groups on the polyacrylonitrile fabric, which offers improved hydrophilicity and dyeability [109]. The polyamides can also be modified by enzymatic treatment using amidases, peroxidases, proteases and cutinase. The nylon 66 fabric treated with laccases showed increased hydrophilicity [110]

2.2.2.8 *Effluent treatment*

The textile processing effluent consists of a variable mixture of both organic and inorganic components, though the concentrations and actual nature depend on the input source. The organic matters generated from textile mills are generally proteins, carbohydrates and fatty acids. Moreover, the inorganic substances are mostly consisting of brine salts, sulphides, sulphites and heavy metals.

However, the bacterial effluent treatment has a prolonged rate of degradation of substances which limits their applications for this purpose. Certain enzymes are considered to confer biological remediation and environmental biotechnology, whether they are anaerobic or aerobic processes. Commercial synthetic colourants used in dyeing and printing during effluent treatment result in the formation of degradative aromatic amines, which are mutagenic, toxic and hazardous to the environment. Enzymes from both aerobic and anaerobic systems are useful in decolourising dyes. The manganese/lignin peroxidases and laccases enzymes are capable of oxidative free radical cleavage of the azo bond [111]. Laccase oxidises both toxic and non-toxic substrates and is used to improve whiteness in conventional bleaching and bio-stoning of cotton. The enzymatic treatment has many advantages, including water, energy and chemicals saving [112].

Currently, worldwide the annual production of dyes is around 800,000 tons/year (10,000 different dyes) and used extensively in the printing and dyeing industries, and at least 10 % of the dyes enter the ecological system via textile effluents. Various methods have been used for the removal of dyes from the effluent; however, most of them are uneconomical and have complex operations. The dyes have a complex chemical structure which offers a resistance to fading when exposed to chemicals, water and sunlight. Therefore, researchers have identified an enzyme, namely laccase, which can degrade the dye structure including synthetic dyes. In the denim industry, laccase is extensively used to degrade indigo to give a stone-wash or abrasion effect to the denim fabric and hence is adequately considered in denim finishing [113].

Concluding remarks and future perspectives

Presently, all over the world, pollution-free industrial processes are gaining momentum, and the use of enzymes in place of hazardous chemicals is an emerging idea to eliminate the polluting processes of textile industries. The use of enzymes can help in achieving overall process economics by reducing water and energy consumption and increasing productivity. Although the applications of enzymes in the textile industry are limited, their applications are increasing and rapidly spreading in all stages of textile processing. Today, companies are continually working on the improvement of their products for more versatile and advanced applications. The use of immobilized enzymes on a commercial scale can be crucial for eco-friendly textile processing.

Additionally, maintaining the stability of the biocatalysts during their application is critical for economic purposes. Moreover, there is a promising future for the reuse of enzymes, which will not only decrease the processing cost drastically but also bring about an extensive renovation in textile wet processing. The textile industry can gain potential benefits from the application of innovative enzymes when their underlying mechanisms are wholly anticipated. The unique characteristics of enzymes are that they act specifically without interpreting the reaction of other components under mild operating conditions. However, simultaneously they are sensitive to pH, temperature, contaminants and humidity. If the cost of enzyme manufacturing can be reduced, enzymes can play an important role in all the textile processing stages.

References

1. Madhu A, Chakraborty JN (2017), 'Developments in application of enzymes for textile processing', *J. Clean. Prod.*, 145, 114–133.

2. Mojsov K (2011), 'Application of enzymes in the textile industry', Int. Congr. "Engineering, Ecol. Mater. Process. Ind., 53, 27–29.

3. Cavaco-Paulo A, Gübitz G (2003). Textile processing with enzymes, Woodhead publishing, Cambridge: pp. 1–228.

4. Tiwari, SP, Srivastava, R, Singh, CS, Shukla, K, Singh, RK, Singh, P, Singh, R, Singh, NL Sharma, R (2015), 'Amylases: an overview with special reference to alpha amylase'. *J. Global Biosci.*, 4, 1886-1901.

5. Chapman J, Ismail A, Dinu C (2018), 'Industrial Applications of Enzymes: Recent Advances, Techniques, and Outlooks, Catalysts', 8, 238.

6. Cavaco-Pauio A (1998). Processing textile fibers with enzymes: an overview, ACS Publications, Washington: pp. 180-189.

7. Roy Choudhury A K (2014). Sustainable textile wet processing: Applications of enzymes. In Roadmap to sustain. text. and cloth., Springer, Singapore: pp. 203-238.

8. Aly AS, Moustafa AB, Hebeish A (2004), 'Bio-technological treatment of cellulosic textiles', *J. Clean. Prod.*, 12, 697–705.

9. Besegatto SV, Costa FN, Damas MSP, Colombi BL, De Rossi AC, de Aguiar CRL, Immich APS (2018), 'Enzyme treatment at different stages of textile processing: A review', *Ind. Biotechnol.*, 14, 298–307.

10. Chatha SAS, Asgher M, Iqbal HMN (2017), 'Enzyme-based solutions for textile processing and dye contaminant biodegradation—A review', *Environ. Sci. Pollut. Res.*, 24, 14005–14018.

11. Nielsen PH, Kuilderd H, Zhou W, Lu X (2009). Enzyme biotechnology for sustainable textiles. In Sustain. Text. Life Cycle Environ. Impact, Woodhead Publishing, Cambridge: pp. 113–138.

12. Shahid M, Mohammad F, Chen G, Tang RC, Xing T (2016), 'Enzymatic processing of natural fibres: White biotechnology for sustainable development', *Green Chem.*, 18, 2256–2281.

13. Diderichsen B (1995), 'A new desizing enzyme shows the changing nature of enzyme production', *J. Biotechnol.*, 38, 196–197.

14. Chen J, Wang Q, Hua Z, Du G (2007), 'Research and application of biotechnology in textile industries in China', *Enzyme Microb. Technol.*, 40, 1651–1655.

15. Patra AK, Madhu A, Chakraborty JN (2010), 'Studies on enzymatic pretreatment of linen', *Indian J. Fibre Text. Res.*, 35, 337-341.

16. Wavhal SD, Balasubramanya RH (2011), 'Role of biotechnology in the treatment of polyester fabric', *Indian J. Microbiol.*, 51, 117–123.

17. Galante Y, Formantici C (2005), 'Enzyme Applications in Detergency and in Manufacturing Industries', *Curr. Org. Chem.*, 7, 1399–1422.

18. Aly AS, Sayed SM, Zahran MK (2010), 'One-step process for enzymatic desizing and bioscouring of cotton fabrics', *J. Nat. Fibers.*, 7, 71–92.

19. Ali S, Khatri Z, Khatri A, Tanwari A (2014), 'Integrated desizing-bleaching-reactive dyeing process for cotton towel using glucose oxidase enzyme', *J. Clean. Prod.*, 66, 562–567.

20. Costa SA, Tzanov T, Paar A, Gudelj M, Gübitz GM, Cavaco-Paulo A (2001), 'Immobilization of catalases from Bacillus SF on alumina for the treatment of textile bleaching effluents', *Enzyme Microb. Technol.*, 28, 815–819.

21. Amorim AM, Gasques MDG, Andreaus J, Scharf M (2002), 'The application of catalase for the elimination of hydrogen peroxide residues after bleaching of cotton fabrics', *An. Acad. Bras. Cienc.*, 74, 433–436.

22. McDonald AG, Tipton KF (2014), 'Fifty-five years of enzyme classification: advances and difficulties', *FEBS J.* 281, 583–592.

23. Araújo R, Casal M, Cavaco-Paulo A (2008), 'Application of enzymes for textile fibres processing', *Biocatal. Biotransformation*, 26, 332–349.

24. Ammayappan L (2013), Application of enzyme on woolen products for its value addition: An overview', *J. Text. Appar. Technol. Manag.*, 8, 1-12.

25. Koshland DE (1995), 'The key–lock theory and the induced fit theory', *Angew. Chemie* Int. Ed. English. 33, 2375–2378.

26. Aniş Davulcu PA, Eren HA (2009), 'Enzymatic pre-treatment of cotton. Part 2: Peroxide generation in desizing liquor and bleaching', *Fibres Text. East. Eur.*, 73, 87–90.

27. Kirk O, Borchert TV, Fuglsang CC (2002), 'Industrial enzyme applications', *Curr. Opin. Biotechnol.*, 13, 345–351.

28. Saini R, Singh Saini H, Dahiya A, Harnek SS (2017), 'Amylases: Characteristics and industrial applications', *J. Pharmacogn. Phytochem.*, 6, 1865–1871.

29. Hossain S, Uddin MK (2011), 'Comparative analysis between conventional pre-treatment and bio-preparation', *Int. J. Eng. Technol.*, 11, 14–19.

30. Sreelakshmi SN, Paul A, Vasanthi NS, Saravanan D (2014), 'Low-temperature acidic amylases from Aspergillus for desizing of cotton fabrics', *J. Text. Inst.*, 105, 59–66.

31. Rehman A, Raza ZA, Masood R, Hussain MT, Ahmad N (2015), 'Multi-response optimization in enzymatic desizing of cotton fabric under various chemo-physical conditions using a Taguchi approach', *Cellulose*, 22, 2107–2116.

32. Chand N, Sajedi RH, Nateri AS, Khajeh K, Rassa M (2014), 'Fermentative desizing of cotton fabric using an α-amylase-producing Bacillus strain: Optimization of simultaneous enzyme production and desizing', *Process Biochem.*, 49, 1884–1888.

33. Lenting HBM, Warmoeskerken MMCG (2004), 'A fast, continuous enzyme-based pretreatment process concept for cotton containing textiles', *Biocatal. Biotransformation*, 22, 361–368.

34. Chand N, Nateri AS, Sajedi RH, Mahdavi A, Rassa M (2012), 'Enzymatic desizing of cotton fabric using a Ca 2+-independent α-amylase with acidic pH profile', *J. Mol. Catal. B Enzym.*, 83, 46–50.

35. Wang WM, Yu B, Zhong CJ (2012), 'Use of ultrasonic energy in the enzymatic desizing of cotton fabric', *J. Clean. Prod.*, 33, 179–182.

36. Hao L, Wang R, Fang K, Liu J (2013), 'Ultrasonic effect on the desizing efficiency of amylase on starch-sized cotton fabrics', *Carbohydr. Polym.*, 96, 474–480.

37. Şahinbaşkan BY, Kahraman MV (2011), 'Desizing of untreated cotton fabric with the conventional and ultrasonic bath procedures by immobilized and native α-amylase', *Starch - Stärke.*, 63, 154–159.

38. Beyler-Çigil A, Çakmakçi E, Danis O, Demir S, Kahraman MV (2013), 'Alpha-amylase immobilization on modified polyimide material', *Chem. Eng. Trans.*, 32, 1687–1692.

39. Talekar S, Joshi A, Joshi G, Kamat P, Haripurkar R, Kambale S (2013), 'Parameters in preparation and characterization of cross linked enzyme aggregates (CLEAs)', *RSC Adv.*, 3, 12485–12511.

40. Agrawal PB, Nierstrasz VA, Klug-Santner BG, Gübitz GM, Lenting HBM, Warmoeskerken MMCG (2007), 'Wax removal for accelerated cotton scouring with alkaline pectinase', *Biotechnol. J.*, 2, 306–315.

41. Karapinar E, Sariisik MO (2004), 'Scouring of cotton with cellulases, pectinases and proteases', *Fibres Text. East. Eur.*, 12, 79–82.

42. Sawada K, Tokino S, Ueda M, Wang XY (2008), 'Bioscouring of cotton with pectinase enzyme', *J. Soc. Dye. Colour.*, 114, 333–336.

43. Lin CH, Hsieh YL (2001), 'Direct scouring of greige cotton fabrics with proteases', *Text. Res. J.*, 71, 425–434.

44. Joshi M, Nerurkar M, Badhe P, Adivarekar R (2013), 'Scouring of cotton using marine pectinase', *J. Mol. Catal. B Enzym.*, 98, 106–113.

45. Rajendran R, Sundaram SK, Radhai R, Rajapriya P (2011), 'Bioscouring of cotton fabrics using pectinase enzyme its optimization and comparison with conventional scouring process', *Pakistan J. Biol. Sci.*, 14, 519–525.

46. Calafell M, Garriga P (2004), 'Effect of some process parameters in the enzymatic scouring of cotton using an acid pectinase', *Enzyme Microb. Technol.*, 34, 326–331.

47. Kalantzi S, Mamma D, Christakopoulos P, Kekos D (2008), 'Effect of pectate lyase bioscouring on physical, chemical and low-stress mechanical properties of cotton fabrics', *Bioresour. Technol.*, 99, 8185–8192.

48. Raza ZA, Rehman A, Hussain MT, Masood R, Ul Haq A, Saddique MT, Javid A, Ahmad N (2014), 'Production of rhamnolipid surfactant and its application in bioscouring of cotton fabric', *Carbohydr. Res.*, 391, 97–105.

49. Csiszár E, Urbánszki K, Szakács G (2001), 'Biotreatment of desized cotton fabric by commercial cellulase and xylanase enzymes', *J. Mol. Catal. B: Enzym.*, 11, 1065–1072.

50. Battan B, Dhiman SS, Ahlawat S, Mahajan R, Sharma J (2012), 'Application of thermostable xylanase of Bacillus Pumilus in textile processing', *Indian J. Microbiol.*, 52, 222–229.

51. Senthil Raja K, Vasanthi NS, Saravanan D, Ramachandran T (2012), 'Use of bacterial lipase for scouring of cotton fabrics', *Indian J. Fibre Text. Res.*, 37, 299-302

52. Hebeish A, Hashem M, Shaker N, Ramadan M, El-Sadek B, Hady MA (2009), 'New development for combined bioscouring and bleaching of cotton-based fabrics', *Carbohydr. Polym.*, 78, 961–972.

53. El Shafie A, Fouda MMG, Hashem M (2009), 'One-step process for bio-scouring and peracetic acid bleaching of cotton fabric', *Carbohydr. Polym.*, 78, 302–308.

54. Preša P, Tavčer PF (2008), 'Bioscouring and bleaching of cotton with pectinase enzyme and peracetic acid in one bath', *Color. Technol.*, 124, 36–42.

55. Pereira L, Bastos C, Tzanov T, Cavaco-Paulo A, Guebitz GM (2005), 'Environmentally friendly bleaching of cotton using laccases', *Environ. Chem. Lett.*, 3, 66–69.

56. Basto C, Tzanov T, Cavaco-Paulo A (2007), 'Combined ultrasound-laccase assisted bleaching of cotton', *Ultrason. Sonochem.*, 14, 350–354.

57. Abou-Okeil A, El-Shafie A, El Zawahry MM (2010), 'Ecofriendly laccase-hydrogen peroxide/ultrasound-assisted bleaching of linen fabrics and its influence on dyeing efficiency, *Ultrason. Sonochem.*, 17, 383–390.

58. Gonçalves I, Herrero-Yniesta V, Perales Arce I, Escrigas Castañeda M, Cavaco-Paulo A, Silva C (2014), 'Ultrasonic pilot-scale reactor for enzymatic bleaching of cotton fabrics', *Ultrason. Sonochem.*, 21, 1535–1543.

59. Buschle-Diller G, Yang XD (2001), Yamamoto R, 'Enzymatic bleaching of cotton fabric with glucose oxidase', *Text. Res. J.*, 71, 388–394.

60. Farooq A, Ali S, Abbas N, Fatima GA, Ashraf MA (2013), 'Comparative performance evaluation of conventional bleaching and enzymatic bleaching with glucose oxidase on knitted cotton fabric', *J. Clean. Prod.*, 42, 167–171.

61. Shin Y, Hwang S, Ahn I (2004), 'Enzymatic bleaching of desized cotton fabrics with hydrogen peroxide produced by glucose oxidase', *J. Ind. Eng. Chem.*, 10, 577–581.

62. Tzanov T, Calafell M, Guebitz GM, Cavaco-Paulo A (2001), 'Bio-preparation of cotton fabrics', *Enzyme Microb. Technol.*, 29, 357–362.

63. Tzanov T, Costa SA, Gübitz GM, Cavaco-Paulo A (2002), 'Hydrogen peroxide generation with immobilized glucose oxidase for textile bleaching', *J. Biotechnol.*, 93, 87–94. doi:10.1016/S0168-1656(01)00386-8.

64. Eren HA, Anis P, Davulcu A (2009), 'Enzymatic one-bath desizing — bleaching — dyeing process for cotton fabrics', *Text. Res. J.*, 79, 1091–1098.

65. Tzanov T, Costa S, Guebitz GM, Cavaco-Paulo A (2001), 'Dyeing in catalase-treated bleaching baths', *Color. Technol.*, 117, 1–5.

66. Cavaco-Paulo A (1998), 'Mechanism of cellulase action in textile processes', *Carbohydr. Polym.*, 37, 273–277.

67. Saravanan D, Vasanthi NS, Ramachandran T (2009), 'A review on influential behaviour of biopolishing on dyeability and certain physico-mechanical properties of cotton fabrics', *Carbohydr. Polym.*, 76, 1–7.

68. Tarhan M, Sarıışık M (2009), 'A comparison among performance characteristics of various denim fading processes', *Text. Res. J.*, 79, 301–309.

69. Ibrahim NA, El-Badry K, Eid BM, Hassan TM (2011), 'A new approach for biofinishing of cellulose-containing fabrics using acid cellulases', *Carbohydr. Polym.*, 83, 116–121.

70. Shen J, Smith E (2015), 'Enzymatic treatments for sustainable textile processing', In Sustain. Appar. Prod. Process. Recycl., Woodhead publishing, Cambridge: pp. 119–133.

71. Esfandiari A, Firouzi-Pouyaei E, Aghaei-Meibodi P (2014), 'Effect of enzymatic and mechanical treatment on combined desizing and bio-polishing of cotton fabrics', *J. Text. Inst.*, 105, 1193–1202.

72. Tavčer PF (2013), 'Effects of cellulase enzyme treatment on the properties of cotton terry fabrics', *Fibres Text. East. Eur.*, 6, 100–106.

73. Pazarlioglu NK, Sariişik M, Telefoncu A (2005), 'Treating denim fabrics with immobilized commercial cellulases', *Process Biochem.*, 40, 767–771.

74. Yu Y, Yuan J, Wang Q, Fan X, Ni X, Wang P, Cui L (2013), 'Cellulase immobilization onto the reversibly soluble methacrylate copolymer for denim washing', *Carbohydr. Polym.*, 95, 675–680.

75. Hao L, Wang R, Liu J, Liu R (2012), 'The adsorptive and hydrolytic performance of cellulase on cationised cotton', *Carbohydr. Polym.*, 89, 171–176.

76. Couto SR, Luis Toca-Herrera J (2006), 'Lacasses in the textile industry', *Biotechnol. Mol. Biol. Rev.*, 1, 115–120.

77. Tavares AP, Cristóvão RO, Loureiro JM, Boaventura RA, Macedo EA (2008), 'Optimisation of reactive textile dyes degradation by laccase-mediator system', *J. Chem. Technol. Biotechnol.*, 83, 1609–1615.

78. Tavares AP, Cristóvão RO, Gamelas JF, Loureiro JM, Boaventura RA, Macedo EA (2009), 'Sequential decolourization of reactive textile dyes by laccase mediator system', *J. Chem. Technol. Biotechnol.*, 84, 442–446.

79. Cristóvão RO, Tavares AP, Loureiro JM, Boaventura RA, Macedo EA (2008), 'Optimisation of reactive dye degradation by laccase using Box–Behnken design'., *Environ. Technol.*, 29, 1357–1364.

80. Tavares AP, Cristóvão RO, Loureiro JM, Boaventura RA, Macedo EA (2009), 'Application of statistical experimental methodology to optimize reactive dye decolourization by commercial laccase', *J. Hazard. Mater.*,162, 1255–1260.

81. Cristóvão RO, Tavares AP, Ribeiro AS, Loureiro JM, Boaventura RA, Macedo EA (2008), 'Kinetic modelling and simulation of laccase catalyzed degradation of reactive textile dyes', *Bioresour. Technol.*, 99, 4768–4774.

82. Dhiman SS, Sharma J, Battan B (2008), 'Pretreatment processing of fabrics by alkalothermophilic xylanase from Bacillus stearothermophilus SDX', *Enzyme Microb. Technol.*, 43 262–269.

83. Losonczi A, Csiszár E, Szakács G, Bezúr L (2005), 'Role of the EDTA Chelating Agent in Bioscouring of Cotton', *Text. Res. J.*, 75, 411–417.

84. Agrawal PB, Nierstrasz VA, Bouwhuis GH, Warmoeskerken MMCG (2008), 'Cutinase and pectinase in cotton bioscouring: an innovative and fast bioscouring process', *Biocatal. Biotransformation*, 26, 412–421.

85. Degani O, Gepstein S, Dosoretz CG (2002), 'Potential use of cutinase in enzymatic scouring of cotton fiber cuticle', *Appl. Biochem. Biotechnol.*, 102, 277–289.

86. Agrawal PB, Nierstrasz VA, Warmoeskerken MMCG (2010), 'Ultrasound-boosted enzymatic cotton scouring process with cutinase and pectate lyase', *Biocatal. Biotransformation.*, 28, 320–328.

87. Agrawal PB, Nierstrasz VA, Warmoeskerken MMCG (2008), 'Role of mechanical action in low-temperature cotton scouring with F. solani pisi cutinase and pectate lyase', *Enzyme Microb. Technol.*, 42, 473–482.

88. Musialak M, Wróbel-Kwiatkowska M, Kulma A, Starzycka E, Szopa J (2008), 'Improving retting of fibre through genetic modification of flax to express pectinases', *Transgenic Res.*, 17, 133–147.

89. Henriksson G, Akin DE, Slomczynski D, Eriksson KEL (1999), 'Production of highly efficient enzymes for flax retting by Rhizomucor pusillus', *J. Biotechnol.*, 68, 115–123.

90. Zhang J, Henriksson G, Johansson G (2000), 'Polygalacturonase is the key component in enzymatic retting of flax', *J. Biotechnol.*, 81, 85–89.

91. Freddi G, Mossotti R, Innocenti R (2003), 'Degumming of silk fabric with several proteases', *J. Biotechnol.*, 106, 101–112.

92. Arami M, Rahimi S, Mivehie L, Mazaheri F, Mahmoodi NM (2007), 'Degumming of Persian silk with mixed proteolytic enzymes', *J. Appl. Polym. Sci.*, 106, 267–275.

93. Gulrajani ML, Agarwal R, Chand S (2000), 'Degumming of silk with a fungal protease, *Indian J. Fibre Text. Res.*, 25, 138-142.

94. Mahmoodi NM, Arami M, Mazaheri F, Rahimi S (2010), 'Degradation of sericin (degumming) of Persian silk by ultrasound and enzymes as a cleaner and environmentally friendly process', *J. Clean. Prod.*, 18, 146–151.

95. Gulrajani ML, Agarwal R (2000), 'Degumming of silk with lipase and protease', *Indian J. Fibre Text. Res.*, 25, 69-74.

96. Heine E, Ruers A, Hocker H (2000), 'Enzymatic degradation of vegetable residues in wool (AiF 11300)', *DWI Reports*, 475–479.

97. Wojciechowska E, Włochowicz A, Wesełucha-Birczyńska A (1999), 'Application of Fourier-transform infrared and Raman spectroscopy to study degradation of the wool fiber keratin', *J. Mol. Struct.*, 511, 307-318.

98. Cegarra J, Gacen J, Cayuela D (2000), 'Wool bleaching with hydrogen peroxide in the presence of proteases', *Congr. Int. Wool Text. Organ.*, 116, 3-15.

99. Jovanĉić P, Jocić D, Molina R, Juliá MR, Erra P (2001), 'Shrinkage properties of peroxide-enzyme-biopolymer treated wool', *Text. Res. J.*, 71, 948–953.

100. Riva A, Algaba I, Prieto R (2002), 'Dyeing kinetics of wool fabrics pretreated with a protease', *Color. Technol.*, 118, 59–63.

101. Kantouch A, Raslan WM, El-Sayed H (2005), 'Effect of lipase pretreatment on the dyeability of wool fabric', *J. Nat. Fibers.*, 2, 35–48.

102. Ibrahim NA, El-Shafei HA, Abdel-Aziz MS, Ghaly MF, Eid BM, Hamed AA (2012), 'The potential use of alkaline protease from Streptomyces albidoflavus as an eco-friendly wool modifier', *J. Text. Inst.*, 103, 490–498.

103. Vílchez S, Jovančić P, Erra P (2010), 'Influence of chitosan on the effects of proteases on wool fibers', *Fibers Polym.*, 11, 28–35.

104. Mei J, Zhang N, Yu Y, Wang Q, Yuan J, Wang P, Cui L, Fan X (2018), 'A novel "trifunctional protease" with reducibility, hydrolysis, and localization used for wool anti-felting treatment', *Appl. Microbiol. Biotechnol.*, 102, 9159–9170.

105. Chen R, Yuan J, Yu Y, Fan X, Wang Q and Zhu Y (2012), 'Preparation of chitosan conjugated protease used for shrink resistance of wool and its properties'. *J food Sci Biotechnolo.*, 31, 63-68.

106. Ribitsch D, Acero EH, Greimel K, Eiteljoerg I, Trotscha E, Freddi G, Schwab H, Guebitz GM (2012), 'Characterization of a new cutinase from Thermobifida alba for PET-surface hydrolysis', *Biocatal. Biotransformation.*, 30, 2–9.

107. Herrero Acero E, Ribitsch D, Dellacher A, Zitzenbacher S, Marold A, Steinkellner G, Gruber K, Schwab H, Guebitz GM (2013), 'Surface engineering of a cutinase from Thermobifida Cellulosilytica for improved polyester hydrolysis', *Biotechnol. Bioeng.*, 110, 2581–2590.

108. Ribitsch D, Herrero Acero E, Greimel K, Dellacher A, Zitzenbacher S, Marold A, Rodriguez RD, Steinkellner G, Gruber K, Schwab H, Guebitz GM (2012), 'A new esterase from thermobifida halotolerans hydrolyses polyethylene terephthalate (PET) and polylactic acid (PLA)', *Polymers (Basel)*, 4, 617–629.

109. Matamá T, Carneiro F, Caparrós C, Gübitz GM, Cavaco-Paulo A (2007), 'Using a nitrilase for the surface modification of acrylic fibres', *Biotechnol. J.*, 2, 353–360.

110. Miettinen-Oinonen A, Silvennoinen M, Nousiainen P, Buchert J (2002), 'Modification of synthetic fibres with laccase', in proceedings of the second international symposium on biotechnology in textiles, 3–6.

111. Khan R, Bhawana P, Fulekar MH (2013), 'Microbial decolorization and degradation of synthetic dyes: A review', *Rev. Environ. Sci. Biotechnol.*, 12, 75–97.

112. Kim S, Moldes D, Cavaco-Paulo A (2007), 'Laccases for enzymatic colouration of unbleached cotton', *Enzyme Microb. Technol.*, 40, 1788–1793.

113. Banat IM, Nigam P, Singh D, Marchant R (1996), 'Microbial decolorization of textile-dye-containing effluents: A review', *Bioresour. Technol.*, 58, 217–227.

3

Natural dyes: Green and sustainable alternative for textile colouration

Luqman Jameel Rather[a*], Mohd Shabbir[b], Showkat Ali Ganie[a], Qi Zhou[a], K. P. Singh[b], Qing Li[a*]

[a]State Key Laboratory of Silkworm Genome Biology, Chongqing Engineering Research Center for Biomaterial Fiber and Modern Textile and College of Sericulture, Textile and Biomass Science, Southwest University, Chongqing, 400715, PR China

[b]Department of Chemistry, Noida Institute of Engineering and Technology, Greater Noida, UP, India -110025.

*Corresponding Author, E-mail: qingli@swu.edu.cn, luqmanjameel123@gmail.com

Abstract: Since time immemorial natural dyes have been explored for the colouration of various textile fibre/fabrics. Recently, natural dyes have been used on an increasing demand due to an increased health and environmental consciousness among the people in response to the hazardous nature of synthetic dyes and their by-products. In this chapter, natural dyes have been described in view of their origin, classification, chemical structures, and, most importantly, their sustainable production and application. Green production and utilisation of natural colourants are supposed to be a permanent alternative to their synthetic counterparts due to high consumer demands for coloured products (textile, food, pharmaceuticals, and cosmetics) and today's environmental challenges. The development of new technologies for dye extraction and dyeing of textiles are a necessity of the present, reviewed in this chapter.

3.1 Introduction

Colour plays an important role in our daily lives as a change in colours serves as an important visual sign. Nature has always captivated mankind with intense and soothing colours from time immemorial [1]. These colourants include dyes as well as pigments. Dyes are coloured compounds having a strong affinity for the textile fibre/fabric via different physical and chemical interactions. Pigments have neither affinity nor interacts with physical or chemical forces with textile fibre/fabrics [2]. People in ancient times used to exploit various plant parts such as roots, stems, barks, berries, leaves, and flower extracts for use in the colouration of textile materials, food, and cosmetics [3,4]. It was 1856 when W. H. Perkin synthesised dye in the laboratory, and the classification of dyes based on the origin has came into existence, i.e., natural and synthetic. Speedy development in synthetic dyes research, production, and application led to a sharp slump in use of natural dyes [5]. However, during the last few decades, researchers around the globe have been motivated to explore new renewable bioresource natural materials for imparting colour which would minimise the negative environmental impacts of the use of azo and benzidine synthetic dyes [6] in response to which many European countries, USA, Germany, and India have imposed Environmental and Ecological Legislations for restricting their uses [7]. As a result of increasing environmental and waste management concerns, natural dyes have re-emerged as potential green chemistry alternatives for synthetic dyes with wide-spread applicability other than textile colouration [8-10].

Enhanced awareness among researchers towards eco-preservation, eco-safety and health concerns, sustainable materials from non-food crops which are environmentally benign, non-toxic and sustainable have revolutionised all industrial sectors, particularly the textile industry [11]. Textile industry has been getting an increased interest all over the globe due to variations in terms of price, durability of end products, design, ease of handling, and product safety [12,13]. Excessive use of water and auxiliary chemicals in dyeing and functional finishing in different textile sectors have led to the development of new dyeing technologies [5,14,15]. Wastewater from wet textile processing contains a wide range of pollutants that need high technical skills for the development of cost-effective cleaner production strategies for their reduction [1,4,16-18]. Therefore, vast research is currently being carried out to minimise the negative environmental impact of synthetic chemical agents through the sustainable harvest of ecofriendly bioresource materials.

This chapter combines the most recent literature regarding the classification, sources, chemistry, extraction and application of natural colourants with a focus on textile colouration and functional finishing properties. Furthermore, analyses regarding the alternative strategies and technical challenges in the processing and development of cost-effective dyeing procedures using renewable agricultural biomass raw materials are also summarised briefly. Lastly, this chapter will discuss the process of modifying the structure of natural colourants to meet the requirements of increasing needs and modern technology, along with the discussion on the sustainability of these products.

3.2 Natural dyes

Dyes/colourants obtained from *flora*, *fauna* and minerals used without chemical processing are renewable and sustainable bioresource materials with no or minimal environmental impact and have been known since antiquity, not only for colouration of textiles but also as food ingredients, cosmetics and medicine [19,20]. Natural flora/fauna provides a wide range of fascinating colours which impart harmonising, elegant and sober shades on different synthetic and natural textile materials. In addition to that, natural colourants impart various functional finishing properties such as insect repellent, deodorising, anti-feedants, antimicrobial, fluorescence, and UV-protective properties, which added more dimensions to the applicability of natural colourants with increasing popularity in the development of systematic, scientific and diversified smart textiles materials [21-25].

3.3 Brief historical aspects of natural dyes

From the well-documented historical records, natural colourants were available to people during the Greco-Roman periods (332 BC - 395 AD) and Egyptians started natural dyeing about 3000 B.C with madder, safflower, alkanet, indigo, bark of the pomegranate as major vegetable dyes [26-28]. Analysis of red fabrics from Tutankhamun's tomb in Egypt has confirmed the presence of Alizarin, an extracted red pigment from madder [29]. Remarks of archaeological monuments reflect the presence of wide spread industrial enterprise of dyeing in Egypt, Mesopotamia and India around the third millennium B.C. It appears that natural dyeing has being practiced in India since the Vedic period. The use of the most important natural dyes of ancient times, such as madder, indigo and henna dates back to Mohanjodaro and Harappa Civilizations, Ajanta Cave Paintings and the Mughal era. Hudud-ul-Alam (982-83) is one of the most vital documents among the contemporary records of the tenth century that reflect the perfect technical hand of Indian dyers in the process of dyeing and bleaching.

Block printing and mordanting techniques are said to be originated from major towns of India like Delhi, Farrukhabad and Lucknow during the Mughal reign (1556-1803) with Catechu, Henna, Dhao, Indigo, Kachnar, Madder, Myrobalan, Pomegranate, Tun, Turmeric, and Patang as major natural dyes. Aluminium, copper, chromium, iron & tin were used for mordanting purposes [30,31]. Till the advent of synthetic counterpart, India used to be one of the largest exporters of natural indigo. Archaeological surveys of different monuments of the Indian subcontinent show that the use of Indigo (*Indigofera tinctoria*) dates back 4000 years ago for dyeing and printing purposes. Woad (*Isatis tinctoria*), another blue dye, has been in use since Bronze Age (2500-800 BC) in Northern Europe. Historical records about Madder red dye indicate its use by ancient Persians and Egyptians and later by the Greeks and Romans.

China has a long history of natural dye extraction from plant sources with potential usage for colouring food and clothing in Dong communities and upland areas. Brilliant scarlet, purple, rose, and violet shades have been obtained on linen and silk in Egypt and China. Moreover, in addition to colouring purposes, Dong communities have used-plant based natural dyes for enhancing the nutritive, medicinal and preservative properties of foods [32,33]. Similarities and differences in textile dyeing all over the world have been studied for a better understanding of the textile dyeing history and mutual influences and exchanges of textile dyeing techniques among different regions with preliminary knowledge gained from the research studies of textile dyeing from Ming (1368-1644) and Qing (1644-1911) dynasties [34]. There are a lot of historical pieces of evidences about vegetable dyes being suitable co-partner to the increasing present-day demand of customers for fascinating and cheerful colours.

3.4 Classification of natural dyes

Natural dyes are classified based on origin, method of application, chemical structure, and colour. However, the earliest known classification of natural dyes was based on alphabetical order. There are also some other methods of classification, such as on the basis of botanical and common names [35]. In Colour Index, natural dyes have been classified based on the mood of applications and the nature of the chemical components, and within the application classes, dyes are classified according to their hues. Detailed classification of natural dyes is discussed below in detail:

3.4.1 Classification based on the origin/source

On the basis of origin, natural colourants are classified into plant/vegetable origin, insect/animal origin, mineral origin and microbial origin (Fig. 3.1). Natural colourants of plant origin can be extracted from the leaves, flowers, fruits, seeds, bark, trunk and roots. Among different plant parts, leaves have been widely used as natural renewable biomass from which pigments are being extracted at a large scale. Encouraging results have been already documented by various researchers utilising tannin-rich extracts of different dye yielding trees such as *T. chebula*, *P. granatum*, *Q. infectoria*, *A. nilotica*, tannic acid, tartaric acid, guava and banana leaves ash in the development of eco-friendly and bioactive shades of different hue and tone [9,10,36-38]. However, it is noteworthy to mention that most of the plant dyes require mordants for their better performances and will be discussed in the latter part of this chapter.

Natural dyes belonging to the animal or insect class include the reds from the exudation of dried Cochineal, Kermes, lac, and molluscs bodies (Fig. 3.1). These red dyes have a long history of usage for the dyeing of different types of fibres [1,39,40]. Pigments comprising of inorganic metal salts and metal oxides belong to the mineral origin category. These include cinnabar, red ocher, yellow ocher, raw sienna, malachite, ultramarine blue, azurite, gypsum, talc, charcoal black, and so on (Fig. 3.1). Mineral origin colourants have also been used in ancient paintings and murals [1,41].

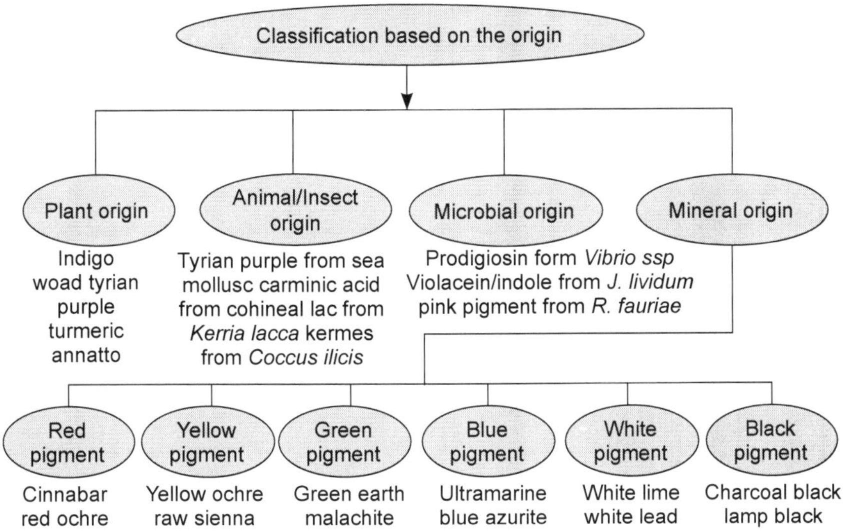

Figure 3.1 Classification of natural dyes based on the origin

Carotenoids, flavonoids, quinones, riboflavin, and prodigiosin are some of the natural pigments which have been isolated from certain microorganisms such as bacteria, algae, and fungi and subsequently applied on various kinds of textile fibres and fabrics for colouration and bioactive fabric production purposes [42]. Table 3.1 summarises some of the microbial pigments, their colour, and their sources with their potential applications. Owing to their potent biological activities, such as anticancer activity, these types of pigments have been frequently used in the pharmaceutical, food, and cosmetics industries [43]. While applying to textiles, these pigments show excellent fastness characteristics along with antimicrobial and antioxidant activities [42]. The colour of the extracted pigments ranges from bright red to bluish-purple to violet and pink. The extracted pigments are stable to light, temperature, and changing pH conditions. Poor light fastness properties widened their dimensions of usage in the printing of reusable and recyclable papers [44].

Table 3.1 Sources and applications of microbial pigments [18]

S. No.	Source	Nature of Pigment	Colour	Applications
1	*Vibrio ssp*	Prodigiosin	Red	Antimicrobial finishing
2	*J. lividum*	Violacein/indole	(Blue-purple)	-do-
3	*S. marcescens*	Prodigiosin	Red	-do-
4	*C. violaceum*	Violacein/indole	Violet	-do-
5	*R. fauriae*	-	Pink	-do-
6	*F. oxysporum*	Anthraquinones	Yellow-orange	Colouration
7	*T. viride*	Anthraquinones	Yellow-orange	Antifungal
8	*Alternaria spp*	Anthraquinones	Reddish brown	Antibacterial and antifungal
9	*D. sanguinea*	Anthraquinones	Red	Colouration
10	Brevibacterium KY -4313	Canthaxanthin and β-carotene	Brown	Poultry feed, fish feed, and Food colourant
11	*Flavobacterium* sp.	Zeaxanthin and lutein	-	Poultry feed
12	*C. michigannise*	Canthaxanthin	Greyish to creamish	Poultry feed and fish feed
13	*H. gramineum*	-	Red	-
14	*H. cynodontis*	-	Bronze colour	-
15	*M. purpureus*	Rubropunctatin	Red	-
16	*P. rhodozyma*	β-carotene	Red	Food colourant
17	*B. trispora*	Lycopene	Cream	-
18	*P. nalgeovensis*	-	Yellow	-
19	*Aspergillus* sp.	-	Orange, Red	-

3.4.2 Classification on the basis of application

Bancroft, in his book "Philosophy of Permanent Colours" classified natural dyes into two groups based on their application methods. The colourants such as indigo, turmeric, orchil, etc., which dye textiles directly, are classified as **substantive** dyes (Fig. 3.2). Another class of dyes that requires a metal salt/mordant to increase the interaction between dye and textiles is called as **adjective** dyes. Adjective dyes were further classified into monogenetic (only one colour irrespective of mordant; annatto) and polygenetic (depends upon mordant; logwood, kamala, lac, and cochineal).

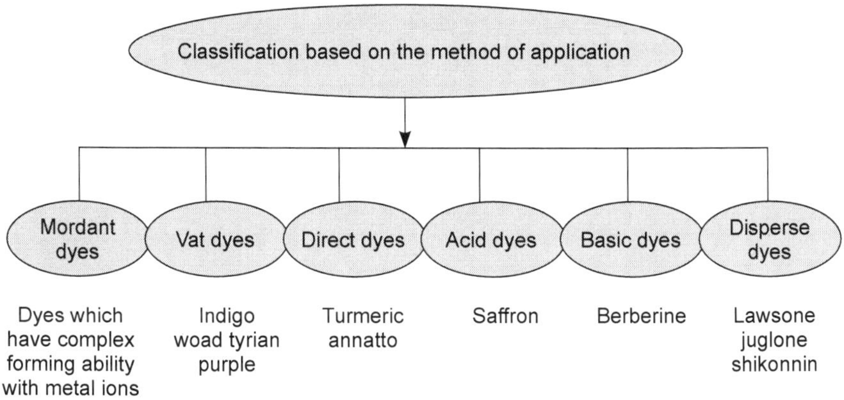

Figure 3.2 Classification of natural dyes based on the method of application

3.4.3 Classification based on chemical structure

Chemical-structure-based classification is widely accepted as it identifies dyes on the nature of chemical constituents present in them [45] (Fig. 3.3):

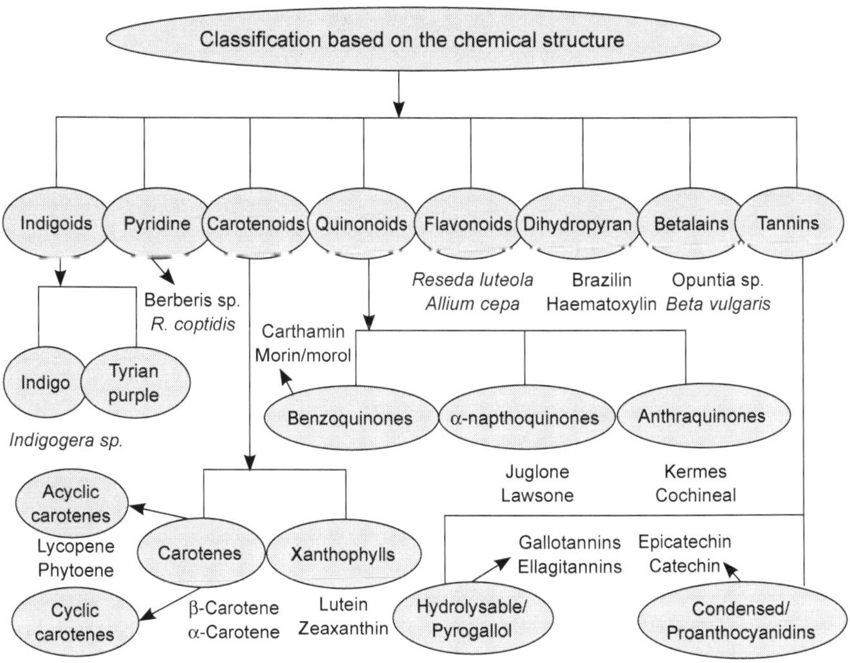

Figure 3.3 Classification based on the type of chemical nature

Indigoids are blue pigments extracted from the leaves of *Indigofera* species (*I. tinctoria*) and mollusc. Indigo from mollusc is industrially not feasible due to the high cost of processing. Indigo and tyrian purple are representative examples of indigoid colourants which have a very high affinity for both cellulosic and protein fibres [46]. However, the advent of synthetic counterparts (Several thousand tons per year) declined the use of natural indigo from *I. tinctoria* and *P. tinctorium* [47].

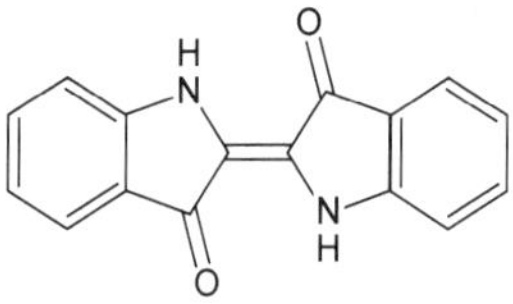

Indigo

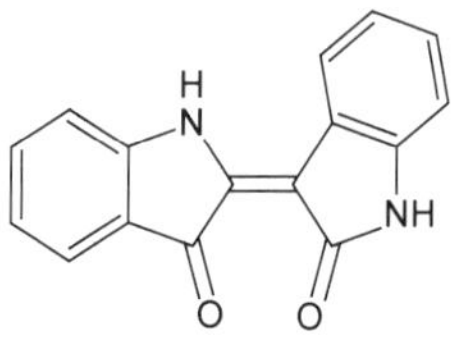

Indirubin

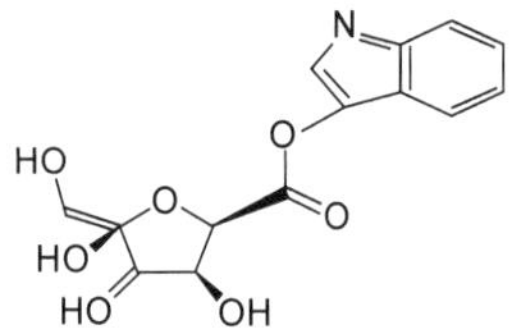

Indoxyl ketogluconate

Pyridine-based dyes consist of isoquinoline alkaloid units with bright yellow colour. Berberine (natural yellow 18; C.I. 75160) is the only natural dye from this class and is being extracted from *B. aristata*, *B. vulgaris*, *P. amurense*, and *R. coptidis* [48,49].

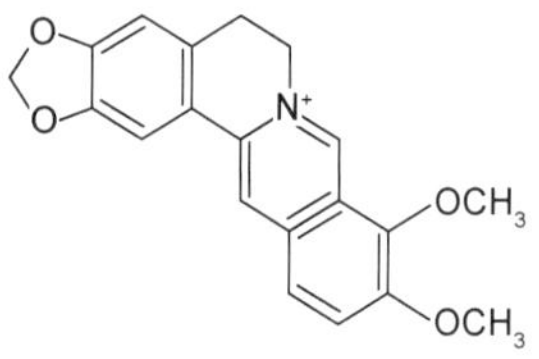

Berberine

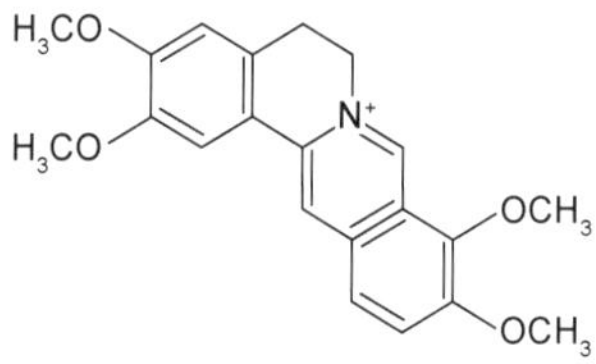

Palmatine

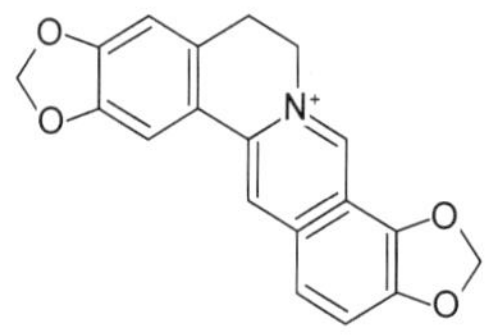

Coptisine

Natural carotenoid compounds are extracted from various plants such as carrots and tomatoes. Carotenoid dyes have a wide range of applicability as food colours due to the presence of vitamin A where they function as antioxidant scavengers [50]. Over 750 carotenoids have been isolated, and structures of which are confirmed through various spectral characterisations. Carotenoids are divided into two groups: carotenes and xanthophylls [51]. Carotenoid dyes impart antioxidant properties to the textiles, which can be further modified by introducing new functionalities into the isoprenoid structure. Further, surface modifications can be introduced by the use of metal mordants.

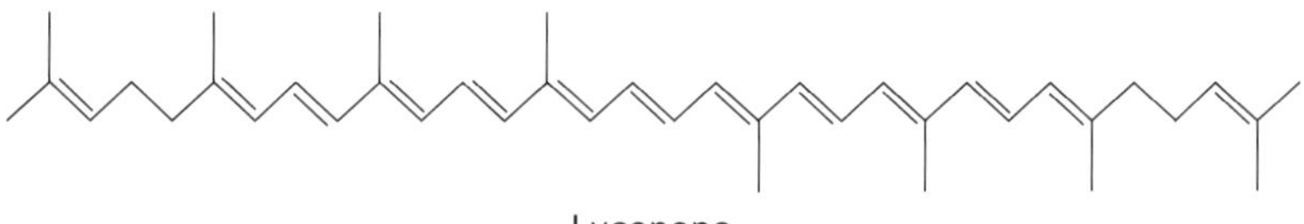

Lycopene

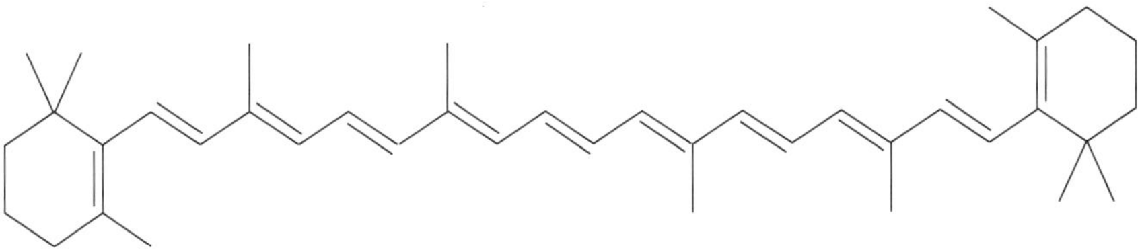

β-Carotene

Natural quinonoids are mostly found in flowering plants, and their colour ranges from yellow to red. Chemically quinonoid colourants are based on benzenoid ring systems with highly diverse structural variations and are further classified into benzoquinones, α-naphthoquinones, and anthraquinones. Some common examples are lawsone, jaglone, emodin, carminic acid, and kermesic acid. Quinonoid dyes are being used to impart antibacterial characteristics to the textile materials in addition to the normal colour [52,53].

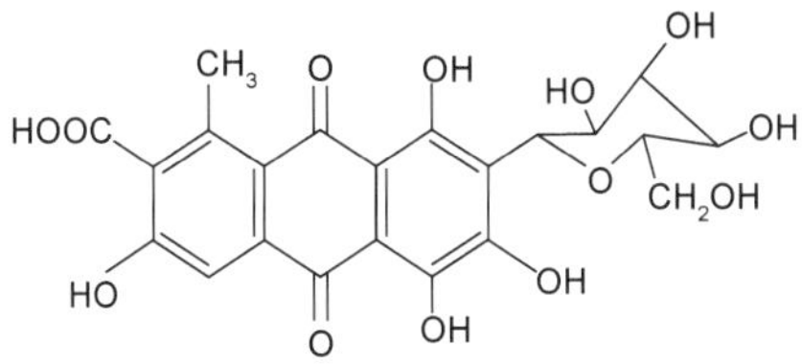

Carminic acid

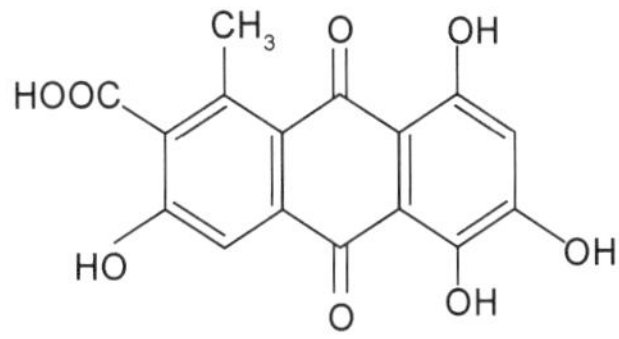

Kermesic acid

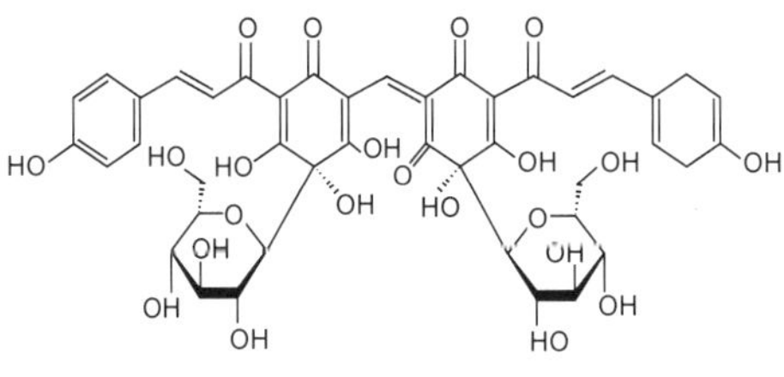

Carthamin

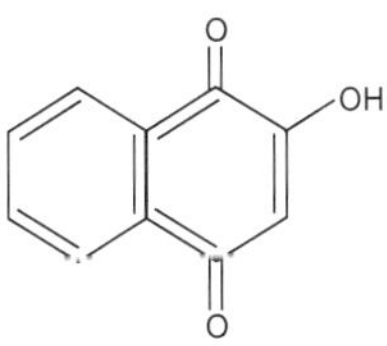

Lawsone

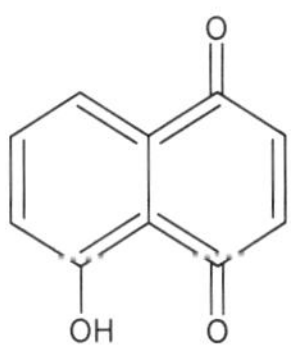

Juglone

Flavonoids are heterocyclic compounds with a 2-phenyl-1,4-benzopyrone structural unit. Flavonoids provide very soothing colours ranging from pale yellow (isoflavones) to reds and blues (anthocyanins) through deep yellow (chalcones, flavones, flavonols, aurones) and oranges (aurones) [54]. Chalcone is thought to be the first flavonoid initially formed in the biosynthetic pathways, by combining 4-coumaroyl-CoA with malonyl-CoA. The most common examples of plant flavonoids are luteolin, quercitrin, apigenin, kaempferol, acacetin, and moral. The performance of these colourants is also enhanced by the use of mordants [55].

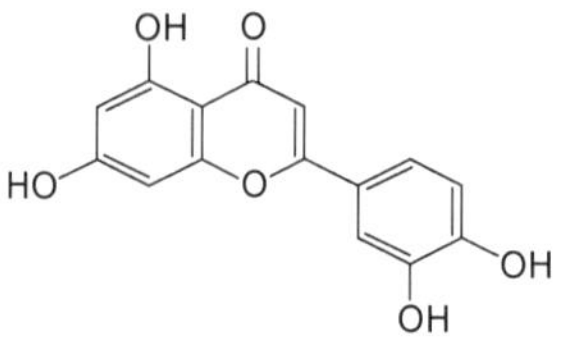

Luteolin

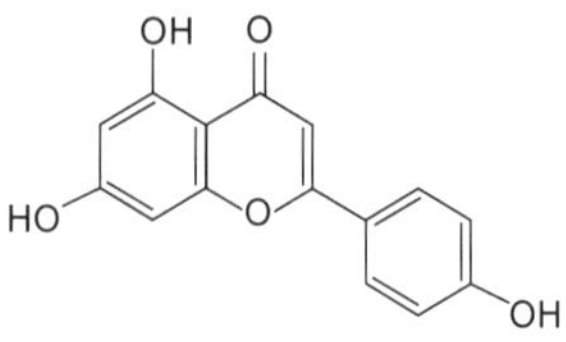

Apigenin

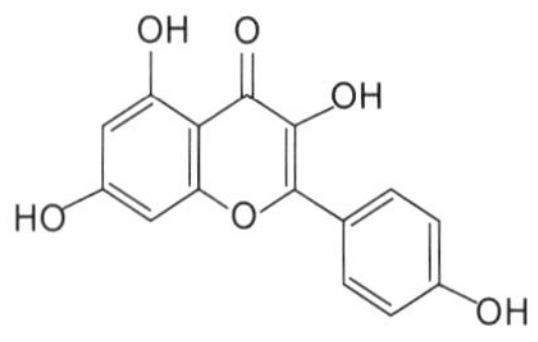

Kaempferol

Dihydropyran pigments include brazilin (C.I. 75280) from brazilwood and haematoxylin (C.I. 75290) from logwood. These dyes also show high affinities for various textile materials in the presence of mordants.

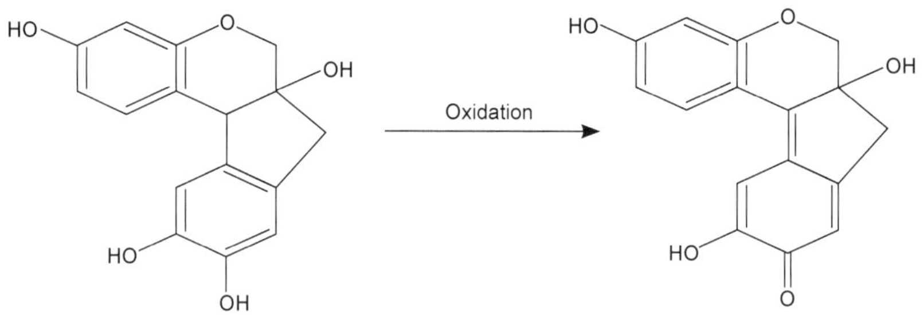

Brazilin Haematoxylin

Betalains are nitrogen-containing pigments comprising of yellow betaxanthins and violet betacyanins. Owing to the presence of bioactive components, these pigments find widespread applications in the food and pharmaceutical industries [56]. However, very few reports are available in the literature concerned with the textile colouration effects of betalains with simultaneous antibacterial effects [57,58].

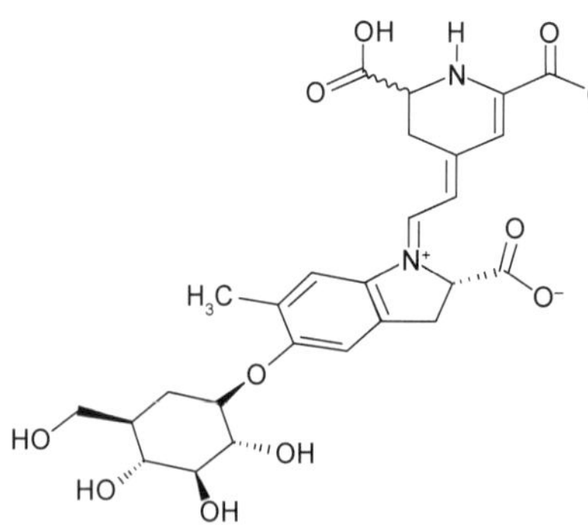

Betanin

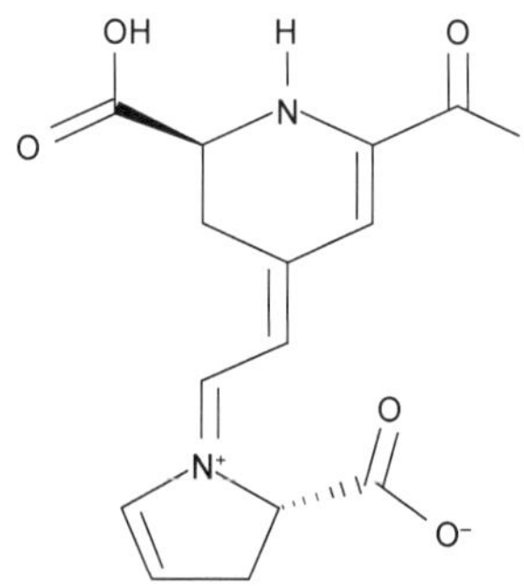

Yellow indicaxanthin

Tannins are high molecular weight water-soluble phenolic compounds found in various plant parts such as fruit, pods, plant galls, leaves, bark, wood, and roots [59,60]. The most common examples of tannins are gallotannis and ellagitannins. When an acid component is gallic acid, the compound is called gallotannins. The simplest and most widespread gallotannin is pentagalloylglucose. Ellagitannins are esters of hexahydroxydiphenic acid, which on complete hydrolysis yields ellagic acid. Tannin extracts have been used to dye wool fibre/fabric samples in the recent past, whose performance (fastness properties and antimicrobial finishing) has been increased by using different metal salts and natural mordants [24,25]. More recently, tannin extract has been used to enhance dye characteristics in conjunction with some previously established natural dyes such as *T. arjuna*, *A. vasica*, etc [10].

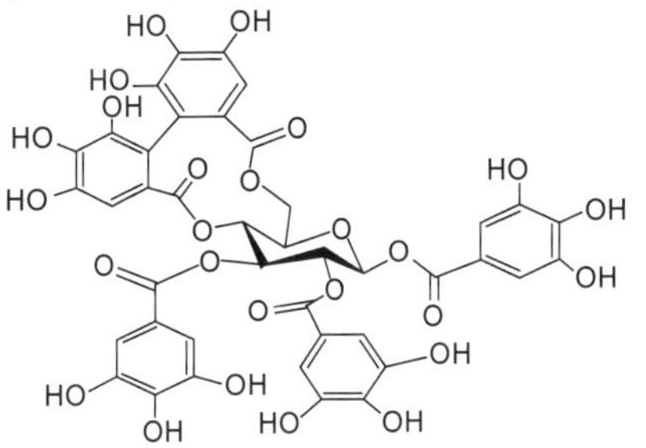

Gallotannins

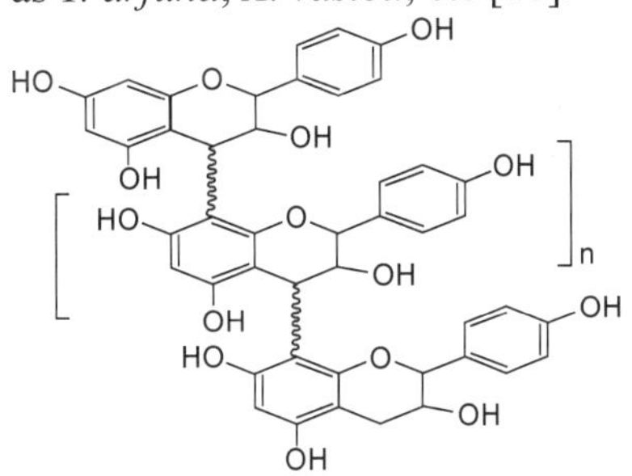

Ellagitannins

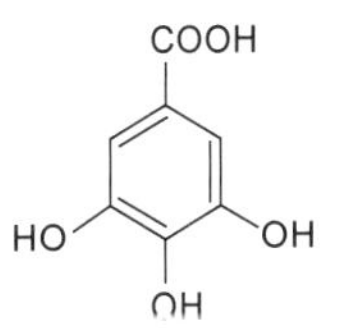

Gallic acid

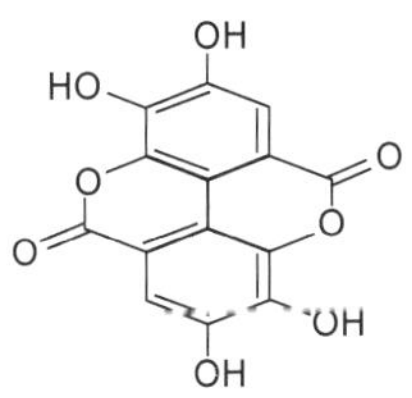

Ellagic acid

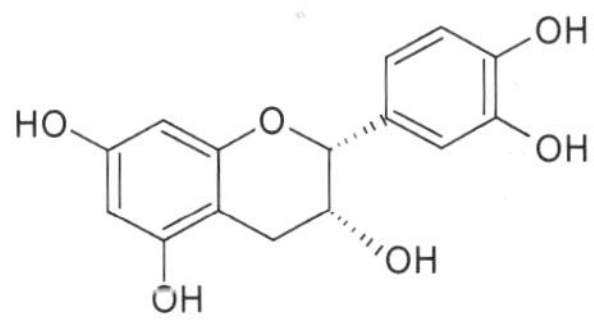

(-)-epicatechin

3.5 Sustainable dye harvesting (Extraction, mordanting, and dyeing)

Dyeing with natural colourants has been in trend since prehistoric times, when mankind started wearing clothes. Natural dyes for textile materials are of great interest to both textile industries as well as research fields due to their eco-safety attributes. Use and production of eco-friendly coloured textiles on customer's repeated demands led to the revival of natural dyes with newer and energy-efficient extraction and dyeing processes for the production of acceptable and reproducible shades. Conventional and non-conventional methods have been employed in dye extraction processes, mordanting, and dyeing involving the use of modern technologies such as microwave energy, ultrasonic energy, plasma, high temperature high pressure (HT-HP), and pad-dyeing techniques [61,62]. As a result, a lot of research findings have been done with regard to the chemical analysis of dyes, improvement in the extraction techniques, re-establishment of traditional dyeing techniques, physico-chemical studies of the dyeing process, dyeing kinetics, development of newer shades, analysis in terms of colour parameters, test for compatibility of binary dye mixtures, chemical modification of textiles, and use of modern technologies at different stages of textile dyeing process [24,63-65].

3.5.1 Extraction of natural colourants

Extraction is the first step that is thought to consume energy along with tedious monitoring and isolation of colouring components, but technological advancements and research studies have made it easy in terms of energy consumption and laborious work. The increasing need for natural dyes in the textile industry demands the use of efficient dye extraction techniques, which not only increases the extraction yield but also saves energy, time, and amount of chemicals. Solid-liquid partitioning extraction techniques may enhance natural dye usage in textile, cosmetics, and food industries due to the increased extraction efficiency. The use of benzene, chloroform, dichloromethane, ethyl acetate, ethanol, methanol, etc., or their binary combinations in solid-liquid partitioning methods helps in increasing the dye extraction yield [66]. However, the use of excessive organic solvents poses environmental and human health threats in addition to increasing the overall cost of natural dyeing processes [67]. To overcome the above-mentioned limitation of using organic solvents, newly discovered extraction methods such as the use of pressurised fluids, gamma radiations, ultrasound, and microwave techniques seem to be the most favorable and sophisticated, with an increased extraction efficiency [68,69].

Since there are abundant amounts of natural resources, aqueous extraction of dyes is highly studied in literature owing to the characteristics of water as a solvent, such as solubility and easy and cheap availability [70-72]. Extraction of coloured compounds from various plant parts using an environment-friendly aqueous solvent is one of the best alternatives with the least side effects. The use of small amounts of polar organic solvents such as ethanol, methanol and ethyl acetate in conjunction with water can increase the partition coefficient, which will enhance the dye transfer from a solid mass to a liquid surface. Also, higher solubility of colour components in the organic part will further help in transferring more dye from solid to liquid surface [73]. Additionally, alkalinity and acidity are important for the extraction of colourants from any dye source because it helps in releasing dye components by enhancing the penetration of solvent and disruption of plant tissues [74]. The use of citric acid (pH = 2.5) has improved the extraction of 3-deoxyanthocyanins from sorghum brans compared to a simple aqueous solvent [75]. Similar results of a high extraction efficiency of colour from sorghum leaves were reported by adding HCl to the ethanol/water combinations [76].

The efficiency of aqueous extraction can be improved by introducing enzymes in the extraction medium. But it takes a longer time to extract a specified amount of dye under enzymatic and fermentation methods compared to other recent sophisticated methods such as ultrasonic and microwave assisted extraction [77]. Several reports are available in the literature that have explored the use of ultrasonic and microwave energy for increasing the extraction efficiency in terms of less solvent consumption, decrease in extraction time, and low temperature [78]. Experimental results from the study the Gong et al. (2019b) showed that changes in ultrasonic frequencies and time accelerated the extraction process with better yield of chlorophyll and anthocyanin under acidic conditions from fresh and old leaves of *C. camphora* [68]. Supercritical fluid extraction uses new, non-toxic, and environmentally-friendly alternative solvents such as carbon dioxide, which under reduced pressure, liquefies and acts as a solvent [79].

3.5.2 Mordanting

Protein fibres such as wool and silk are believed to have good substantivity towards natural dyes owing to the abundant functional groups on their surface, but some synthetic fibres and cellulose show difficulty in dyeing with some natural dyes. Various ecofriendly pretreatments such as natural polymers (chitosan, cyclodextrin) and metal salts (alum, iron) have been developed to solve this low substantivity of natural dyes on some textile fibres [80,81]. Such pretreatments chemicals resulting in high fastness and simultaneously altering the colour characteristics are known as mordants. However, to increase the special performance of textile materials using natural dyes, scientists all over the globe are working on enhancing the intrinsic low light fastness properties of natural dyes, which restricts their end uses [71, 82]. Metal salts and their combinations, known as mordants, have been used to overcome this problem over a wide range of natural and synthetic fibres [83,84].

The colour characteristics modifications depend on the factors such as the chemical nature of mordant, mordant combinations or their concentration ratios, and mordant-dye-fibre interactions. Significant research work has been done regarding the types of mordants used, mordanting methods, and their effects on photofading characteristics of naturally dyed textile materials. Some of the commonly studied metallic salt mordants as well as biomordants used to alter the colour properties of textile materials dyed with natural dyes are enlisted in some review papers published earlier [1,4].

Various metallic salts have been used in natural dyeing processes to change the colour characteristics and fastness properties of dyed textiles, including less toxic (up to a certain limit) metals such as alum, ferrous sulphate, stannous chloride [16] and rare earth metals (on ramie fabrics) [21]. Satisfactory results were obtained by using different metallic salts, among which copper sulphate and ferrous sulphate gave light-resistant shades. But the use of metal salts for mordanting processes possesses environmental and human health concerns as a significant amount of metal ions remains unused in the residual dye baths, due to which discharge of textile industries containing the concentration of metal ions beyond a certain limit is strictly banned [37,85]. Alternative use of natural mordants, which are biologically and environmentally safe, has become a new trend in modern-day natural dyeing research and is thought that they will eliminate the hazardous effects of previously used metal salts [86-88].

Mordants of biological origin such as myrobalans (*T. chebula*), pomegranate rinds extract (*P. granatum*), tannin, tannic acid, tartaric acid, guava, and banana leaves ash have been used increasing dye-fibre interactions. Phenolic hydroxyl groups of tannins help in effective crosslinking between dye components and fibre which eventually improves colour fixation. Tannin-based mordants are found to be most effective in cotton dyeing for most of the natural dyes. Some plant materials (leaves, bark, ash, or fruits) with high tannin or metal content (Chlorophyll) may also be used for mordanting purposes depending upon the nature of phytoconstituents, structure and amount of metal present in them. Polyphenolic compounds from gallnut, pomegranate peel, and babool have been explored as biomordants with *Adhatoda vasica* and *Alkanna tinctoria* natural dyes for wool dyeing to establish a correlation between colour characteristics and biomordants. Various biomordants such as chlorophyll-a, valex, rosemary, *Eurya acuminate,* and tamarind seed coats have been explored with different natural dyes in literature. Some of the natural mordants are listed in Table 3.2. Aluminium-rich or hyperaccumulating plants are also a potential source of natural metallic mordants and have been used in traditional Indonesian textiles, and some Al-rich plants are also listed in Table 3.2, which have been previously used for bio-mordanting processes [10,36,89-93].

Table 3.2 List of plants used as source of biomordants in natural textile dyeing (Table extended from Shahid et al., 2013) [4]

Botanical Name	Reason for use as mordant	References
Eucalyptus spp.	Used as sources of tannin mordant	[85]
Entada spiralis		[52]
Acacia catechu		[38]
Acacia nilotica		[10,86]
Emblica officinalis		[87]
Memecylon scutellatum		[88]
Punica granatum		[10]
Quercus infectoria		[71]
Quercus ithaburensis ssp. macrolepis		[95]
Rosmarinus officinalis		[95]
Thuja orientalis		[95]
Rhus coriaria		[96]
Rumex hymenosepolus		[36]
Tamarindus indica		[97]
Prosopis spp.		[98]
Terminalia bellerica		[98]
Terminalia chebula		[35]
Terminalia arjuna		[10]
Enterolobium cyclocarpum	Al-hyperaccumulating plants used as substitute of alum mordant	[89]
Caesalpinia coriaria		[89]
Symplococcus spp.		[99]
Aporusa spp.		[99]
Baccaurea racemosa		[99]
Xanthophyllum lanceatum		[99]
Eurya acuminate		[100]
Pyrus pashia	Cu rich plants	[90]
Sasa veitchii		[91]
Cinnamomum camphora	Chlorophyll rich plants	[17]
Xylocarpus moluccensis		[101]

There are three types of mordanting methods based on the time of their application during the process of dyeing.

In a pre-mordanting method, mordants are applied prior to dyeing. It is most commonly employed for cotton and cellulosic textile materials as they have the least affinity for most of the natural dyes. The advantage of this mordanting method is that the mordant baths can be used many more times continuously. This makes the pre-mordanting method an economical and sustainable process with large-scale applications along with the reduction of pollution load. Metal ions form a coordination complex with the functional groups of wool, silk, etc., and act as a bridge between textile and dye components involving complex formation during the pre-mordanting method (Fig. 3.4). Pre-mordanting technique is superior for stannous chloride and alum mordant [94].

In the post-mordanting method, mordanting of fabric/fibre/yarn is done after the dyeing process in a separate bath, and the final colour is developed in the last phase (Fig. 3.5) [93]. Post-mordanting is recommended in the case of iron and copper mordants for producing grey and black colours, and it improves light fastness to a greater extent [94].

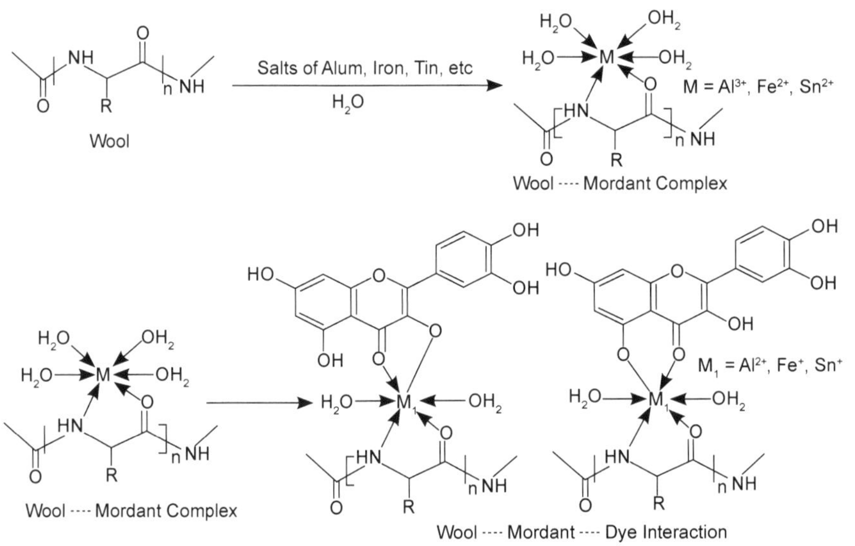

Figure 3.4 Probable schematic representation of Wool-mordant-Dye Interaction in pre-mordanting method

In the meta-mordanting method, mordanting and the dyeing processes are simultaneously carried out in the same bath. The advantage of this mordanting method is the reduction in the dyeing time due to a smaller number of steps involved. Best results have been achieved for ecru denim dyeing with a harda-tartaric acid combination using onion extract [102]. For animal fibres, mordant or mordant combinations may be introduced into the dye bath solutions towards the end of the dyeing process, when dye exhaustion will be near equilibrium. Through this method, darker shades have been produced for some dyes, while the colour yield for some other dyes may decrease due to the interaction between the dye and mordant (dye-mordant complex formation) and may also cause uneven dyeing. The drawback of this mordanting method is that the unused mordant in the dye bath cannot be reused.

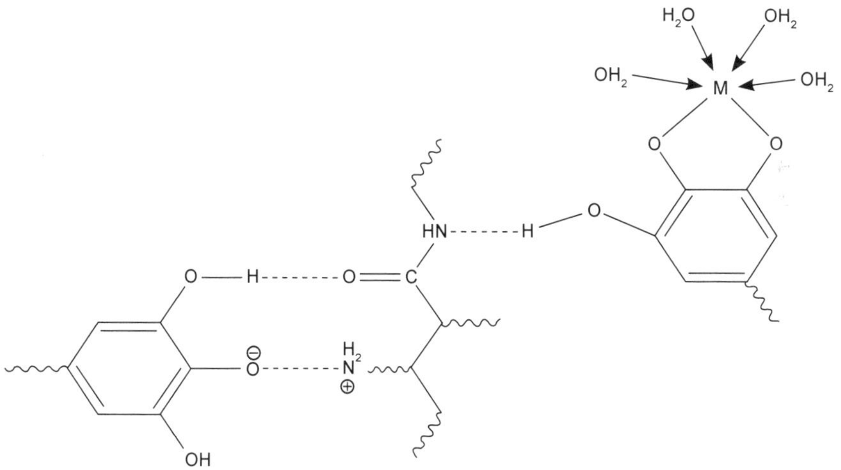

Figure 3.5 Probable schematic representation of Wool-Dye-Mordant Interaction in post-mordanting method

3.5.3 Simultaneous dyeing and functional finishing of textile materials using natural dyes

Significant advancements have been achieved in the last few decades in natural dye applications producing coloured textile materials with additional environmental and aesthetic benefits, but insufficient scientific studies and systematic reports render full exploitation of these dye yielding flora and fauna with a lot of natural products still untouched. Eventually, a large number of factors involved in natural dyeing processes need proper standardisation of dyeing variables. Researchers all over the globe have done enormous investigations on dyeing and subsequent functional finishing [68,103-108]. Exploration and fitting of thermodynamic and kinetic models on the dye adsorption data help in better understanding of dye chemistry and helps in enhancing the dyeing performance of natural dyes [55,105-108].

Conventional and non-conventional dyeing methods have been used to impart colour to textile fibres and fabric. Conventional dyeing methods involve adsorption of colourants through padding or exhaustion technique involving only conventional energy in the form of heat energy [4]. Conventional dyeing methods use large amounts of solvents, energy, and auxiliary chemicals such as salts. In comparison, non-conventional dyeing techniques involve the use of modern technologies such as ultrasonic, microwave, and supercritical CO_2. These methods involve a different kind of energy and require less chemicals and other auxiliaries. The results of ultrasonic dyeing have been found to be much better compared to conventional dyeing [68]. Non-conventional dyeing technique such as supercritical CO_2 is environmentally favourable as it uses less or no chemical auxiliary and does not release any wastewater from the dyeing bath [109]. In view of the full exploitation of natural dyes, optimisation of dyeing conditions/variables needs to be done in response to the type of natural dye, type of mordant, pH, temperature, M:L ratio, etc [103,104].

Textile materials, such as wool, cotton and silk, are ideal substrates for the growth and proliferation of microorganisms having a negative impact on hygiene and other properties. Research in the area of production of hygienic clothing via imparting antimicrobial, antifungal, and antioxidant properties has been under way from the last few decades. Bioactive components from natural plant extracts, in addition to the normal colouration, are used to impart functional finishing properties to coloured fabric/fibre besides being biocompatible, non-toxic, and renewable [1,3,68]. Rather et al. (2016b, 2017b) explored the potential of tannin-rich extract of *Acacia nilotica* and *Terminalia arjuna* for the production of antioxidant and antimicrobial wool fibres with acceptable and satisfactory wash durability [24,25]. UV-protection properties of wool, silk, and cotton fabric were investigated by simple adsorption of *Cinnamomum camphora* leaves extract from fresh and old leaves [68]. Gomes da Silva et al. (2018) evaluated the antimicrobial and UV protection properties of cotton fabric dyed with eucalyptus leaves extract which were enhanced by the use of chitosan as a mordant [110]. The pad-steam process was used to impart antimicrobial functionality in addition to normal colour with tea stem extract without using chemical mordants. Tannin-rich extract from gallnut was used to prepare chitosan/gallnut tannin composite fibre with 99.7% reduction in *S. aureus* bacterial colonies in addition to the double green and red fluorescent finishing [111]. Simultaneous colouration and flame retardant property with bioactive functionalisation (Antibacterial and antioxidant) of silk fabric was carried out using functional dyes extracted from tea steam wastes [112]. Natural dyes extracted from *Acacia auriculiformis* imparted excellent antibacterial resistance to wool and silk fabric against *S. aureus* and *E. coli*. UPF values were found to increase with an increase in the dye concentration [113]. Pomegranate peel and walnut green husks have been explored as green antimicrobial agents, and the results were compared with that of the antimicrobial action of inorganic nanoparticles ($Ag/Cu_2O/ZnO$) on wool yarns [114].

3.6 Natural dye printing

A lot of work has been carried out for the dyeing of textiles using natural dyes, but the work on printing is very limited. Natural pigment printing involves the use of natural dyes and a thickener to develop colourful shades. Hakeim et al. (2005) used curcumin natural dye to print chitosan pre-treated cotton fabric [115]. An increase in colour yield was found with increasing chitosan concentration. Hebeish et al. (2006) used henna as a natural dye in presence of reactive cyclodextrin in cotton printing [116]. Karolia and Buch (2008) documented the magnificent resistant block cotton fabric (Ajrakh) printed with natural dyes (*Terminalia chebula*, *Curcuma domestica*, *Punica granatum*, and *Tamarindus indica*) [117]. However, to cut down the time of printing production, natural indigo was replaced by synthetic indigo in due course of time. Rekaby et al. (2009) used *Alkalna tinctoria* and *Rheum emodi* natural dyes for printing cotton, wool, silk and flax fabrics with sodium alginate and British gum as thickening agents using the pigment printing technique [118]. The effect of different printing styles on cotton fabrics was studied by Mukherjee et al. (2011) [119]. The flatbed screen-printing technique was used to print water-based digital printing ink of annatto dye to get wash-fast cotton fabric with improved physical properties such as conductivity, pH, and surface tension [120]. Printed jute fabric with diversified value-added applications had been produced by eco-friendly printing with *Rubia cordifolia*, *Onosma echoides*, and *Bixa orellana* in presence of natural thickeners such as sodium alginate, gum, and gum arabic [121].

Ink-jet printing is one of the emerging fields in textile colouration with many benefits over conventional printing, such as simplicity, low cost, waste management, and low water and energy consumption. Recently, there has been growing interest in ink-jet inks for textile printing, and the use of annatto natural dye is considered a novelty in the race of replacing synthetic dyes [120]. A series of natural dyes, pomegranate peel, cutch, golden dock, and annatto was used to prepare water-based ink for printing of cotton fabric and evaluated over a period of 90 days in terms of stability, wash, light, and rub fastness [122].

3.7 Sustainability and environmental prospects of natural dyeing and finishing

Natural dyes obtained from plants and animals are considered eco-friendly and biocompatible since these are produced from never ending-natural resources. These compounds of natural origin for dyeing and finishing of textiles prove themselves sustainable based on the following characteristics:

- Replacement to synthetics
- Easy availability
- Economically cheap
- Environment friendly

Sustainable development is defined as the development which occurs keeping in mind the betterment and fulfilment of the needs of coming generations. Sustainable development mainly stood on three pillars of economy, environment and society of any state (Fig. 3.6).

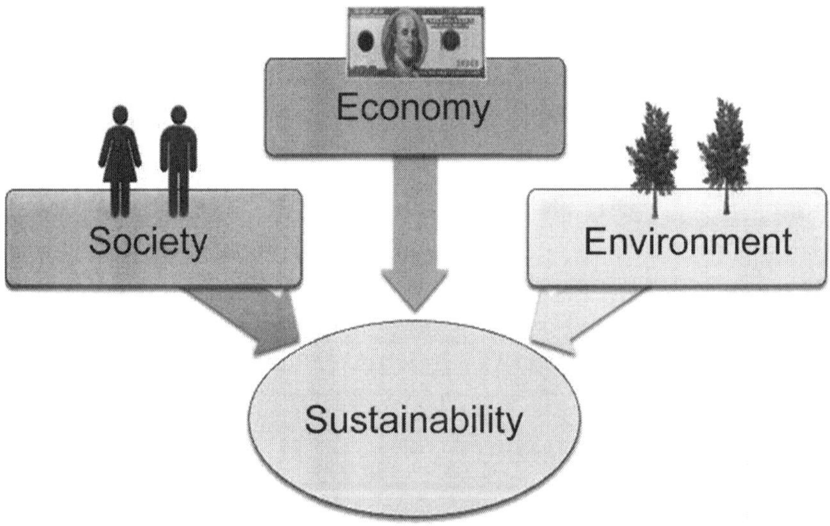

Figure 3.6 Sustainability pillars i.e. society, economy and environment

Above mentioned characteristics of natural dyes exactly fit into the definition of sustainable development in every aspect. Synthetic dyes have replaced natural dyes for easy dyeing, a wide range of shades and high fastness properties. But, since people have been known about the associated hazards of the use of synthetic dyes, interest towards natural dyes has increased in the last years. Advanced eco-friendly methods of extraction, dyeing and finishing of textiles have emerged for the revival of natural dyes. Mordants are generally used to enhance substantivity of the natural dyes on textiles, and previously many heavy metal ions had been used for dyeing, but due to their harmful effects on the environment, their use had to be discontinued, and now some mordants with negligible or no toxicity are used within the eco-permissible limits. Within the last few years mainly, biomordants (discussed in detail earlier in the chapter) have emerged as an alternative to metal mordants, and many research works have been reported in recent years.

Modern techniques such as radiation treatment have emerged as non-conventional techniques to improve dyeing effects of natural dyes and the dyeing behavior of textile substrate towards natural dyes. This technique has been proved highly beneficial, and there is a lot to explore in this area. As per our knowledge, these radiations have no known effects on the environment till now. Previously, a lot of research work has been carried out by using radiation treatments on textile substrates for the improvement of dyeing behavior [123,124]. Several techniques regarding dyeing and functional finishing are gaining popularity owing to their associated benefits. UV, gamma, ion-beam irradiation, laser, plasma, microwave and ultrasound energy treatments are nonconventional methods for textile dyeing and finishing, which are in accordance with sustainability aspects. Solvent-free and heat treatment dyeing are also in the same line of sustainability.

Concluding remarks and future perspectives

Natural dyes, owing to their ecological and environmental benefits in addition to being cheap, non-toxic, and derived from a sustainable and renewable resource, have attracted the attention of the people in the recent past in order to explore them for their potential use in a variety of application disciplines. Although, in the past few years, research in the area of natural dyes has helped scientists to understand the technological details required for their efficient utilisation. However, there are still technical challenges that are to be overcome before these technologies can be adopted on an industrial scale. Enhancing the extraction of colourants in terms of percentage yield and their subsequent applications with cost-effective processing is a challenging job which needs further deep research studies. Well-documented research studies and their enthusiastic results, which have been cited in this chapter, may help in exploring further research with improving results, eventually leading us towards the economic viability of natural dye production at an industrial scale for sustainable utilisation of bioresources.

References:

1. Yusuf M, Shabbir M, Mohammad F (2017), 'Natural Colorants: Historical, Processing and Sustainable Prospects', *Nat Products Bioprospect*, 7, 123-145.
2. Gürses A, Açıkyıldız M, Güneş K, Gürses MS (2016), 'Classification of Dye and Pigments. In: Dyes and Pigments, SpringerBriefs in Molecular Science. Springer, Cham.
3. Ramamoorthy S, Mudgal G, Rajesh D, Nawaz Khan F, Vijayakumar V, Rajasekaran C (2009), 'Characterisation of novel pH indicator of natural dye Oldenlandia umbellata L', *Nat Prod Res*, 23, 1210-1217.
4. Shahid M, Shahid-ul-Islam, Mohammad F (2013), 'Recent advancements in natural dye applications: a review', *J Clean Prod*, 53, 310-331.
5. Holme I. (2006), 'Sir William Henry Perkin: a review of his life, work and legacy', *Color Technol*, 122, 235-251.
6. Ahlström LH, Sparr Eskilsson C, Björklund E (2005), 'Determination of banned azo dyes in consumer goods', *TrAC Trends Anal Chem*, 24, 49-56.
7. Rather LJ, Shahid-ul-Islam, Akhter S, Hassan QP, Mohammad F (2017), 'Chemistry of Plant Dyes: Applications and Environmental Implications of Dyeing Processes', *Curr Environ Eng*, 4, 103-120.
8. Kumar JK, Sinha AK (2004), 'Resurgence of natural colourants: a holistic view', *Nat Prod Res*, 18, 59–84.
9. Mirjalili M, Nazarpoor K, Karimi L (2011), 'Eco-friendly dyeing of wool using natural dye from weld as co-partner with synthetic dye', *J Clean Prod*, 19, 1045-1051.
10. Rather LJ, Shahid-ul-Islam, Shabbir M, Bukhari MN, Shahid M, Khan MA, Mohammad F (2016), 'Ecological dyeing of Woolen yarn with *Adhatoda vasica* natural dye in the presence of biomordants as an alternative copartner to metal mordants', *J Environ Chem Eng*, 4, 3041-3049.
11. Shahid-ul-Islam, Shahid M, Mohammad F (2013), 'Green Chemistry Approaches to Develop Antimicrobial Textiles Based on Sustainable Biopolymers-A Review', *Ind Eng Chem Res*, 52, 5245–5260.
12. Gorjanc M, Gorensek M, Jovancic P, Mozetic M (2013), 'Multifunctional Textiles - Modification by Plasma, Dyeing and Nanoparticles. Eco-Friendly Textile Dyeing and Finishing Melih Günay, IntechOpen.
13. Shahid-ul-Islam, Shahid M, Mohammad F (2013), 'Perspectives for natural product based agents derived from industrial plants in textile applications - a review', *J Clean Prod*, 57, 2-18.
14. Shahid-ul-Islam, Mohammad F (2015), 'Natural Colorants in the Presence of Anchors So-Called Mordants as Promising Coloring and Antimicrobial Agents for Textile Materials', *ACS Sustain Chem Eng*, 3, 2361-2375.
15. Yusuf M, Shahid M, Khan MI, Khan SA, Khan MA, Mohammad F (2015), 'Dyeing studies with henna and madder: A research on effect of tin (II) chloride mordant', *J Saudi Chem Soc*, 19, 64-72.

16. Samanta AK, Agarwal P (2009), 'Application of natural dyes on textiles', *Indian J Fibre Text Res*, 34, 384-399.

17. Gong K, Rather LJ, Zhou Q, Wang W, Li Q (2019), 'Natural dyeing of merino wool fibers with *Cinnamomum camphora* leaves extract with mordants of biological origin: a greener approach of textile coloration', *J Text Inst*, https://doi.org/10.1080/00405000.2019.1698228

18. Nambela L, Haule LV, Mgani Q (2020), 'A review on source, chemistry, green synthesis and application of textile colorants', *J Clean Prod*, 246, 119036.

19. Kadolph SJ (2008), 'Natural dyes: a traditional craft experiencing new attention', *Delta Kappa Gamma Bull*, 75, 14-17.

20. Taylor GW (2008), 'Natural Dyes in Textile Applications', *Rev Prog Color Relat Top*, 16, 53-61.

21. Guesmi A, Hamadi N Ben, Ladhari N, Saidi F, Maaref H, Sakli F (2013), 'Spectral characterization of wool fabric dyed with indicaxanthin natural dye: Study of the fluorescence property', *Ind Crops Prod*, 46, 264-267.

22. Hwang E-K, Lee Y-H, Kim H-D (2008), 'Dyeing, fastness, and deodorizing properties of cotton, silk, and wool fabrics dyed with gardenia, coffee sludge, Cassia tora. L., and pomegranate extracts', *Fibers Polym*, 9, 334-340.

23. Specos MM, García JJ, Tornesello J, Marino P, Vecchia MD, Tesoriero MV, Hermida LG (2010), 'Microencapsulated citronella oil for mosquito repellent finishing of cotton textiles', *Trans R Soc Trop Med Hyg*, 104, 653-658.

24. Rather LJ, Shahid-ul-Islam, Azam M, Shabbir M, Bukhari MN, Shahid M, Khan MA, Haque QMR, Mohammad F (2016), 'Antimicrobial and fluorescence finishing of woolen yarn with *Terminalia arjuna* natural dye as an ecofriendly substitute to synthetic antibacterial agents', *RSC Adv*, 6, 39080-39094.

25. Rather LJ, Akhter S, Padder RA, Hassan QP, Hussain M, Khan MA, Mohammad F (2017), 'Colorful and semi durable antioxidant finish of woolen yarn with tannin rich extract of *Acacia nilotica* natural dye', *Dye Pigm*, 139, 812-819.

26. Abdel-Kareem O (2012), 'History of Dyes Used in Different Historical Periods of Egypt', *Res J Text Appar*, 16, 79-92.

27. Kuswandi B, Jayus, Larasati TS, Abdullah A, Heng LY (2012), 'Real-Time Monitoring of Shrimp Spoilage Using On-Package Sticker Sensor Based on Natural Dye of Curcumin', *Food Anal Methods*, 5, 881-889.

28. Zyoud A, Zaatar N, Saadeddin I, Helal MH, Campet G, Hakim M, Park D, Hilal HS (2011), 'Alternative natural dyes in water purification: Anthocyanin as TiO2-sensitizer in methyl orange photo-degradation', *Solid State Sci*, 13, 1268-1275.

29. Loyson P (2011), 'Chemistry in the Time of the Pharaohs', *J Chem Educ*, 88, 146-150.

30. Vankar PS, Shanker R, Srivastava J (2007), 'Ultrasonic dyeing of cotton fabric with aqueous extract of *Eclipta alba*', *Dye Pigm*, 72, 33-37.

31. Siva R (2007), 'Status of natural dyes and dye-yielding plants in India', *Curr Sci*, 92, 916-925.

32. Chen W (1984), '*Zhongguo Fangzhi Kexue Jishushi, Gudai Bufen* (*History of the Science and Technology of Chinese Textiles, the Ancient Part*)', Beijing: Kexue chubanshe.

33. Huang N, Chen (1995), '*Zhongguo Fuzhuang Shi* (*History of Chinese Costumes*) (Beijing: Zhongguo lüyou chubanshe)'.

34. Zhu P (2003), 'Mingdai Yu Qingdai Qianqi Guangdong De Haishang Sichou Maoyi (Guangdong's Maritime Silk Trade during the Ming and Early Qing Dynasties)' (PhD thesis, Jinan University).

35. Samanta AK, Konar A, Chakraborti S, Datta S (2011), 'Dyeing of jute fabric with tesu extract: Part 1 — Effects of different mordants and dyeing process variables', *Indian J Fibre Text Res*, 36, 63-73.

36. Aminoddin H (2010), 'Functional dyeing of wool with natural dye extracted from *Berberis vulgaris* wood and *Rumex hymenosepolus* root as biomordant', *Iran J Chem Chem Eng*, 29, 55-60.

37. Guesmi Ahlem, Ladhari N, Hamadi N Ben, Msaddek M, Sakli F (2013), 'First application of chlorophyll-a as biomordant: sonicator dyeing of wool with betanin dye', *J Clean Prod*, 39, 97-104.

38. Mansour HF, Heffernan S (2011), 'Environmental aspects on dyeing silk fabric with sticta coronata lichen using ultrasonic energy and mild mordants', *Clean Technol Environ Policy*, 13, 207-213.

39. Dapson R (2007), 'The history, chemistry and modes of action of carmine and related dyes', *Biotech Histochem*, 82, 173-187.

40. Marzano A (2013), 'Harvesting the Sea: The Exploitation of Marine Resources in the Roman Mediterranean', Oxford University Press.

41. Yuan L, Han A, Ye M, Chen X, Yao L, Ding C (2018), 'Synthesis and characterization of environmentally benign inorganic pigments with high NIR reflectance: Lanthanum-doped BiFeO3', *Dye Pigm*, 148, 137-146.

42. Tuli HS, Chaudhary P, Beniwal V, Sharma AK (2015), 'Microbial pigments as natural color sources: current trends and future perspectives', *J Food Sci Technol*, 52, 4669-4678.

43. Ahmad WA, Wan Ahmad WY, Zakaria ZA, Yusof NZ (2012), 'Application of Bacterial Pigments as Colorant, Springer Berlin Heidelberg.

44. Agboyibor C, Kong W-B, Chen D, Zhang A-M, Niu S-Q (2018), 'Monascus pigments production, composition, bioactivity and its application: A review', *Biocatal Agric Biotechnol*, 16, 433-447.

45. Perkin AG, Everest AE (1918), 'The Natural Organic Colouring Matters. London, New York Longmans, Green and Co.

46. Clark RJ, Cooksey CJ, Daniels MA, Withnall R (1993), 'Indigo, woad, and Tyrian Purple: important vat dyes from antiquity to the present', *Endeavour*, 17, 191-199.

47. Garcia-Macias P, John P (2004), 'Formation of Natural Indigo Derived from Woad (*Isatis tinctoria* L.) in Relation to Product Purity', *J Agric Food Chem*, 52, 7891-7896.

48. Ahn C, Zeng X, Obendorf SK (2012), 'Analysis of dye extracted from Phellodendron bark and its identification in archaeological textiles', *Text Res J*, 82, 1645-1658.

49. Leona M, Lombardi JR (2007), 'Identification of berberine in ancient and historical textiles by surface-enhanced Raman scattering', *J Raman Spectrosc*, 38, 853-858.

50. Rodriguez-Amaya DB (2016), 'Natural Food Pigments and Colorants', *Curr Opin Food Sci*, 7, 20-26.

51. Jackson H, Braun CL, Ernst H (2008), 'The Chemistry of Novel Xanthophyll Carotenoids', *Am J Cardiol*, 101, S50-S57.

52. Park JH, Gatewood BM, Ramaswamy GN (2005), 'Naturally occurring quinones and flavonoid dyes for wool: Insect feeding deterrents', *J Appl Polym Sci*, 98, 322-328.

53. Singh R, Jain A, Panwar S, Gupta D, Khare SK (2005), 'Antimicrobial activity of some natural dyes', *Dye Pigm*, 66, 99-102.

54. Giusti MM, Wallace TC (2009), 'Flavonoids as Natural Pigments, *Handbook of Natural Colourants, 1st ed., John Wiley & Sons, West Sussex,* 257-275.

55. Rather LJ, Shahid-ul-Islam, Mohammad F (2015), 'Study on the application of *Acacia nilotica* natural dye to wool using fluorescence and FT-IR spectroscopy', *Fibers Polym*, 16, 1497-1505.

56. Azeredo HMC (2009), 'Betalains: properties, sources, applications, and stability - a review', *Int J Food Sci Technol*, 44, 2365-2376.

57. Ali NF, El-Mohamedy RSR (2011), 'Eco-friendly and protective natural dye from red prickly pear (*Opuntia Lasiacantha* Pfeiffer) plant', *J Saudi Chem Soc*, 15, 257-261.

58. Ganesan P, Karthik T (2017), 'Analysis of colour strength, colour fastness and antimicrobial properties of silk fabric dyed with natural dye from red prickly pear fruit', *J Text Inst*, 108, 1173-1179.

59. Haslam E (1996), 'Natural Polyphenols (Vegetable Tannins) as Drugs: Possible Modes of Action', *J Nat Prod*, 59, 205-215.

60. Serrano J, Puupponen-Pimiä R, Dauer A, Aura A-M, Saura-Calixto F (2009), 'Tannins: Current knowledge of food sources, intake, bioavailability and biological effects', *Mol Nutr Food Res*, 53, S310-S329.

61. Kamel MM, El-Shishtawy RM, Youssef BM, Mashaly H (2007), 'Ultrasonic assisted dyeing. IV. Dyeing of cationised cotton with lac natural dye', *Dye Pigm*, 73, 279-284.

62. Nourmohammadian F, Gholami MD (2008), 'An Investigation of the Dyeability of Acrylic Fiber Via Microwave Irradiation', *Prog Color Color Coat*, 1, 57-63.

63. Bechtold T, Turcanu A, Geissler S, Ganglberger E (2002), 'Process balance and product quality in the production of natural indigo from *Polygonum tinctorium* Ait. applying low-technology methods', *Bioresour Technol*, 81, 171-177.

64. Borges ME, Tejera RL, Díaz L, Esparza P, Ibáñez E (2012), 'Natural dyes extraction from cochineal (*Dactylopius coccus*). New extraction methods', *Food Chem*, 132, 1855-1860.

65. Sinha K, Saha P Das, Datta S (2012), 'Response surface optimization and artificial neural network modeling of microwave assisted natural dye extraction from pomegranate rind', *Ind Crops Prod*, 37, 408-414.

66. Adedokun O, Sanusi YK, Awodugba AO (2018), 'Solvent dependent natural dye extraction and its sensitization effect for dye sensitized solar cells', *Optik (Stuttg)*, 174, 497-507.

67. Luque de Castro MD, Priego-Capote F (2010), 'Soxhlet extraction: Past and present panacea', *J Chromatogr A*, 1217, 2383-2389.

68. Gong K, Pan Y, Rather LJ, Wang W, Zhou Q, Zhang T, Li Q (2019), 'Natural pigment during flora leaf senescence and its application in dyeing and UV protection finish of silk and wool – a case study of *Cinnamomum Camphora*', *Dye Pigm*, 166, 114-121.

69. Duval J, Pecher V, Poujol M, Lesellier E (2016), 'Research advances for the extraction, analysis and uses of anthraquinones: A review', *Ind Crops Prod*, 94, 812-833.

70. Canche-Escamilla G, Colli-Acevedo P, Borges-Argaez R, Quintana-Owen P, May-Crespo JF, Cáceres-Farfan M, Puc JAY, Sansores-Peraza P, Vera-Ku BM (2019), 'Extraction of phenolic components from an Aloe vera (*Aloe barbadensis* Miller) crop and their potential as antimicrobials and textile dyes', *Sustain Chem Pharm*, 14, 100168.

71. Grifoni D, Bacci L, Zipoli G, Albanese L, Sabatini F (2011), 'The role of natural dyes in the UV protection of fabrics made of vegetable fibres', Dye Pigm, 91, 279-285.

72. Shabbir M, Rather LJ, Mohammad F (2018), 'Economically viable UV-protective and antioxidant finishing of wool fabric dyed with *Tagetes erecta* flower extract: Valorization of marigold', *Ind Crops Prod*, 119, 277-282.

73 Wizi J, Wang L, Hou X, Tao Y, Ma B, Yang Y (2018), 'Ultrasound-microwave assisted extraction of natural colorants from sorghum husk with different solvents', *Ind Crops Prod*, 120, 203-213.

74. Vatai T, Škerget M, Knez Ž (2009), 'Extraction of phenolic compounds from elder berry and different grape marc varieties using organic solvents and/or supercritical carbon dioxide', *J Food Eng,* 90, 246-254.

75. Barros F, Dykes L, Awika JM, Rooney LW (2013), 'Accelerated solvent extraction of phenolic compounds from sorghum brans', *J Cereal Sci*, 58, 305-312.

76. Kayodé APP, Bara CA, Dalodé-Vieira G, Linnemann AR, Nout MJR (2012), 'Extraction of antioxidant pigments from dye sorghum leaf sheaths', *LWT - Food Sci Technol*, 46, 49-55.

77. Armenta-Lopez R, Guerrero IL, Huerta S (2002), 'Astaxanthin Extraction From Shrimp Waste by Lactic Fermentation and Enzymatic Hydrolysis of the Carotenoprotein Complex', *J Food Sci*, 67, 1002-1006.

78. Chuyen HV, Nguyen MH, Roach PD, Golding JB, Parks SE (2018), 'Microwave-assisted extraction and ultrasound-assisted extraction for recovering carotenoids from Gac peel and their effects on antioxidant capacity of the extracts', *Food Sci Nutr*, 6, 189-196.

79. Herrero M, Mendiola JA, Cifuentes A, Ibáñez E (2010), 'Supercritical fluid extraction: Recent advances and applications', *J Chromatogr A*, 1217, 2495-2511.

80. Kampeerapappun P, Phattararittigul T, Jittrong S, Kullachod D (2010), 'Effect of chitosan and mordants on dyeability of cotton fabrics with *Ruellia tuberosa* Linn.', *Chiang Mai J Sci*, 38, 95-104.

81. Ghouila H, Meksi N, Haddar W, Mhenni MF, Jannet HB (2012), 'Extraction, identification and dyeing studies of Isosalipurposide, a natural chalcone dye from *Acacia cyanophylla* flowers on wool', *Ind Crops Prod*, 35, 31-36.

82. Ibrahim NA, El-Gamal AR, Gouda M, Mahrous F (2010), 'A new approach for natural dyeing and functional finishing of cotton cellulose', *Carbohydr Polym*, 82, 1205-1211.

83. Chairat M, Bremner JB, Chantrapromma K (2007), 'Dyeing of cotton and silk yarn with the extracted dye from the fruit hulls of mangosteen, *Garcinia mangostana* linn.', *Fibers Polym*, 8, 613-619.

84. Zheng GH, Fu H Bin, Liu GP (2011), 'Application of rare earth as mordant for the dyeing of ramie fabrics with natural dyes', *Korean J Chem Eng*, 28, 2148-2155.

85. Sachan K, Kapoor VP (2007), 'Optimization of extraction and dyeing conditions for traditional turmeric dye', *Indian J Tradit Knowl*, 6, 270-278.

86. Rather LJ, Khan MA, Mohammad F (2019), 'Biomordanting potential of *Acacia nilotica* (Babul) in conjunction with *Kerria lacca* and *Rheum emodi* natural dyes', *J Nat Fibers*, 16, 275-286.

87. Prabhu KH, Teli MD, Waghmare NG (2011), 'Eco-friendly dyeing using natural mordant extracted from *Emblica officinalis* G. fruit on cotton and silk fabrics with antibacterial activity', *Fibers Polym*, 12, 753-759.

88. Chairat M, Darumas U, Bremner JB, Bangrak P (2011), 'Dyeing of cotton yarn with the aqueous extract of the leaves of *Eupatorium odoratum* L. in Thailand and associated extract toxicity studies', *Color Technol*, 127, 346-353.

89. Arroyo-Figueroa G, Ruiz-Aguilar GML, Cuevas-Rodriguez G, Sanchez GG (2011), 'Cotton fabric dyeing with cochineal extract: influence of mordant concentration', *Color Technol*, 127, 39-46.

90. İşmal ÖE, Yıldırım L, Özdoğan E (2015), 'Valorisation of almond shell waste in ultrasonic biomordanted dyeing: alternatives to metallic mordants', *J Text Inst*, 106, 343-353.

91. Park SJ, Park YM (2010), 'Eco-dyeing and antimicrobial properties of chlorophyllin copper complex extracted from *Sasa veitchii*', *Fibers Polym*, 11, 357-362.

92. Vankar PS, Shanker R, Dixit S, Mahanta D, Tiwari SC (2008), 'Sonicator dyeing of natural polymers with Symplocos spicata by metal chelation', *Fibers Polym*, 9, 121-127.

93. Shabbir M, Rather LJ, Bukhari MN, Shahid-ul-Islam, Khan MA, Mohammad F (2019), 'First-time application of biomordants in conjunction with the *Alkanna tinctoria* root extract for eco-friendly wool dyeing', *J Nat Fibers*, 16, 846-854.

94. Gupta D, Gulrajani ML, Kumari S (2004), 'Light fastness of naturally occurring anthraquinone dyes on nylon', *Color Technol*, 120, 205-212.

95. İşmal ÖE, Yıldırım L, Özdoğan E (2014), 'Use of almond shell extracts plus biomordants as effective textile dye', *J Clean Prod*, 70, 61-67.

96. Bruni S, Guglielmi V, Pozzi F, Mercuri AM (2011), 'Surface-enhanced Raman spectroscopy (SERS) on silver colloids for the identification of ancient textile dyes. Part II: pomegranate and sumac', *J Raman Spectrosc*, 42, 465-473.

97. Prabhu KH, Teli MD (2014), 'Eco-dyeing using *Tamarindus indica* L. seed coat tannin as a natural mordant for textiles with antibacterial activity', *J Saudi Chem Soc*, 18, 864-872.

98. Shekhawat MS (2012), 'Studies on the use of in vitro synthesized red dye of *Arnebia hispidissima* L. in textile industries', *Pharma Innov*, 1, 63-68.

99. Cunningham AB, Maduarta IM, Howe J, Ingram W, Jansen S (2011), 'Hanging by a Thread: Natural Metallic Mordant Processes in Traditional Indonesian Textiles', *Econ Bot*, 65, 241-259.

100. Vankar PS, Shanker R, Mahanta D, Tiwari SC (2008), 'Ecofriendly sonicator dyeing of cotton with Rubia cordifolia Linn. using biomordant', *Dye Pigm*, 76, 207-212.

101. Rahman NAA, Tajuddin R, Tumin SM. Optimization of Natural Dyeing Using Ultrasonic Method and Biomordant. Int J Chem Eng Appl 2013:4:161-164.

102. Deo HT, Paul R (2000), 'Dyeing of ecru denim with onion extract as a natural dye using potassium alum in combination with harda and tartaric acid', *Indian J Fibre Text Res*, 25, 217-220.

103. Zhou Q, Rather LJ, Ali A, Wang W, Zhang Y, Haque QMR, Li Q (2020), 'Environmental friendly bioactive finishing of wool textiles using the tannin-rich extracts of Chinese tallow (*Sapium sebiferum* L.) waste/ fallen leaves', *Dyes Pigm*, 176, 108230.

104. Rather LJ, Zhou Q, Ali A, Haque QMR, Li Q (2020), 'Valorization of natural dyes extracted from mugwort leaves (*Folium artemisiae argyi*) for wool fabric dyeing: Optimization of extraction and dyeing processes with simultaneous coloration and biofunctionalization', *ACS Sustain Chem Eng*, 8, 2822-2834.

105. Chairat M, Rattanaphani S, Bremner JB, Rattanaphani V (2005), 'An adsorption and kinetic study of lac dyeing on silk', *Dye Pigm*, 64, 231-241.

106. Hou X, Yang R, Xu H, Yang Y (2012), 'Adsorption Kinetic and Thermodynamic Studies of Silk Dyed with Sodium Copper Chlorophyllin', *Ind Eng Chem Res*, 51, 8341-8347.

107. Rather LJ, Shahid-ul-Islam, Khan MA, Mohammad F (2016), 'Adsorption and Kinetic studies of *Adhatoda vasica* natural dye onto woolen yarn with evaluations of Colorimetric and Fluorescence Characteristics', *J Environ Chem Eng*, 4, 1780-1796.

108. Sun S-S, Tang R-C (2011), 'Adsorption and UV Protection Properties of the Extract from Honeysuckle onto Wool', *Ind Eng Chem Res*, 50, 4217-4224.

109. Shah DS, Shah J (2013), 'Unconventional Techniques for Energy Conservation in Textile Wet Processing', Formatex.

110. Gomes da Silva M, SD de Barros MA, Ribeiro de Almeida RT, Pilau EJ, Pinto E, Soares G, Santos JG (2018), 'Cleaner production of antimicrobial and anti-UV cotton materials through dyeing with eucalyptus leaves extract', *J Clean Prod*, 199, 807-816.

111. Zhu X, Hou X, Ma B, Xu H, Yang Y (2019), 'Chitosan/gallnut tannins composite fiber with improved tensile, antibacterial and fluorescence properties', *Carbohydr Polym*, 226, 115311.

112. Cheng T-H, Liu Z-J, Yang J-Y, Huang Y-Z, Tang R-C, Qiao Y-F (2019), "Extraction of Functional Dyes from Tea Stem Waste in Alkaline Medium and Their Application

for Simultaneous Coloration and Flame Retardant and Bioactive Functionalization of Silk', *ACS Sustain Chem Eng*, 7, 18405-18413.

113. Chakraborty L, Pandit P, Maulik SR (2020), '*Acacia auriculiformis* - A natural dye used for simultaneous coloration and functional finishing on textiles', *J Clean Prod*, 245, 118921.

114. Sadeghi-Kiakhani M, Tehrani-Bagha AR, Gharanjig K, Hashemi E (2019), 'Use of pomegranate peels and walnut green husks as the green antimicrobial agents to reduce the consumption of inorganic nanoparticles on wool yarns', *J Clean Prod*, 231, 1463-1473.

115. Hakeim OA, Abou-Okeil A, Abdou LAW, Waly A (2005), 'The influence of chitosan and some of its depolymerized grades on natural color printing', *J Appl Polym Sci*, 97, 559-563.

116. Hebeish AA, Ragheb AA, Nassar SH, Allam EE, El Thalouth JIA (2006), 'Technological evaluation of reactive cyclodextrin in cotton printing with reactive and natural dyes', *J Appl Polym Sci*, 102, 338-347.

117. Karolia A, Buch H (2008), 'Ajarkh, the resist printed fabric of Gujarat', *Indian J Tradit Knowl*, 7, 93-97.

118. Rekaby M, Salem AA, Nassar SH (2009), 'Eco-friendly printing of natural fabrics using natural dyes from alkanet and rhubarb', *J Text Inst*, 100, 486-495.

119. Mukherjee A, Nandi N, Mukherjee A (2011), 'Discharge printing with natural dyes', *Text Trends*, 45, 33-35.

120. Savvidis G, Karanikas E, Nikolaidid N, Eleftheriadis I, Tsatsaroni E (2014), 'Ink-jet printing of cotton with natural dyes', *Color Technol*, 130, 200-204.

121. Chattopadhyay SN, Pan NC, Roy AK, Saxena S, Khan A (2013), 'Development of natural dyed jute fabric with improved colour yield and UV protection characteristics', *J Text Inst*, 104, 808-818.

122. Savvidis G, Zarkogainni M, Karanikas E, Lazaridis N, Nikolaidid N, Tsatsaroni E (2013), 'Digital and conventional printing and dyeing with the natural dye annatto: optimisation and standardisation processes to meet future demands', *Color Technol*, 127, 55-63.

123. Zia KM, Adeel S, Rehman F, Aslam H, Khosa MK, Zuber M (2019), "Influence of ultrasonic radiation on extraction and green dyeing of mordanted cotton using neem bark extract', *J Ind Eng Chem*, 77, 317-322.

124. Adeel S, Gulzar T, Azeem M, Fazal-ur-Rehman, Saeed M, Hanif I, Iqbal N (2017), 'Appraisal of marigold flower based lutein as natural colourant for textile dyeing under the influence of gamma radiations', *Radiat Phys Chem*, 130, 35-39.

4

Microbial colourants – Future of sustainable colouration of textiles

Akankshya Panda, Pallavi Madiwale, Saptarshi Maiti, Madhura Nerurkar, Aranya Mallick, Ravindra Adivarekar*

Department of Fibres and Textile Processing Technology, Institute of Chemical Technology, Matunga (E), Mumbai- 400019, India

*Corresponding Author, Email: rv.adivarekar@ictmumbai.edu.in

Abstract: Today's textile dyeing industry is dominated by synthetic dyes, which got an initiation by the accidental discovery of a dye by William Henry Perkin in 1869. The gradual development in the potent chemical colours overtook the industry era, thereby affecting the environment and humans and hence led to the intense research in natural colours and dyes from flora and fauna. Though these conventional natural dyes are environment friendly, they can only supplement but an not substitute synthetic dyes due to various limitations. Microbial pigments are coloured metabolites secreted by microorganisms in stress conditions which can be an attractive source of non-toxic, biodegradable and non-carcinogenic alternatives for human use with the potential to substitute synthetic colours.

In this chapter, we discuss; the evolution of these microbial pigments, the current technology involved, the production stages, the metabolic mechanism of biosynthesis, the economics involved in pigment production, application of pigments in the textile colouration industry for therapeutic and other commercial finishes and novel strategies attempted by the various researchers. It also touches upon the bulk production of microbial pigments and their future challenges and further opportunities.

4.1 Introduction

Colour is the most astonishing feature of any material. It determines their characteristics as well as their intensity of attractiveness to the viewers. Colour strongly contributes to the aesthetics of any product. It is because its appearance is considered as the primary aspect that is touched upon by our senses, whereas colour helps to determine its acceptance. In this age of fast consumerism, this factor has become more crucial, and major industries like food, cosmetics and textiles have enhanced the usage of colouring components for their products. This increased demand for colourants caused the acceptability of synthetically produced colours over natural colourants for their reliability, reproducibility and economics [1]. The history of synthetic dyes is not a long one, and it can be said to have started in 1856. In this year, aniline purple was synthesised by William Henry Perkin, and the colour was named as Perkin's Mauve or Mauveine. Since then, innumerable colourants have been chemically synthesised, mainly utilising petroleum resources. Natural colourants, on the other hand, have been utilized since the dawn of human civilisation. The wall paintings of the cave of Altamira, Spain, are one of the bright examples of it. Scientists have found out that these were painted in Upper Paleolithic Age, almost 36,000 years ago. The colouration of clothing material is also a part of human history. Coloured fabrics were found in the Indus valley civilisation (2600-1900 B.C.), where the clothes show traces of madder dye in the remains at Harrapa and Mohenjodaro. The attraction of the human eye towards colour and the art of dyeing also have pieces of evidence in the Bronze Age in Europe. Traces of blue pigment on 6000-year-old cotton fabrics from Huaca Prieta on the north coast of Peru was positively identified to be an indigoid dye (indigotin), making it the earliest known use of indigo in the world, derived from *Indigofera spp.* native to South America [2–4]. In the year 1884, the cultivation of *Monascus sp.* was seen and utilised for manufacturing Chinese rice and rice wine, both of which were red in colour. The commercialisation of the first carotenoid pigment from *Cryptococcus* happened in 1954, followed by a production of the carotenoid pigment from other species like *Rhodotorula sp.* Later on, the production and development of astaxanthin from *Phaffia rhodozyma* and beta carotene from *Dunaliella salina* using natural products containing beta carotene were reported [5]. There are many other examples that can be drawn out from the history of colour production from natural resources, and its utilisation has gradually grown with the civilisations. The various natural sources like plants, animals, microorganisms and minerals were used to achieve colours for various types, and they were the only source of colourants. However, with the advent of steam engines followed by the industrial revolution, the amount of yarn and fabric production increased many folds. In order to suffice the need for the coloured fabric of this amount, natural sources became scarce and synthetic colours created their place in textile colouration. Their brightness, wide range of colour gamut, colour reproducibility, ease of application and performance properties made them a favourite of most of the dyers [6].

4.1.1 Synthetic dyes and their limitations

The utilisation of synthetic colourants has increased many folds since their discovery. Along with their positive aspects, synthetic dyes have certain limitations which are causing havoc to the environment. Firstly, these dyes are mainly petroleum-based, and therefore their resource is limited and depleting at a very fast rate. Secondly, the textile industry stands in the second position globally in the category of polluting industry, where the major pollution is due to the dyeing process. It is estimated that nearly 20% of the global water pollution is related to textile dyeing processes. These petroleum-based colourants, the fixation agents (e.g., formaldehyde and chromium) and finishing agents like flame retardants are the main contributors to this pollution [7,8]. The release of huge proportions of colourants and fixing agents after treatment in water streams leads to major destruction in the surrounding ecosystems due to toxicity and non-biodegradability [9,10]. Therefore, the world is experiencing a shift towards the use of biodegradable and eco-friendly commodities, and the necessity is increasing for nature-based colourants day by day. Different natural sources such as plants, insects, microorganisms and mineral ores are explored for alternatives to synthetic colourants. The environmental demand necessitates the utilisation of natural dyes, which pose no to minimum risk to the ecology, and colours obtained from these sources possess characteristic properties like UV protection, anti-bacterial, anti-oxidant and anti-carcinogenic due to their parent source [7,9–12].

The textile industry widely uses synthetic dyes because they are easily synthesised at a low cost, highly photostable, thermally stable and cover a wide colour spectrum. However, it has caused a discharge of a high amount of coloured waste-water. Due to this, water transparency gets affected in water bodies, and it also generates trouble for photosynthetic plants and algae as synthetic dyes restrict the absorption of light. Additionally, synthetic dyes are usually harmful due to their toxicity, mutagenicity, as well as carcinogenic nature, which leads to several human health hazards. Again the irreducible synthetic colours are a major obstacle in the effluent treatment plants and lead to blockage of the membranes.

These drawbacks of synthetic colours have diverted a huge interest toward natural colourants.

4.1.2 Natural colourants

Natural colouring agents are commonly (majorly known as pigments) recovered from a plant, animal, or microbial sources to impart distinct and different colours. These pigments have been in use as colours since time immemorial. Synthetic dyes have provided a full range of colours but at the cost of threat to the environment and our health. Therefore, the sustainable textile market has again demanded for diverting towards bio dyes. In practice, so far, plants are found to be the major sources of natural dyes, and their extracts from seeds, fruits, flowers, leaves, stems, and roots of different plant species give colours along with therapeutic properties. The demand for natural dyes has increased phenomenally to many folds. It is impossible to meet this demand from only plant-based dyes due to shrinking land. Thus, there is a need to explore alternative sources of natural dyes. Insects and mineral sources also have limited applicability as of now, as availability in bulk is majorly missing [13]. Additionally, these sources though natural, can eventually strain the eco-system due to the very high requirement of quantity to achieve the same tinctorial strength of the synthetic dyes [14]. The quality and colourfastness of such colourants, in most cases, are not at par with the synthetic colours. Large agricultural land and water are required to grow the plant material, and colour extraction also requires an ample amount of water and energy. Along with it, many of the plant sources are seasonal. The major limitation of these natural dyes is their low yield after extraction.

Altogether, an alternative colouring source is of great interest to the textile and fashion industry. This has led the research arena to focus on microbial pigments to overcome these issues. Generally, US and EU consumers appear to be well informed of those problems. For example, the commercial success of underwear having natural dyes onto it indicates a new positive outlook towards the usefulness of natural colourants, though their synthetic counterparts are cheaper. In order to evaluate the techno-economic feasibility of present natural dye sources, several research projects have been undergone.

4.1.3 Microbial pigments

Microorganisms have recently gained much interest as a colour source since they are known to generate coloured pigments. Fermentation of various organic sources using specific microorganisms is carried out to obtain certain colours. Microorganisms are known to be potential sources of bio-pigments as these form a wide range of pigments, namely melanins, carotenoids, quinines, monascins, flavins, violancein, etc.

The takeover benefits of microbial sources include availability in abundance, environmental stability, cost efficiency, less labour intensiveness, high yields and ease in downstream processing [1]. The techniques include culture/strain development and cultivation, their engineering, and the establishment of textile dyeing process. Possibility of providing exceptional medicinal properties in addition to desired colouration interests the fashion and the textile industry to move towards microbial dyes. With the strong movement of fashion towards eco-footprint and sustainable fashion, the researchers are working on various methods of dyeing textiles using colours obtained from natural or engineered microorganisms. Pigments obtained from microorganisms are useful because they can sustain any weather conditions, grow very easily and rapidly on low-cost substrates, and provide different shades of pigments.

4.2 Pigment production

The process of production of pigments from microorganisms is not the invention of 20th-century research. Many traditional cultural processes have utilised it, although without going in much deep to check out the science behind it. Two of the most common ancient and traditional microbial colouration processes are the preparation of red yeast rice in China and Japan and the preparation of black colour paste used in Kalamkari printing, India.

In ancient China and Japan, some rice types were cultivated along with *Monascus purpureus* yeast. As a result, the rice grains were used to collect the red-purple pigment from the yeast and turn it into that colour. The inception of this process is dated as back as 300 BC. Along with its use in culinary, ancient Chinese applied this in many traditional medicines.

Kalamkari printing is one of the many traditional and beautiful hand-printing processes of India. The name means craftmanship using pens. The black colour utilised in this process is prepared by fermenting jaggery in the presence of iron rust in an earthen vessel for a fortnight. The extract is then applied on naturally mordanted cotton fabric using wooden pens. It produces a bright black colour, and this process is mainly based upon the utilisation of microorganisms to produce colours.

Systematic production of pigments from microorganisms can become a significant phenomenon for the colouration of textiles in today's era. In this process, the first step is always to identify a microorganism that can produce colourant. This is generally done in two ways. The culture can be isolated from the nature or any existing microorganism can be genetically modified for the same purpose. This process is also known as screening, and the main objective of this step is the identification and isolation of colour producing microorganisms from non-producing ones. This procedure involves a culture enrichment technique at an optimised incubation condition followed by isolation of microorganism as per interest. When the pure microorganisms are obtained through screening and isolation, their gene sequencing analysis is carried out and was then deposited at a registered bank [15]. The production of crude pigment can be carried out by growing microbial inoculums at required conditions and nutrients. As mentioned earlier, growing these microorganisms in agricultural wastes, e.g., in sugar cane bagasse, due to higher nutrient content is an effective and economic step for industrial-level production [15–18]. The amount of product generated depends on the fermentation step. The required pigment is then extracted as per their solubility level and characterised for further application in various sectors. The process of production of microbial colourants is explained hereafter.

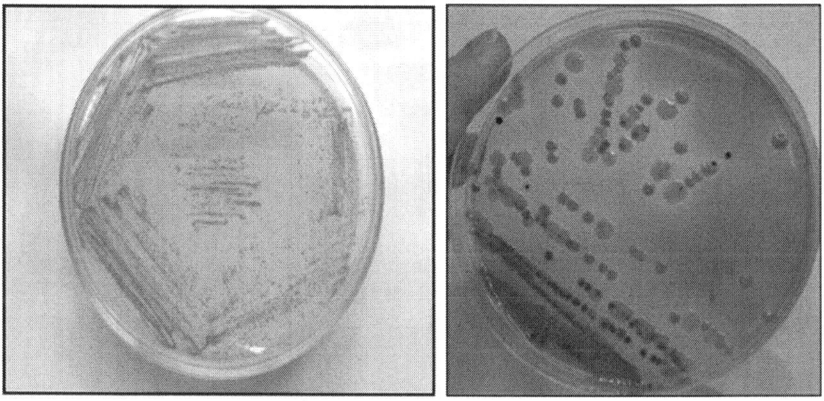

Figure 4.1 Colour producing microorganism

4.2.1 Strain collection

Microbial organisms can grow in almost any environmental condition. According to their food source, aerobic nature, the requirement of light, environmental stimuli like pH, temperature, salinity, humidity, etc., they can adapt and flourish in various situations. In their favourable conditions, they thrive. In adverse conditions, however, it has been seen that they tend to produce more pigments as an action of shock and protection. An ideal microorganism that can produce pigment should be familiar with carbon and nitrogen sources, has tolerance to pH minerals and temperature, and produce good colours. Such microorganisms having non-hazardous and non-pathogenic nature that can be easily separated from the cell biomass, are always opted for successful large-scale production.

Fungi, bacteria and algae are three major sources of pigments [2,18]. Among these sources, fungi are the well-explored source of microbial pigments. Many fungal pigments are chemically stable, non-toxic, and stable over wide pH and temperature range [20–25]. However, the pigment developed by the fermentation process requires high capital for setting up in terms of components for growth media. Along with that, the contamination rate of fungi is higher and they are highly contagious. Therefore, the isolation and growth of these microorganisms are difficult. High production costs are also a matter of concern. Therefore, researchers are attempting to explore the possibility of industrial side-streams and wastes for fermentation processes to counter balance the production cost [26]. There are many fungi available in the literature that produce pigment in a low-cost substrate, mainly in agro-residual wastes [27,28]. Different fungi sources have been found out for industrial production of pigments of red, yellow and orange tint [29].

Bacteria are also an important source of microbial pigments which can be utilised in textile industries. *Actinobacteria* is the highest pigment-producing group of bacteria. Some bacteria that produce different colours are *Flavobacterium sp.*, *Bacillus spp.*, *Serratia marcescens, Chromobacterium violaceum, Staphylococcus aureus, Agrobacterium aurantiacum,* etc. Colours produced by these organisms are Yellow, Creamy, Red, Purple, Golden Yellow, and Pink-red, respectively [30]. Production of these pigments consists of shorter life cycles as compared to fungi. Furthermore, the genetic modification of bacteria is easier. The focus of the research fraternity is on exploring the bacteria for the production of colour pigment and its application [12]. Fungal pigments stand in a higher position than bacterial pigments in terms of commercial outlook. [31].

Algae are another type of widely available microorganisms and can be a source for colour pigments. It has been seen that macroalgae may contain carotenoids and phycoerythrins, which can become important pigments for the textile finishing processes [32,33]. Many algal pigments are soluble in water or organic medium and, therefore, can be extracted without much effort. extraction of algal pigment on large scale results in biomass disintegration, followed by organic solvent mixture treatment. The supernatant of the solvent mixture can be utilised further to extract pigment using various methods that include phase separation and chromatographic techniques.

The colour pigments extracted from these sources have great structural diversity. The derivatives of carotenoids, phenazine dyes, pyrrole dyes, azaquinones, anthraquinone, violacein, melanin, pyocyanin, prodiginines, and benzoquinone are among the few colourants already extracted from microbial sources and have potential usage in textile processing field [34–42]. A further boost to the production of these natural pigments has been created by their outstanding pharmacological characteristics like antioxidant, anti-inflammatory, anti-bacterial and anti-carcinogenic which gives an added advantage along with the basic aesthetics [43].

4.2.1.1 *Pigments from wild bacteria source*

Marine water has been a good source of the microorganisms, and a novel rose red pigment has been extracted from *Serratia marcescens subsp. Marcescens* collected from mangrove forest soil of Mumbai, India [44]. *P. aeruginosa* (GS-33) from a marine source produced Phenazine-1-carboxylic acid (PCA), a lemon yellow pigment that also has an anti-carcinogenic property [45].

4.2.1.2 *Pigments from wild fungi source*

Fungus tends to grow in almost every atmospheric condition, preferably in damp weather. In many cases, it has been observed that fungi grow in wood and cause its colouration. This is also known as spalting, and researchers are utilising it to find pigment-producing fungi. *Chlorociboria aeruginosa* (green), *Scytalidium ganodermophthorum* (yellow) and *Scytalidium cuboideum* (red) were found to be producing pigments in different types of plants like bamboo, and pinus wood in the North Pacific Region, USA [46–49].

Healthy roots of trees and plants are another good source of dark pigment-producing endophyte fungi. These endophytes can produce pigments which are bioactive secondary metabolites. *Phyllosticta capitalensis*, melanin-producing fungi, has been seen in different types of vegetation like both dry and wet deciduous forests, mangroves and semi-evergreen forests [50–52].

Biochar is charcoal produced from plant matter and stored in the soil. It is generally generated by biomass pyrolysis. Biochar and similar chemical compounds are found to be a booster for fungal growth, and it has been reported that aspergillin pigment (black) producing *Aspergillus niger* grows very well in this type of soils [27,52,53].

4.2.2 Strain development

The science and technology of engineering and enhancing microbial strains, in order to improve the capacities of metabolism for applications in biotechnology is known as strain improvement/development.

It has been reported in a recent article that genetically developed strains are able to produce pigments in a shorter period of time with severalfold increase in pigment generation [54,55]. With this genetic modification, the mutation process can be sped up using certain mutagens like ethyl methane sulphonate (EMS), 1-methyl-3-nitro-1-nitrosoguanidine (NTG) and ultraviolet light [56,57]. In another report, microwave radiation was used successfully for mutation of *Serratia marcescens* which caused considered augmentation of production in prodigiosin pigment [58].

4.2.3 Fermentation

Fermentation is the process of chemical breakdown of substances by microorganisms, typically involving effervescence and the generation of heat as a by-product. Fermentation can take place in the presence or absence of air. Pigments form as a secondary metabolite during fermentation. Fermentation can be classified into two types: submerged fermentation and solid-state fermentation.

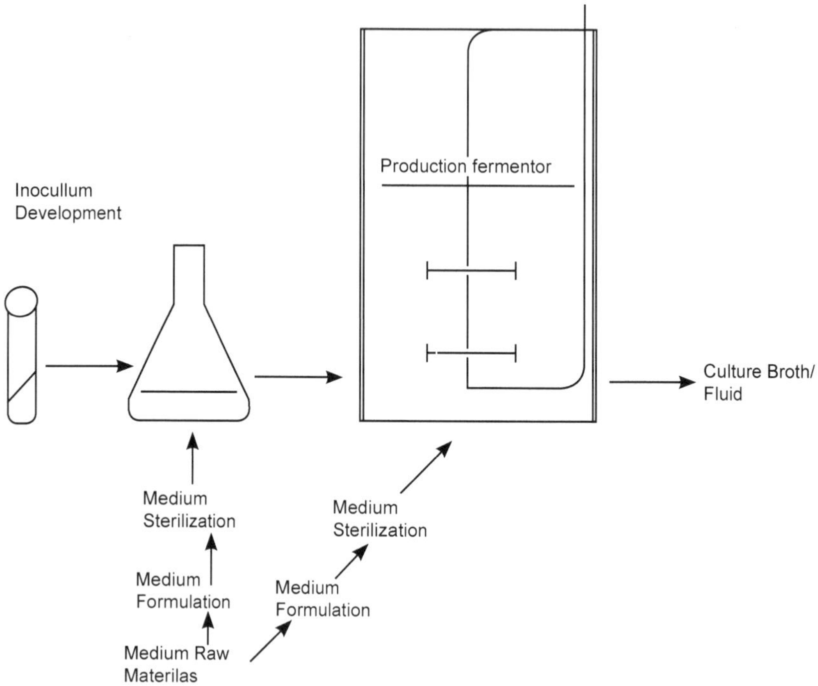

Figure 4.2 Fermentation process of microbial extract

4.2.3.1 *Solid state fermentation*

This type of fermentation process does not involve a liquid medium. Microorganisms are grown in solid media without any water, and only 15% moisture is required for the growth. These solid substrates are sparingly soluble biopolymers and are a rich source of nutrients for microorganisms [59]. It is a traditional process, and the most common reactors utilised for this process are tower reactor, drum reactor and forced aeration reactor. Agricultural materials like jackfruit seed and corn cob are reported to be used for the production of pigments from *Monuscus spp.* [27,60–62].

4.2.3.2 *Submerged fermentation*

Submerged fermentation technology began its course in the early 1930s, and it has become very promising since then. In this process, the nutrients used for fermentation are in a liquid state which allows the growth of the microorganism culture throughout the solution. Both batch-wise and continuous processes are in use for submerged fermentation. This process has been utilised in many microbial colour production processes [1,63,64].

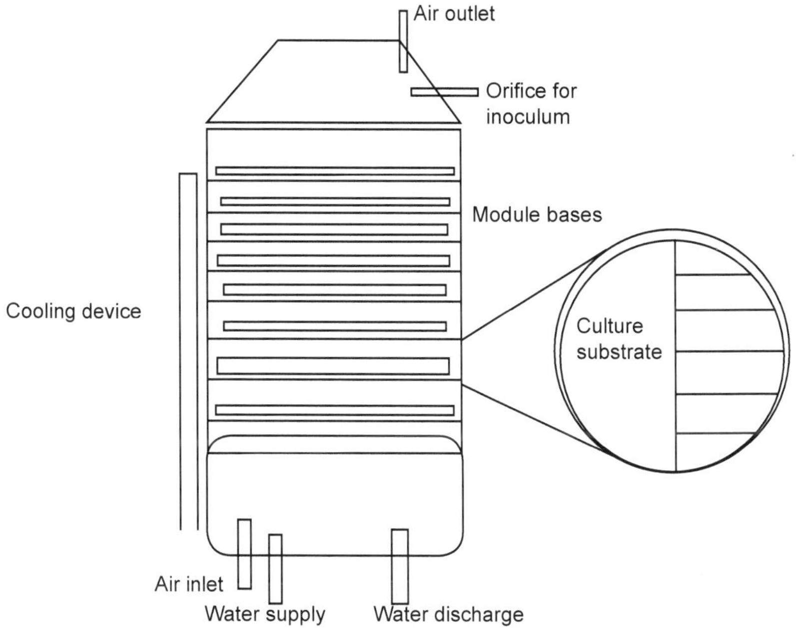

Figure 4.3 Schematic diagram of solid state fermentation

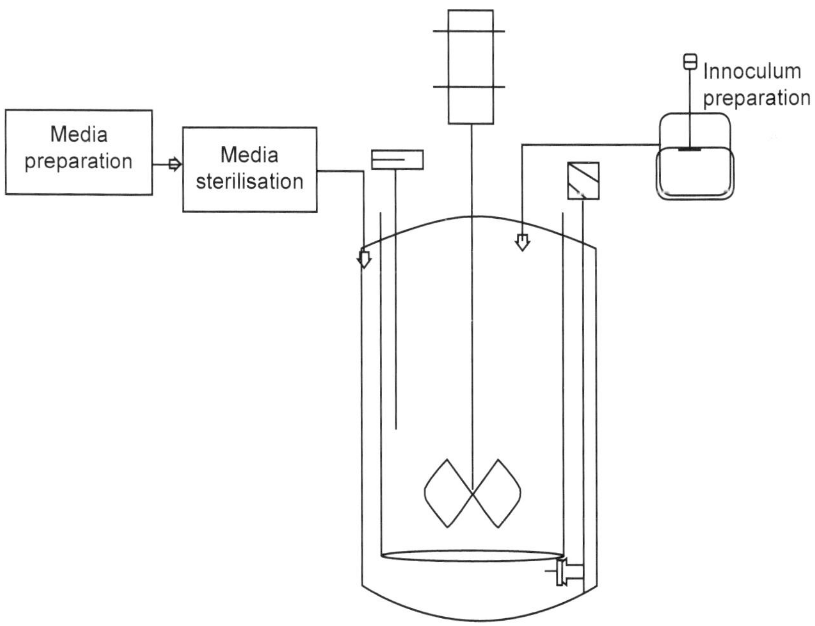

Figure 4.4 Schematic process diagram of submerged fermentation

4.2.4 Extraction

The fermentation process completes the production of pigments, and the next step involves the separation of these pigments from the residue nutrients, dead mass and water. This step is commonly termed as extraction, which finally delivers the purified pigment for further processes. The extraction process of natural colourants can be widely classified into an aqueous, enzymatic, supercritical fluid, alkali or acid assisted, microwave, and ultrasonic method and solvent extraction.

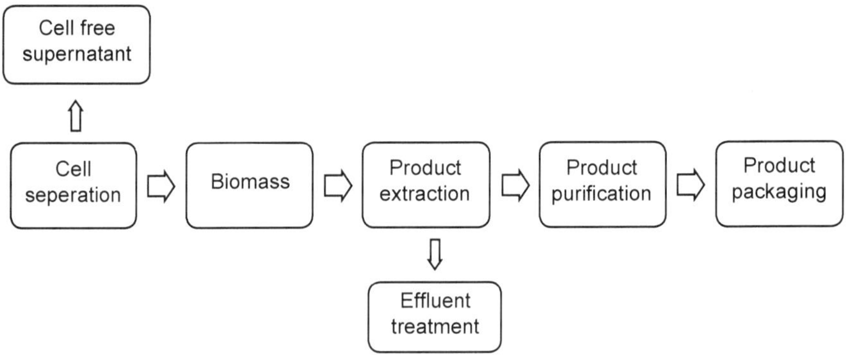

Figure 4.5 Extraction process of the coloured pigments

The aqueous extraction is always a sought-after process as it is an environment-friendly process, and the coloured solution can directly be utilised for the consequent dyeing process. It has to be noted that only the water-soluble colourants are suitable for this type of extraction.

The enzyme extraction process uses enzymes for separating the dye from the place where it is produced. It generally happens by breaking down the bonds between pigments and the holding substrate. As the enzyme is very substrate-specific, the separation happens without any chemical degradation of the pigment, and the pigment thus produced is dissolved in the extraction media.

Fermentation of barks, leaves and roots is carried out to take out the colours from the substrates. A very common example of dye extraction by fermentation is the production of natural indigo. Fresh leaves and twigs are collected in a wooden vat and are first kept in the warm water of around 32°C temperature. Then those are kept for fermentation. The cellulose part of the leaves degrades, and the pigment comes out of it. Both the enzyme and fermentation-based processes are mainly used for the extraction of dyes from plant sources. Nevertheless, they can also be utilised to extract microbial colours from spalted wood-based microorganisms [65].

Alkali or acid-assisted extraction is very much suitable for extracting polar colour compounds from the fermented media. Alkaline extraction is useful, especially for colourants that contain polyphenolic compounds. Acidic extraction, on the other hand, is utilised for flavones-based dyes as it prevents the oxidative degradation of the molecules. This process is comparatively more effective and economical. The only drawback is the removal of acid or alkali in the later stages.

The solvent-assisted extraction process is very effective in most cases as it can be utilised for both the polar and non-polar colourants. It is economical and efficient, and if the solvents are recollected and reused, it is quite environment-friendly. Most of the microbial pigments are extracted by this process. Some of the common organic solvents are acetone, methanol, ethanol, PET ether, chloroform, and water-alcohol mixture. This water-alcohol mixed system is applied to extract both water-soluble and insoluble pigments [66,67]. Acid or alkali is also added in certain cases to hydrolyse the glycosides and bring about a higher yield in extraction.

Ultrasound and microwave can be incorporated during the extraction process to reduce the time and increase the yield. These processes cause agitation and collision at a molecular level which causes the separation of pigment from the holding substrate much more effectively than normal mechanical agitation processes [65].

Utilisation of supercritical fluids for colour and other extraction from plant sources has been reported [68]. Although this process is very efficient and environment friendly, the drawback lies in the high cost and selectivity of raw materials.

4.3 Recent trends to overcome the limitations of production of microbial dyes

While discussing about microbial pigments, microorganisms found in the surrounding nature are the actual inspiration and source of various colours. It has been discussed throughout the chapter about those various microorganisms which bear beautiful colours. However, the important challenge is to enhance the pigment production from these microorganisms so that the process becomes economically viable. One of the obvious solutions might be using a large fermenter; however, the required option is to utilise molecular biology for producing high-yielding microorganisms from the existing ones. The genes which are responsible for the biosynthesis of various pigments can be either cloned/mutated, and recombinant DNA technology can be applied to find out different species with far better pigment yield capacity. Amgen Inc.

scientists engineered a non-hazardous *E. coli* strain whose yield of producing indigo pigment (exclusively obtained from the woad plant) was very high [69,70]. Polyketide, like kalafungin, which is generally bright yellow in colour, is a genetically modified version of *Streptomyces coelicolor*, a blue pigment actinorhodin. Alternatively, biosynthesis of actinorhodin can also produce orange or yellow-red anthraquinones by genetic engineering [71].

4.3.1 Characterisation

The colours thus obtained need to be characterised to know their structure, functional groups and chemical stability. This is important for their final application. Sometimes, a combination of pigments is produced by the microorganisms, and those are needed to be separately identified for the same purpose. However above all, it is an absolute necessity to find out whether the colour has any affinity towards textile materials. The colours may not have any attraction towards textiles, which may be the case when the colourant does not have any functional group or are non-polar in nature. In that case, there might need an additional step to carry out fixation of the colourant onto textile materials physically or chemically. Therefore, it is necessary to check their dyeability before application to textiles. The extracted pigments can be tested under NMR, Raman spectroscopy, and FTIR to find out the functional groups in the pigment required for colouration. Table 4.1 shows the list of pigments majorly applied and experimented along with their chemical structures and functional groups.

Table 4.1 Reported microbial pigments and their chemical structures

1	Astaxanthin (Purple)	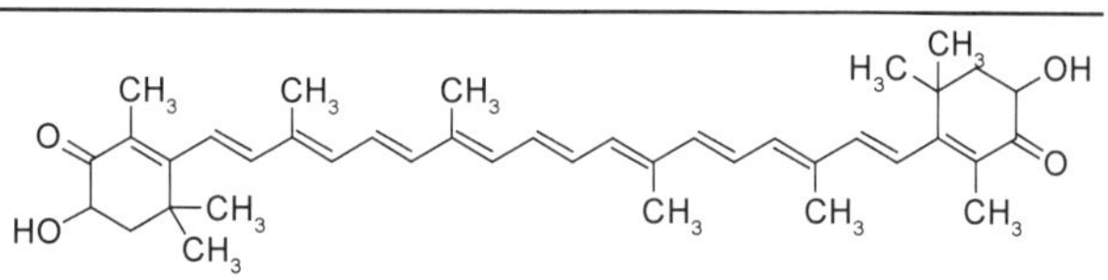
2	Flavoglaucin (Pellow)	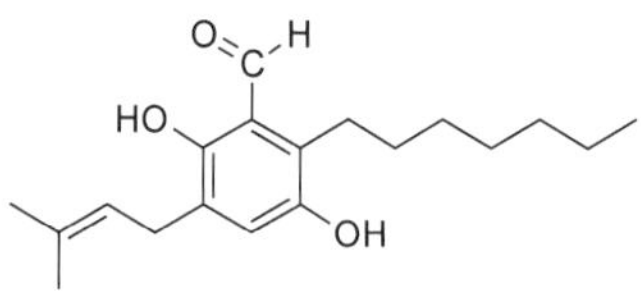
3	Indigo (Blue)	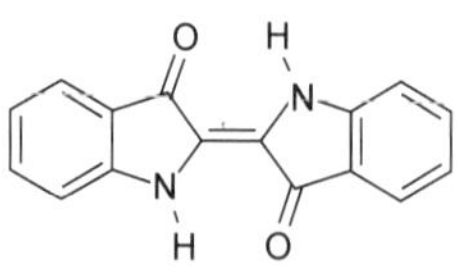

Cont...

Cont...

4	Riboflavin (Yellow orange)	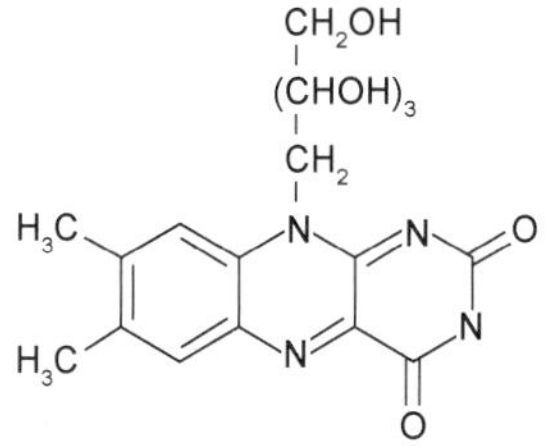
5	Anthraquinone (Red)	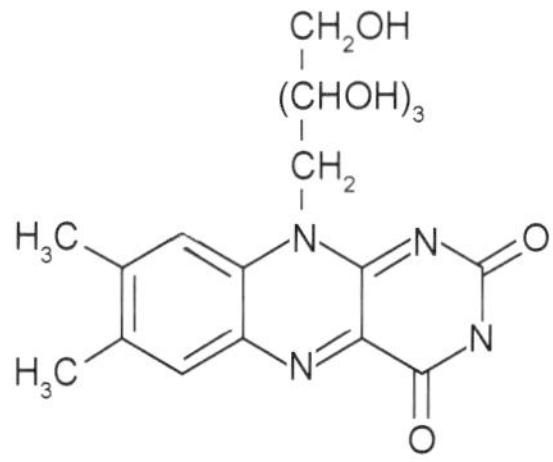
6	Violacein (Purple)	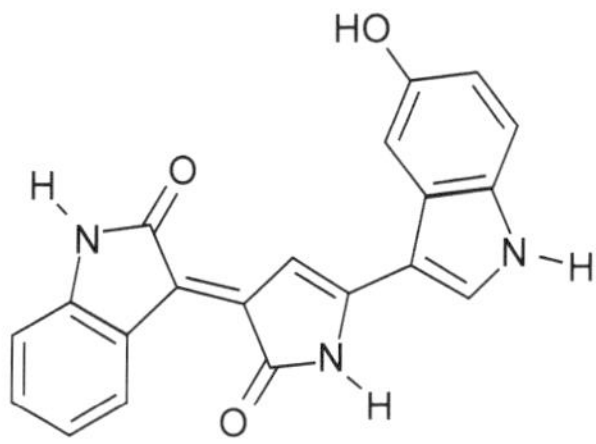
7	Napthaquinone (Deep blood red)	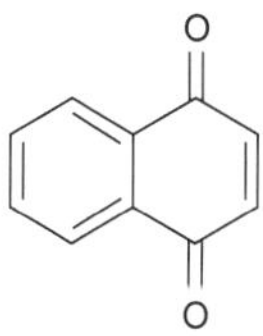
8	Prodigiosin (Red)	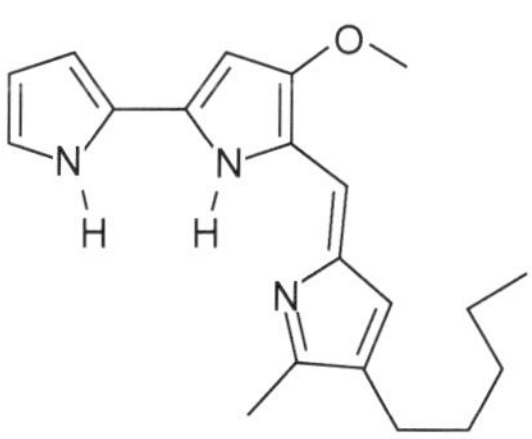
9	Monascorubamine (Red)	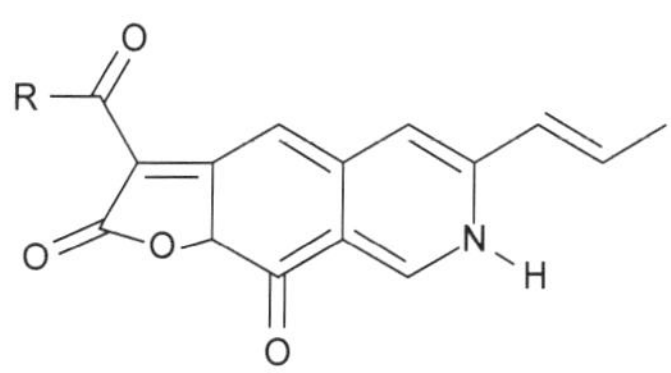

Cont...

Cont...

10	Monascorubin (Orange)	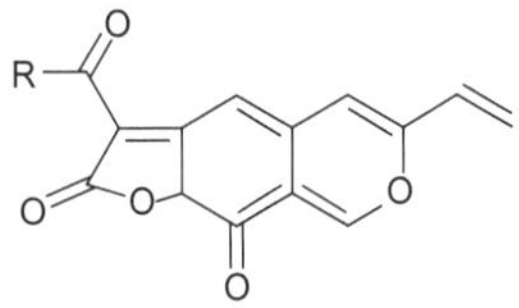
11	Carotenoid (Yellow)	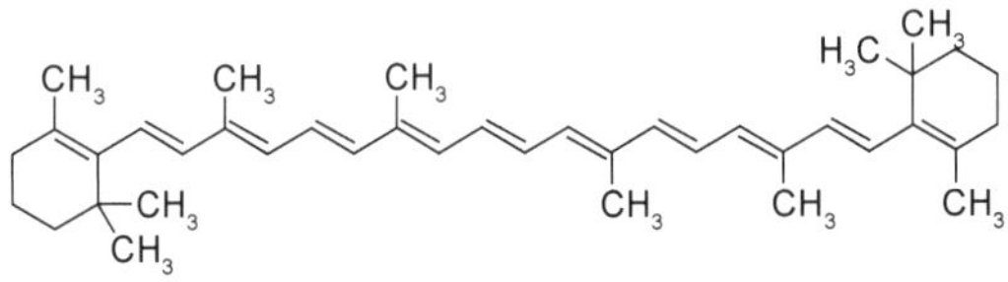

Many researchers have reported the physico-chemical properties and structures of metabolite obtained by the fungal strains using those characterisation processes. The microbial pigments used for the colouration of a textile substrate are studied for the presence of required functional groups, and their dyeing characteristics were analysed. The performance properties like washing, rubbing and light fastness were also evaluated to determine their significance in dyeability [14,72]. As filamentous fungi accumulate a complex mixture of compounds as a pigment, further purification is generally required for its clear analysis. In some cases, micro-spectrometric measurements, ultraviolet excitation and infrared irradiation demonstrate the effective study of the type of compound [73].

4.4 Application of pigments

Pigments are molecules of higher importance to many industries. Their usage varies from a wide range of colouring agents to additives and antioxidants. These colours and all the other pigments are mostly applied in areas of cosmetics, pharmaceuticals, food, textiles and plastics. Pigments that are applied in textile colouration are summarised below.

4.4.1 Textile colouration

Beautiful coloured fabrics are always demanded. It requires skill and knowledge of science, technology and craftsmanship for textile colouration. The colour choice of both designer and customer can be challenging at times, and in order to fulfil the demand, a dyer has to have different colours which can cover the complete colour gamut when used in self or combination. These dyes should have fair thermal and chemical stability because, during dyeing, temperature and different chemicals like salt, acid, alkali and many other formulated chemicals are utilised. There are different classes of dyes in existence, and those are applied to specific textile materials using variety of distinc processes. To use the colourants in combination, it is absolutely necessary to have compatibility within them and get onto fabric by a single process. All these variables make the process complicated. This is the main challenge for colouration using natural colourants. At the same time, it is also true that natural colours are better in many aspects than synthetic colourants. Within natural colours, microbial dyes and pigments are slowly finding their niche and have the potential to become a commercial success. The need is to undertake intensive research to discover more and more microbial colourants to complete the shade gamut, classify as per user's convenience for easy selection and develop user-friendly processes of application adhering to present-day industrial set-up to be suitable for the bulk colouration of textiles.

The researchers have produced violacein by the local isolation of *Chromobacterium violaceum* (Gen Bank accession no. HM132057), which is mainly found in different agricultural waste materials like brown sugar, molasses, solid pineapple waste, sugarcane bagasse as a substitute to the traditional rich medium [15]. This highly competitive, acetone and methanol soluble pigment is stable over a wide range of temperatures and pH.

Marine bacteria were bio-prospected as an attempt towards the isolation of a feasible dye pigment source. *Serratia sp.* [44]. BTWJ8 was isolated which secreted prodigiosin-like pigment. This red-yellow natural pigment was evaluated for colouration in the textile industry and showed stable performance [74]. Research has been conducted on the isolation of *Vibrio sp.* strain from marine sediments that resulted in huge production of bright red pigments for dyeing fibres like silk, wool, acrylic, and nylon [40,75]. Ahmad et. al. have evaluated the impact of various pigments like violacein and prodigiosin on dyeing of cotton and silk fabrics [15,66].

A study was conducted for the isolation of three fungi, namely *Curvularia lunata, Alternaria alternate,* and *Trichoderma virens* to retrieve pigments for their application in textile dyeing. It gave good results for dyeing wool and silk in the absence of mordant. Pigment from *Trichoderma virens* also demonstrated antifungal property. As the research was conducted and molecules were analysed, it was found that the pigments were multi-component in nature [76]. Several *Monascus purpureus species* were cultivated on rich starch substrates and were well known to produce monascorubramine and rubropunctamine, which are red and yellow in colour, respectively [19].

Many works have been carried out regarding colouration of textile materials to explore promising future for these colourants in textile industry, and some of those are summarised below;

Table 4.2 List of microorganism, produced coloured pigment and their application on specific substrate

Sl. No.	Microorganism	Type	Pigment	Colour	Application	References
1	*Chromobacterium violaceum*	Bacteria	violacein	purple	Pure cotton, pure silk, pure rayon, acrylic, polyester	[15]
2	*Serratia sp.*	Bacteria	prodigiosin	Red-yellow	Wool, nylon, acrylics and silk	[44,77]
3	*Vibrio sp.*	Bacteria	prodiginines	Red	Wool, Silk, Nylon and Acrylic	[40,74]
4	*Chryseobacterium sp.*	Bacteria	violacein	Yellow	Natural silk, Linen, Cotton	
5	*Trichoderma virens*	Fungus	multi-component	Yellow	Wool and silk	[78]
6	*Alternaria alternate*	Fungus	multi-component	Brilliant olive	Wool and silk	[78]
7	*Curvularia sp.*	Fungus	multi-component, anthraquinone	Skin melanin, Blue	Wool and silk	[78,79]
8	*Dermocybe sanguinea*	Fungus	anthraquinone	Yellow/ Red	Wool/ Cellulosics	[80]
9	*Chlorociboria aeruginosa*	Fungus	xylindein	Green	Polyester	[81][82][83]
10	*Scytalidium cuboideum*	Fungus	not reported by author	Red	Polyester	[81][83]
11	*Scytalidium ganodermorphthorum*	Fungus	not reported by author	Yellow	Polyester	[81][83]
12	*Monascus purpureus*	Fungus	monascorubramine, rubropunctamine	Orange, Red, Yellow	Wool, Cotton	[84–86]
13	*P. purpurogenum*	Fungus	not reported by author	Yellow	Cotton	[85,86]
14	*Thermomyces sp.*	Fungus	not reported by author	Yellow	Silk	[87]
15	*Phymatotrichum sp. (NRC 151)*	Fungus	not reported by author	Brown	Wool	[88]
16	*Drechslera sp.*	Fungus	anthraquinone	Reddish	Polyamide	[79]
17	*Trichodenna sp.*	Fungus	anthraquinone	Reddish	Polyamide	[79]
18	*Aspergillus sp.*	Fungus	anthraquinone	Reddish	Polyamide	[79]
19	*Janthinobacterium lividum*	Bacteria	violacein	Bluish-Purple	Cotton, Wool, Silk and Nylon	[89]

In a nutshell, it can be said that many fungal and bacterial pigments show promise for textile colouration. Research is on throughout the world, and more microbial colours are likely to be discovered in the near future.

4.5 Future challenges and limitations

The introduction of microbial colourants is very promising and eco-friendly, but there are still many hurdles that need to be overcome. The textile industry, being a major manufacturing sector, works under tremendous pressure where there is all time battle for textile production at the lowest possible time using minimum resources. This makes the colouration tougher than it should be in terms of time. This may become a hindrance for using these colours. For a textile dyer, it is necessary to know the following criteria of a colourant;

- Classification of the dye
- Substrates that can be dyed
- Compatibility when used in combination
- A simple application procedure
- Fixation of the dyes

Microbial colourants contain various functional groups, and many times, they do not possess any functional groups (pigments). Therefore, using them in a single process becomes difficult.

Along with it, there are certain limitations in pigment production itself which are preventing the industrial-scale production and colouration using microbial pigments. The main limitations are,

- Limited colour gamut
- Compatibility of one pigment with the other during colouration
- Solubility in water and other solvents
- Chemical and thermal stability
- Initial high setup cost
- Maintaining cultural purity

In short, it can be said that, although there are challenges, microbial colouration has a huge potential in the future for cleaner and sustainable production.

4.6 Economics for pigment production

Public interest in microbial colourants has increased due to the awareness of health safety and eco-friendliness, and their growth is predicted to increase by 7% annually [21,90]. Microbial source of pigment is an important source of natural colourants, but huge efforts are needed to generate low-cost organic substrate to grow pigment-producing organisms. The bio processing of microbial pigments is a tedious job as it requires a higher level of sterility and nutrient enrichment. On one end, the natural/microbial colourant is gaining higher interest, but on an other end, these may be manifolds more expensive than synthetic ones. It can be expected, in a few years, that the application of microbial pigments will increase many folds, and the consumers will be ready to accept it with its cost factor.

It has been reported by Venil et., al. that microbial pigments are comparatively costlier than synthetic colours and plants/insects-based natural colours. A standard synthetic violet (Brilliant blue FCF, E133) costs around 42 – 60 US \$/kg while violacein costs around 5000x10^5 \$/kg; Allura Red AC, E129 comes at a price of 24-42 \$/kg, annatto extract (plant-based red colour) is of 80 \$/kg while prodigiosin costs around 5000x10^5 \$/kg. In case of yellow colour, FD&C Yellow No. 5 is comparatively costly (500 \$/kg), but still is far cheaper than saffron (1400 \$/kg) and carotenoid (1000 \$/kg) [66]. This high cost includes the cost of pure nutrition source, extremely carefully carried out fermentation, the extraction, purification and proper packaging of the final material. Along with it, the low to medium yield of pigment production also adds up to it. However, these unjustifiably higher costs may be due to the novelty of the technology and high purity supply of these colours as per the requirement of other industries. More research is likely to bring down the cost to an affordable level. In addition to the present high cost, the major limitation of its use in textile colouration is that it requires an ample amount of dyes or pigments to colour the substrate in comparison to its application in other industries. This brings the current market value amounting to nearly US\$ 1/g, which restricts their application mainly to naturally coloured garments with high value addition. In order to face these challenges, there is a requirement for the development of an economical process to produce pigments that can successfully substitute the stand of synthetic dyes and pigments. The use of abundantly available agro residues in solid-state fermenters might help in reducing production costs to a greater extent [91].

Summary

Sustainability is the buzzword of today's society. To attain it, a balance between economy, environment and society is absolutely necessary. The textile industry, being an integral part of the economy and society, requires more attention to follow eco-friendly processes. Finding renewable sources, utilising optimum resources and creating environment-friendly processes are the basics to reach sustainability in textile. In this chapter, the production, characterisation and application of the pigment from microbial organisms in an effective and economical way to achieve sustainability have been reviewed.

The usefulness of microbial colour over other natural colourants has been discussed. The typical sources of microbial colours have been summarised along with the chemical structures of many of these microbial colourants. Furthermore, reported colourants that are used in textile materials are also been listed. The advantages, limitations and prospects of textile colouration using microbial colourations have been explained.

References

1. Joshi V K, Attri D, Bala A, Bhushan S (2003), *Microbial pigments*, India, NISCAIR-CSIR.
2. Mallick A (2018), *Interconnection Between Biotechnology and Textile: A New Horizon of Sustainable Technology*, Handb. Renew. Mater. Color. Finish. Wiley online lib. 549–573.
3. Cardon D (2007), 'Natural dyes', *Sources, Tradit. Technol. Sci.*, 268.
4. Siva R (2007), 'Status of natural dyes and dye-yielding plants in India', *Curr. Sci.*, 92, 916-925.
5. Venil K C, Zakaria Z A, Ahmad W A (2013), 'Bacterial pigments and their applications', *Process Biochem.*, 48, 1065–1079.
6. Lomax S Q, Learner T (2006), 'A review of the classes, structures, and methods of analysis of synthetic organic pigments', *J. Am. Inst. Conserv.*, 45, 107–125.
7. Chattopadhyay S N, Pan N C, Roy A K, Saxena S, Khan A (2013), 'Development of natural dyed jute fabric with improved colour yield and UV protection characteristics', *J. Text. Inst.*, 104, 808–818.
8. Singh O P (2000), 'Natural dyes: the pros & cons', *Indian Text. J.*, 110, 42–47.
9. Gupta D, Khare S K, Laha A (2004), 'Antimicrobial properties of natural dyes against Gram-negative bacteria', *Color. Technol.*, 120, 167–171.
10. Jimenez-Gonzalez O, Ruiz-Espinosa H, Luna-Guevara J J, Ochoa-Velasco C E, Luna V D, Luna-Guevara M L (2018), 'A potential natural coloring agent with antioxidant properties: Microencapsulates of Renealmia alpinia (Rottb.) Maas fruit pericarp', *NFSJ.*, 13, 1–9.
11. Tuli H S, Sandhu S S, Sharma A K, Gandhi P (2014), 'Anti-angiogenic activity of the extracted fermentation broth of an entomopathogenic fungus, Cordyceps militaris 3936' *Int J Pharm Pharm Sci,* 6, 581-583.
12. Venil C K, Lakshmanaperumalasamy P (2009), 'An Insightful Overview on Microbial Pigment , Prodigiosin', *Electronic J. Biology,* 5, 49–61.
13. Neves M I L, Silva E K, Meireles M A A (2019), 'Trends and Challenges in the Industrialization of Natural Colorants', *Food Public Heal.,* 9, 33–44.
14. Saxena S, Raja A S M (2014), 'Natural dyes: sources, chemistry, application and sustainability issues', *Roadmap to sustainable textiles and clothing. Springer, Singapore,* 37-80.

15. Ahmad W A, Ahmad W Y W, Zakaria Z A, Yusof N Z (2012), *Application of Bacterial Pigments as Colorant,* Application of Bacterial Pigments as Colorant, Berlin, Heidelberg, Springerbriefs in Molecular Science, Springer.

16. Nigam J N, Kakati M C (2002), 'Optimization of dilution rate for the production of value added product and simultaneous reduction of organic load from pineapple cannery waste', *World J. Microbiol. Biotechnol.,* 18, 303–308.

17. Tanaka K, Hilary Z D, Ishizaki A (1999), 'Investigation of the utility of pineapple juice and pineapple waste material as low-cost substrate for ethanol fermentation by Zymomonas mobilis', *J. Biosci. Bioeng.,* 87, 642–646.

18. Creczynski-Pasa T B, Antônio R V (2004), 'Energetic metabolism of Chromobacterium violaceum', *Genet. Mol. Res.* 3, 162–166.

19. Heer K, Sharma S (2017), 'Microbial Pigments As a Natural Color: a Review', *Int. J. Pharm. Sci. Res.* 8, 1913–1922.

20. Poorniammal R, Gunasekaran S (2015), 'Physical and Chemical Stability Analysis of Thermomyces Yellow Pigment for Food Application', *Int. J. Food Ferment. Technol.,* 5, 47-52.

21. Mapari S A, Nielsen K F, Larsen T O, Frisvad J C, Meyer A S, Thrane U (2005), 'Exploring fungal biodiversity for the production of water-soluble pigments as potential natural food colorants', *Curr. Opin. Biotechnol.,* 16, 231–238.

22. Morales-Oyervides L, Oliveira J C, Sousa-Gallagher M J, Méndez-Zavala A, Montañez J C (2015), 'Effect of heat exposure on the colour intensity of red pigments produced by Penicillium purpurogenum GH2', *J. Food Eng.*, 164, 21–29.

23. Morales-Oyervides L, Oliveira J, Sousa-Gallagher M, Méndez-Zavala A, Montañez J C (2017), 'Assessment of the Dyeing Properties of the Pigments Produced by Talaromyces spp.', *J. Fungi,* 3, 38-46.

24. Hamlyn P F (1998), 'Fungal biotechnology', *Br. Mycol. Soc. Newsl.,* 17–18.

25. Gill M, Steglich W (1987), 'Pigments of fungi (Macromycetes)', *Fortschritte Der Chemie Org. Naturstoffe/Progress Chem. Org. Nat. Prod., Springer*, 1–297.

26. Panesar R, Kaur S, Panesar P S (2015), 'Production of microbial pigments utilizing agro-industrial waste: a review', *Curr. Opin. Food Sci.,* 1, 70–76.

27. Nigam P S N, Pandey A (2009), *Biotechnology for Agro-Industrial Residues Utilisation: Utilisation of agro rsidues*, Springer Science & Buisness Media.

28. Brar S K, Dhillon G S, Soccol C R (2014), *Biotransformation of waste biomass into high value biochemicals*, USA, Springer.

29. Narsing Rao M P, Xiao M, Li W J(2017), 'Fungal and Bacterial Pigments: Secondary Metabolites with Wide Applications', *Front. Microbiol.,* 8, 1-13.

30. Usman H M, Abdulkadir N, Gani M, Maiturare H M (2017), 'Bacterial Pigments and its Significance', *MOJ Bioequiv Availab*, 4, 1-5.

31. Dufossé L (2006), 'Microbial production of food grade pigments', *Food Technol. Biotechnol*, 44, 313–323.

32. Rossano R, Ungaro N, D'Ambrosio A, Liuzzi G M, Riccio P (2003), 'Extracting and purifying R-phycoerythrin from Mediterranean red algae Corallina elongata Ellis & Solander', *J. Biotechnol*, 101, 289–293.

33. Takaichi S (2011), 'Carotenoids in algae: distributions, biosyntheses and functions', *Mar. Drugs*, 9, 1101–1118.

34. Dharmaraj S, Ashokkumar B, Dhevendaran K (2009), 'Food-grade pigments from Streptomyces sp. isolated from the marine sponge Callyspongia diffusa', *Food Res. Int.*, 42, 487–492.

35. Saranya R, Jayapriya J, Tamilselvi A (2012), 'Dyeing of silk fabric with phenazine from Pseudomonas species', *Color. Technol.*, 128, 440–445.

36. Nagia F A, EL-Mohamedy R S R (2007), 'Dyeing of wool with natural anthraquinone dyes from Fusarium oxysporum', *Dye. Pigment.*, 75, 550–555.

37. Durán N, Justo G Z, Ferreira C V, Melo P S, Cordi L, Martins D (2007), 'Violacein: properties and biological activities', *Biotechnol. Appl. Biochem.*, 48, 127-133.

38. El-Naggar N E A, El-Ewasy S M (2017), 'Bioproduction, characterization, anticancer and antioxidant activities of extracellular melanin pigment produced by newly isolated microbial cell factories Streptomyces glaucescens NEAE-H', *Sci. Rep*, 7, 1-19.

39. El-Fouly M Z, Sharaf A M, Shahin A A M, El-Bialy H A, Omara A M A (2015), 'Biosynthesis of pyocyanin pigment by Pseudomonas aeruginosa', *J. Radiat. Res. Appl. Sci.*, 8, 36-48.

40. Alihosseini F, Ju K, Lango J, Hammock B D, Sun G (2008), 'Antibacterial colorants: characterization of prodiginines and their applications on textile materials', *Biotechnol. Prog.*, 24, 742–747.

41. Zheng L, Cai Y, Zhou L, Huang P, Ren X, Zuo A, Meng X, Xu M, Liao X (2017), 'Benzoquinone from Fusarium pigment inhibits the proliferation of estrogen receptor-positive MCF-7 cells through the NF-κB pathway via estrogen receptor signaling', *Int. J. Mol. Med.*, 39, 39–46.

42. Krishnamurthy K V, Siva R, Senthil T K (2002), 'Natural dye-yielding plants of Shervaroy Hills of Eastern Ghats', *Proc. Natl. Semin. Conserv. East. Ghats, Environ. Prot. Train. Res. Institute, Hyderabad,* 151-153.

43. Tuli H S, Chaudhary P, Beniwal V, Sharma A K (2015), 'Microbial pigments as natural color sources: current trends and future perspectives', *J. Food Sci. Technol.*, 52, 4669–4678.

44. Nerurkar M, Vaidyanathan J, Adivarekar R, Langdana Z B (2013), 'Use of a natural dye from Serratia marcescens subspecies Marcescens in dyeing of textile fabrics', *Octa J. Environ. Res.,* 1, 129-135.

45. Patil S, Paradeshi J, Chaudhari B (2016), 'Anti-melanoma and UV-B protective effect of microbial pigment produced by marine Pseudomonas aeruginosa GS-33', *Nat. Prod. Res.*, 30, 2835–2839.

46. Hernandez V A, Galleguillos F, Robinson S (2016), 'Fungal pigments from spalting fungi attenuating blue stain in Pinus spp.', *Int. Biodeterior. Biodegrad.,* 107, 154–157.

47. Robinson S C, Weber G, Hinsch E, Vega Gutierrez S M, Pittis L, Freitas S (2014), 'Utilizing Extracted Fungal Pigments for Wood Spalting: A Comparison of Induced Fungal Pigmentation to Fungal Dyeing', *J. Coatings*, 1–8.

48. Vega Gutierrez S, Vega Gutierrez P, Godine A, Pittis L, Huber M, Stanton S, Robinson S (2016), 'Feasibility of Coloring Bamboo with the Application of Natural and Extracted Fungal Pigments.', *Coatings,* 6, 37-47.

49. Robinson S C, Hinsch E, Weber G, Freitas S (2014), 'Method of extraction and resolubilisation of pigments from C hlorociboria aeruginosa and S cytalidium cuboideum, two prolific spalting fungi', *Color. Technol*, 130, 221–225.

50. Suryanarayanan T S, Ravishankar J P, Venkatesan G, Murali T S (2004), 'Characterization of the melanin pigment of a cosmopolitan fungal endophyte', *Mycol. Res*, 108, 974–978.

51. Addy H D, Piercey M M, Currah R S (2005), 'Microfungal endophytes in roots', *Can. J. Bot.,* 83, 1–13.

52. Pagano M C, Dhar P P (2015), 'Fungal pigments : an overview', *Fungal Biomolecules: Sources, Appli. Rec. Develop*, 173-181.

53. Ladygina N, Rineau F (2013), *Biochar and soil biota*, New York, CRC Press.

54. Nigam P S, Luke J S (2016), 'Food additives: Production of microbial pigments and their antioxidant properties', *Curr. Opin. Food Sci.,* 7, 93–100.

55. Sonderegger M, Sauer U (2003), 'Evolutionary engineering of Saccharomyces cerevisiae for anaerobic growth on xylose', *Appl. Environ. Microbiol.*, 69, 1990–1998.

56. Tripathi U, Venkateshwaran G, Sarada R, Ravishankar G A (2001), 'Studies on Haematococcus pluvialis for improved production of astaxanthin by mutagenesis', *World J. Microbiol. Biotechnol.*,17, 143–148.

57. Chen Y, Li D, Lu W, Xing J, Hui B, Han Y (2003), 'Screening and characterization of astaxanthin-hyperproducing mutants of Haematococcus pluvialis', *Biotechnol. Lett.,* 25, 527–529.

58. Venil C K, Aruldass C A, Dufossé L, Zakaria Z A, Ahmad W A (2014), 'Current perspective on bacterial pigments: emerging sustainable compounds with coloring and biological properties for the industry–an incisive evaluation', *RSC Adv.*, 4, 39523–39529.

59. Pandey A (2003), 'Solid State Fermentation', *Biochem. Engg. J.*, 13, 81-84.

60. Nimnoi P, Lumyong S (2011), 'Improving Solid-State Fermentation of Monascus purpureus on Agricultural Products for Pigment Production', *Food Bioprocess Technol.*, 4, 1384–1390.

61. Velmurugan P, Hur H, Balachandar V, Kamala-Kannan S, Lee K J, Lee S M, Chae J C, Shea P J, Oh B T (2011), 'Monascus pigment production by solid-state fermentation with corn cob substrate', *J. Biosci. Bioeng.*, 112, 590–594.

62. Babitha S, Soccol C R, Pandey A (2006), 'Jackfruit seed-a novel substrate for the production of Monascus pigments through solid-state fermentation', *Food Technol. Biotechnol.*, 44, 465–471.

63. Domínguez-Espinosa R M, Webb C (2003), 'Submerged fermentation in wheat substrates for production of Monascus pigments', *World J. Microbiol. Biotechnol.*,19, 329–336.

64. Gunasekaran S, Poorniammal R (2008), 'Optimization of fermentation conditions for red pigment production from Penicillium sp. under submerged cultivation', *African J. Biotechnol.* 7, 1894-1898.

65. Saxena S, Raja A S M, Mutthu S (2014), 'Roadmap to Sustainable Textile and clothing' *Textile Sci. Cloth. Techno., Springer*, 978-81.

66. Venil C K, Nordin N, Zakaria Z A , Ahmad W A (2014). 'Chryseobacterium artocarpi sp. nov., isolated from the rhizosphere soil of Artocarpus integer' *International journal of systematic and evolutionary microbiology,* 64, 3153-3159.

67. Gmoser R, Ferreira J A, Lennartsson P R, Taherzadeh M J (2017), ' Filamentous ascomycetes fungi as a source of natural pigments', *Fungal Biol. Biotechnol.* 4, 1–25.

68. Baysal T, Ersus S, Starmans D A J (2000), 'Supercritical CO_2 extraction of β-carotene and lycopene from tomato paste waste', *J. Agric. Food Chem.,* 48, 5507–5511.

69. Pfeifer B A, Khosla C (2001), 'Biosynthesis of polyketides in heterologous hosts', *Microbiol. Mol. Biol. Rev.*, 65, 106–118.

70. Hong S T, Carney J R, Gould S J (1997), 'Cloning and heterologous expression of the entire gene clusters for PD 116740 from Streptomyces strain WP 4669 and tetrangulol and tetrangomycin from Streptomyces rimosus NRRL 3016', *J. Bacteriol.,* 179, 470–476.

71. Mcdaniel R, Ebert-khosla S, Fu H, Hopwoodt D A, Khosla C (1994), 'Engineered biosynthesis of novel polyketides: influence of a downstream enzyme on the catalytic specificity of a minimal aromatic polyketide synthase', *Proceedings of the National Academy of Sciences,* 91, 11542–11546.

72. Durán N, Teixeira M F S, De Conti R, Esposito E (2002), 'Ecological-Friendly Pigments From Fungi Ecological-Friendly Pigments From Fungi', *Critical reviews in food science and nutrition*, 42, 53-66.

73. Garcia-Pichel F, Castenholz R W (1991), 'Characterization and biological implications of scytonemin, a cyanobacterial sheath pigment', *J. Phycol.*, 27, 395–409.

74. Krishna J G, Basheer S M, Beena P S, Chandrasekaran M (2008), 'Marine Bacteria As Source of Pigment for Application As Dye in Textile Industry', *Proc. Int. Conf. Biodivers. Conserv. Manag.,* 1, 743–750.

75. Shahitha S, Poornima K (2012), 'Enhanced production of prodigiosin production in Serratia marcescens', *J. Appl. Pharm. Sci.,* 2, 138-140.

76. Usman H M, Farouq A A, Baki A S, Abdulkadir N, Mustapha G (2018), 'Production and characterization of orange pigment produced by Halophilic bacterium Salinococcus roseus isolated from Abattoir soil', *J Microbiol Exp,* 6, 238-243.

77. Yusof N Z (2008), *Isolation and Applications of Red Pigments from Serratia Marcescens.*, Malaysia, Univ. Tekno. Doc. Diss.

78. Sharma D, Gupta C, Aggarwal S, Nagpal N (2012), *Pigment extraction from fungus for textile dyeing,* India, NISCAIR-CSIR.

79. Hobson D K, Wales D S (1998), 'Greeny dyes', *J. Soc. Dye. Color.*, 114, 42–44.

80. Räisänen R (2009), *Anthraquinones from the fungus Dermocybe sanguinea as textile dyes.*, Helsingin yliopisto.

81. Hinsch E M, Weber G, Robinson S C, Chen H L (2015), 'Colorfastness of extracted wood-staining fungal pigments on fabrics: A new potential for textile dyes', *J. Text. Apparel, Technol. Manag,.* 9, 1–11.

82. Hinsch E M, Robinson S C (2016), 'Mechanical Color Reading of Wood-Staining Fungal Pigment Textile Dyes: An Alternative Method for Determining Colorfastness', *Coatings,* 6, 1-12.

83. Weber G, Chen H L, Hinsch E, Freitas S, Robinson S (2014), 'Pigments extracted from the wood-staining fungi Chlorociboria aeruginosa, Scytalidium cuboideum, and S. ganodermophthorum show potential for use as textile dyes', *Color. Technol.*, 130, 445–452.

84. De Santis D, Moresi M, Gallo A M, Petruccioli M (2005), 'Assessment of the dyeing properties of pigments from Monascus purpureus', *J. Chem. Technol. Biotechnol.: International Research in Process, Envi. and Clean Technol.*, 80, 1072–1079.

85. Velmurugan P, Chae J C, Lakshmanaperumalsamy P, Yun B S, Lee K J, Oh B T (2009), 'Assessment of the dyeing properties of pigments from five fungi and anti-bacterial activity of dyed cotton fabric and leather', *Color. Technol*, 125, 334–341.

86. Velmurugan P, Kim M J, Park J S, Karthikeyan K, Lakshmanaperumalsamy P, Lee P, Park Y J, Oh B T (2010), 'Dyeing of cotton yarn with five water soluble fungal pigments obtained from five fungi', *Fibers Polym*, 11, 598–605.

87. Poorniammal R, Parthiban M, Gunasekaran S, Murugesan R,Thilagavathi G (2013), 'Natural dye production from Thermomyces sp fungi for textile application', *Indian J. Fibre Text. Res*, 38, 276–279.

88. Atalla M M, El-khrisy E A M, Youssef Y A, Mohamed A A (2011), 'Production of textile reddish brown dyes by fungi, Malays' *J. Microbiol*, 7, 33–40.

89. Shirata A (2000), 'Isolation of bacteria producing bluish-purple pigment and use for dyeing', *Jpn. Agric. Res. Quar.*, 34, 131–140.

90. Sen T, Barrow C J, Deshmukh S K (2019), 'Microbial Pigments in the Food Industry—Challenges and the Way Forward', *Front. Nutr.*, 6, 7.

91. Babitha S (2009), 'Microbial pigments', *Biotechnol. Agro-Industrial Residues Util., Springer*, 147–162.

5

Functional dyes for simultaneous dyeing and finishing of textiles

Ankit Singh, Indrajit Bramhecha, Mukul Gupta, Javed Sheikh*
Indian Institute of Technology Delhi, India
*Corresponding author, Email: jnsheikh@iitd.ac.in

Abstract: There is a huge demand for functional textiles due to the increased public awareness about special textile products, which can provide various additional functionalities. Imparting functional properties to textiles involves a separate process called finishing. This separate application process needs additional cost, resources, and production time; this leads to wastage of resources and degradation of the base substrate, which makes it unsustainable. The modern-day textile industries are seeking sustainable alternatives to impart functionalities to textiles using dyes at a dyeing stage to avoid finishing treatment. This chapter gives a brief overview of several approaches and reported works in the direction of synthesis of various types of functional dyes and their application to textiles for the development of functional textiles.

5.1 Introduction

The process of imparting uniform colouration, in most cases, to textiles using dyes is known as dyeing. In general, dyes should have substantivity towards fibre which is to be dyed. Dyes have two parts in their chemical structures, i.e., chromophore and auxochrome. The chromophore is the colour-bearing group, and auxochrome modifies the maximum wavelength of the dye to provide the desired colour. Finishing is a process of adding aesthetic and functional values to textiles to make them acceptable in the market. Generally, dyeing is followed by finishing in the conventional textile processing, and both processes require intensive consumption of resources, such as water, chemical, energy, and labour. Combining dyeing and finishing processes is a crucial requirement for modern textile production as it can reduce water consumption, effluent load, and production time.

The application of chemical finishes at the dyeing stage always faces problems due to their incompatibility with dyes and auxiliaries. Functional dyes are the dyes that provide combined dyeing and finishing effects in a single dyeing operation. They contain a functional group, in addition to auxochrome and chromophore, which is responsible for their functional property. They can be synthesised by introducing suitable functional groups in the chemical structures of dyes. The synthesis of functional dyes requires additional knowledge of chemistry, as the synthesis expert should keep in mind the required hue and functionality.

Functionality is a broad term that includes various properties as per the market requirements. Trending functionalities in textiles, like antibacterial activity, UV protection, water repellency, and mosquito repellency, are covered in this chapter. Even a single dye can impart multiple functionalities; such dyes are called multifunctional dyes. Some of the multifunctional dyes are also discussed in this chapter.

5.2 Antibacterial dyes

Bacteria can grow on textiles during their storage and use. Textiles, mainly made from natural fibres, can be easily attacked by bacteria because they hold water, giving a suitable environment for their growth. Such textiles have a bad odor and may cause allergy/infection to the wearer. Various approaches have been attempted for making antibacterial textiles using finishing chemicals. Quantifying market review of these chemicals was reported [1]. Due to the current pandemic situation, the interest in antibacterial textiles is constantly increasing. Some of the natural antibacterial dyes are readily available in nature, while antibacterial synthetic dyes are manufactured through various reactions of intermediates.

5.2.1 Natural dyes

Natural dyes are environment-friendly and abundantly available in nature. Some natural dyes possess pharmaceutical activities, such as antifungal, antibacterial and antioxidant. They are derived from plants, animals, and insects. Natural dyes are hailed as an eco-friendly alternative to synthetic dyes because most of them are non-allergic, non-toxic, and biodegradable. Natural dyes have low substantivity towards textiles, and generally, mordanting is required to fix natural dyes on textiles. Mordants can be classified as synthetic and natural, and most of the metal mordants are not eco-friendly, which necessitates the exploration of natural bio-mordants. This provides an opportunity to impart functional properties through functional bio-mordants.

Acacia catechu was utilised as a herb in the pharmaceutical field. It has numerous pharmaceutical properties, such as antibacterial, antioxidant,and antifungal. Catechu was used as a functional dye in many research works [2–4]. It contains catechin as a principal component (Fig. 5.1). The wool dyed with catechu using metal mordant indicated less antibacterial activity than unmordanted wool; metal mordants interact with a functional group of catechu, such as hydroxyl, which decreases the antibacterial action [5]. Chitosan from waste was utilised as a bio-mordant to dye the cotton with catechu, and an enhancement in antibacterial activity with chitosan as mordant was obtained [6]. Viscose rayon and cotton coloured with catechu demonstrated antibacterial property upto 25 washes [7]. Copper, ferrous,and aluminium salts were used as mordants to dye protein fibres (wool and silk) with catechu dye, which showed a maximum of 99.9% antibacterial activity [8].

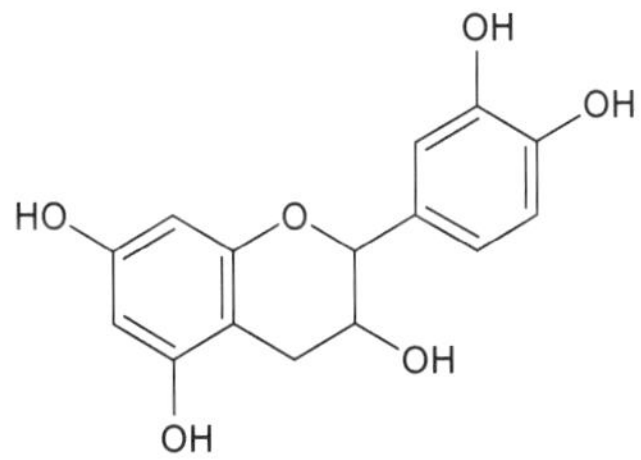

Figure 5.1 Principle component of catechu (catechin)

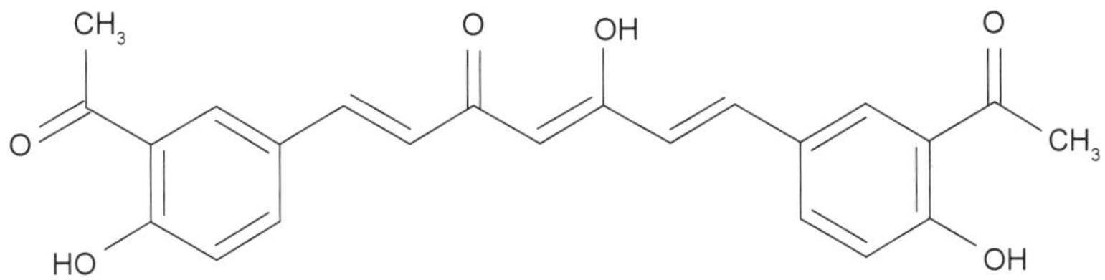

Figure 5.2 Curcumin

Curcumin (turmeric) is a food colourant widely utilised as an antibacterial compound to dye textiles. Curcumin is diferuloylmethane (Fig. 5.2), and studies indicated that curcumin has many functionalities, such as anti-tumor [9], antiviral [10], antiseptic, antioxidant, anti-inflammatory, analgesic, and anticarcinogenic activity. Cotton fabric was dyed with curcumin using Iron(II) sulphate as a mordant; the dyed fabric showed antibacterial activity against Gram-positive and Gram-negative bacteria [11]. The presence of methoxyl and hydroxyl moiety is responsible for its antibacterial action. Wool fabric dyed with curcumin showed 91% antibacterial activity against *S. aureus* [12]. The modified acrylic fibre containing amidoxime groups was coloured with curcumin natural dye using sodium sulphate as a mordant [13]. Turmeric was applied to nylon 6,6 woven fabric utilising potassium aluminum sulphate, cupric sulphate, and ferric sulphate mordants; 100% antibacterial activity against *E.coli* was obtained for samples dyed in conjunction with cupric sulphate mordant [14]. Wool fabric was functionalised with 99.8 % antibacterial activity against *E.coli* using turmeric dye and the aluminium sulphate mordant [15].

Henna (*Lawsonia inermis*) is a red-orange dye used to dye hairs and textiles. The principal component of henna leaves is 2-hydroxy-1, 4-naphthoquinone (Fig. 5.3). Henna leaves possess many pharmaceutical properties, such as antibacterial, anti-inflammatory, anti-hyperglycemic, and anti-apoptotic [16]. Chitosan-treated wool fabric was dyed using henna dye, and the dyed fabric showed more than 80% bacterial colony reduction [17]. Mordants (ferrous sulphate and alum) were utilised to dye wool yarn using henna dye [18]. The improvement in antibacterial activity was obtained using chitosan-pretreatment of jute followed by dyeing with henna dye [19]. The functionalisation of linen fabric was performed using henna dye and copper sulphate mordant; 99% antibacterial activity, excellent UV protection, and antioxidant activity were achieved [20]. Table 5.1 summarises the research works related to the antibacterial functionalisation of textiles using natural dyes.

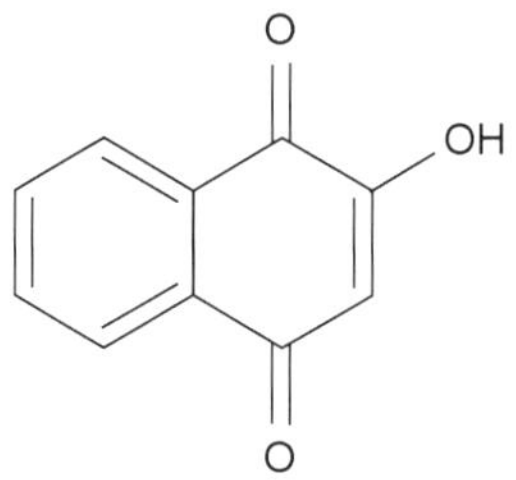

Figure 5.3 Principle component in henna leaves

Table 5.1 Summary of various works related to antibacterial functionalisation of textiles using natural dyes

Dye	Substrate	Mordants	Properties obtained	Reference
Cassia fistula, onion peels	Wool	Tannic acid	Antibacterial activity	[21]
Berberis Vulgaris	Wool	–	Maximum 99.2 % antibacterial activity	[22]
Saraca asoca and *Albizia lebbeck*	Silk	Alum, copper Sulphate, stannous chloride, ferrous sulphate	Antibacterial activity	[23]
Turmeric	Silk	Copper sulphate, ferrous sulphate, and potassium aluminium sulphate	100 % antibacterial activity	[24]

Contd...

Contd...

Dye	Substrate	Mordants	Properties obtained	Reference
Psidium guajuva	Silk	Potassium aluminium sulphate, tannic acid, and tartaric acid	Antibacterial activity	[25]
Clitoria flowers, Marigold, pomegranate	Cotton	Ferrous sulphate and copper sulphate, lemon	70% inhibition rate against *E. coli*	[26]
Madder and safflower yellow	Polyamide-6	Alum, Zn-sulphate, and tannic acid	UV-protection and anti-bacterial properties	[27]
Curcumin	Cotton	-	90% antibacterial property	[28]
Punica granatum, Diospyros peregrine, Terminalia chebula	Cotton	Aluminum potassium sulphate	Antibacterial property	[29]
Araucaria Columnaris	Cotton		Antibacterial activity	[30]
Tamarindus indica	Cotton, wool, and silk	Copper sulphate	Antibacterial activity	[31]
Quince (Cydonia oblonga) leaves	Wool	Zinc chloride and silver nitrate	Antibacterial activity	[32]
Melia composita leaves	Silk, wool, and cotton	-	Antibacterial property	[33]
Crataegus Elbursensis Fruit	Wool	Aluminum sulphate, copper sulphate, and tin chloride	Antioxidant and antibacterial properties	[34]

5.2.2 Chitosan-based antibacterial dyes

Chitosan is a biopolymer with inherent antibacterial properties and is widely explored in textile finishing. Chemical modification in synthetic dyes through various reactions is a simple way to achieve antibacterial functionality. The amino group of chitosan acts as a site to prepare the antibacterial dye. Chitosan requires an acidic medium for its dissolution in water. A reactive dye needs an alkaline medium for its fixation on cellulose through covalent bonding. Thus, it is difficult to obtain the solubility of reactive dye and chitosan in a medium. Tang et al. synthesised a reactive dye based on chitosan [35]. N-(2-hydroxyl) propyl-3-trimethyl ammonium chitosan was prepared by reacting chitosan with a quaternary ammonium salt. The reaction of chitosan quaternary ammonium salt was further carried out with reactive red dye (Figs 5.4 and 5.5). Chitosan quaternary ammonium salt-based dye showed more antibacterial activity than the reactive red dye. The developed dye demonstrated a good solubility at pH 11 and was used as a reactive dye.

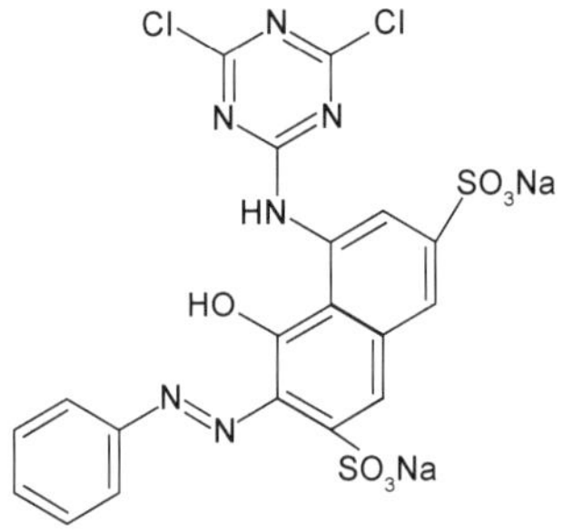

Figure 5.4 Unmodified reactive red dye

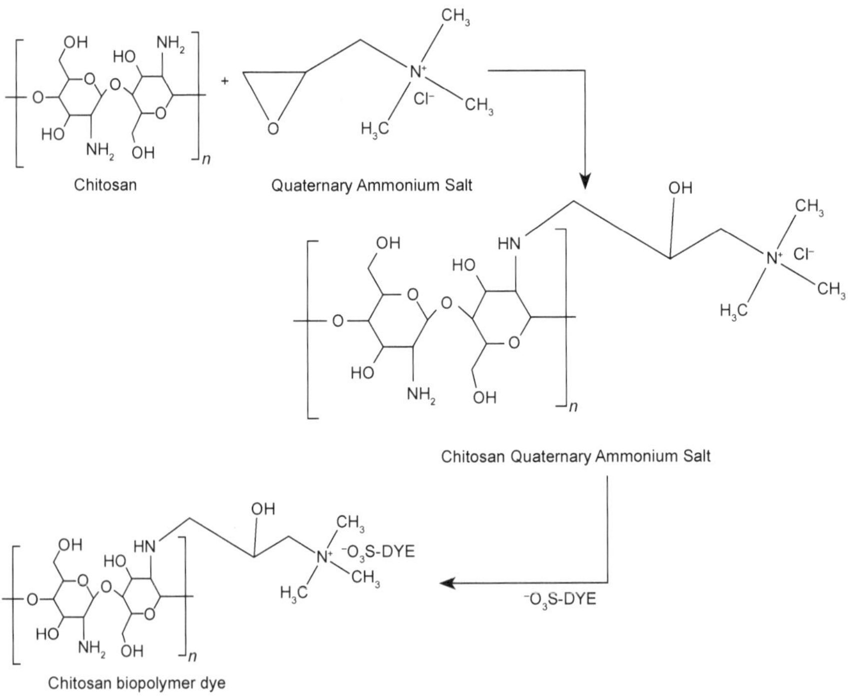

Figure 5.5 Chitosan-based modified reactive dye

Reactive blue 19 dye was altered to make it antibacterial with the assistance of chitosan [36]. The modified dye showed an improvement in wash fastness as well as light fastness in comparison to the unmodified dye (Figs. 5.6 and 5.7).

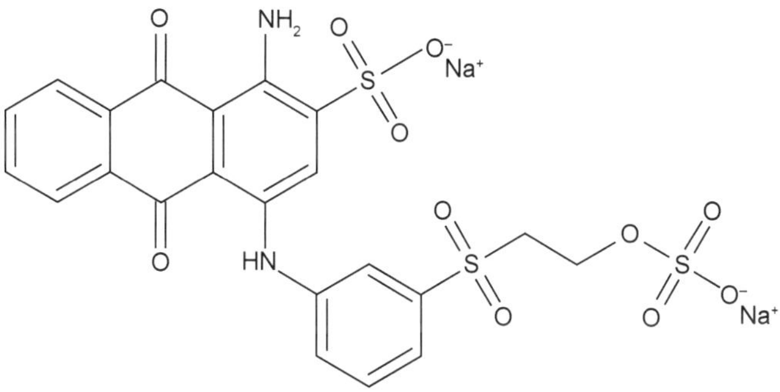

Figure 5.6 Reactive blue 19

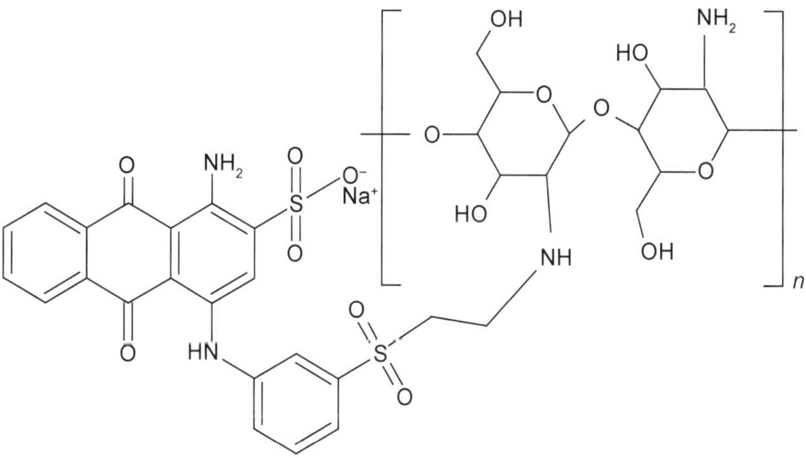

Figure 5.7 Modified reactive blue 19

5.2.3 Antibacterial azo dyes based on curcumin

As discussed in the previous section, curcumin is a compound known for its antibacterial activity. It can also be used to synthesise azo dyes for the simultaneous colouration and antibacterial finishing of textiles. Antibacterial azo dyes based on curcumin have been synthesised by Gaffer et al. [37]. These colourants showed a maximum of 88 % antibacterial activity on silk due to the availability of curcumin in the structure of the dyes (Fig. 5.8).

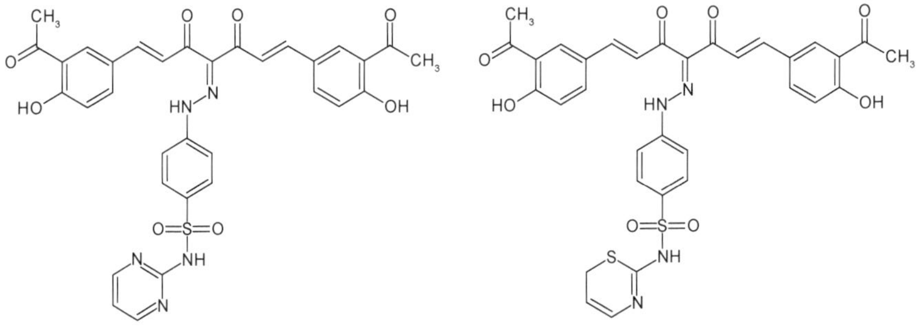

Figure 5.8 Curcumin-based azo dyes

5.2.4 Naphthalamide dyes

Antibacterial dyes can be synthesised by incorporating napthalamide functional moiety in the structure of the dyes. The structures of monoazo naphthalimide-based dyes are shown in Fig. 5.9. These naphthalamides dyes showed excellent levelness and fastness. The dyes also demonstrated a maximum of 25 mm zone inhibition against *S. aureus* bacteria and a good colour range [38].

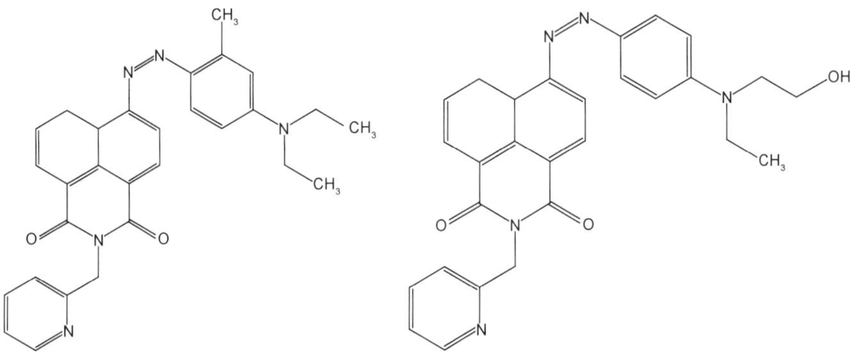

Figure 5.9 Monoazo naphthalimide-based dyes

Two dyes based on naphthalimide were synthesised and applied to nylon 6 [39]. Fig. 5.10 represents the structure of dyes. Nylon 6 dyed with thiol-based dye indicated more antibacterial activity than hydroxyl group-based dye.

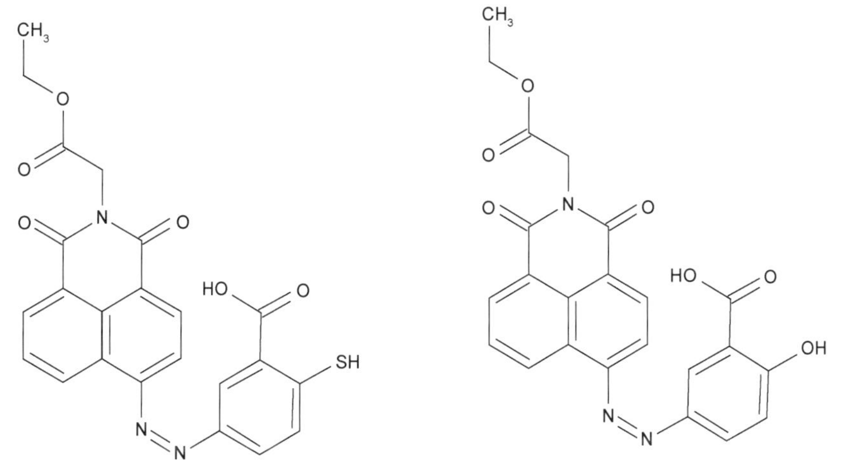

Figure 5.10 Naphthalimide-based disperse dyes

5.2.5 Phthalazinedione dyes

Numerous methodologies were explored by researchers to make phthalazinedione-based dyes for imparting antibacterial properties to textiles. Phthalazinedione moiety is responsible for the antibacterial activity of such dyes. Dyes derived from 2-(thiazol-2-yl)phthalazine-1,4-diones showed good light, washing, heat, and perspiration fastness and antibacterial activity on polyester fabric [40]. Fig. 5.11 represents the structures of some antibacterial phthalazinedione azo dyes [41].

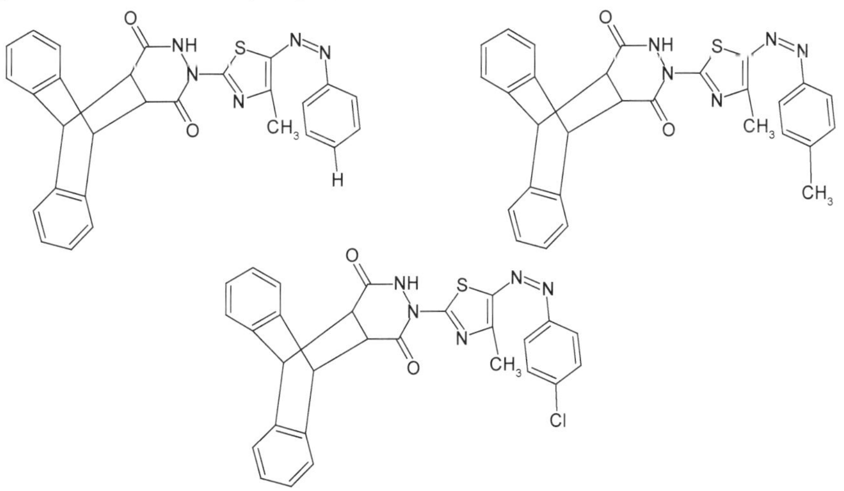

Figure 5.11 Phthalazinedione-based disperse dyes

5.2.6 Quinazolinone dyes

Quinazolinone is a heterocyclic compound widely used in the medical field, and it has various functional properties, such as antibacterial and antifungal activities [42]. A reactive dye based on the quinazolinone was synthesised, which showed antibacterial and antifungal activities on nylon, wool, and cotton [43]. Chloro-subbed quinazolinone colours demonstrated more antibacterial properties than bromo-subbed quinazolinone colours [44]. Parekh & Maheria reported 3-(4-Aminophenyl)-2-phenylquinazolin-4(3h)-based antibacterial dyes; Fig. 5.12 indicates the chemical structure of one such antibacterial dye [45].

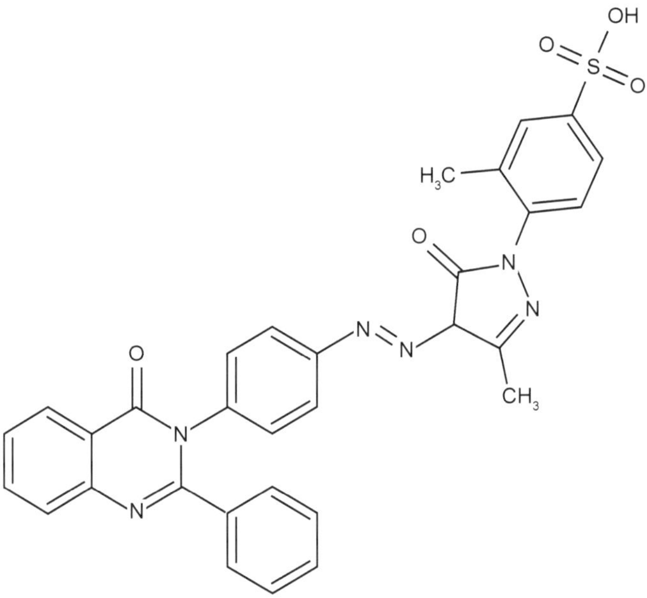

Figure 5.12 Quinazolinone-based antibacterial dye

5.2.7 Pyrazole dyes

Pyrazole-based dyes are notable for their antibacterial and anticancer properties. The polyester fabric was dyed with azo-pyrazole disperse dyes, and the dyed fabric displayed antibacterial properties [46]. Fig. 5.13 shows the structures of the reactive dyes based on pyrazole. These reactive dyes can give good wash fastness on cotton because of covalent bonding and magnificent antibacterial activity because of the presence of pyrazole moiety [47]. Pyrazole triazine-based disperse dyes were synthesised and applied to polyester; the dyes indicated a maximum of 32 mm inhibition zone against *S. aureus* [48].

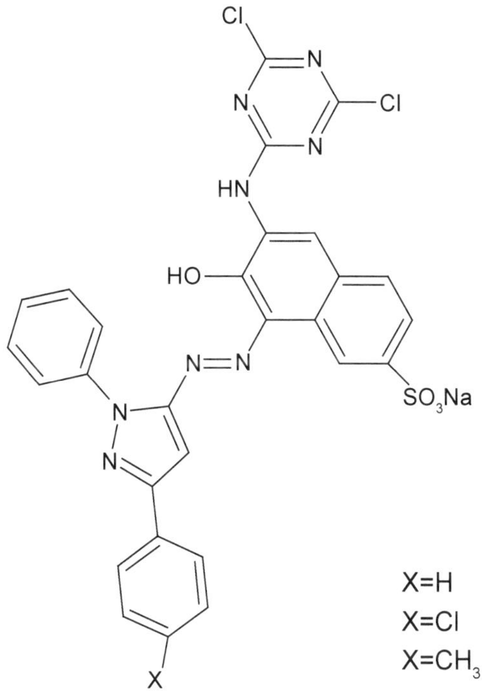

Figure 5.13 Pyrazole-based reactive dyes

5.2.8 Sulfonamide dyes

Sulfonamides are used to treat bacterial infections (Fig. 5.14). The excellent performance of sulfonamide makes it an attractive chemical for the synthesis of antibacterial compounds. Dye chemists used the proficiency of this drug to synthesise antibacterial dyes. Sulfonamide-based dyes were simultaneously applied with the finishing agent DMDHEU (Dimethylol dihydroxy ethylene urea) on viscose fabric; the treated viscose fabric demonstrated UV protection, crease-resistance, and antibacterial activity [49].

The synthesis route of sulphonamide moiety-containing disperse dye is mentioned in Fig. 5.15. 4-Amino-N-(2-pyrimidinyl) benzene sulphonamide was diazotised using sodium nitrite and hydrochloric acid. The previously diazotised base was reacted with a coupling component for a disperse dye synthesis. The polyester dyed with synthesised disperse dye showed wet rubbing fastness rating of 4-5 and antibacterial activity [50].

Fig. 5.16 indicates the synthesis strategy for the sulfonamide-based reactive dye. The reaction of J-acid was performed with cyanuric chloride using acetone as the reaction solvent at 0-5 °C. Cyanurated J-acid was further reacted with sulfonamide. The resultant product was used as a coupler and reacted with diazotised sulfonamide derivative base. The synthesised dye was applied to cotton fabric, and their interaction was through a covalent bond [51].

Chemical groups, such as -NH_2, -OH, and -SO_3Na, can enhance the antibacterial action of sulfonamide-based dyes [52]. The Thiazole-based azo dyes with sulfonamide moiety were cross-linked with cotton cellulose using DMDHEU. The treated fabric indicated UV protection along with antibacterial activity [53].

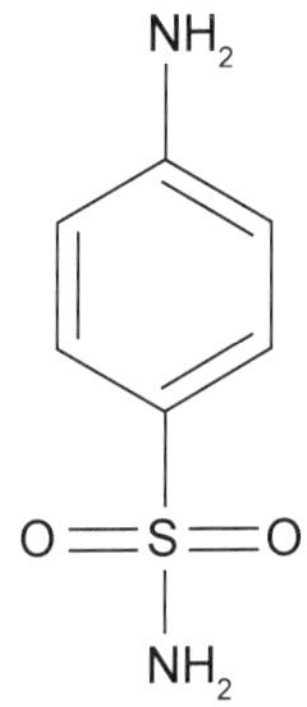

Figure 5.14 Sulphonamide

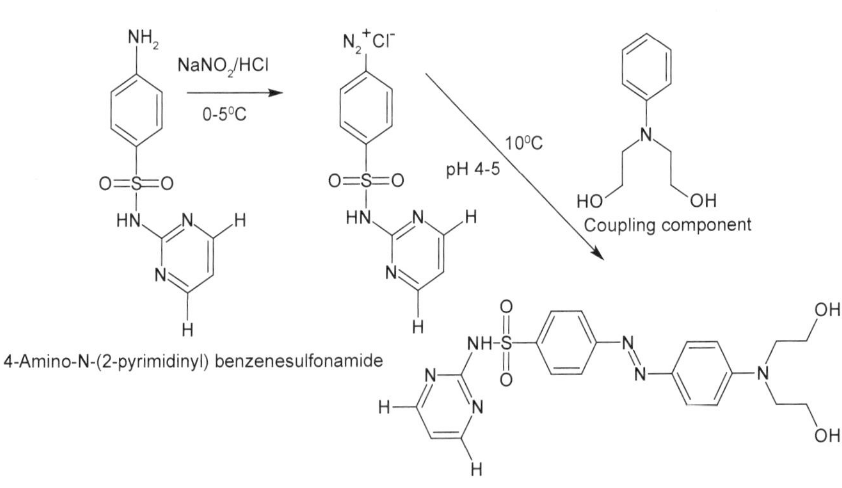

Figure 5.15 Synthesis of sulphonamide-based disperse dye

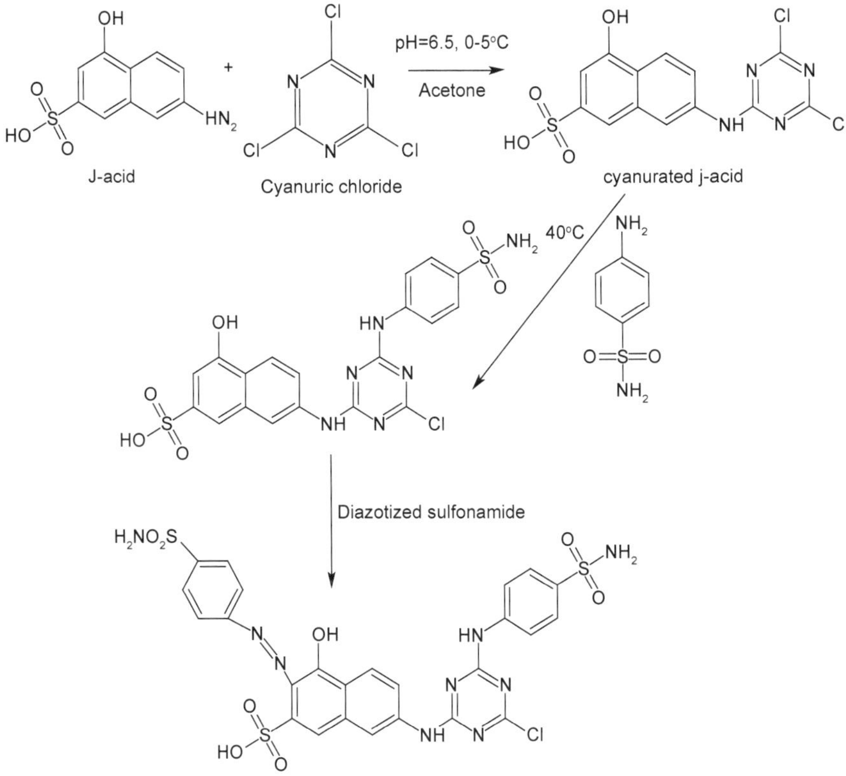

Figure 5.16 Synthesis of Sulphonamide-based reactive dye

5.3 UV-protective dyes

Nowadays, protection from UV rays is a requirement because of the gradually depleting ozone layer. Long-term exposure to UV rays can cause skin problems, such as sunburn, premature ageing, allergies, and skin cancer. UV rays are electromagnetic radiation of a wavelength 100- 400 nm. UV rays are classified into three categories; UV-A (315–400 nm), UV-B (290–315 nm), and UV-C (100–290 nm). The ozone layer absorbs UV-C rays; only UV-A and UV-B incident on earth surface. Textile is a basic need of human beings, and UV-protective textile is an earnest necessity of the modern customer. A UV protection offered by a fabric is a combined effect of yarn structure, fibre type, fabric structure, dye structure, and finishing treatments. UPF (UV protection factor) value is the indicator that defines UV protection provided by the clothing.

When UV radiation falls on the fabric surface, it can be scattered, absorbed, reflected, and transmitted [54]. The transmitted rays interact with the human body (Fig. 5.17). The UV-protective finish on textiles can act as a barrier to reduce the transmission of UV rays. The protection provided by dyes depends on the type and concentration of dyes on textiles. The dye with an excellent UV absorbing nature can offer excellent UV protection properties. A widely researched organic UV absorbers are O-hydroxybenzo-phenones, O-hydroxy phenyl triazines, and O-hydroxy phenyl hydrazines.

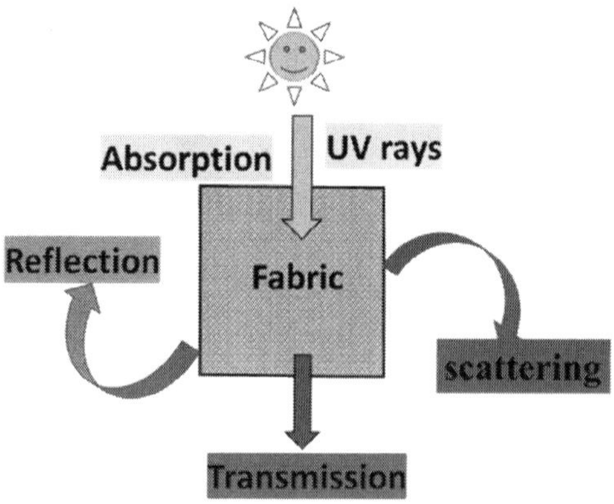

Figure 5.17 Interaction of UV rays with fabric

Natural colourants, such as madder, cochineal, and Indigo, can act as UV-protective eco-friendly colourants for textiles [55]. A mordanting can also enhance UPF values [56]. Cotton and silk were dyed by *Rheum and Lithospermum erythrorhizon* natural dyes. The coloured samples demonstrated excellent UPF values because of the UV protection property provided by dyes. Fig. 5.18 indicates the possible mechanism of UV-protective action. The absorption behavior might be due to the breakage of hydrogen bonding [57].

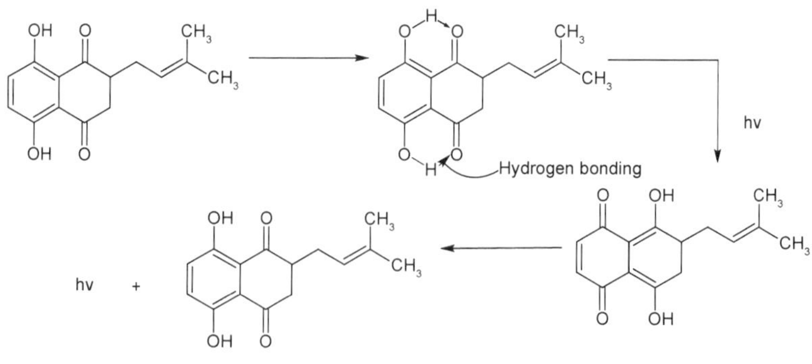

Figure 5.18 Mechanism of UV protection

Eucalyptus leaf extract (composition is shown in Fig. 5.19) can act as a good UV absorber. Fig. 5.20 represents the UV absorption spectrum of eucalyptus leaf extract, which suggests its excellent UV absorption (Fig. 5.20) [58].

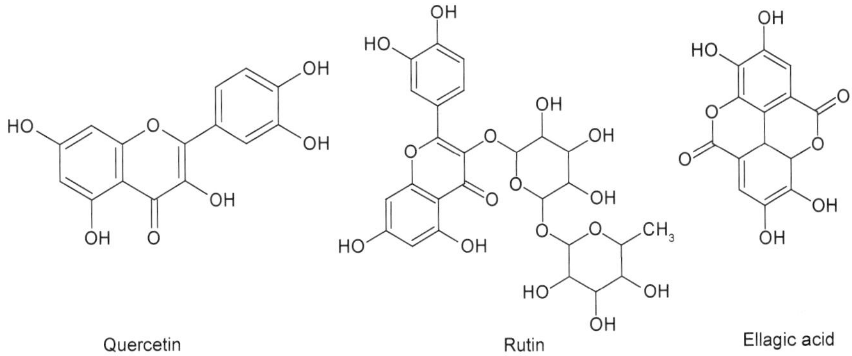

Figure 5.19 Composition of Eucalyptus leaf extract

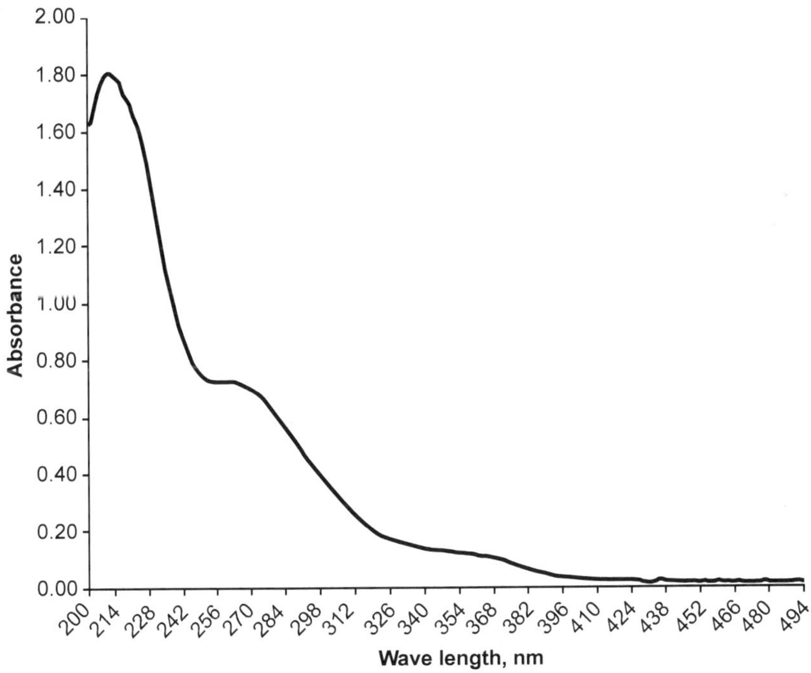

Figure 5.20 UV-Vis absorption spectrum of eucalyptus leaf extract [58]

An aqueous extract of honeysuckle was applied to wool for imparting UV protection. Chlorogenic acid (Fig. 5.21) is the active substance in the honeysuckle, which is responsible for UV protection [59]. *Mediterranean* flora (Helichrysum) plant extract was applied to cotton and flax, and excellent UV protection was obtained [60].

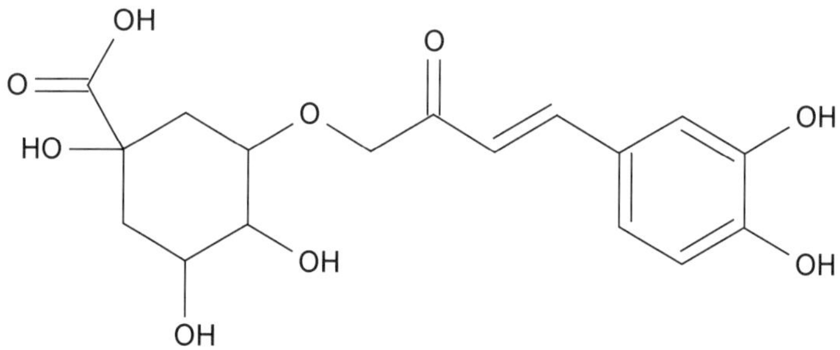

Figure 5.21 Chlorogenic acid

The diamine UV absorbers were used to synthesise UV-protective reactive dyes (Fig. 5.22). These reactive dyes were applied to the cotton fabric to obtain UV-protective dyed cotton [61].

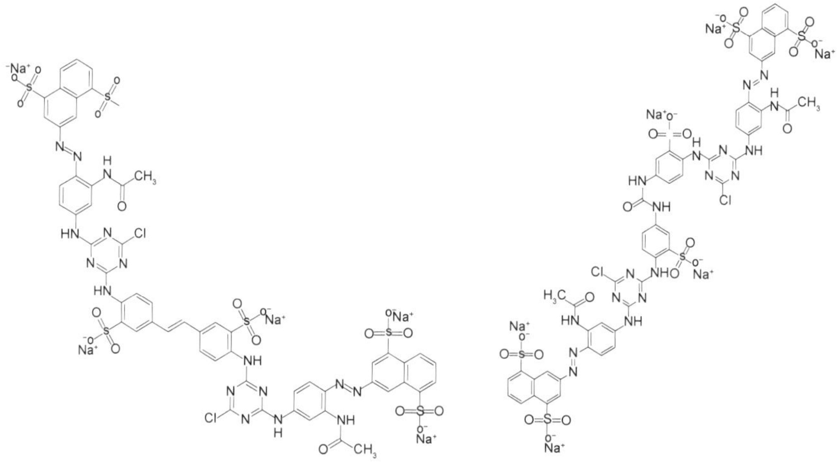

Figure 5.22 Bifunctional UV-protective reactive dyes

The benzophenone UV absorber-based reactive dye was prepared, and Fig. 5.23 indicates the structure of the reactive dye [62].

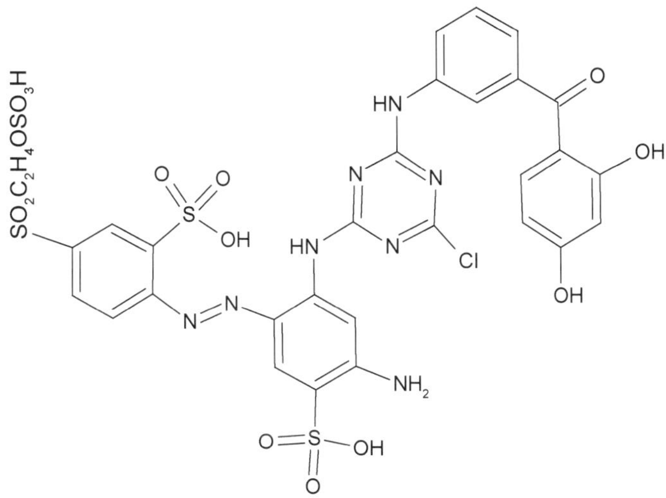

Figure 5.23 Benzophenone UV absorber-based reactive dye

Three monoazo dyes based on 2-hydroxy-4-methoxybenzophenone as a coupler were synthesised [63]. Fig. 5.24 indicates the methodology of preparation for these dyes. The broad absorption in the range of 280-420 nm confirmed the UV-protective action of dyes. The silk dyed with these acid dyes showed UPF values of 39.7, 43.7, and 27.8 for Dye 1, Dye 2, and Dye 3, respectively.

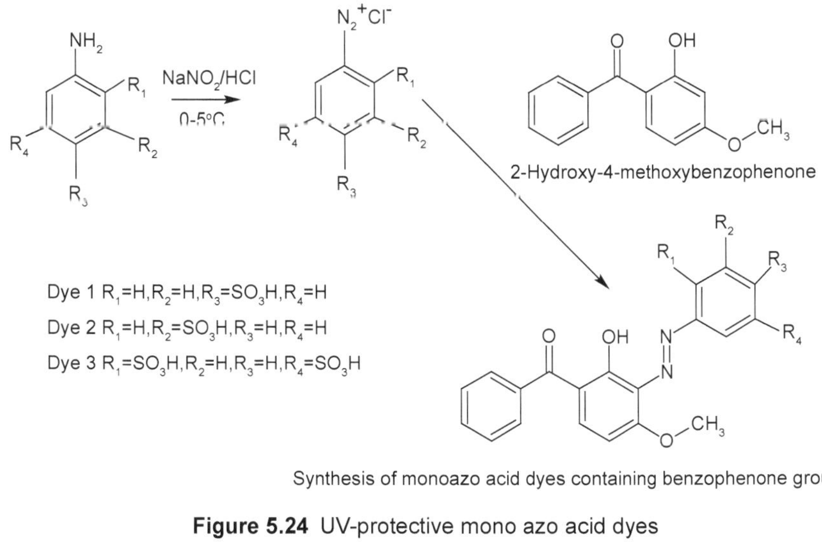

Synthesis of monoazo acid dyes containing benzophenone group.

Figure 5.24 UV-protective mono azo acid dyes

UV absorbers, like resorcinol, 2,4-dihydroxybenzophenone and 1-hydroxy benzotriazole sulfonic acid were utilised for the synthesis of azo dyes [64]. A recently reported study revealed the synthesis of heterocyclic azo disperse dye with UV protection-cum-antibacterial functionalities [65].

5.4 Water repellent dyes

Water repellent textiles are used to prepare products, like raincoats, umbrellas, tarpaulins, and swimming suits. Water repellent dyes can be an excellent option to provide wash-fast water repellent properties. The preparation of water repellent acid dyes was reported by Teli et al. [66]. Fig. 5.25 shows the synthesis route for these water-repellent acid dyes. Firstly, pentadecafluoroctanoyl chloride was prepared by refluxing pentadecafluorooctanoic acid with thionyl chloride. The reactions of pentadecafluoroctanoyl chloride with γ-acid and H-acid were further carried out. The resultant compounds were used as couplers in azo dye synthesis. Four acid dyes were synthesised using bases, like m-trifluoromethylaniline and m-toluidine. The improvements in water repellency were observed after the dyeing of silk fibre with these dyes. The dyed silk additionally displayed a decrease in the wicking height. Similarly, disazo water repellent dyes were also synthesised (Fig. 5.26). The dye-6 imparted a better water repellency than dye-5 due to the presence of the -NO_2 group in the para-position in dye-6, unlike the -CH_3 group in dye-5 at m-position [67].

A new emerging research area is the development of textiles with water-repellent, self-cleaning, and stain-resistant properties. The surface self-cleaning effect is displayed when the water contact angle exceeds 150 °. A moving drop of water on superhydrophobic textiles removed the soil from the textile surface. Alkyl and fluorinated alkyl groups having very little interaction with water can enhance the contact angle of textile surfaces. A recently reported study discloses a method to make superhydrophobic cotton through dyeing [68]. Two kinds of dyes were prepared: (A) dyes that fix through impregnation and (B) the dye which makes stain; however, is covalently connected to cotton (Fig. 5.27). Fig. 5.28 represents the contact angle between textiles dyed with superhydrophobic dyes, and water. The self-cleaning effect is clearly visible in Fig. 5.29.

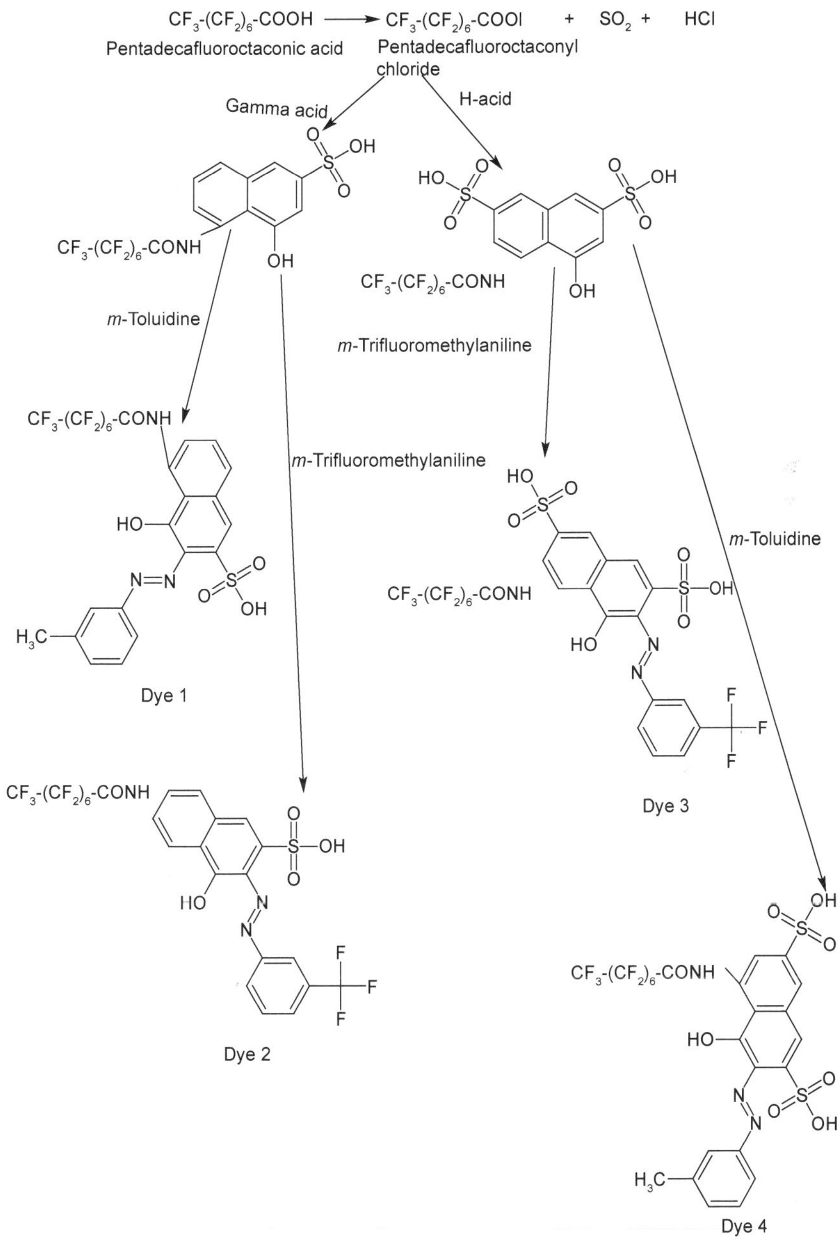

Figure 5.25 Monoazo water repellent acid dyes

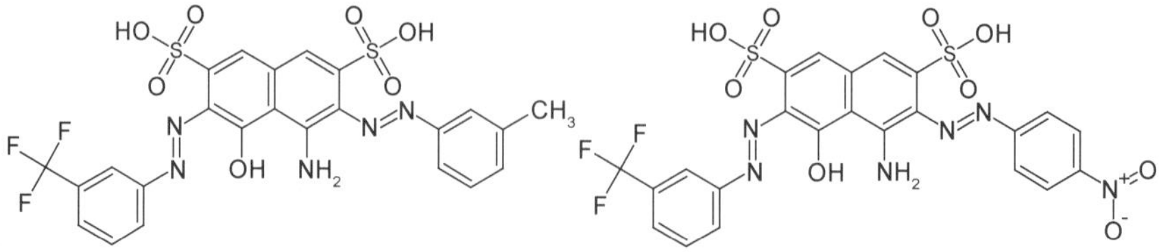

Water repellent Dye 5

Water repellent Dye 6

Figure 5.26 Bisazo water repellent dyes

Stain by dye impregnation

Stain by covalent attachment of the dye

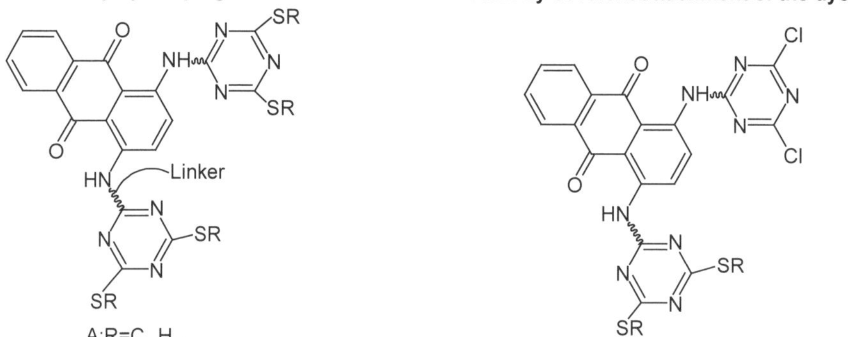

A:R=$C_{12}H_{25}$

B:R=$CH_2CH_2C_8F_{17}$

C:R=$C_{12}H_{25}$

D:R=$CH_2CH_2C_8F_{17}$

Figure 5.27 Superhydrophobic anthraquinone dyes

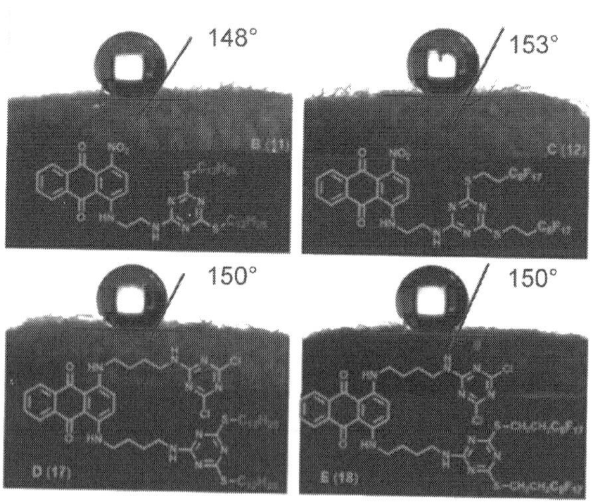

Figure 5.28 Angle of contact between dyed cotton and water droplets [68]

Figure 5.29 Self-cleaning effect [68]

5.5 Mosquito repellent dyes

Mosquitoes are insects of concern as mosquito-bite is a cause of several diseases to humankind. Mosquitoes act as vectors in the spread of diseases like malaria, dengue, chickengunya, zika, and yellow fever. The population of mosquitoes is increasing day by day due to climate change and urbanisation. Several products, including creams, lotions, liquids, roll-ons and patches, are available in the market for human protection against mosquito-bites. However, these products have limitations in terms of poor retention of efficacy over a longer time, requirement of repeated applications, localised protection in the area of application, loss of effectiveness after washing, and allergic reactions in some cases. Textile materials are one of the basic needs of humankind and cover most of the body parts; this provides a clear advantage to mosquito repellent clothing. The products that are ideal to be imparted with mosquito repellency include anti-mosquito garments, bed nets, head nets, bed sheets, window curtains, home textiles, protective clothing, and repellent ropes. DEET (N, N-Diethyl-meta-toluamide) is the most widely utilised synthetic mosquito repellent. Out of 6,241 mosquito repellent compounds tested against *Aedes aegypti*, DEET was found to be the most effective [69]. The researchers tried several methods to apply mosquito repellents on textiles, such as direct padding, coating, and microencapsulation. Microencapsulation of mosquito repellent compounds to generate mosquito repellent microcapsules for textile finishing is one of the most widely used techniques. The primary issue with existing technologies is the lack of efficacy and limited durability of mosquito repellency on textiles.

The term mosquito repellent dyes refer to those dyes which have inherent repellent property against mosquitoes. Textiles dyed with such dyes become unfavorable for the attack of mosquitoes. Although various chemicals were studied as mosquito repellent finishes, mosquito repellent dyes are beneficial as they can impart efficient mosquito repellency with better durability. Mostly, azo chromophore-based mosquito repellent dyes have been reported. Akbarzadeh et al. reported the synthesis of a mosquito repellent reactive dye [70]. A typical route followed by them to synthesise the reactive dye is given in Fig. 5.30.

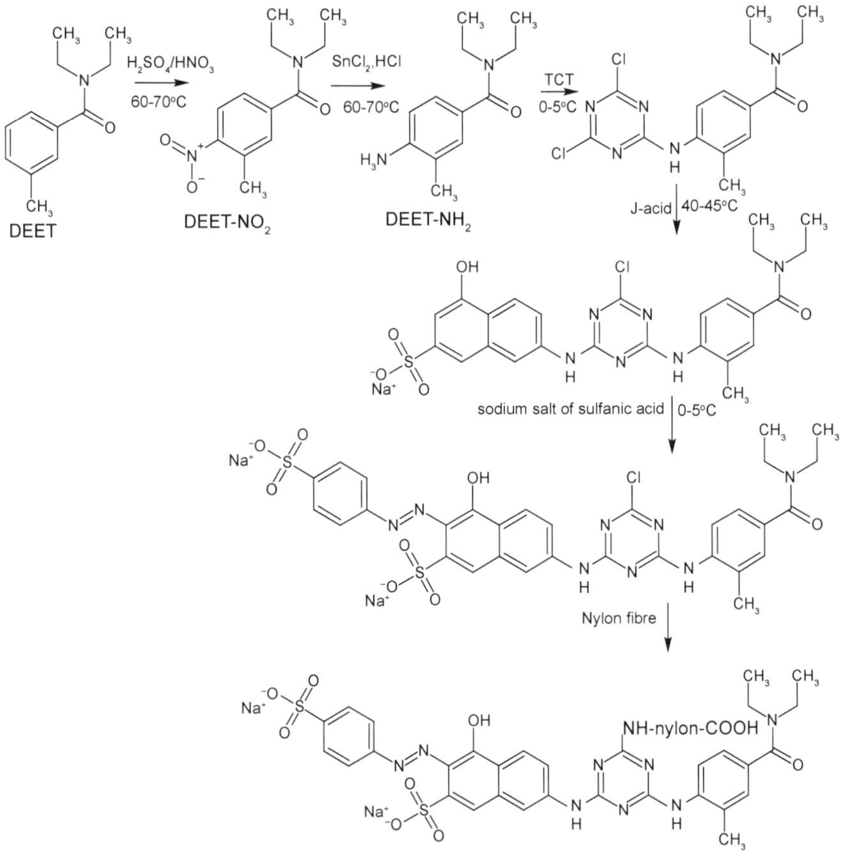

Figure 5.30 Mosquito repellent reactive Dye 1

Firstly, the nitration of DEET was done by using a nitrating mixture of sulphuric acid and nitric acid. The nitrated product of DEET (DEET-NO_2) was further subjected to reduction using $SnCl_2$ and hydrochloric acid. The reduction product of DEET-NO_2 (DEET-NH_2) was reacted with 2,4,6-Trichloro-triazine at 0-5 °C. The condensation of the previous compound was done using J-acid. The resultant compound was used as a coupler, and the coupling reaction was performed with the diazotised base of sulfanilic acid sodium salt. The dye obtained (Dye 1) was applied to nylon-6, making a covalent linkage with nylon fibre. The dyed nylon displayed 90% mosquito repellency.

Teli & Chavan synthesised a mosquito repellent reactive dye (Fig. 5.31) using a different procedure [71]. The nitration of DEET was performed using sulphuric acid and potassium nitrate. The reduction of DEET-NO_2 was done using a reducing system (Zn/HCl). The reactive group in DEET-NH_2 was introduced by reacting it with 2,4,6-Trichloro-triazine. Further, the reaction of the previously synthesised compound was done with Bronner's acid. The prepared dye was applied to nylon fabric, which displayed mosquito repellent action till at least ten washes.

Mosquito repellent and antibacterial azoic colourants were synthesised by *in situ* dyeing method on cotton fabric. The dyed cotton demonstrated a maximum of 100% mosquito repellency and 99.6 % antibacterial activity [72].

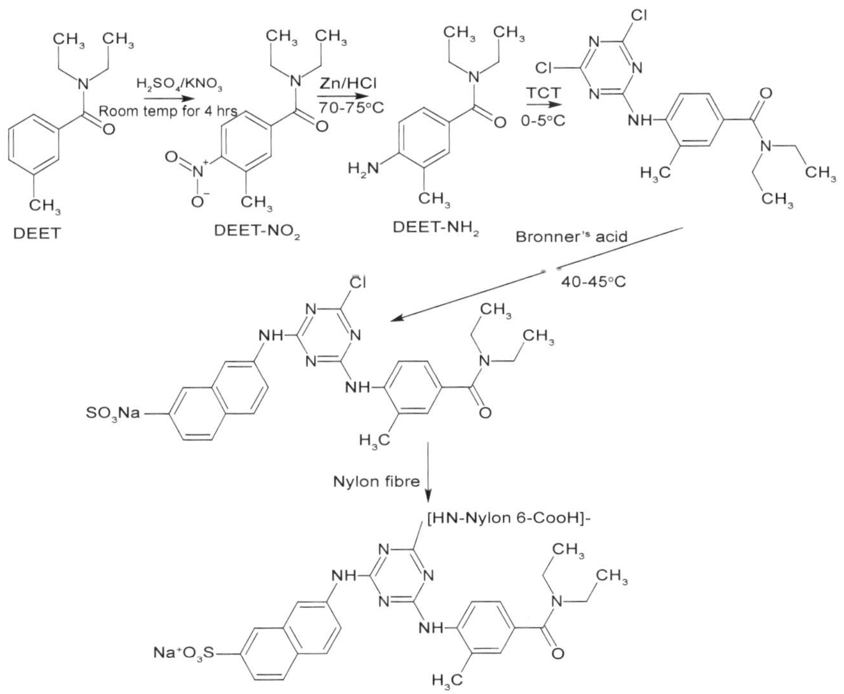

Figure 5.31 Mosquito repellent reactive Dye 2

Concluding remarks and future perspectives

The development of short routes for getting the final products is the need of the hour in all fields. Short routes ensure better ecology and savings in resources. Thus, the route to obtain a finished product can be shortened by using functional dyes. The development and growth of technical textiles are expected to drive growth in the sector of functional dyes. The focus of the researchers is expected to shift from traditional dyeing and functional finishing to functional dyeing, especially for obtaining functional textile products. The limitations of lack of durability and ecological perspectives in traditional dyeing and finishing processes can be thus solved. Even though most of the reported functional dyes are not commercialised yet, their usefulness and sustainability will lead to a revised interest in the development of such functional colourants on a commercial scale. Joint efforts from the researchers working in various fields like chemistry, chemical engineering, textile chemical processing, and material science are required. A huge market for such functional colours is expected.

References

1. Antimicrobial Medical Textiles Market Size | Global Industry Report, 2024, https://www.grandviewresearch.com/industry-analysis/antimicrobial-medical-textiles-market (accessed April 8, 2022).
2. Bhattacharya SD, Shah AK (2000), 'Metal ion effect on dyeing of wool fabric with catechu', *J. Soc. Dye. Colour.*, 116,10–12. https://doi.org/10.1111/j.1478-4408.2000.tb00002.x.
3. Yeasmin F, Al Mamun MA, Rokunuzzaman MD (2018), 'Comparative Study of Cotton Knitted Fabric Dyed with Natural Dyes Extracted from Catechu V / S Reactive Dyes', *Int. J. Text. Sci.,* 7,65–68. https://doi.org/10.5923/j.textile.20180703.01.
4. Md S, Kabir M, Dhar AK, Bhattacharjee M (2020), 'The use of natural Areca catechu dyes for silk and nylon and its halochromic effect', *J. Text. Inst.*, 0,1–8. https://doi.org/10.1080/00405000.2019.1674542.
5. Khan MI, Ahmad A, Khan SA, Yusuf M, Shahid M, Manzoor N, Mohammad F, (2011), 'Assessment of antimicrobial activity of catechu and its dyed substrate', *J. Clean. Prod.*,19,1385–1394. https://doi.org/10.1016/j.jclepro.2011.03.013.
6. Teli MD, Sheikh J, Pradhan C (2014), 'Simultaneous natural dyeing and antibacterial finishing using chitosan from bio-waste', *Melliand International.*, 20,171–172.
7. Ibrahim HM, Hamada MM, El-Hawary NS, Kamel MM, El-Alfy EA (2017), 'Effect of Catechu Natural Dye Extracts on Coloration and Antibacterial Protection Factor for Different Cellulosic Fabrics', *Der Pharma Chemica.* ,9,84–88.
8. Ki-Yeon N, Jung-Soon L (2010), 'Dyeability and Functionality of Catechu (Part II) -Dyeing Properties of Protein Fiber with Catechu', *Korean Journal of Human Ecology,*19, 709-717.

9. Tomeh MA, Hadianamre R, Zhao X (2019), 'A review of curcumin and its derivatives as anticancer agents', *Int. J. Mol. Sci,*. 20,1033. https://doi.org/10.3390/ijms20051033.

10. Moghadamtousi SZ, Kadir HA, Hassandarvish P,Tajik H, Abubakar S, Zandi K (2014) , 'A review on antibacterial, antiviral, and antifungal activity of curcumin', *Biomed Res. Int.*,186864,1-12. https://doi.org/10.1155/2014/186864.

11. Gawish SM, Mashaly HM, Helmy HM, Ramadan AM, Farouk R (2017), 'Effect of Mordant on UV Protection and Antimicrobial Activity of Cotton, Wool, Silk and Nylon Fabrics Dyed with Some Natural Dyes', *J. Nanomed. Nanotechnol.*,08,1–9. https://doi.org/10.4172/2157-7439.1000421.

12. Han S, Yang Y (2005), 'Antimicrobial activity of wool fabric treated with curcumin', *Dyes. Pigment..*, 64,157–161. https://doi.org/10.1016/j.dyepig.2004.05.008.

13. El-Shishtawy RM, Shokry GM, Ahmed NSE, Kamel MM (2009), 'Dyeing of modified acrylic fibers with curcumin and madder natural dyes', *Fibers Polym.,* 10, 617–624. https://doi.org/10.1007/s12221-010-0617-4.

14. Mirjalili M, Karimi L (2013), 'Antibacterial dyeing of polyamide using turmeric as a natural dye', *Autex Res. J.*, 13,51–56. https://doi.org/10.2478/v10304-012-0023-7.

15. Ghaheh FS, Mortazavi SM, Alihosseini F, Fassihi, A, Nateri AS, Abedi D (2012), 'Assessment of antibacterial activity of wool fabrics dyed with natural dyes', *J. Clean. Prod.* 72,139–145. https://doi.org/10.1016/j.jclepro.2014.02.050.

16. Al-snafi AE (2019), 'A Review on Lawsonia Inermis: a Potential Medicinal Plant', *Int. J. Curr. Pharm. Res.* 11, 1–13. https://doi.org/10.22159/ijcpr.2019v11i5.35695.

17. Giri Dev VR, Venugopal J, Sudha S, Deepika G, Ramakrishna S (2009),' Dyeing and antimicrobial characteristics of chitosan treated wool fabrics with henna dye', *Carbohydr. Polym.*, 75,646–650.

18. Yusuf M, Ahmad A, Shahid M, Khan MI, Khan SA, Manzoor N, Mohammad F (2012), 'Assessment of colorimetric, antibacterial and antifungal properties of woollen yarn dyed with the extract of the leaves of henna (Lawsonia inermis)', *J. Clean. Prod.*, 27,42–50. https://doi.org/10.1016/j.jclepro.2012.01.005.

19. Bhuiyan MAR, Islam A, Islam S, Hossain A, Nahar K (2017), 'Improving dyeability and antibacterial activity of Lawsonia inermis L on jute fabrics by chitosan pretreatment', *Text. Cloth. Sustain.,* 3,1–10. https://doi.org/10.1186/s40689-016-0023-4.

20. Yadav R, Anubhav, Mathur P, Sheikh J (2019), 'Antibacterial, UV protective and antioxidant linen obtained by natural dyeing with Henna', *Cellul. Chem. Technol.* 53,357–362. https://doi.org/10.35812/cellulosechemtechnol.2019.53.36.

21. Ali NF, Mohamedy RSREL, El-Khatib EM (2011), 'Antimicrobial Activity of Wool Fabric Dyed with Natural Dyes', *Res. J. Text. Appar.,* 15,1–10. https://doi.org/10.1108/RJTA-15-03-2011-B001.

22. Haji A (2013), 'Eco-friendly dyeing and antibacterial treatment of cotton', *Cellul. Chem. Technol.*,47,303–308.

23. Baliarsingh S, Panda AK, Jena J, Das T, Das NB (2012), 'Exploring sustainable technique on natural dye extraction from native plants for textile: Identification of colourants, colourimetric analysis of dyed yarns and their antimicrobial evaluation', *J. Clean. Prod.* 37,257–264. https://doi.org/10.1016/j.jclepro.2012.07.022.

24. Ghoreishian SM, Maleknia L, Mirzapour H, Norouzi M (2013), 'Antibacterial properties and color fastness of silk fabric dyed with turmeric extract', *Fibers Polym.*, 14 ,201–207. https://doi.org/10.1007/s12221-013-0201-9.

25. Narayanaswamy V, Gowada KNN, Sudhakar R (2013), 'Dyeing and Color Fastness of Natural Dye from Psidium guajuva on Silk', *J. Nat. Fibers,*. 10,257–270. https://doi.org/10.1080/15440478.2013.797948.

26. Kanchana R, Fernandes A, Bhat B, Budkule S, Dessai S, Mohan R (2013), 'Dyeing of textiles with natural dyes - An eco-friendly approach', *Int. J. ChemTech Res.* ,5,2102–2109.

27. Ibrahim NA, El-Zairy WM, El-Zairy MR, Ghazal HA (2013), 'Enhancing the UV-protection and Antibacterial Properties of Polyamide-6 Fabric by Natural Dyeing', *Text. Light Ind. Sci. Technol.* 2,36-41.

28. Reddy N, Han S, Zhao Y, Yang Y (2013), 'Antimicrobial activity of cotton fabrics treated with curcumin', *J. Appl. Polym. Sci.* 127,2698–2702. https://doi.org/10.1002/app.37613.

29. Datta S, Uddin MA, Afreen KS, Akter S, Bandyopadhyay A (2013), 'Assessment of antimicrobial effectiveness of natural dyed fabrics', *Bangladesh J. Sci. Ind. Res.*,48,179–184. https://doi.org/10.3329/bjsir.v48i3.17327.

30. Devi S, Rathinamala J, Jayashree S (2014), 'Study on antibacterial activity of natural dye from the bark of Araucaria Columnaris and its application in textile cotton fabrics', *J. Microbiol. Biotechnol. Res. Sch.* , 4,32–35.

31. Prabhu KH, Teli MD (2014), 'Eco-dyeing using Tamarindus indica L. seed coat tannin as a natural mordant for textiles with antibacterial activity', *J. Saudi Chem. Soc.* 18,864–872. https://doi.org/10.1016/j.jscs.2011.10.014.

32. Cerempei A, Mureşan EI, Cimpoeşu N, Carp-Cărare C, Rimbu C (2016), 'Dyeing and antibacterial properties of aqueous extracts from quince (Cydonia oblonga) leaves', *Ind. Crops Prod.* 94,216–225. https://doi.org/10.1016/j.indcrop.2016.08.018.

33. Pal A,Tripathi YC, Kumar R, Upadhyay L (2016), 'Antibacterial Efficacy of Natural Dye from Melia composita Leaves and Its Application in Sanitized and Protective Textiles', *J. Pharm. Res.* ,10(4),154–159.

34. Safapour S, Sadeghi-Kiakhani M, Eshaghloo-Galugahi S (2018), 'Extraction, Dyeing, and Antibacterial Properties of Crataegus Elbursensis Fruit Natural Dye on Wool Yarn', *Fibers Polym.* 19,1428–1434. https://doi.org/10.1007/s12221-018-7643-z.

35. Tang R, Zhang Y, Zhang Y, Yu Z (2016), 'Synthesis and characterization of chitosan based dye containing quaternary ammonium group', *Carbohydr. Polym.* 139,191–196. https://doi.org/10.1016/j.carbpol.2015.12.047.

36. Tang R, Yu Z, Zhang Y, Qi C (2016), 'Synthesis, characterization, and properties of antibacterial dye based on chitosan', *Cellulose.*, 23,1741–1749. https://doi.org/10.1007/s10570-016-0935-9.

37. Gaffer H, Mashaly H, Abdel-Rhman SH, Hammouda M (2017), 'Synthesis of novel dyes based on curcumin for the creation of antibacterial silk fabrics', *Pigment Resin Technol.* 46, 478–484. https://doi.org/10.1108/PRT-11-2016-0112.

38. Shaki H, Gharanjig K, Rouhani S, Khosravi A, Fakhar J (2012), 'Synthesis and application of some novel antimicrobial monoazonaphthalimide dyes: Synthesis and characterisation', *Color. Technol.* 128, 270–275. https://doi.org/10.1111/j.1478-4408.2012.00374.x.

39. Mohammadkhodaei Z, Mokhtari J, Nour M (2010), 'Novel anti-bacterial acid dyes derived from naphthalimide: Synthesis, characterisation and evaluation of their technical properties on nylon 6', *Color. Technol.* 126 ,81–85. https://doi.org/10.1111/j.1478-4408.2010.00230.x.

40. Khalil AM, Berghot MA , Gouda MA, El Bialy SA (2010), 'Synthesis and antibacterial studies of azodispersed dyes derived from 2-(thiazol-2-yl)phthalazine-1,4-diones', *Monatshefte Fur Chemie.* ,141,1353–1360. https://doi.org/10.1007/s00706-010-0392-3.

41. Khalil AEGM, Berghot MA, Gouda MA (2011), 'Design, synthesis and antibacterial activity of new phthalazinedione derivatives', *J. Serbian Chem. Soc.* 76,329–339. https://doi.org/10.2298/JSC091122028K.

42. Prabhakar V, Babu KS, Ravindranath LK , Latha J, Nagamaddaiah KH (2015), 'Quinazoline Derivatives and its Biological Significance', *International Journal of Current Trends in Pharmaceutical Research* . 3, 997–1010.

43. Patel DR, Patel KC. (2010), 'Synthesis, characterization and application of quinazolinone based reactive dyes for various fibers', *Fibers Polym.* ,11,537–544. https://doi.org/10.1007/s12221-010-0537-5.

44. Patel DR, Patel KC. (2012), 'Synthesis, Antimicrobial Activity and Colorimetric Studies of Some New Bromo-quinazolinone Derivative as Potential Reactive Dyes', *Arab. J. Sci. Eng.* 37, 1347–1368. https://doi.org/10.1007/s13369-012-0241-2.

45. Parekh N, Maheria K (2012), 'Studies on antimicrobial activity for multidrug resistance strain by using phenyl pyrazolones substituted 3-(4-aminophenyl)-2-phenylquinazolin-4(3H)-one derivatives in vitro condition and their dyeing performance', *Fibers Polym.* 13 , 162–168. https://doi.org/10.1007/s12221-012-0162-4.

46. Elmaaty TA , Elnagar K , Hassan S, Gamal H (2014), 'Antibacterial activity and dyeing characteristics of some azo-pyazole disperse dyes using eco- friendly ultrasound energy for PET fabric', *International Journal of Scientific and Engineering Research*, 5,1156–1161.

47. Rizk HF, Ibrahim SA, El-Borai MA (2015), 'Synthesis, fastness properties, color assessment and antimicrobial activity of some azo reactive dyes having pyrazole moiety', *Dyes. Pigment..* ,112,86–92. https://doi.org/10.1016/j.dyepig.2014.06.026.

48. Rizk HF, Ibrahim SA, El-Borai MA (2017), 'Synthesis, dyeing performance on polyester fiber and antimicrobial studies of some novel pyrazolotriazine and pyrazolyl pyrazolone azo dyes', *Arab. J. Chem.* ,10, S3303–S3309. https://doi.org/10.1016/j.arabjc.2014.01.008.

49. Gaffer HE , Fouda MMG, Fahmy HM (2013), 'Enhancing Functional Properties of Viscose Fabric by Using Novel Aryl-Azo sulfonamide derivatives ', *J. Appl. Sci. Res.*, 9,4059–4067.

50. Mokhtari J, Babaii B, Akbarzadeh A (2017), 'Synthesis and evaluation of novel antibacterial monoazo disperse dyes based on sulfonamide derivatives on polyester', *Pigment Resin Technol.*, 46, 440–448. https://doi.org/10.1108/PRT-10-2016-0093.

51. Gaffer HE, R. Elgohary M, Etman HA, Shaaban S (2017), 'Antibacterial evaluation of cotton fabrics by using novel sulfonamide reactive dyes', *Pigment Resin Technol..*, 46, 210–217. https://doi.org/10.1108/PRT-08-2015-0080.

52. Zahran MK, Gaffer HE, Mashaly HM (2019), 'Dyeing and Antibacterial Properties of Some Newly Synthesized Sulfonamide Reactive Dyes', *Macromol. Symp.*,384,1–8. https://doi.org/10.1002/masy.201800166.

53. Mohamed HA, Abdel-Wahab BF, Fahmy HM (2020), 'Thiazole Azodyes Containing Sulfonamide Moiety for UV Protection and Antimicrobial of Cotton Fabrics Thiazole Azodyes Containing Sulfonamide Moiety for UV', *Polycycl. Aromat. Compd.*,42,195-203. https://doi.org/10.1080/10406638.2020.1723655.

54. Alebeid OK, Zhao T (2017), 'Review on: developing UV protection for cotton fabric', *J. Text. Inst.* ,108, 2027–2039. https://doi.org/10.1080/00405000.2017.1311201.

55. Sarkar AK (2004), 'An evaluation of UV protection imparted by cotton fabrics dyed with natural colorants', *BMC Dermatol.*, 8 ,1–8. https://doi.org/10.1186/1471-5945-4-15.

56. Gupta D, Jain A, Panwar S (2005), 'Anti-UV and anti-microbial properties of some natural dyes on cotton', *Indian J. Fibre Text. Res.*, 30,190–195.

57. Feng XX, Zhang LL, Chen, JY, Zhang, JC (2007), 'New insights into solar UV-protective properties of natural dye', *J. Cleaner Prod.*,15, 366–372. https://doi.org/10.1016/j.jclepro.2005.11.003.

58. Mongkholrattanasit, R, Krystufek J, Wiener J, Vikova M (2011), ' Dyeing, fastness, and UV protection properties of silk and wool fabrics dyed with eucalyptus leaf extract by the exhaustion process ', *Fibres Text. East. Eur.*,19(3),94-99.

59. Sun SS, Tang RC (2011), 'Adsorption and UV Protection Properties of the Extract from Honeysuckle onto Wool', *Ind. Eng. Chem. Res* ,50,4217–4224. https://doi.org/10.1021/ie101505q.

60. Grifoni D, Bacci L, Lonardo SD, Pinelli P, Scardigli A, Francesca C, Francesco S, Zipoli G, Roman A (2014), 'UV protective properties of cotton and flax fabrics dyed with multifunctional plant extracts', *Dyes. Pigment..*,105, 89–96. https://doi.org/10.1016/j.dyepig.2014.01.027.

61. Czajkowski W, Paluszkiewicz J (2008), 'Synthesis of Bifunctional Monochlorotriazine Reactive Dyes Increasing UV-Protection Properties of Cotton Fabrics', *Fibres Text. East. Eur.*,16,122-26

62. He L, Gong G, Freeman HS, Jian W, Chen M, Zhao D (2011), 'Studies involving reactive dyes containing a benzophenone ultraviolet absorber', *Color. Technol,* 27,47–54. https://doi.org/10.1111/j.1478-4408.2010.00277.x.

63. Wang Y, Tang B, Ma W, Zhang S (2011), 'Synthesis and UV-protective properties of monoazo acid dyes based on 2-hydroxy-4-methoxybenzophenone', *Procedia Eng.*,18,162–167. https://doi.org/10.1016/j.proeng.2011.11.026.

64. Youssef BM, Ahmed MHM, Arief MMH, Mashaly HM, Abdelghaffar RA, Mahmoud SA. (2014), 'Synthesis and application of functional (Anti-UV) Azo-dyes based on γ-acid on wool fabrics', *Indian J. Sci. Technol*,7, 1005–1013.

65. Mishra VR, Ghanavatkar CW, Sekar N. (2019), 'UV protective heterocyclic disperse azo dyes: Spectral properties, dyeing, potent antibacterial activity on dyed fabric and comparative computational study, *Spectrochim. Acta A Mol. Biomol. Spectrosc.* 223,117353. https://doi.org/10.1016/j.saa.2019.117353.

66. Teli MD, Sekar N, Prabhu KH, Chaudhari JB (2008), 'Synthesis and application of monoazo acid dyes containing long chain perfluoroalkyl group for water repellent polyamides', *Asian J. Chem.*,20,4212–4220.

67. Teli MD, Sekar N, Prabhu KH (2009), 'Synthesis and application of bisazo acid dyes for water repellent polyamides', *Asian J. Chem*, 21,3411–3418.

68. Salabert J, Sebastian RM, Vallribera A (2015), 'Anthraquinone dyes for superhydrophobic cotton', *Chem. Commun*, 51, 14251–14254. https://doi.org/10.1039/c5cc06028a.

69. Degennaro M (2015), 'The mysterious multi-modal repellency of DEET', *Fly*, 9 , 45–51. https://doi.org/10.1080/19336934.2015.1079360.

70. Mokhtari J, Nateri AS, Sarli MA (2012), ' Imparting insect repellency to nylon 6 fibers by means of a novel MCT reactive dye', *J. Appl. Polym. Sci.* 126,1097–1104. https://doi.org/10.1002/app.36816.

71. Teli MD, Chavan PP (2017), 'Synthesis of reactive dye to impart mosquito repellency to nylon', *J. Text. Inst*, 108 ,226–232. https://doi.org/10.1080/00405000.2016.1161695.

72. Teli MD, Chavan PP (2018), 'Dyeing of cotton fabric for improved mosquito repellency', *J. Text. Inst,* 109, 427–434. https://doi.org/10.1080/00405000.2017.1351066.

6

Ink-jet printing onto textiles

Kushal Sen
Department of Textile and Fibre Engineering
Indian Institute of Technology, Delhi, India
Email: kushal@textile.iitd.ac.in

Abstract: The chapter briefly describes the conventional textile printing process and its challenges, and how the transfer printing process served as a pre-cursor to the development and growth of inkjet printing of textiles. The fundaments of inkjet printing, the types of inks and the complexities involved, the dynamics of drop formation and the importance of pre-printing treatment and post-printing requirements have been briefly covered. Also discussed are the print head complexity and essential machine elements.

6.1 An overview of conventional printing and limitations

Which attribute of a textile product attracts a consumer besides its need? It is the visual appeal. The colour and the design play a very important role in the selection of a garment or household textile. Textile printing naturally creates an opportunity for the manufacturer and thus has a distinct advantage. The touch and the feel come next. Printing, therefore, is one of the most important unit operations in the chemical processing of textiles as it not only creates visual impact but also adds tremendous value to all textile products. It undoubtedly makes a great economic sense.

However, printing, by far, is the most complex and technologically challenging process. From the management perspective, the printing process has to be right the first time. The possibility of post-printing corrections is near zero. Usually, more than one colour is used to create a design. And to top it all, the non-experts can easily find the faults. One cannot afford to make mistakes. That is a challenge for an expert.

There are many technological considerations. It starts with the obvious requirement of containment of colours within the boundaries of the design besides the creation of a pleasing design. Therefore, one needs high viscosity pastes. The printing paste requires high molecular weight compounds called the thickening agents. Their chemistry and interaction with various dyestuff and essential additives have to be clearly understood. The temperature dependence of viscosity as also the effect of ageing on the stability of stock thickening poses a challenge. One must know the rheological behaviour of the print paste. The print paste must demonstrate shear thinning characteristics.

Next comes the selection of print-worthy dyestuff. The dye and fibre chemistry come into play. The dye fixation processes are different for different dye-fibre combinations. One must appreciate that printed design has many colours, tones, and shades. It is assumed that the properties of all the dye molecules used, such as fastness to washing, light, perspiration, etc., are the same. Invariably the desired shades are obtained by mixtures of dyes and not by a single dye. If the rates of fading of different dyestuff in the mixture are slightly different, the changes in the tone would be easily discernible with time.

After the fixation of dyes, washing is equally challenging. The unfixed dyes, thickening agents and auxiliaries have to be washed off. It is expected that during washing, the run-off colours don't stain the other part of the fabric, whether printed or unprinted.

In a nutshell, the printing process is more complex and needs careful consideration and optimisation at all steps. Irrespective of the style and method used for printing, the following are the essential elements of a printing process;

1. The design
    a. Selection of suitable design
    b. Colour separation
    c. Preparation of tracings; if design has eight colours one needs eight tracings
    d. Preparations of screens or rollers; eight colours means eight screens or rollers
2. Printing process
    a. Selection of dyes and pigments
    b. Selection of thickener
    c. Selection of auxiliaries, binders (in the case of pigment printing)
    d. Preparation of colour paste
    e. Printing
    f. Drying

3. Fixation
    a. Steam
    b. Dry heat
4. Elaborate Washing
    a. Batch or
    b. Continuous
5. Drying

All of these steps are time and energy-consuming and need expert printing technologists. Satisfying the customer in terms of accuracy of design, shade matching, and acceptable fastness is a challenge for the most learned. Add to it the complications related to the printing of blended fabrics; no one would envy the job of a textile printer.

6.2 Transfer printing: A pre-curser to digital printing of textiles

It is now clear that the printing of textiles is a very complex and tedious process. Can we do things differently? Can we make the life of the textile printer simple? Is it possible to recreate the design anytime, anywhere and with the same accuracy? Can someone else print the design at a different location on a different medium and the textile printer simply transfers the same onto the fabric whenever needed? The answer is yes.

This was the beginning of the transfer printing of textiles. The paper printing technology was well established. The smoothness and dimensional stability of the paper are definitely better compared to those of a textile fabric. The print effects are always crisp and more accurate. The expectation from printed papers was obviously different. The requirements of fastness to washing, perspiration, rubbing, etc., are never a consideration in the printed paper.

The first serious attempts were made around 1929-1930, when the British Celanese patented a process for transfer printing [1]. The process was suitable for printing on cellulose acetate. With the advent of polyester, about 20 years later, the designs printed on paper with disperse dyes were transferred onto the polyester fabric. The combination of disperse dye and polyester was almost a God-given opportunity. Disperse dyes would sublime and diffuse into polyester at the sublimation temperatures under suitable pressure. The process was quick, crisp and dry. All that the textile print house needed was a suitable thermal transfer calender. There is neither a need to prepare the print paste nor any worry about design accuracy, and on top of this, there is no requirement for washing. It is definitely more environment-friendly; it requires less energy and almost no water. Of course, it helped in the reduction of process steps, manpower and space requirements. The turnaround time to get approvals and completion of orders too is reduced significantly. There was an air of excitement. One could print any type of design, even the photographic prints. Transfer printing was an important landmark in the history of printing.

Paper printing technology was very advanced. The off-set paper printing technology could create photographic prints using just four colours, namely, CMYK (cyan, magenta, yellow and black). So whatever one could print onto paper, the same could be transferred onto textiles. It appeared the sky was the limit. As long as the dye could sublime and the textile substrate has an affinity for the same, one could do the transfer printing. Some enthusiasts printed the photographic worsted design on paper and later transferred the same onto synthetic fabrics to give the impression of worsted suiting. This was stretching it too far. However, thermal transfer printing was exciting. Printing paper with sublimable disperse dyes, which had no affinity for this cellulosic substrate, and transferring the design onto polyester fabrics became a commercial reality. This process is referred to as dry transfer or thermal transfer process. Thermal transfer printing, though it is possible on nylon, acrylic or even polypropylene fabrics, its success has been mainly due to polyester. Transfer printing was no more a complicated process for the textile printer.

Of course, the paper printing process had to be tweaked, particularly with respect to the ink. The requirement of fastness, never before concern of the paper printing industry, is obviously a requirement of the textile industry. Selection of dyes, auxiliaries and transfer paper became critical. Inks had to be made from sublimable disperse dye. The paper industry responded well, and the textile industry lapped it up. The paper printing later became digital. Inkjet printers are now a household name. These are used to print transfer papers as well.

6.2.1 Advantages of sublimation transfer printing process

Sublimation transfer printing technology offers the following advantages;

- It is a simple process; the textile printer does not have to go through the cumbersome processes of conventional printing,
- Designs may be printed and stored on a cheap and non-bulky substrate such as paper,

- Design can be printed onto textiles quickly based on the sales demand,
- The production of short-run repeat orders is easy,
- This technology requires relatively low skill input and results in no or low rejects,
- Stocking and storage costs are lower,
- Inexpensive printing equipment with less space requirements,
- No effluent as there is no need for washing-off,
- Photographic effects can be realised which is not possible with the conventional textile printing processes, and
- It is possible to print on garments or garment panels.

6.2.2 Limitation of sublimation transfer printing

From the commercial point of view, the transfer printing has only a limited share of the total textile printing business. What is holding transfer printing technology to be the lead technology? The answer is the technology itself. Thermal or sublimation transfer printing process can be used for polyester fabrics and, maybe to some extent, other synthetic fabrics. This process is not suitable for cotton, viscose, silk, wool, and other non-synthetic textiles. One needs different classes of dyes for different substrates which range from, reactive, acid, mordant, metal complex, basic and of course disperse. Except for the disperse dyes, others do not sublime. While a lot of research on wet transfer printing of textiles has been done, the commercial success has been miniscule. This, therefore, has been a major limitation of this technology, although the market share of sublimation transfer printing has been projected to increase. According to the Smithers survey and projections, the worldwide market for sublimation printing will grow to USD 13.42 billion by 2023 [2].

6.3 Inkjet printing of textiles: The potential

It is evident from the previous discussion that although the sublimation transfer printing technology offered tremendous advantages over the conventional printing technology, its application, however, is limited to the printing of synthetic textiles; in fact, to the printing of polyester textiles. This has been its biggest drawback. Its share in the textile printing business is, therefore, limited. The paper printing technology kept on developing, be it the inkjet or the laser printing. The flexibility that digital printing technology offers has made it very popular, handy and cheap. Paper printing is a child's play so much so that inkjet and laser printers are today found in almost every household, attached to a desktop computer for printing documents and photographs on paper. Just a click of a mouse, and out comes the beautiful coloured print. Sublimation transfer printed papers are now produced using inkjet printing technology.

Considering the limitation of transfer printing, the logical question was can one use digital technology to directly print the designs on textiles. The extended question was, can digital printing technologies be used to apply other dyes, besides the disperse dyes, to print textiles. Naturally, this would increase the range of textiles substrates that could be printed. The digital printing machinery manufacturers responded, why not? This opened a plethora of opportunities. Which were the new names in the textile printing machine manufacturing? Naturally, those who were successful in the digital paper printing business, the likes of Epson. While the paper industry uses both laser and inkjet printing technologies, only inkjet printing technology has been used for printing textiles till now.

The world of textile printing is fashion-dependent and changing rapidly. The fashion seasons are becoming shorter. Customers demand more variety of colours, shades and unique designs. The chances of repeat orders are getting lesser. Average print lengths are rapidly reducing. Globalisation of business needs a quick response, quicker sampling and turnarounds. Today, flexibility and versatility are the need of the hour. Mass customisation rather than mass production is the buzzword. On top of these, requirements of waste minimisation and reduced load on the environment pose substantial challenges to businesses. Many regulations related to these are mandatory in many countries. The question is, 'do we have a printing technology that can address all these issues?' Yes, inkjet technology is the answer.

The inkjet printing technology is clean and crisp. All you need is

- A computer
- The software
- An inkjet printer, and
- The textile substrate

Putting it in perspective, the inkjet printing technology involves

- Selection of a master design
- Uploading the design file onto a computer; it may simply be a scanned image or a clicked photograph
- Transfer of design data to an inkjet printer with the help of software; colour separation takes place automatically into the four colours (CMYK)
- Printing onto a textile surface

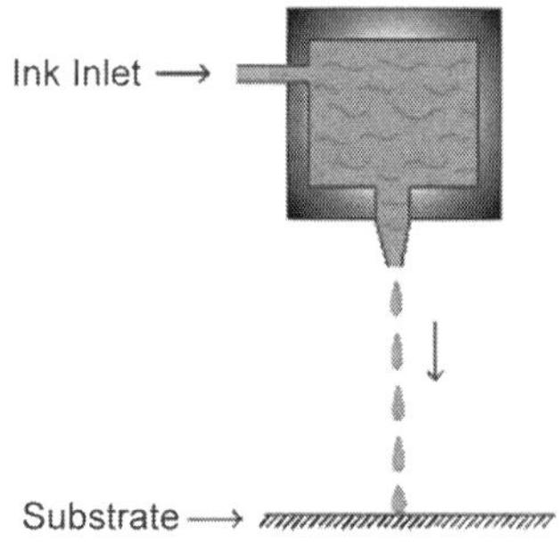

Figure 6.1 Principle of inkjet printer

Inkjet printing is essentially a non-contact printing process, as shown in Fig. 6.1. Recall that the conventional printing processes are contact processes. The screen, the roller and the block necessarily touch the fabric's surface. It is appropriate at this juncture to introduce two terms, viz., spot colours and process colours. Visualise the conventional screen printing process. If there are eight different colours/shades in the design, one needs eight tracings representing the eight colours; made after careful separation of colours, and eight different screens for the eight colours. How are these colours/shades produced? These are produced by judicious mixing of various dyes and pigments to match the required shade. These are known as *spot colours*.

Inkjet printing technology does not rely on the pre-mixing of colours. The perceived colour is generated during the printing process itself. Colours produced during the printing process are called the *process colours*. This is done by printing very small dots from the four primary colours, at the designated places, as required by the design and determined by the appropriate software and hardware. The dots are created by ejecting very small droplets of coloured ink (Fig. 6.1). As the name suggests, inks are low viscosity fluids. High viscosity colour pastes are not suitable for inkjet printing. Like one refers to pixels on a monitor or display units, it is the dots that are used to create the design on the textile substrate. This technology thus allows millions of shades to be produced by just four colours. That is why one can create photographic prints. Different colours or shades are produced by judicious placement of dots in the right proportion without actually mixing.

In principle, inkjet printing technology is very simple. Can one say that inkjet textile printing is exactly the same as that of paper printing? Not at all. The printed paper is not washed; however, textiles have to be washed again and again. Penetration and fixation of dyes are expected. The fastness to light, rub, perspiration, etc., have to be met too. Unlike paper, the fabric surface is much rougher. We shall learn in detail about the stringent requirements that entail textile printing.

Suffice to say that we do need *pre-treatments* to make fabrics suitable for inkjet printing. It may be noted that some necessary chemicals/auxiliaries cannot be incorporated in ink. It would restrict the free flow and even corrode the jets. These are applied on the fabric by coating or padding. Another objective of pre-treatment is also to restrict the spread of colours as the viscosity of the printing ink is low. Similarly, textile printing would require *post-treatments* also to ensure fixation of colours and washing-off of the unfixed colour. In these respects, the inkjet printing of textiles is different from the printing of paper.

Inkjet printing technology, nevertheless, is very attractive and has rightly caught the fancy of one and all. The current share of inkjet printed fabrics is very low. Screen printing still dominates. This is mainly due to the high cost of printing machines and inks and lower speeds of production. However, in the ITMA 2019, there were sufficient machinery manufacturers who displayed the latest technologies with printing speeds comparable to the conventional rotary screen printing machine. The future is promising. According to the survey and projections from Smithers, inkjet printing of textiles remains one of the fastest-growing technologies worldwide and is likely to rise up to about USD 6.13 billion by 2023 [3].

6.4 Inkjet printing technology: Technological edge

It is clear from the discussion in the previous section that inkjet printing technology is a non-contact printing process where in small droplets of ink from a nozzle are directed onto the textile substrate to create dots of desired colours. So what? What is the edge? There are two important features of this technology,

(a) Creation of photographic prints onto textiles and

(b) High resolution leading to remarkable design accuracy and quick reproduction of prints at any time and every time. Conventional printing is incapable of doing so.

6.4.1 Photographic prints

The technology primarily uses inks with four basic colours, CMYK, to generate drops and print dots on the textile surface and thus helps to create photo-quality images. It must, however, be kept in mind that accurate placement of droplets is the key. This process is controlled by high-frequency digital signals and involves the application of a controlled pressure on the ink fluid at the pre-determined frequency to generate droplets of the required size.

The inkjet technology allows the generation of drops of volume ranging from 3-40 pL. The smaller the size of the drop, the smaller would be the dot, and thus more accurate reproduction of the design. It may be recalled that this technology used process colours, i.e., there is no mixing of colours. If the dot size is small enough, the human eye cannot resolve the dots as separate entities. Therefore, the perceived colour appears to be a mix of the different coloured dots in that area. Different proportions of the CMYK dots would create different colour perceptions. This helps to create millions of shades. How many shades can one produce with just one colour? Well, it depends on the resolution. Say, there is only one 1×1 square matrix; this square may either be filled or be empty. So one can have two grey levels (Table 6.1). If same area is divided into four say, in a 2×2 matrix, either one or all smaller squares can be filled or all could be empty. So one can get 5 grey levels. Right?

Table 6.1 Possible arrangements of single colour dots

1×1	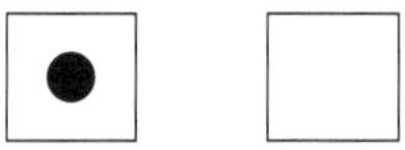	2 shades
2×2	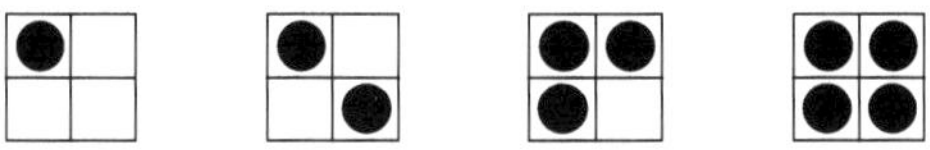	5 shades

If there are three colours at our disposal, one can generate 125 colours for a 2×2 matrix. Now think of reducing the dot size further, and thus increasing the possibility of more dots being accommodated in the same area. The number of possible colours keeps rising (Table 6.2). When the dot size is further reduced to fit in a 16×16 matrix, millions of colours can be produced. The key concept is that the dot size should be too small to be visible to the naked eye. This helps inkjet technology to create photographic images, which is certainly not possible with conventional textile printing technologies.

Table 6.2 Possible grey levels with just three colours

Matrix resolution	Grey levels including white	Number of colours
2x2	5	125
4x4	17	4913
8x8	65	274625
16x16	257	16.9 million

6.4.2 Quality of print and resolution

It is understood that if the dot size is reduced, millions of shades can be obtained with only four colours. What about the quality of the print? Whenever one talks about the quality of an image, the term that is generally referred to is resolution. It is pixels in display monitors and dots in the inkjet-printed designs. The smaller are the drops, the crispier is the image. The printheads with higher nozzles per unit area and with the capability of delivering smaller droplets would give the truer colour representation and accurate images. In the earlier versions of inkjet machines, the drop volumes were ~100 pL. These are significantly smaller now, as low as 1.5 -2.0 pL. One can print more dots in the designated print area to get better resolution. A printed design of 600×600 dpi means that there are 600 dots per inch in the warp direction and 600 dots per one inch in the weft direction. For a given print head, the resolution may not be uniform in both directions; it may have a resolution of 2400×1200 dpi. In any case, higher resolutions mean better print quality. Photographic prints would need high-resolution printers.

What happens if the dot size is reduced at the same level of resolution? Fig. 6.2 is a case in point. How would the two be perceived by the human eye? The perceived depth of shade in (b) is expected to be less. On the other hand, what happens if the dot size increases such that the diameter of the dot is more than the length of the square? The dots would overlap, and the quality would deteriorate.

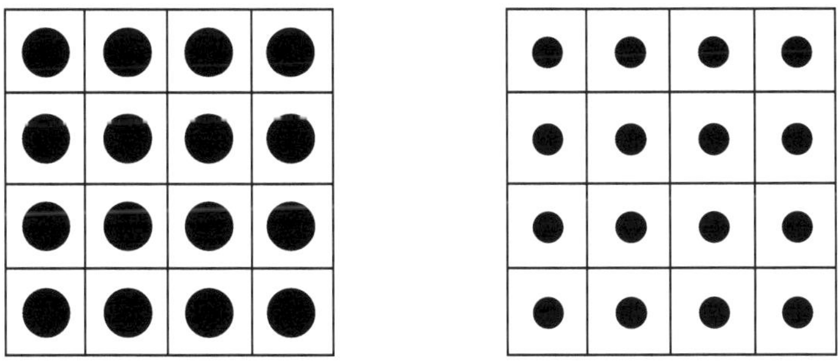

(a) (b)

Figure 6.2 Effect of dot size

6.5 Jetting principles

Having learnt the importance of small droplets and small dots, it would be interesting to learn how the drops are generated. There are two jetting principles in use for generating drops, namely, Continuous Inkjet (CIJ) and Drop-on-demand (DoD).

6.5.1 Continuous Inkjet (CIJ) technology

In this, a high-pressure pump pushes the liquid ink from the reservoir through the microscopic nozzle creating a continuous stream of droplets. A high-frequency signal to a piezoelectric crystal transducer helps to break the stream into droplets, which are then suitably directed to the substrate. The frequency of excitation controls the rate of production. The higher the excitation frequency, the lower will be the average drop size. At a higher frequency, the wavelength of the jet is shortened; thus, the volume of the broken stream, which evolves into a droplet, reduces, and as a consequence, the drop size becomes smaller [4]. In some CIJ systems, excitation frequencies of over 1MHz are employed. The drop size may vary from 3-40 pL. What is the role of drop size? A smaller drop size produces a higher resolution and sharper design. CIJ uses different inkjet systems for design creation. One of these is called binary. The schematic of a *binary* type jet is shown in Fig. 6.3. A piezo-electric system is used for generating drops continuously while a charging device and drop deflecting system places the drop appropriately. The name binary suggests that the drop either goes straight to the substrate or to the drain or gutter as per the requirement of the design. Why is the charging done? This helps in deflecting the drops to the desired spot. Of course, this needs accurate control. It may happen that a smaller proportion of drops are used for the design, and the major proportion goes to the gutter and is recycled.

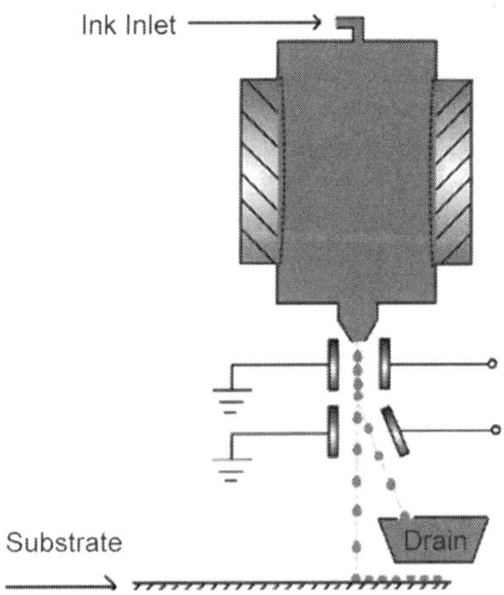

Figure 6.3 Binary CIJ

The other jetting system for CIJ technology is a *multiple deflection* type jet. The schematic of this is shown in Fig. 6.4. In this system, the drops can be deflected in multiple directions and are placed on the substrate, while those which are not required are sent straight to the gutter. What is the advantage of this system? Printing speeds can be increased. And the disadvantage? The dots would not be circular in shape. It may be interesting to note that charged droplets are separated by one or more uncharged *guard droplets* to minimise electrostatic repulsion between neighbouring droplets.

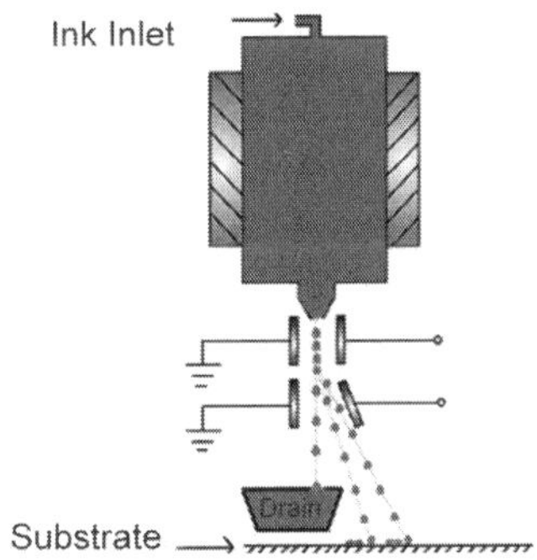

Figure 6.4 CIJ multiple deflection jet

In general, because of the complexities associated with CIJ (charging of droplets and deflection, ink recirculation, pressurisation) print heads tend to be costly. The ink chamber needs to be actively refilled by the positive pressure.

6.5.2 Drop-on-demand (DoD) technology

As the name suggests, this technology creates a drop only when there is a demand. No unwanted drops are to be directed to drain. Here too, piezoelectric crystal-based transducers are the major drivers; however, others such as thermal transducers are also used for creating DoD printheads. The frequency of excitation here is of a magnitude lower as compared to CIJ.

6.5.2.1 Piezoelectric crystal based DoD inkjet technology (PIJ)

In these jets, the piezoelectric crystals are used as a displacement tool by impressing suitable voltages. The transducer is generally attached to a membrane that forms the ink chamber wall, or in some cases, this itself may act as a wall. On application of an electric field, the chamber volume is proportionally reduced, and the ink drop is ejected. Note that the drop is generated when needed. This transducer can be placed on the top, on the sides and may be a part of multiple chambers. Fig.6.5 is a schematic of a DoD jet with the crystal on top, also known as the shear mode type system. In this mode, the electric field is perpendicular to the poling direction of the crystal. It may here be mentioned that the dipoles in unprocessed piezoelectric crystals are present in a random order. For these crystals to function as effective transducers, these dipoles need to be aligned in a suitable direction. This is called *poling*. In the squeeze mode, a tubular piezoelectric transducer may be placed around the tubular ink chamber (Fig. 6.6). On the application of an electric field, the tube squeezes, reduces the volume, and the drop is ejected.

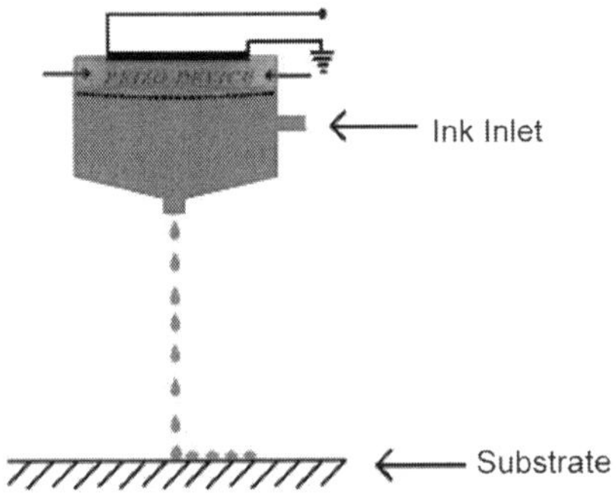

Figure 6.5 DoD with crystal on top

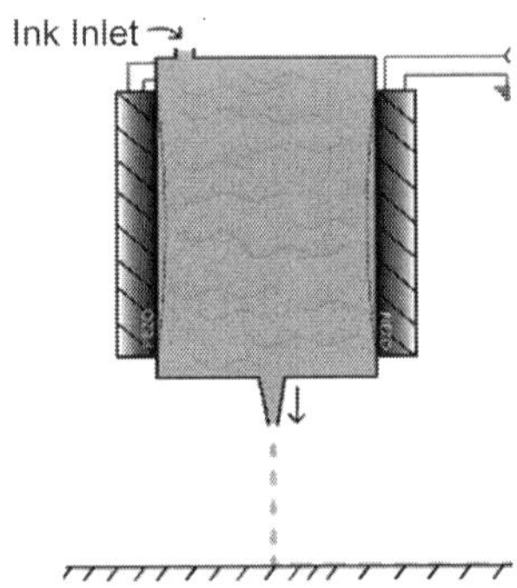

Figure 6.6 DoD tubular crystal; squeeze mode

To increase the rate of production, multiple chambers can be used in the squeeze mode, where the transducer may serve as the wall (Fig. 6.7). Different transducers may get different signals and thus give more flexibility. Theoretically, many configurations can be designed to suit the requirements of the machinery manufacturers.

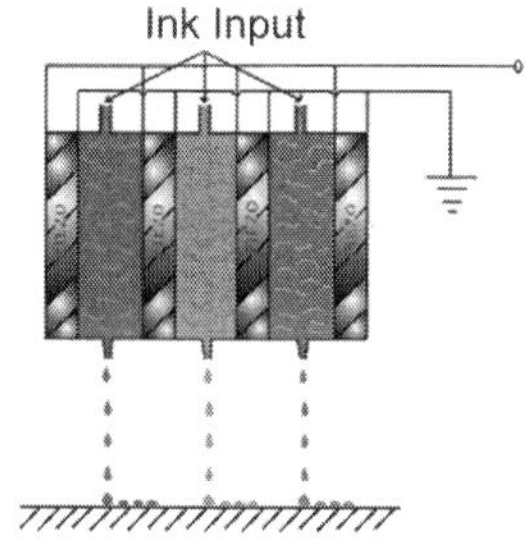

Figure 6.7 DoD jet with multiple chambers

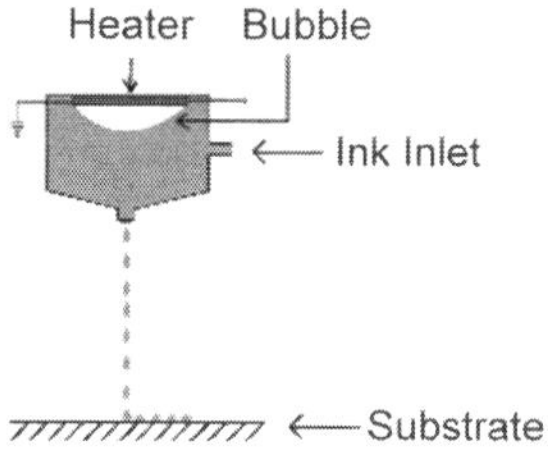

Figure 6.8 Schematic of thermal DoD jet; top shooter

6.5.2.2 Thermal Drop-on-demand (DoD) inkjets (TIJ)

Another very interesting type of DoD inkjet belongs to the category of thermal jets. Printers based on these are sometimes referred to as bubble jet printers. In such inkjets, each nozzle houses a tiny heating element, rather than a piezo crystal. Inks are generally aqueous-based, and the quick heating helps to generate localised steam, creating a bubble. This ejects the equivalent volume of ink from the nozzle. It is interesting to note that the temperature can quickly rise to 300°C, and the pressure in the localised zone may be over 70 atmosphere. As the drop is ejected, temperature and pressure quickly go back to normal. If the heating element is placed on top of the chamber, the arrangement is called top shooter (Fig. 6.8). The heater could be placed on the side or in the middle of the chamber and thus would be termed as a side shooter or middle shooter, respectively. There can be two heaters within the same chamber, and this would be termed a double shooter. Once the principle is understood and mastered, many possible combinations can be employed to increase the efficacy and, sometimes, to beat or protect intellectual property rights.

The area of piezoelectric transducer required for ejecting the same volume of ink is an order higher than that required for thermal transducers. This means more nozzles can be housed in a smaller area. Also, the fabrication of thermal jets is easy and less complex. But the bubble formation and bubble collapse are violent processes. The trapped bubbles on rough surfaces and in the corners of the ink chamber absorb the pressure created by the bubble and thus alter the magnitude of the pressure pulse. Drop volume control is thus compromised.

The control on the volume of the generated drop is higher in piezoelectric systems. Also, for pigment inks, one would prefer to use piezo-based inkjets rather than thermal jets.

What about the robustness of printheads? TIJ may fail due to a) deposits on the heater surface that may reduce the thermal conductivity and b) corrosion of the heater surface. On the other hand, the PIJ has a longer useful life and offers better control.

6.6 The ink

It is clear that with just four primary colours (CMYK), millions of colours and shades can be produced. The question is, which are those four molecules whose chemistry would truly represent these primary colours. It needs a higher order of standardisation. True colour or shade will not be obtained even if deviations in the visible spectra of molecules are very little. The chemical architecture of the synthesised dyes has to be such that it truly translates the image from the display monitor to the textile fabric. Of course, it is expected that the image of the design captured by a camera or a scanner is a true representation and is displayed as such on the monitor. This is a different domain and is being addressed by hardware and software engineers. It has reached a fairly high degree of standardisation across the tech community. Similar standards are expected in the designing of the dye molecules. This is not an easy task. In addition, it must have an affinity for the substrates, be it cotton, silk, polyester, nylon or blends thereof, and should have the required fastness properties. The storage stability of inks is always a challenge, and one would like the inks to be stable for six months to years.

Table 6.3 Commercially available inks

Ink type	Fibre type	Essential pre-printing ingredients	Essential post-printing processes
Reactive	Cotton, viscose rayon, silk, wool, nylon	Alkaline/ neutral	Steam, wash
Acid	Silk, wool, nylon	Acidic	Steam, wash
Disperse	Polyester	Thickener	High temperature steam /thermosol , wash, reduction clear
Pigment	All	None	Dry heat

In ink jet printing, none of the conventional printing chemicals, such as alkali, urea and thickener can be directly incorporated into the ink formulations [10-11]. Therefore, the pre-printing formulations may have a suitable thickening agent to make the fabric surface suitable for inkjet printing. This enhances the print quality by reducing the spread of the ink on the fabric by blocking the capillaries and also by suppressing the protruding hair. A word of caution, the presence of thickener affects the penetration. Optimisation of the thickener concentration is required.

6.6.1 Pigment inks

Pigment inks appear to be an attractive option, and in the early days, these were the natural choice for inkjet printing. The process was simple; print the design and fix it by hot press. No need to wash. Simple and crisp. These inks are available as aqueous dispersions. As the pigments do not have any affinity to the substrate, these need to be fixed with the help of a binder. The question is, at which stage the binder should be added, at the pre-printing stage or in the ink itself? If one adds the binder at the pre-printing stage, the binder film will develop all over the fabric surface, even in non-design areas. Not a very attractive proposition. Can the binder be added in the ink itself? What if it polymerises in the ink and chokes the printhead nozzles? Ink manufacturers have chosen the latter option. Considering the inks are stored and used at temperatures much below the curing temperatures, the chances of it polymerizing are negligible. The catalyst may be there in the ink or the pre-treatment formulation. As the pigments need to be fixed with binders adding thickener in the pre-treatment recipe for the pigment inks is a challenge. The debate still goes on. One thing, however, is clear that if the binder is in the ink, thermal drop-in-demand inkjets are not suitable. Piezoelectric-based inkjet printheads are therefore recommended for pigment inks. Pre-padding with binders having very low T_g can give the machinery manufacturer the flexibility to use thermal inkjet printers.

6.6.2 Reactive inks

The next historical step was to design and develop the reactive inks mainly for cellulosic textiles. These inks are available as aqueous solutions. The reactive dyes form covalent bonds with cellulose in alkaline conditions. These can be used for protein-based and polyamide-based textiles as well. The reactive dyes, interestingly, can be fixed onto silk, wool and nylon at neutral pH. A neutral medium has an added advantage. It eliminates the formation of the hydrolyzed dye. This is one reason why alkali is not incorporated in the ink.

This has to be added during the pre-printing treatment, if needed. In this respect, the reactive inks are versatile. For cotton and cellulosics, the pre-treatment recipe may have,

- Medium viscosity sodium alginate
- Urea
- Sodium carbonate
- Resist salt

For silk, wool, and nylon one may use,

- Guar gum
- Urea
- Maintain pH at slightly less than neutral

Of course, the fixation in steam followed by careful thorough washing is needed.

6.6.3 Acid inks

The acid inks are designed specifically for silk, wool or polyamide fibres. These, too, are available as aqueous solutions. Naturally, these need an acidic environment for fixation. Suitable acids can be a part of the pre-printing treatment. Acids are not added in the ink as these can adversely affect the useful life of the printhead. The following may be part of the pre-treatment recipe for wool and silk or even nylon,

- Guar gum
- Urea
- Ammonium tartrate

Fixation in steam and washing is required.

6.6.4 Disperse inks

The disperse inks are suitable for polyester. These can be used for other synthetics textiles as well. These are also available as aqueous dispersions. Sublimable disperse inks are used for printing paper to be used for thermal transfer printing. High energy disperse dyes are used for direct printing of polyester fabric. Any suitable thickening agent other than synthetic thickeners can be used for maintaining a slightly acidic pH. The fixation can be done in high-temperature steam or by dry thermosol treatment. Reduction clearing may be required.

In all these ink types, the four standard primary colours (CMYK) are available. To increase the colour gamete, some machinery manufacturers recommend and use six or eight colours, e.g., an eight colour system may use black, cyan, magenta, yellow, orange, red, blue and violet. It gives a truer colour representation. Considering the complexity of the selection of dye, ink formulation and compatibility with printhead technologies, there are only a few players in the world who manufacture the ink [12].

6.6.5 Dynamics of drop formation

Drop formation from an ink stream happens on a microsecond time scale because of the high operating frequencies. On the other hand, placement of dots, spreading and penetration happen on a millisecond time scale. Surface energy plays a very important role. Process dynamics and fluid flow characteristics are very critical to the science behind ink formulation. Other necessary ingredients required in the ink are bound to change this dynamics. The selection of ingredients and the optimisation is a complex process. The performance of the printhead depends on ink formulation. Ink, as expected, will have a high concentration of dyes as well.

The viscosity of the inks is very low, around 20 cps, as compared to that of the conventional print paste. The drop formation happens at very high frequencies and at a high velocity of a fluid. These create stress on the printhead. Inks should have shear thinning behaviour so that stresses on the transducers are low and don't damage the printhead. At high frequency, the fluid flow shows a low level of non-Newtonian behaviour, which strongly influences ink jettability. Therefore, the rheology of the ink has to be properly managed. Viscosity changes every time a component is added, be it the solvent, dispersing agent, humectant or biocide. Viscosity modifiers are among many ingredients of the ink. Ink viscosity is measured on piezo-axial vibrators (PAV) or high-frequency rheometers to mimic the actual jetting conditions.

Viscosity at low shear rates determines,

- The ink fill-up behaviour,
- The cleaning and priming of the nozzle; priming refers to initiating a process of running ink through a printhead to expel the air from the chamber or from the manifold, and
- The ready-to-jet condition

The surface tension of the ink and drop formation are linked as well. The higher surface tension helps in the formation of spherical drops. This reduces the surface energy. Surface active agents are added to control the surface tension of the ink. Surfactants with High HLB values help to maintain the colloidal stability, and those with low HLB values help to a) wet the nozzle capillary, b) maintain the meniscus at the nozzle exit, c) facilitate jettability and interaction with the substrate and d) improve dot formation and image quality.

In any case, the fluid dynamics is affected by both the viscosity and surface tension. The Weber number (We) and Reynold number (Re) are important quantities that relate to inertial and viscous forces as given under, and both are dimensionless quantities,

$$We = \frac{(\rho v^2 L)}{\sigma}$$

where, ρ is the density, v the velocity, L the diameter of nozzle, and σ the surface tension.

Weber number characterises the atomizing quality of the spray and, therefore, the size of the droplet. Simply speaking, it's a ratio between the inertial forces and stabilizing forces, such as surface tension, which represents a cohesive force. Higher velocity and lower surface tension help in drop formation and spray.

$$Re = \frac{vL}{\nu}$$

where, v is the velocity, L the diameter of nozzle, and ν the dynamic viscosity. Higher R_e means inertial forces dominate the cohesive forces. The higher is the Reynolds number, the higher is the turbulence in the flow.

6.6.5.1 *Drop formation*

Drop ejection takes place as a result of the pressure impressed by the actuators at pre-determined frequencies. As the drop exits the nozzle, the shape is near-spherical with a long thin trailing ligament. This ligament breaks due to instability and perturbance into several small droplets (Fig. 6.9), causing unwanted splash onto the substrate, thereby reducing the print quality. How does one control this phenomenon? The formation of satellite drops can be managed by incorporating a suitable amount of viscosity builders, e.g., polymeric substances, also referred to as rheology modifiers. This helps to build elastic stresses in the ligament and pulls it back to the meniscus after the drop separates. PEG with molecular weights ranging from several hundred to several thousand can be used as rheology modifiers. A small concentration of these compounds modifies the in-flight fragmentation and drop detachment phenomenon. More than optimum concentration would adversely affect the jettability and increase the stress on the actuators. Surface tension also plays a role in the formation of satellite drops. Higher surface tension leads to creation of spherical drops quickly, and thus the ligament abruptly fragments into smaller satellite drops.

One would like the machine to start printing immediately. It does not happen. Why? The failure to immediately start is related to ink deposits at the nozzle which blocks the orifice. Evaporation of water leads to the deposition of solids. Ink formulation, therefore, must aim that the rheology facilitates longer idling periods without affecting the jetting efficiency. The rate of evaporation can be reduced by adding humectants such as diethylene glycol or covering the nozzle plate, say during homing, to avoid contact with air. A term that is often used in inkjet technology is latency. Latency of a nozzle is sometimes described as the number of firings needed before a first droplet is jetted as the printing is resumed. It is basically the time needed to start the printing process. Kye-Si Kwon *et al.* in their study, have studied the formation of the first drop after an idle state using a high-speed camera before a steady state is achieved. They suggest that the printhead performance and print quality can be related to the first drop formation [13]. C. Leigh Herran *et al.*, in their study, have shown that the diameter of the drop a) decreases as the glycerol concentration in the ink increases, b) decreases as the excitation frequency increases, c) increases with excitation pressure amplitude, and d) decreases as the carrier stream velocity increases [4].

Figure 6.9 In-flight fragmentation and drop detachment

6.6.6 Ink formulation

A significant number of auxiliary chemicals are added to create a balance between viscosity and surface tension, printhead sturdiness and ink performance, ink type and the colour gamut, as also stable dispersion and particle size. It may again be mentioned that the textile inks are aqueous solutions or dispersions. Following are some of the necessary ingredients that find their way into the inks,

- The dye or pigment particles; acid and reactive dyes are used as such, but pigments and the disperse dyes are used in aqueous *dispersions*. The solid content may be around 20%, while the particle size may vary from 0.05-0.15 μm. Higher particle size would lead to settling and clogging, while the lower size would result in aggregation.
- In the case of pigments, the *binders* are a part of the ink. Film formed by these should have very T_g.

- Water is the *solvent.*
- *Co-solvents* enhance jetting performance, reduce the rate of evaporation and ink deposits, facilitate re-dissolving the deposits and cleaning, and help in wetting and adhesion to the substrate.
- *Surfactants* are added to manage the surface tension of the inks.
- *Rheology modifiers* to reduce the formation of satellite drops.
- *Humectants* for absorbing water from the atmosphere whenever needed.
- There are surfactants in the ink, and the jets function at very high frequencies. Foaming is expected. *Antifoaming* agents are required to control the foam formation.
- To improve the shelf life, *biocides* are also a part of the ink.
- The selection of the chemicals has to be done as per the requirements of the substrate and that of the printhead technology. Needless to say that all these should be compatible.

6.7 Inkjet printing machines

It is clear that inkjet printing of textiles is a different ball game compared to the printing of paper. The textile fabrics and garments have a rough surface with protruding hairs. Therefore these cannot be fed to the inkjet printing machines as such and need to be prepared for inkjet printing. It must be mentioned that this is in addition to other preparatory processes such as singeing, desizing, scouring, bleaching, etc. The fabric to be printed may be white or dyed fabric. In conventional printing, one would feed these directly to the printing machines. But not here. Inkjet printing is a non-contact printing, where the drops are fired from the nozzle and travel through the space between the printhead and fabric surface to form the dot. Any obstacle in between would change the flight path and thus would reduce the quality of the print.

In addition to the design center, a typical inkjet printing house would have

- Pre-printing processing machine/s
- Inkjet printing machine
- Post-printing processing machine/s

6.7.1 Pre-printing processing unit

The goal of a pre-processing unit is to suitably prepare the fabric for printing. Such a unit may have a padding or coating unit followed by a drying unit. The coating or padding is needed to

- Smoothen the surface by suppressing the protruding hair, remove creases by stretching, and partially fill the surface undulations.
- Provide dimensional stability to knitted fabrics or hard twisted fabrics such as crepes and georgettes or otherwise lightweight fabrics to avoid skewing.
- Improve feed performance by controlling the slippage.
- Add certain essential chemicals required to fix the dyes or inks on to the textiles.

Typical padding or coating solution would have a thickener, such as, sodium alginate, guar gum, PVA, etc., in low concentrations mainly to control the spreading of colour by capillary forces. In addition, it may have necessary dye-fixing auxiliaries such as acids or alkalis, as per the chemistry of the substrate and dyestuff. It is not desirable to add such chemicals in the ink as these may damage the printheads and also reduce the stability of inks. The padded or coated fabrics are invariably dried on a stenter or any other suitable drier. Fabric, after drying, is sent to the inkjet printing unit.

6.7.2 Inkjet printing assembly

A typical inkjet printing machine would essentially have the following subunits, namely

- Feeder unit
- Printing unit
- Drying unit
- Folding unit

The layout of the printing system is shown in Fig. 6.10.

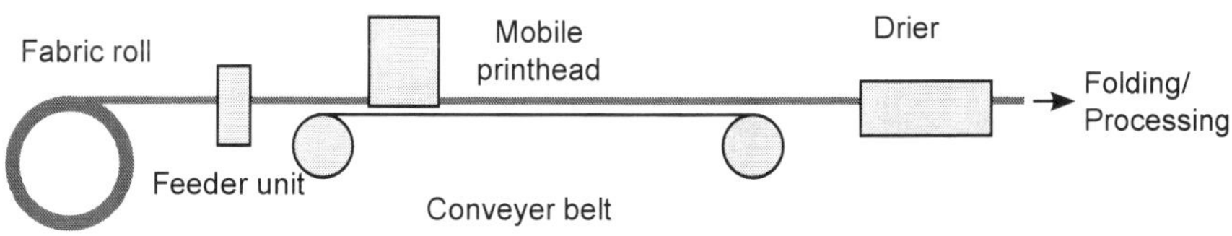

Figure 6.10 Layout of inkjet printing machine

The role of the feeder unit is to pick up the fabric from the selvedges, control alignment and fix the fabric to the conveyer belt by gluing it. Why? So that there is no fabric movement or slippage during the printing process. This is similar to the conventional printing process. The success of printing depends on the controlled entry of the fabric into the inkjet printing machine. This is done by proper

- Spreading of fabric to remove creases
- Tension control
- Gluing to the non-deformable endless blanket
- Washing and cleaning of the blanket during return
- Fabric speed control; done by precision stepper motors. This is the most important step and is synchronised with the printhead motion. A mismatch means inaccurate design translation.

6.7.2.1 Printing unit

The inkjet printing unit has a carriage assembly. The most important element is, of course, the printhead. The printhead moves over the carriage assembly and across the fabric width from one end to the other and returns to the base. It can print either in one direction or in both directions. It carries with it all the colour tubes and electronic flexible flat cable (FFC) for conveying the information from the computer to the actuators, e.g., the piezoelectric transducers sitting on top of the ink chamber. Without a doubt, the printhead is the heart of the machine, by far the most complex and costly element of the inkjet printing machine. Printhead is a complex assembly that houses a huge number of piezoelectric drivers, the ink chambers and the nozzle plate. The size of the orifice is very small and is difficult to see by the naked eye. One may be surprised to know that the thickness of the printhead may be as small as 1.5 mm. It operates at high frequencies and is expected to fire the right coloured drops and direct those to the right areas. Simultaneously, it replenishes the ink from the reservoirs through the flexible tubes.

A typical printing machine may have two carriages, each carriage carrying 4-8 printheads, each printhead has more than 250 active nozzles capable of delivering up to 30 pL of drops at 20000 drops per second and is expected to achieve a print resolution of 600 dpi or more. Wow! Complex assembly and huge expectations.

The productivity of the printhead (P) depends on the operating frequency of the printhead, f, the number of nozzles per printhead and the volume of the drop, V. In other words,

$$P = Vnf$$

The DoD inkjets function around 10 kHz while the CIJ inkjets function around 100 kHz, about an order higher and thus CIJ inkjets have higher productivity. The machine printing speed (m^2/h) depends on the print quality or the resolution set on the printer. For the same machine and design, if the resolution is doubled, say from 360 dpi to 720 dpi, the machine speed would reduce to half. In general, the consumption of ink is relatively small. Typically for 100% coverage, about 15-20 ml of ink per square meter may be needed. While four primary colours (CMYK) can give satisfactory results, however, these days, six or eight colour machines are also in vogue for textile printing. This is done to increase the colour gamut and produce a better colour rendition on the fabric.

The printing cost per unit area on inkjet machines is higher as compared to that on rotary screen printing. The inkjet machines are costly, the machine speeds are slower, and the cost of ink is higher. However, if the print lengths are smaller, the comparative cost for inkjet turns out to be lower. If the print run lengths are higher, the rotary screen printing cost is lower. The advantages of printing short lengths by inkjet printing have long been realised. The main contribution to the recurring cost comes from the printhead replacement besides the cost of ink. The robustness of the printhead is a prime concern of the printing managers and machinery manufacturers. It may be noted that the piezoelectric-based printheads are more robust and have a longer life. The thermal inkjet-based ones are more prone to corrosion and deposits on the heater surface affecting their useful life; the bubble formation and collapse are very violent processes and pose a major challenge.

However, the inkjet printing technology is experiencing a quantum jump. In the ITMA 2019, Barcelona, advanced inkjet printing machines were demonstrated where the machine speeds are comparable to the rotary screen printing machines. This is likely to bring down the cost of printing. The new technology employs the concept of the stationery printhead and is known as *single-pass machines* (Fig. 6.11)). One may recall that in the existing inkjet machines, also known as multi-pass machines, the printheads move across the fabric from one end to the other over a carriage to create the design. This compromises the machine speed to a large extent. In single-pass machines, printheads are stationary. One printhead array is used for each colour. The schematic for such machine is depicted in Fig. 6.11.

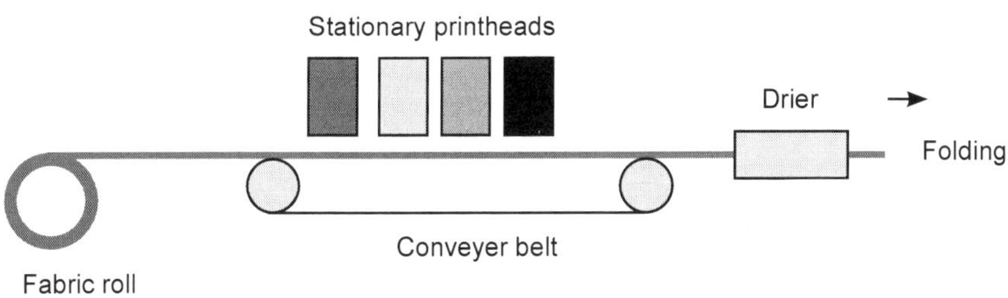

Figure 6.11 Layout of single-pass inkjet printing machine machine

These machines demonstrate not only higher speeds but also better quality of print as the drop trajectory is not affected. These machines also may use six or eight colours to increase the colour gamete. In addition, with a better understanding of the chemistry of dyes and an increase in demand, the cost of inks is also reducing. This will make inkjet printing more cost-competitive and attractive.

While the printheads are the heart of the machines, the importance of the sophisticated software which does colour separation in 4, 6, or 8 colours and actuates the drop generation, drop size and manipulates the drop trajectory, cannot be undermined. The software and hardware design are beyond the scope of this chapter, but it must be appreciated that these need the contributions from computer and electronics engineers, mathematicians, and material scientists.

6.7.3 Post-printing processing unit

Unlike the paper printing, post-processing of inkjet printed fabrics is a must. The machinery required is very much the same as is needed for conventional printing. These may include

- Steaming/backing
- Washing
- Drying
- Folding

Without a doubt, the fixation of colours and washing are very crucial steps, and all care must be taken to ensure the print quality. The pigment-printed fabrics are backed or cured to fix the binder. These are not washed. However, fabrics printed with reactive, acid and disperse inks would require thorough washing after fixation by steam. Fabrics printed with disperse inks may also require reduction clearing.

6.7.4 Collaborative strategy

The technology of inkjet printing of textiles is complex and highly sophisticated. Understanding the colour coordinates of the design, separation of colours in 4, 6, or 8 colours based on the machine considerations, generation of drops and manipulation of their trajectory, and of course, standardisation of colour chemistry and other related issues are not the forte of a single company or one specialist. Special domain knowledge is required at every level. Collaboration is, however, the key. The hardware specialist must collaborate with software designers for customised software. The machinery manufacturers must take the ink formulators in confidence for ink formulation suitable for the specific printheads. Together all must ensure standardisation in evaluation and manufacture for satisfactory printing and, of course, strive to reduce the overall printing cost.

6.8 Quality attributes in inkjet printing

Finally, when all the parameters are set, one must evaluate the output. How much is the deviation from the desired goal? The term quality by itself appears to be a loose term. How does one measure the quality? Inkjet printing of a textile, which is flexible, has rough surfaces with fibres and yarns assembled in three dimensions, has pores and capillaries, and also has an appearance that is not truly white, is definitely not the same as printing of a photographic paper. Therefore, logically one expects the printed design may have some variance from what one sees on the computer monitor. Nevertheless, some parametric attributes that are suggested for evaluation and monitoring the quality of the print are related to the attributes of the

- Dots,
- Lines, and
- Printed Area

The dot is the building block of the image. One may like the measure the deviation in the *dot placement*. This may happen due to obstruction in the jet, obstruction in flight path by protruding hair, or simple variation in the actual gap between the printhead surface and the surface to be printed. Another measurable parameter is *dot gain*. The size of the dot may increase because of the in-plane spreading of ink because of capillaries. The *dot shape* may change, i.e., the major and minor axes of the may be different. This may result because of differential absorptive characteristics of warp and weft. *Satellite drops* formed due to rheology mismatch with printhead is also a performance metric.

Line width may be higher than intended, also due to capillary forces. The effect may also be due to fabric structure, i.e., plain or twill weave. The edge sharpness may get affected, and the optical density may be different because of out of plain absorption.

The printed area can be evaluated in terms of the colour (Lab value), and noise (graininess). The image is sensitive to variations in the brightness of the fabric (bleached / OBA treated) and varies with the fabric and the yarn structure, e.g., spun yarn or filament yarn. The overall quality may be less on textile fabrics as compared to the one on photo-quality paper. Overall reflectance may get affected as well. *Drop coalescing* may happen due to slow penetration leading to an increase in dot size and shape. All these lead to the noise. In deep shades, some of these may get obscured.

Good pre-printing processing, to a great extent, can improve these attributes. Will the print have the same quality as that obtained on the photographic paper? No. But why would someone want this? Textile has its own attraction. The colour reproduction and the quality of design obtained by inkjet printing are always superior to the conventional printing processes. Nevertheless, for successful inkjet printing, standardisation at all the stages is needed, be it the machinery and printheads, the ink formulation, fabric pre-processing and post-processing.

6.9 Sustainability and inkjet printing

The term sustainability of any process includes its economic benefits, the ecological effects accompanying this, and also the assessment of overall impact of the process on the society. It encompasses the ethical jurisprudence as well. Modern successful businesses often include sustainability in their vision documents. Industry 4.0 and circular economy are not just the buzz words. Emancipated entrepreneurs and businessmen are slowly and steadily inching in this direction and would like to get assessed and be certified, e.g., get a bluesign certificate.

The textile wet processing industry, including the printing, is often cited as the biggest polluter and consumes very large amounts of water and energy. The recent focus on inkjet technology is justified not only in terms of the reproduction efficacy and the design accuracy, but also in terms of the sustainability of the process. It is often referred to as the green technology.

In this respect, the inkjet textile printing process consumes less colour, less water, less energy, and requires almost no thickening agents. The comparative advantages of inkjet printing of textiles have been discussed in an article [14] and are depicted in Fig. 6.12.

The manufacturers of the modern inkjet machines insist on a much-optimised maintenance protocol, and these machines don't require frequent replacement of structural parts. These measures definitely reduce the carbon footprint, lead to a circular economy and offer environmental benefits. The machine produces less noise and requires less production space.

Printing with reactive, acid or disperse inks, involves pre-processing, printing, fixing, washing and drying. As mentioned before, the amount of ink used in inkjet printing is very low; almost completely utilised. It minimises pollution. How? This technology does not need thickener paste and requires very less auxiliaries, and therefore leads to efficient use of chemicals. It means a reduced amount of water required for washing. The associated energy required for drying the fabric as and electrical energy needed to run the machine is also less.

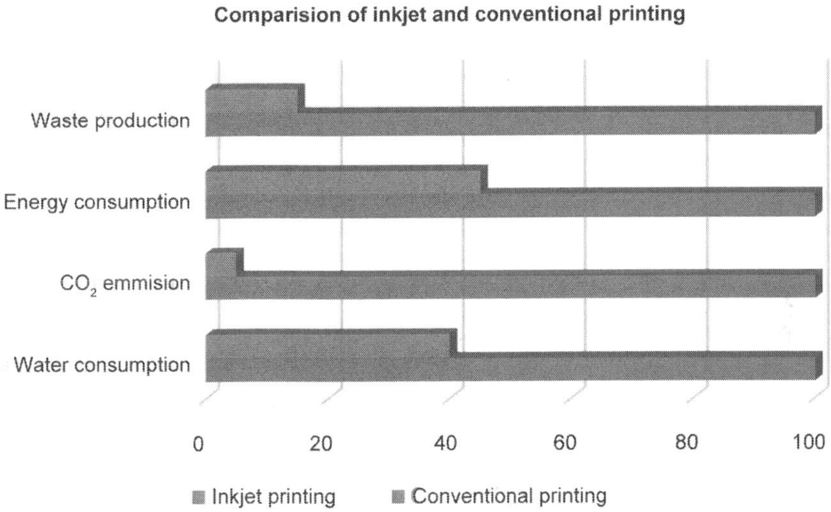

Figure 6.12 Ecological advantages of inkjet printing of textiles

Printing with pigment ink, on the other hand, simply involves, printing and hot pressing. That's it. No washing. No drying. No pollution. It is, therefore, a very eco-friendly process.

The modern inkjet technology, such as the single-pass, surely addresses economic and ecological aspects of sustainability. In this respect, it is far ahead of conventional printing technologies. The important societal impact issues have to be addressed, however, by the concerned industry which wishes to use this technology. However, it does address this to some extent, in that the process by itself is highly standardised and automated and requires very little human intervention or exploitation.

References

1. British Celanese, BP 293 022 (1929), 349 683 (1931).
2. https://www.smithers.com/services/market-reports/printing/the-future-of-dye-sublimation-printing-to-2023. (Accessed 30 September 2019).
3. https://www.smithers.com/services/market-reports/printing/the-future-of-digital-textile-printing-to-2023. (Accessed 30 September 2019).

4. Herran CL, Wang W, Huang Y, Mironov V, Markwald R (2010), 'Parametric study of acoustic excitation-based glycerol-water microsphere fabrication in single nozzle jetting', *J. Manuf. Sci. Eng.*, 132, 1-11.

5. Fashandi S (2017), 'Digital printing of acrylic fabric with cationic dyes using conventional inkjet printer', *Color Res. Appl.*, 42, 244-249.

6. Zhang Y, Cheung V, Westland S, Beverley KJ (2009), 'Colour management of a low-cost four-colour ink-jet printing system on textiles', *Color. Technol.*, 125, 29-35.

7. Kanik M, Hauser PJ (2003), 'Ink-jet printing of cationised cotton using reactive inks', *Color. Technol.*, 119, 230-234.

8. Savvidis G, Karanikas E, Nikolaidis N, Eleftheriadis I, Tsatsaroni E (2014), 'Ink-jet printing of cotton with natural dyes', *Color. Technol.*, 130, 200-204.

9. Aldib M (2015), 'Photochromic ink formulation for digital inkjet printing and colour measurement of printed polyester fabrics', *Color. Technol.*, 131, 172-182.

10. Aston SO, Provost JR, Masselink H (1993), 'Jet printing with reactive dyes', *J. Soc. Dye. Colour.*, 109, 147-152.

11. Fan Q, Kim YK, Perruzzi MK, Lewis AF (2003), 'Fabric pretreatment and digital textile print quality', *J. Imaging Sci. Technol.*, 47, 400-407.

12. Gupta S (2001), 'Inkjet printing-a revolutionary ecofriendly technique for textile printing', *Indian J. Fibre Text. Res.*, 26, 156-161.

13. Kwon KS, Kim HS, Choi M (2016), 'Measurement of inkjet first-drop behavior using a high-speed camera', *Rev. Sci. Instrum.*, 87, 035101.

14. https://catnewtech.com/digital-printing-a-positive-step-towards-sustainability-in-textiles/(Accessed 30 September 2019).

Suggested reading

1. Miles L W C, *Textile Printing,* Society of Dyers and Colourists, Bradford, 2003.

2. Ujiie H., *Digital Printing of textiles,* Woodhead Publishing Limited, 2006.

3. Zapka W. (ed), *Handbook of Industrial Inkjet Printing,* Volume 1 & 2, John Wiley & Sons, 2017.

7

Sustainable textile finishing using natural materials

S. Wazed Ali*, Sourav Banerjee and Satyaranjan Bairagi
Department of Textile and Fibre Engineering, Indian Institute of Technology Delhi, India
*Corresponding Author, Email: wazed@textile.iitd.ac.in

Abstract: The use of chemicals in the textile processing sector, starting from preparatory to dyeing and finally finishing, is vital for the production of value-added materials for fashion and intrinsic end-use with longevity. The use of bioactive agents is one such hopeful pathway having the potential to tidy up environmental pollution arising due to the inordinate use of hazardous synthetic dyes and auxiliaries. As a result, the textile finishing segment is also in a great crisis and is continuously looking for eco-friendly alternatives. In this connection, various biomolecules have been explored in recent eras. So many compounds from natural extracts have exhibited excellent UV protective, antimicrobial, mosquito repellent, and aroma functionality, and these have been integrated into the textiles to get those functional properties. This chapter elaborates on implementing and adopting such greener alternatives to make sustainable value-added textile substrates.

7.1 Introduction

In a recent trend, a revolutionary movement has been started towards the use of green chemistry in every field, and it is a very important approach for sustainable solutions. Environmental impact can be reduced by the use of biomolecules, and textile researchers are greatly involved in it with much interest. Heavy industries like textiles consume a significant amount of energy and water. Some serious environmental threats are encountered by the use of many chemicals that belong to this sector. Textile researchers are engaged in exploring and applying various new materials to deal with the aforesaid issues. The real need is to cut down the huge amount of pollution from the textile sector by banning some toxic chemicals, and on the other hand, there is a need to explore natural alternative materials that can fulfill the requirements of existing chemicals [1]. Textile industries and researchers are in continuous search to find out green technologies to produce finished value-added products to meet the worldwide demands. There is a substantial customer demand for ultraviolet protective, fire-retardant, hygiene, and antimicrobial textiles, and these products are increasing their market share day by day. To meet these requirements, in the last few years, research has been carried out on various natural materials and their integration into textiles to impart desirable functional properties.

7.2 Textiles and issues of sustainability

A composite textile industry comprises various departments like spinning, weaving, chemical processing, and, ultimately, garment manufacturing. There are huge amounts of hazardous materials produced globally, and a large proportion is related to the textile sector. Mainly in the processing hub, textile substrates are treated with a range of chemicals like synthetic colours, auxiliaries and finishing agents, which can lead to a serious overall pollution level in the existing ecology. The environment is greatly affected by the release of various toxic substances, and we adopt various effluent treatment technologies to mitigate this effect. Nowadays, sustainable process houses adopt different research and development strategies to replace hazardous chemicals with greener alternatives. Environmental issues of synthetic fibres can be replaced by natural fibres such as protein-based fibres (silk, wool), cellulosic fibres (cotton, hemp, jute, bamboo fibre), etc. Various eco-friendly approaches of using natural fibres, natural dyes, natural finishing agents, and recycling of fibres are the recent trends to ensure sustainability [2]. Sustainability is the process that measures the degree of protection of the environment for future coevals without hampering the resources of their uses. The requirement of water is always high for textile processing units. Enormous energy consumption at each step to make a final value-added product is again a big concern. There is an awful requirement to practise sustainable processes of quality improvement of materials and conservation of energy for the betterment of environmental and economic impact.

7.3 Green extract application for textile finishing

Several studies have been carried out on natural extracts that exhibit multifunctional properties due to the presence of various functional components like tannins, carotenoids, anthocyanins, naphthoquinones, etc. In this context, various herbal ingredients have been studied for application to natural and synthetic textile materials to impart multifunctional effects. Different parts of plants, like fruits, barks, flowers, and shells, contain various secondary metabolites that can be utilised in finishing operations. Various works in the literature claimed environment-friendly dyeing and finishing of textile substrates using extract of tamarind, amla flower petal waste, tea extract, tulsi, aloe vera, and neem. There are bulk requirements of materials for various segments of textiles, such as home textiles, hygiene textiles, and automotive cum aerospace textiles for the incorporation of green technology. Many researchers are engaged in demonstrating various functional properties (such as UV protection, antibacterial and fire-retardant properties, etc.) on textile substrates using bio-resources. The recent research outcomes of natural resource-treated textile materials based on the usage of different plant extracts or protein waste extracts (derived natural agents) have been accumulated and systematically presented in this chapter.

7.3.1 Antimicrobial finishing of textile using natural bio-extract

Textiles act as a medium to alleviate the growth of microbes, and their number can increase quickly after getting basic ontogenetic conditions like humidity, nutrients, and temperature. This growth of microbes on textiles can create problems of reduction in tensile properties, staining, unpleasant odour, infections, etc. A range of synthetic antimicrobial agents such as metals and their salts, triclosan, and quaternary ammonium-based compounds have been successfully applied as finishing agents for making products with excellent antimicrobial activity. But these agents have important issues such as their perniciousness associated with concentration and retention in non-target microorganisms, increased effluent loads, water contamination, etc. Recently, customers are more aware of the eco-friendliness of functional textiles. A lot of research work has been carried out on producing hygienic clothing using bioactive agents. Many biomolecules have been studied by many research groups for producing multifunctional value-added textile products. Such biomolecules like neem, tulsi leaf, aloe vera, coconut shell extract (CSE), and chitosan have already been used in textile substrates to impart an antimicrobial effect [3]. Antimicrobial agents inhibit or kill microbes by destroying the cell wall or changing membrane permeability, denaturing proteins, or suppressing lipid synthesis or enzyme activity.

7.3.1.1 Active ingredients of plants responsible for antimicrobial activity

Major components for plant-based biomolecules responsible for imbuing bioactive functionalities have been represented in the below text [3].

Phenolic compounds

This is a diversified group of secondary metabolites. Phenolic compounds have hydroxyl groups (-OH) in their structure which are responsible for inhibition by interacting with the cell membrane to rupture it and cause leakage of plasmic components. Also, the position and numbers of hydroxyl groups on the extracted molecules are responsible for their activity toward organisms. Phenolic compounds like thymol, resorcinol, pyrogallol, catechol, eugenol, caffeic acid, etc. have been explored for microbial destruction on textile substrates.

Quinones

These are the aromatic compounds having high reactivity with two ketones substitution as anthraquinones and naphthoquinones. Quinones form complex with nucleophilic amino acids in microbes and inactivate the protein portion. Heena and madder extracts contain quinone-based components like lawsone and alizarin compounds.

Flavonoids

Flavonoids are hydroxylated phenolic substances. Their activity is mainly due to complex formation with extracellular soluble protein and disruption of membrane. Catechin, quercetin, chrysin, galangin, etc. are extracted flavonoids used in textiles.

Tannins

Tannins are polyphenolic substances capable of astringency and are obtained in almost every portion of a plant's barks, leaves, roots, fruits, etc. The antimicrobial activity of tannin is mainly driven by the inhibition of extracellular enzymes, and also it inhibits oxidative phosphorylation (microbial metabolism). It is often used as a bio-mordant for anchoring natural dyes with textiles. Hydrolysable tannins and condensed tannins are the common examples that are widely used.

Essential oils

The fragrance is the identification of essential oil. This is enriched by isoprene structures called terpenes. If any extra oxygen is added to that structure, then it is termed as terpenoids. Membrane disruption of the microorganisms is the main way of inhibition. Neem or clove oils are examples, in this category, known for applying to textiles.

Polysaccharides

Aloe leaf contains different polysaccharides like glucomannan, galactogalacturan, etc. These active ingredients destruct the cell membrane of the microorganisms and lead to the leaking of cytoplasmic liquid out of the cell.

7.3.2 Green extracts as bioactive agents

In a study, Teli et al. [4] reported the multifunctional activity of the jute fabric treated with green coconut shell extract (CSE). The flame-retardant properties of the CSE-treated jute fabric in terms of limiting oxygen index (LOI) have been improved up to 81% as compared to the untreated jute fabric. At the same time, this CSE treatment on the jute fabric imparted natural colour and a very good antibacterial behavior against both the Gram-positive and Gram-negative bacteria. The antibacterial effect of treated jute was studied based on different pH conditions. The authors concluded that jute fabric treated with CSE in alkaline conditions contains different phosphorus and nitrogenous compounds, which help to impart better flame retardancy and antibacterial properties. CSE contains phenolic substances and tannin-metal ion complexes, which have a potential contribution to antibacterial efficacy. In another study, cotton fabric was finished with herbal extracts of chamomile, sage, and green tea [5] without using any source of chemicals. Extracted tea leaves have an active phenolic component called catechin. Sage is a characteristic source of different polyphenolic flavonoids, and chamomile also has flavonoids called Apigenin. Researchers proposed a mechanism involving cross-linking of these active polyphenolic components with cotton fabric. Only sage and green tea components showed the antibacterial activity against Gram-positive bacteria. Different aqueous extracts of the plant [6], such as *bael* leaves exhibited zone of inhibition around 16 and 12 mm against *S. aureus* and *E. coli* bacteria, respectively. Also, zone of inhibition against *S. aureus* and *E. coli* bacteria are reported for *aonla* leaves (12 and 11mm), *safeda* leaves (16 and 09 mm), and curry leaves (12 and 07 mm). Also, in the case of ethanolic extraction, the *safeda* leaves extract showed a higher zone of inhibition, i.e., 23 and 20 mm against *S. aureus* and *E. coli* bacteria, respectively, followed by *bael* leaves (20 mm each), *curry* leaves (20 and 18 mm), *aonla* leaves and *neem* leaves exhibited the same zone of inhibition, i.e., 18 and 16 mm, respectively. The lowest antimicrobial activity was found in lemon (nimboo) leaves extract against both the Gram-positive and Gram-negative bacteria strains. Das et al. [7] have studied the antibacterial activity of *aloe barbadensis* biomolecule-treated cotton fabric. Besides antibacterial properties, these *aloe barbadensis* biomolecules on the cotton fabric showed other properties such as natural colour and fragrance. An increased zone of inhibition was found in the aloe barbadensis biomolecule-treated cotton fabric sample compared to control cotton. These antibacterial activities also got changed by varying concentrations of the biomolecules applied. Antibacterial properties of the treated cotton fabric were evaluated against both the Gram-positive and Gram-negative bacteria. Similarly, Ali et al.[8] studied the antibacterial activity of the *aloe vera* gel finished cotton fabric. In their study, they developed *aloe vera* gel finished cotton fabric using carboxylic acid (BTCA) as a cross-linking agent. This cross-linking agent helped to create bonds between the hydroxyl groups of the cotton fabric and *aloe vera* gel. The *aloe vera* gel treated cotton fabric showed higher antibacterial properties as compared to the untreated cotton fabric. Antibacterial behavior of the *aloe vera* treated cotton substrate was evaluated against both the Gram-positive and Gram-negative bacteria. Teli et al.[9] have reported the ultraviolet protection and natural colouration properties of the linen fabric treated with the waste material (*Sterculiafoetida* fruit shell). The linen fabric was treated with the *Sterculiafoetida* shell extract by using mordant (alum and harda) or without mordant. In both the cases (with mordant and without mordant), waste shell extract-treated linen fabric showed higher washing, light, and rubbing fastness properties as compared to the untreated linen fabric. The optimised concentration of shell extract was 30% which was applied at 90°C for 30 min. This research team has also reported the antibacterial activity of the *Sterculiafoetida* fruit shell waste on dyed silk fabric against both the Gram-positive and Gram-negative bacteria[10]. The dyed silk fabric showed enhanced antibacterial properties against Gram-positive bacteria (93.87%-97.33%) as compared to the Gram-negative bacteria (82.63%-97.63%). The authors have concluded that this increased antibacterial activity of the dyed silk fabric may be due to the presence of flavonoids, terpenoids, saponin, and tannins agents in the waste shell extract. In addition, they have also studied the antibacterial properties of the natural dyed cotton, wool, and silk fabric using turmeric and pomegranate rind extract [11], where they used *Tamarindus indica L.* seed coat tannin as a natural mordant. The dyed fabrics followed by pre-mordanting showed better washing, light, and colour strength when compared with the simple dyed fabrics (without mordanting). The antibacterial properties of the treated and untreated fabrics were also evaluated against Gram-positive and Gram-negative bacteria. Finally, they concluded that natural mordant-based dyed fabrics showed acceptable light, washing, colour strength, and antibacterial activity even after 20 washing cycles. Chitosan is very much effective for producing value-added cellulosic textiles. Many researchers tried to incorporate this novel biomaterial into textile substrates by using cross-linking agents. Chitosan has a wide spectrum of antibacterial activity against both Gram-positive and Gram-negative bacteria. From various pieces of literature, it is evident that mainly three mechanisms work behind its antibacterial effect, such as breaking the cell membrane by charged or ionic surface engagement, inhibiting the action of protein synthesis, and chelating with any metal to create a barrier for bacteria. Recently, Sheikh et al. have reported the antimicrobial activity of chitosan on linen fabric [12,13]. They modified linen with two-stage processes: first, the fabric was passed through chitosan and citric acid solution, and finally, the fabric was treated with thiourea and phytic acid. The antibacterial effect of the treated fabric is represented in Table 7.1, whereas washing durability results are tabulated in Table 7.2. The main reason behind these two-stage processes was proposed as the phosphorylation of chitosan and the act of synergistic effect on the substrate. The mechanism behind antimicrobial efficacy is the interaction of positively charged chitosan (having -NH_2 group) and negatively charged bacteria which results in leakage of nutrients followed by destroying the cell membrane. They also proposed that quenching of DNA may be another reason for the bacterial reduction.

Table 7.1 Different natural materials used for antimicrobial finishes of textiles

Natural material	Source of the active agent for antimicrobial activity	Reference
Neem	Seed extract, Leaf extract	[14-16]
Chitosan	Shrimp and other Crustacean shells	[12,13,16]
Silk sericin	Raw silk	[16]
Eucalyptus	Eucalyptus oil	[16,17]
Azuki beans	Azuki beans extract	[16]
Tea	Tea leaf extracted oil	[16]
Tulsi	Tulsi leaves extract	[16,18]
Pomegranate	Pomegranate peel extract (rind), seed extract	[16]
Clove	Clove oil	[16]
Onion	Onion skin and pulp extracts	[16]
Aloe vera	Leaf extract	[7,8,16]
Tridax procumbens	Whole plant extract	[16]
Coconut	Coconut shell extract	[4]

Table 7.2 Antimicrobial activity of different natural materials

Agent	Fabric	Concentration	Bacterial reduction	Reference
Neem seed extract	PET/cotton (48/52)	5% (w/v)	95% against *Bacillus subtilis* and 70% against *Proteus vulgaris.*	[14]
	PET/cotton (67/33)	10% (w/v)	98% against *Bacillus subtilis* and 65% against *Proteus vulgaris.*	
Neem leaf extract	Cotton fabric	50 gpl	98% against *Staphylococcus aureus* and 96% against *Escherichia coli*	[15]
Chitosan	Linen fabric	10 gpl	83% against *Staphylococcus aureus* and 84% against *Escherichia coli*	[12]
Aloe vera	Cotton fabric	5% (w/v)	98.5%against *Staphylococcus aureus* and 98% against *E. Coli*	[8]
Coconut shell extract	Jute fabric	1:15 MLR in pH 7	97% against *Staphylococcus aureus* and 96% against *Escherichia coli*	[4]
Tulsi	Cotton fabric	5% owf	93% against *Staphylococcus aureus* and 75% against *Escherichia coli*	[16,18]
	PET/cotton fabric	5% owf	90% against *Staphylococcus aureus* and 70% *Escherichia coli*	

Ancient people were dependent on natural materials which were used in their daily life. Tulsi or *Ocimum sanctum* plant extracts from various parts have been reported as insecticidal, antimicrobial, antioxidant, anti-inflammatory, and antibiotic in the traditional Ayurveda system. Researchers have reported that it can be applied on the cellulosic substrate by direct coating, cross-linking or encapsulation processes. Ravindra et al. have reported in their work that dried tulsi leave powder can be mixed with methanol for soxhlet extraction, and the extract can be applied on cotton and cotton/polyester blend employing glutaraldehyde and sodium hypophosphite [16,18]. A noticeable bacterial reduction was observed against both the Gram-positive and Gram-negative bacteria. 10% concentrated extract exhibited a maximum reduction of 95% and 82% against *S. aureus and E. coli*, respectively. Also, they have mentioned that an almost 40% reduction of activity for both types of bacteria was noticed even after washing up to 20 cycles. Overall, blend fabric showed comparatively less activity than a pure one. Eugenol, methyl eugenol, and carvacrol are the main responsible components for showing antibacterial effects. Many researchers have found that various kinds of secondary metabolites like gallic acid, ellagic acid, rutin, and quercetin are actively present in eucalyptus leaves which can be used as an effective finishing agent on textiles. Silva et al. have reported that aqueous extraction of eucalyptus leaves has an antibacterial effect on cotton knitted fabric against *E. coli, S. aureus, and C. albicans* [19]. The treated fabric showed better results against *S. aureus and C. albicans.* Antimicrobial activity got increased with an increase in the concentration of the extract. It has also been concluded that the effect is not so much lucrative against *E. coli* bacteria. The fabric treated with chitosan was chosen for the pre-treatment process to get a better rubbing and washing fastness value. Also, chitosan helped to get higher bacterial reduction against Gram-positive and Gram-negative bacteria. The synergistic effect of eucalyptus component and chitosan provided higher antibacterial efficacy. Balamurugan et al. have reported the antimicrobial coating of cotton fabric by eucalyptus oil extracted from steam distillation of leaves [17]. Therapeutically used eucalyptus oil has extensively been used in pharmaceutical firms as it contains 1.8-cineole. The authors coated the cotton fabric with extracted oil by pad-dry-cure methods. The treated fabric was evaluated to find out the zone of inhibition (around 18, 15 and 11 mm for *B. subtilis*, *P. aeruginosa* and *S. aureus,* respectively). After the first wash, a 12 mm zone of inhibition was observed for *B. subtilis* bacteria, and the rest of them showed no such effect. Neem has been used in the Indian traditional medicinal system from the Vedic era for its bioactive enriched metabolites. Every portion of neem plant has many pharmacological activities which mainly play as protecting agent from oxidation, malarial infection, diabetics, inflammation, ulcer, etc [16]. The most common active components of neem are azadirachtin, nimbin, nimbidin, salannin, nimbidol, gedunin, sodium nimbinate, queceretin, etc. Joshi et al. have stated the potentiality of neem extract as an antibacterial agent on polyester/cotton blend fabric [14]. Neem seed extract has no affinity for fabric. So it was integrated into blend fabric by padding and curing using glyoxal/glycol, aluminium sulphate, and tartaric acid. Mainly neem terpenoids are responsible for bonds created between the cellulosic hydroxyl groups. The antibacterial efficacy was checked against *Bacillus subtilis* and *Proteus vulgaris* bacteria, and for both cases, it showed above 95% bacterial reduction. After the first wash, it got reduced up to 75% and 57%, respectively.

7.4 Natural materials used for fire retardant textiles

Recently there has been a trend to make sustainable fire-resistant finishing. For a long time, inorganic salts, metal oxide (aluminum oxide, calcium carbonate), halogen-based and nitrogen, sulphur derivatives, pyrovatex CP, tetrakis hydroxymethyl phosphonium chloride (THPC), etc. have been successfully explored as a promising fire protective agent for the cellulosic textiles [20]. Due to the synergistic effect, phosphorus and nitrogen-based compounds together show better fire-retardant effects. The most usable textile fabric is cotton-based; therefore, different processes have been tried to impart flame retardancy on cotton-based textiles by using different suitable flame-retardant agents. For instance, antimony in combination with halogen (mixture of ATO/BrFr) exhibits very good flame-retardant properties. On the other hand, halogen-based materials are cheap, easily available, and show promising flame-retardant properties when applied to the cellulosic textile substrates [21]. Various researchers have recently started their exploratory works to evaluate the fire retardant action of natural materials on textile, paper, and other polymeric substrates. A comprehensive study is revealed by building up all the recent advancements in flame retardancy to the textile materials based on using different bio-macromolecules.

7.4.1 Fire retardant biomolecules from various plant sources

From various literatures, it has been summarised that plant extracts have minerals, secondary metabolites, phosphorous, various salts, etc. Recently, different plant-based secondary metabolites have been studied to evaluate their fire-retardant efficacy on various textile substrates. In this context, waste coconuts shell, lignin sulfonate, banana pseudo stem sap, spinach juice, starch, pomegranate rind extract, etc. have been presented in detail.

7.4.1.1 Lignosulphonate

Lignin is one of the natural amorphous polyaromatic compounds and generally originated from wood pulp. Lignin has been explored as a filler, plasticiser, hygroscopic agent, binding material, flocculating agent, etc. Various research works have been executed to study the flame retardant property of lignin. Lignosulphonates-based materials have also been revealed for applications in broad areas [22]. Shukla et al. [23] have reported that cotton fabric implemented by sodium lignin sulfonate (SLS) showed adequate fire retardancy on cotton textiles. No residue was found in the case of untreated fabric, whereas, in the case of treated fabric, 28.5% of residual char was observed even after burning.

The authors demonstrated the fire retardant efficacy of SLS-treated cotton fabric using thermo-gravimetric analysis (shown in Fig. 7.1). The researchers explained that an additional char layer increases the flame retardancy, which works in a condensed phase mechanism and prevents flammable gases from releasing for further combustion. The volatile gases that evolved during burning were evaluated by GC-MS characterisation. Along with fire retardant action, SLS-treated cotton fabric could also show a natural colour and UV protective properties compared to untreated cotton fabric. From their study, it was revealed that cotton fabric attained fire-retardant properties with a higher add-on percentage of SLS, and the optimal concentration of SLS, as stated, was 30%.

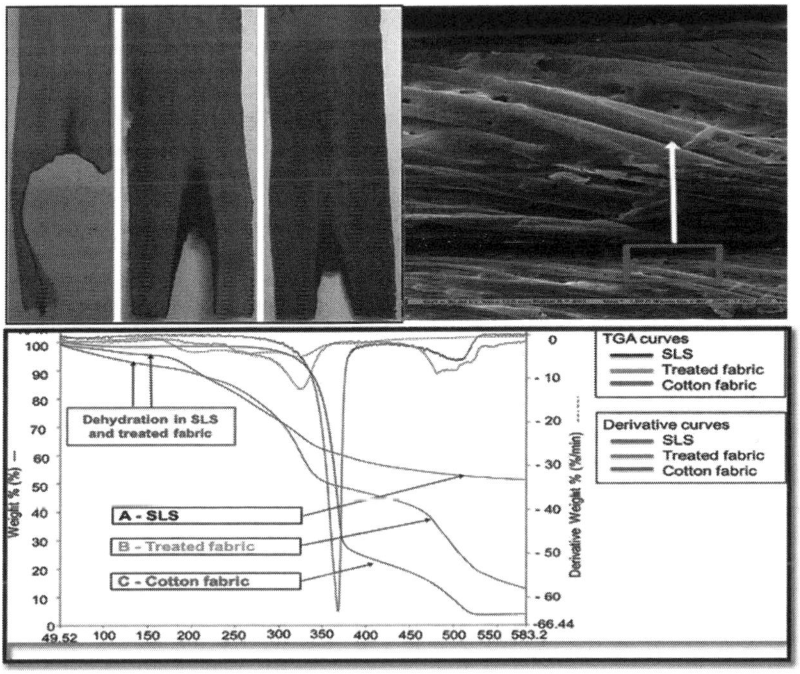

Figure 7.1 SEM analysis image along with TGA curves of SLS treated fabric [23]

7.4.1.2 *Banana Pseudostem Sap (BPS)*

Banana Pseudostem Sap (BPS) extract plays an important role as a natural fire retardant ingredient for a textile substrate. The reason behind this resistivity is due to the existence of different flame-retardant agents, such as inorganic salts, calcium, phosphate, sodium, potassium, etc., in sap [20]. Basak et al. [24] have reported that BPS extract-treated cotton fabrics showed good fire-resistant properties (Fig. 7.2). In a comparison of limiting oxygen index (LOI) value, it was observed that sap-treated cotton fabric (LOI approx. 30) showed a very good improvement as compared to the untreated cotton fabric (LOI approx 20). They also studied the flame retardant efficacy of banana pseudostem sap-treated jute fabric [20]. It was observed that the treated jute fabric showed 1.9 times higher LOI value than the control jute fabric.

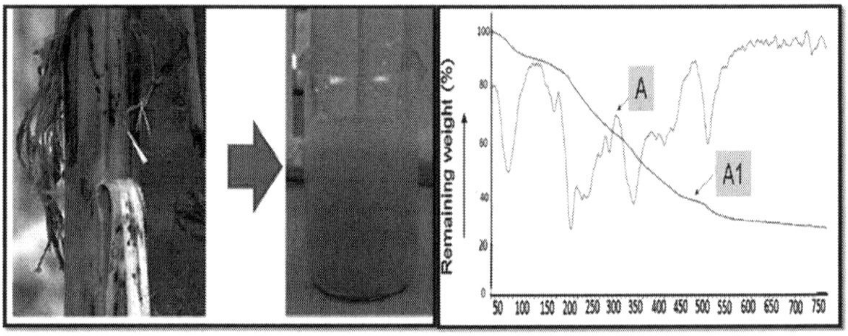

Figure 7.2 Banana pseudostem sap along with TG and DTG curves of BPS [20]

7.4.1.3 *Spinach Juice (SJ)*

As per the reported study, spinach juice also plays an important role in imparting flame-retardancy to textile fabrics as the juice is composed of different types of inorganic salts, sodium, potassium, iron, and so many other elements. Basak et al. [25,26] have reported that spinach juice-treated cotton fabric exhibited reasonable flame-retardancy (as shown in Fig. 7.3). The study was carried out by applying spinach juice to the bleached and fully mercerized cotton fabrics by exhaust method. The study demonstrated the fire retardant efficacy of the SJ-treated cotton fabric in terms of LOI, vertical and horizontal flammability test, and radiant heat testing. Improved flame-retardant property was observed after treatment of SJ on cotton fabric. The treated sample showed 1.6 times higher LOI value than the control cotton fabric.

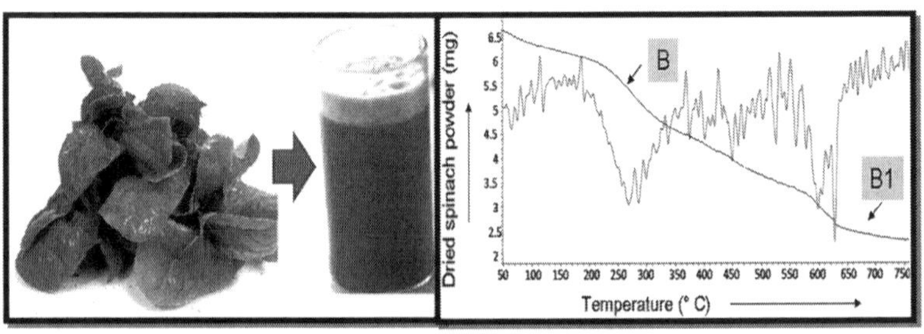

Figure 7.3 Spinach juice from green spinach leaves and TG and DTG curves of dried SJ powder[20]

7.4.1.4 Pomegranate Rind Extract (PRE)

Extracted juice of pomegranate rind has been studied in detail for its efficacy in imparting flame retardancy to textile substrates. Basak et al. [27] have explored the fire-retardant action of PRE extract-treated lignocellulosic jute fabric by applying the active ingredients by the exhaust method. In their study, the control jute sample burned out within 3 min, whereas the treated jute showed a significant improvement as evaluated by the vertical flammability test. During the evaluation of afterglow of treated fabric, it was observed that an increase in pH value of extract solution reduced the afterglow as compared to control. The char structure of treated jute fabric (as shown in Fig. 7.4) had multiple pores and bubbles, which are indications of intumescence characteristics [28]. Catalyzed dehydration indicated by the thermogravimetric (TG) profile of the treated fabric and further formation of heavy char after burning were the proofs of the condensed phase mechanism of fire retardant action. In the case of the control jute sample, isothermal TGA analysis at 500 °C showed more than 30% weight loss, whereas in the case of alkaline PRE treated jute showed only 4%–5% weight loss after 20 min of heat exposure. The rind extract is composed of different polyphenols, nitrogen, and phosphorus contents. These lead to the formation of char, help in catalyzed dehydration of the structure, and oppose the formation of flammable gases.

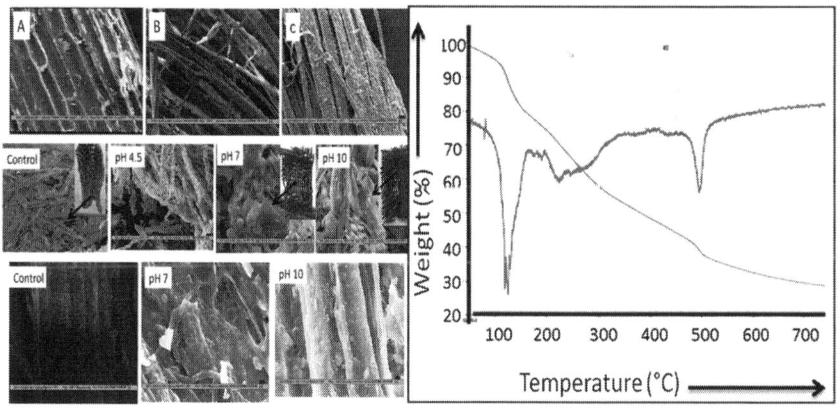

Figure 7.4 SEM analysis of PRE treated fabric along with TGA and DTG curve [27]

7.4.1.5 Green Coconut Shell Extract (CSE)

Coconut (*Cocos nucifera*) is an important fruit-related plant for its everyday use and has almost been discovered everywhere, including the coastal areas of India. The nutritional water of coconut is used as a natural drink with more nutritional value. In a study, the coconut was cut into small parts, and then an extractor was used to extract the fresh sap[29]. The appearance of fresh extracted CSE solution was dark yellowish in colour with acidic nature. It was made neutral and alkaline as per the requirement for the experiment by the addition of mild alkali sodium carbonate (Na_2CO_3). Teli et al. have reported that textile substrates were padded with different concentrations of CSE at 90°C with a material to liquor ratio of 1:15 for 60 min. Treated cotton and jute samples were dried in open condition. A significant increment of LOI value was found after CSE application on the fabric samples. Jute got swelled from acidic to alkaline conditions, which increased the LOI value with an add-on percentage [4]. The standard value of LOI for wool is 25, having some inbuilt flame-retardant properties. However, the wool polymers are not able to withstand high-temperature. After the application of the CSE, LOI values of the treated wool samples were found to be increased to 36. Authors reported that the control wool fabric got burnt within 90 sec, whereas CSE-treated wool fabrics showed no afterglow effect and formed a char length of 8 cm. Different kinds of metal were noticed in FTIR (Fourier transform infrared spectroscopy) analysis of shell extract. The alkaline shell extract finished cotton and jute textiles reported greater thermal stability than the acidic biomolecules treated material, as the content of phosphorus and silicon is more. These phosphates and chlorinated compounds can work as flame retardants for the jute and cotton substrate. Also, a synergistic effect is reported due to the presence of metal, phosphorous, nitrogen, silicate bonded with water molecules, and various phyto-compounds. Thermogravimetric (TG) analysis provided information related to char formation and non-flammable gases like CO_2, H_2O, etc. Researchers reported that a thick black coloured char was formed after burning, which was also reaffirmed by SEM analysis. It was concluded that CSE works based on the principle of condensed phase intumescent mechanism.

7.4.2 Animal-based natural flame retardants

7.4.2.1 Chitosan

Chitosan is a biopolymer derived by deacetylation of chitin, a bio-based glycosidic component obtained from crabs, shellfish, insects' cuticles, and fungi cells. Chitosan has polyamine groups with reactive amino components and available hydroxyl groups, etc. A group of researchers claimed that chitosan has the ability to increase the wash durability of soluble phosphate-treated cotton textiles. Fang et al. have reported that phosphorylated chitosan enhanced the flame retardancy of cotton fabric. Diammonium hydrogen phosphate gets its nitrogen source from chitosan. Wash durability was improved along with thermal stability by incrementing chitosan concentration. Researchers have recently reported the intumescent effect of fire retardancy on cotton fabric once it was coated with the mixture of chitosan and polyphosphate compound by layer-by-layer technique, as shown in Fig. 7.5 [30].

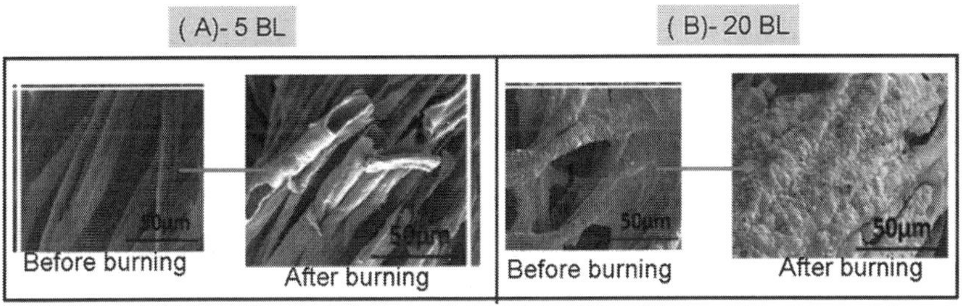

Figure 7.5 Surface and char morphology of the chitosan and ammonium polyphosphate treated fabric, 5 bi layer (A) and 20 bi layer (B) treated fabric [20]

7.4.2.2 Casein

Casein is a major phosphorous-containing milk protein and generally gets separated when preparing cream from skimmed milk. Milk protein is composed of various kinds of amino acids. Carosio et al. have revealed a casein biomolecule to improve the thermal properties of cotton. The thermal constancy of the casein molecule is mainly due to the presence of nitrogenous and phosphorous components in its structure. The researcher reported the coating of cotton sample with alkaline casein solution[31]. The treated fabric

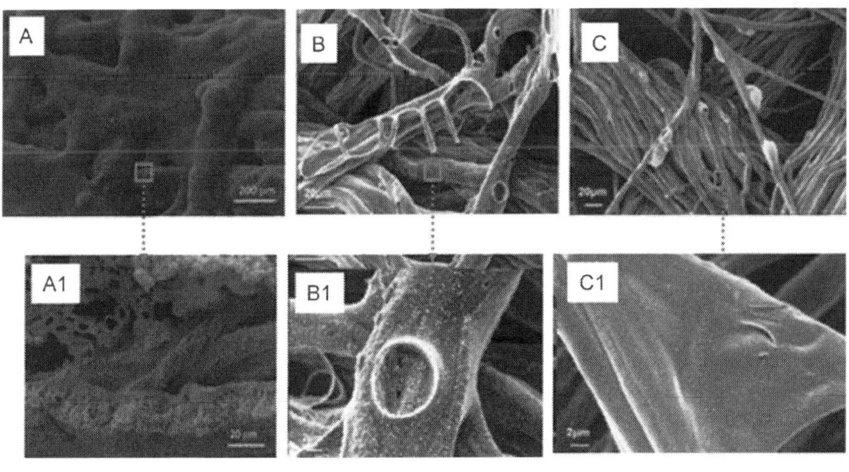

Figure 7.6 Surface morphology of the char mass generated after burning of casein treated PC blend (A), cotton (B) [20]

registered a remarkably lower burning rate and increased final residue. Also, the burning time was increased by 30% more than the control fabric. Research teams also studied the fire retardancy of casein powder on synthetic and blended materials. After completion of burning, char morphology of casein finished fabrics showed lighter, voluminous, globular structures as shown in Fig. 7.6, indicating an intumescent mechanism by bubble formation of phosphorus-rich components.

7.4.2.3 Whey Protein

Whey protein is also a milk protein, generally having a globular α-helix structure. Whey protein has a sulphur composition in its amino acids, which differs from casein. Bosco et al. have reported the integration of whey protein on cotton fabric to enhance the thermal behavior of the fabric. The main compositions of whey protein powder are protein (93.5%), fat, glucose, ash, and water molecules. This powder was applied to the substrate as an

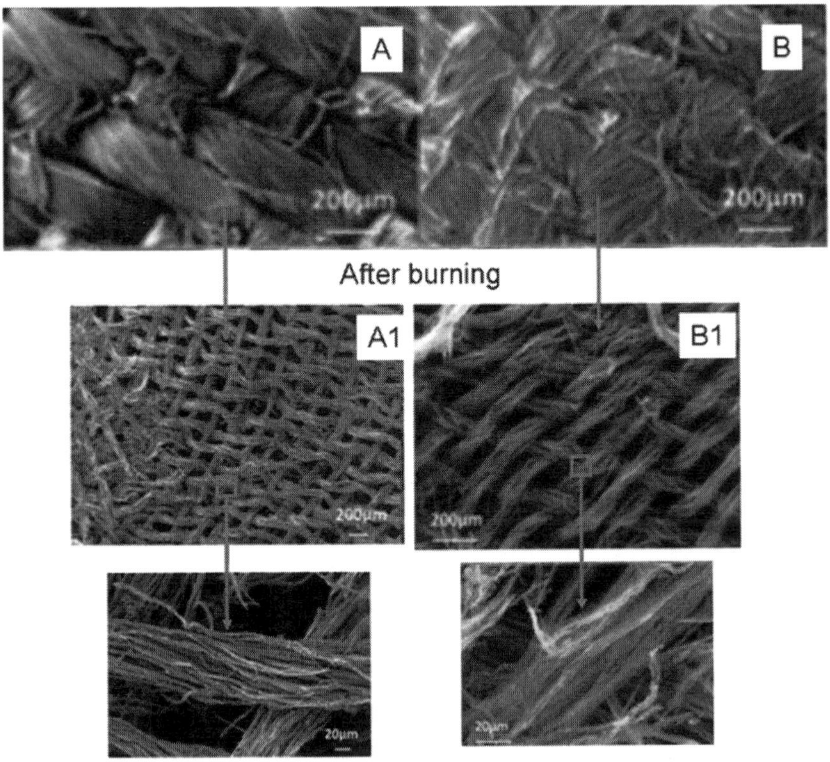

Figure 7.7 SEM morphology of the WP and DWP treated cotton fabric (A and B), char morphology of the A (A1) & B (B1)[20].

isolated coating comprising folded or unfolded chains. As per result, folded protein molecules were suspended. Separately alkalisation was done into the fold configured whey solution that sets its pH. Unfolded and denatured whey proteins were suspended at high temperatures. The study revealed that the reduction of the rate of burning of the sample was around 25% and 22.5% in comparison with the control samples. In case of burning time, the treated fabric was burnt in 126 sec (for whey protein) and 133 sec (for denatured protein), respectively. Control cotton fabric was burnt in 70 sec. Whey protein treatment reduced the time of combustion of the finished material. TGA profile in the air atmosphere denoted that whey protein application reduced thermal degradation due to the formation of oligopeptides. The amount of char mass production also got increased [32], as shown in Fig. 7.7.

7.4.2.4 Deoxyribonucleic Acid

Deoxyribonucleic acid (DNA) is a helical structured long chain consisting of nitrogen-containing molecules like adenine, guanine, cytosine, and thymine. In deoxyribose units (penta carbon sugar), nitrogenous bases and polyphosphates are mainly present in the DNA skeleton by ester bonds. Bosco et al. have reported TG analysis of DNA molecules from herring sperm. They mentioned their weight reduction in three stages attributed to more thermal stability due to phosphorus and di-ester bonds or phosphate breakage in the DNA structure [33]. DNA can form thick and denser char. Phosphoric acid is responsible for catalytic desiccation and NH_3 formations, reducing the available oxygen to provide high thermal stability. DNA molecule is the most efficient for forming flame retardant treatment on the cotton substrate. This was reported that cotton fabric was padded through the solution formed by dissolving DNA powder in de-ionized water in acidic pH at room temperature for over half an hour, under continuous mixing. The LOI values of the DNA-treated cotton samples were 23, 25, and 28 at different concentrations of 5%, 10%, and 19%, respectively. This thermal stability was due to phosphorous groups in DNA that got decomposed to H_3PO_4 and initiated the char formation process. Also, some theories suggest that nucleic acid behaves as a fire-retardant on principle based on nitrogen–phosphorous synergism, which increases the efficacy of finished cellulosic fabric.

7.4.2.5 Hydrophobin

Hydrophobins are non-hazardous cysteine-based amino acids that are produced from the fungal cell. These also assist in the growth of the cell membrane and spore dispersion into the surroundings. In a study, Alongi et al. have applied proteinous solution on cotton textile, keeping approx. 1:26 material to liquor ratio by a dipping technique [34]. In such cases, the add-on of the finished fabric was maintained up to 20%. Thermo-oxidative profile of the hydrophobin treated fabric showed the reduction of pyrolytic conditions and a lower weight reduction rate at elevated temperatures. They also highlighted that the presence of cysteine bonds in the hydrophobin layer enhances the dehydration at 200°C with a formation of H_2S and H_2SO_4. TGA pattern of the hydrophobin-treated cotton was almost the same as the degradation of the crude protein casein padded cotton substrate in an air atmosphere.

7.4.2.6 Chicken Feather

In recent days bird plumage has been a hot topic of research due to its composition of many hygroscopic and lyophilic amino acids, among which serine is the most predominant. It has been reported that moisture retention is higher for feather protein than in wool. Keratin is mainly found in feathers containing nitrogen, sulfur, and ash material. Wang *et al.* have reported that the addition of feather protein increased the flame resistivity of the cotton textiles by a synergistic mechanism of phosphorus–nitrogen-based compounds. Raw feather molecules were accumulated from the waste of animal firms and added to alkali and urea at a high temperature. In the LOI test, the chicken feather protein treated cotton (250 gpl) showed a value of around 30 and eradicated the fire in 7 sec in the vertical flame tester. Even though the total sample got burnt with afterglow for 180 sec, the burning rate was less. Treated cotton substrate exhibited a voluminous and compact coating on the surface of the fabric. Researchers have reported that the fabric treated with a mixture of boric acid and borax with feather protein evinced tiny uniform grain dispersed on the top with more intended charred fibre morphology, as shown in Fig. 7.8. This absorbed heat in the presence of phosphorous, nitrogen, and sulphur and cut back the entrance of the inflammable gases into the molecular construction and enhanced the char generation [35].

Figure 7.8 Char morphology of the control (A), CFP treated (B) and (CFP + BA + BX) treated (C) cotton fabric [20]

Table 7.3 Overall comparison of flammability parameters [20]

Fabric details	Flammability parameters			
	LOI	Burning rate (mm/s)	Burning time (s)	Char length (mm)
Control cotton	18	1.3	78	Nil
Control PET	21	1.8	57	Nil
Casein (20% add on)-cotton	24	0.4	75	30
Casein (20% add on)-PET	26	0.6	54	30
Casein (20% add on)-PC blend	21	0.7	171	Nil
DNA-cotton (19% add on)	28	1.2	76	89
Chicken feather-cotton (5.72% add on)	30.1	1.3	7	Nil

7.5 UV-protective textiles using natural materials

It is known that different types of radiation, such as infrared (IR), ultraviolet (UV), and visible light (VL), are emitted by the sun. IR and UV radiation are not good for the human body since they cause different health issues. Ultraviolet radiation can be classified into three groups: UVA, UVB, and UVC, depending upon their wavelength. UVA has a wavelength ranging from 320-395 nm, whereas 280-320 nm and 200-280 nm wavelengths range for UVB and UVC, respectively. The ozone layer imbibes UVC. The ozone concentration (O_3) in the ozone layer is not high. Ozone molecules get splitted into molecular oxygen (O_2) and oxygen atoms (O) by UV radiation. Again, oxygen molecules absorb UV rays and get splitted into atomic oxygen, and this atomic oxygen reacts with the oxygen molecule to form ozone. This process is known as the ozone-oxygen cycle. UVA is the most concerning to the human body because this category of UV reaches the earth with the maximum amount that causes different problems in our body. Different problems such as skin cancer, eye damage, skin aging, and so forth can be observed due to excessive exposure to UVA radiation. Therefore, protection from this radiation is an important task to solve the above-said problems. Research has drawn great interest in developing effective UV protective fabrics in recent days [36,37].

In the past few decades, different organic (o-hydroxy benzophenones, o-hydroxyphenyl benzotriazoles, o-hydroxyphenyl hydrazine, salicylic acid derivatives, and aromatic conjugated systems) and inorganic (silver, titanium, zinc nanoparticles, graphene, etc.) finishing agents have been used to develop UV protective fabrics. But these finishing agents are not eco-friendly, causing environmental pollution, and the aquatic life gets hampered too. To resolve these environmental and aquatic problems, researchers are trying to discover different eco-friendly natural finishing agents for UV-protective fabrics.

Kim et al. have reported the dyeing and UV characteristics of the cotton fabric dyed with natural green tea extract [38]. Green tea extract has different phenolic ingredients. One of them is catechin, which has excellent UV protection properties. To compare their UV resistance, the authors developed four different samples (control cotton fabric, only chitosan treated cotton fabric, green tea extract-treated cotton fabric, and chitosan and green tea extract-treated cotton fabric). It was seen that there was no such significant change in UV protection properties (84.4%, 85.5%, and 84.7% against UV-A and 86.6%, 87.0%, and 87.4% against UV-B) of the control, chitosan treated and green tea extract-treated cotton fabric. But the combined chitosan-green tea extract treated cotton fabric showed higher UV protection values, with 91.3% protection against UV-A and 92.8% protection against UV-B rays. Rungruangkitkrai et al. [39] have studied the UV protection properties of the eucalyptus leaf extract dyed wool fabric. They treated the fabric by two techniques: pad-batch and pad-dry process, where they used different metal mordants for better dye uptake by the wool fabric. The finished fabric showed better UV protection properties than the control wool fabric. It was found that the UV protection property got enhanced when the dye concentration was increased. The UPF value for the 5 g/L dye concentration was 81.8, whereas the UPF value was 104.2 for the 20 g/L dye concentration. This concept has also been established by other researchers [40, 41].

In recent days, the concept of multifunctional finishing (antibacterial, fire retardant, antioxidant, UV protection, and so on) of the textile fabric by using natural finishing agents has taken a new momentum among the research communities. For instance, Saini et al. [13] have studied the multifunctional properties of linen fabric which was treated with a combination of chitosan and green tea extract. The finishing agents (chitosan and green tea extract) were applied on the linen fabric using simple layer-by-layer (L-B-L) techniques. After the treatment of the linen fabric, the authors evaluated different properties (antibacterial, antioxidant, and UV protective). As compared to the control linen fabric, the finished fabric showed antibacterial activity in the range of 75–97% and antioxidant property of around 58–96%. In addition, this fabric showed enhanced UV protective property (UPF > 22.5). This multifunctional effect was achieved due to the synergistic effect of the chitosan and green tea extract together. Specifically, the UV protection factor of the linen fabric was improved as a consequence of the barrier layer made by the L-B-L technique and the UV adsorption capability of the different polyphenolic chemicals (catechin, ferulic, etc.) present in the green tea extract. In another study, Teli et al. [42] studied the UV protection properties of wool fabric treated with coconut shell extract (CSE) biomolecules, as shown in Table 7.4. They concluded that the CSE finished wool fabric showed a higher UPF (~189.47) value than the untreated wool fabric (~20.38) which is due to the presence of different colouring components in the extract.

Table 7.4 UPF values of treated and untreated wool fabric [42]

Parameters	UVA	UVB	UPF (290-400 nm)
Untreated wool	14.11	3.64	20.38
Only CSE-A	1.55	0.80	116.61
Only CSE-B	0.94	0.51	189.47
Acid dye in CSE-A	1.10	0.59	165.81
Acid Dye in CSE-B	0.55	0.27	364.97

Pucciarini et al. [43] have observed natural colouration, antioxidant and UV protective properties of the wool fabric finished with onion skin extract biomolecules. Different antioxidant components in the onion skin extract help to reduce the oxidation and decomposition of the lipids present in the human skin. An experiment was conducted using 2-thiobarbituric acid (TBA) assay to prove the UV protection capability of the biomolecules. From this experiment, lipoperoxidation (LPO) was measured to quantify the lipid oxidation. Malonaldehydebis (dimethyl acetal) (MDA) is one of the byproducts of this photo-oxidative decomposition method. It was found that the LPO concentration was decreased in the treated wool sample compared to the untreated wool fabric. This was due to the presence of different antioxidant components in the onion skin extract. Iyer et al. [44] have reported UV-protective properties of the viscose fabric treated with vitamin 2 (Flavin and its coenzymes such as Riboflavin (RF) and Flavin Mononucleotide (FMN)). It was found that this biomolecule-finished viscose fabric could provide multifunctional properties such as natural colouration, photoluminescence,

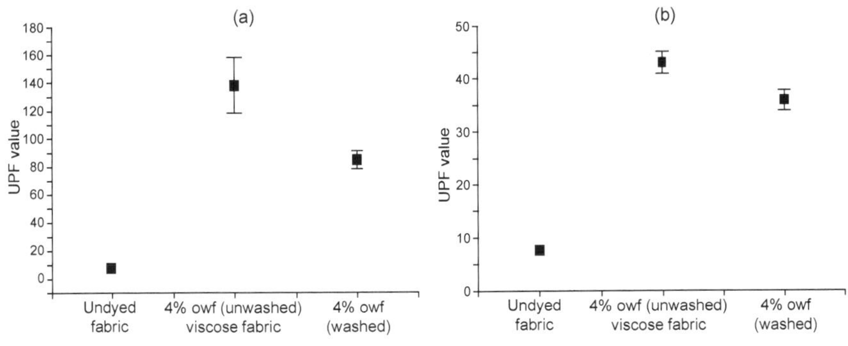

Figure 7.9 UPF factor of RF (a) and FMN (b) viscose dyed fabric samples [44].

and UV protection. Specifically, it was observed from their comparative results that UV blocking efficiency was improved much in the case of treated viscose fabric. The UPF value of the RF and FMN treated viscose fabric was reported as >50 and 35, respectively, while the value was only 5 for the control viscose fabric, as shown in Fig. 7.9.

In a study, Shukla et al. [23] have reported the flame-retardant property of the cellulosic fabric (cotton), where the authors finished the cotton fabric using sodium lignin sulfonate. Besides fire retardancy, the treated fabric showed various other functionalities such as antibacterial and UV protection effects. The finished fabric attained a higher UPF (> 50) value than the untreated fabric, as shown in Table 7.5. This is due to the presence of different high molecular weight phenolic compounds, carbonyl compounds, and other chromophore components in sodium lignin sulfonate.

Table 7.5 UV resistant parameters of the control and treated fabric [23]

Sample	UPF	Transmission (UV-A) (%)	Transmission (UV-B) (%)	Blocking (UV-A) (%)	Blocking (UV-B) (%)
Control	20.8	12.61	3.44	87.39	96.56
Treated	630.3	0.14	0.20	99.86	99.80

7.6 Mosquito repellent finish by natural materials

With the growing technological advancements, textile substrates have also started to provide much-needed characteristics other than their usual use, and many of the credits go to value addition by finishing. Mosquitoes are one of the threats to people, and about 3 million people die every year from mosquito bites. Due to increased public awareness about health and hygiene issues in the future, researchers have shown a deep interest in developing mosquito repellent textiles. As an effect of global warming, the birth rate of mosquitoes is rapidly increasing day by day, and as a result of viral infection, complicated diseases are also spreading rapidly in an alarming manner. Mosquito-borne diseases like malaria, dengue fever, chikungunya, Japanese encephalitis, and filariasis are the most commonly transmitted diseases in India. Mosquito repellants are insecticides used to kill or control mosquitoes and destroy them. Many products are available in various forms like spray, soap, oil, powder, cream repellant, coil, etc. DEET (N, N diethyl-3-methylbenzamide) is the most commonly used synthetic material which has been proven to obstruct a wide spectrum of insects, including mosquitoes. The main disadvantage of DEET is its high toxicity. Till now, many natural biomolecules have been identified for their repellency against mosquitoes, but nothing has proved as durable or effective as DEET. The most widely studied natural repellents are based on oils extracted from various plants. These natural ingredients effectively repel mosquitoes but often require frequent reapplication. Various natural oils like lemongrass oil, citronella oil, cinnamon oil, castor oil, eucalyptus oil, rosemary oil, clove oil, cedar oil, garlic oil, peppermint oil, etc. are the few best examples that can impart a very good mosquito repellency effect on textile substrates. Anitha et al. have reported mosquito repellent properties of microencapsulated lemongrass oil on polyester textiles [45]. Microencapsulation is a rapidly growing technique of micro-scale packaging of solids or liquids in the form of a thin coating applied to a substrate. Lemongrass is a traditional aromatic plant used in pharmacological applications. Lemongrass oil is a cyclic terpene and a natural source of vitamin A. The authors experimented with aqueous, methanolic, and their microencapsulated forms. Anopheles mosquitoes were selected for the repellency test, and the highest mosquito repellency activity was observed in the case of microencapsulated extracts. Among them, 92% repellency was reported in the case of microencapsulated aqueous extract, whereas the efficacy was 80% in the case of the methanolic process. They had also studied the washing durability up to 15 washes, and 84% repellency was achieved in the case of microencapsulated aqueous extract. Banupriya et al. have reported the mosquito repellency activity of rosemary extract-treated cotton fabric [46]. Rosemary powder was mixed with methanol overnight. Then the filtered extract was used as a herbal finishing agent. The treated fabric showed approximately 92% mosquito repellency. Cineole is the main component responsible for this activity. Sumithra et al. have elaborately expressed the mosquito repellency effect of a combination of Ricinuscommunis, Senna auriculata, and Euphorbiahcrita leaf extracts [47].

The cotton denim fabric was pad-dry-cured with a methanolic extract of leaf powder. Approximately 68% repellency was achieved for medicinal herb-treated fabric. After 30 industrial washes, the effect decreased to 50%. Mero Specos et al. have illustrated the mosquito repellent effect of microencapsulated citronella oil on cotton fabric [48]. Citronellol, geraniol, and other terpene contents are the main responsible active components for mosquito repellency action. Encapsulation is an important technique to provide shield-like protection against moisture or oxygen, and during rupture or damage, it releases the active components for proper action. They had studied this experiment on female A. aegypti species. They proved that fabric sprayed with citronella rendered protective action up to 16 days. Still, in the case of microencapsulated oil-treated fabric, it showed about 90% repellency effect for up to 21 days.

7.7 Fragrance finished textiles by natural materials

Recently, fragrance finishing has been a new trend and is growing rapidly in the emerging markets. Fabrics are treated with various techniques by different types of essential oils from aromatic plants. Natural aromas are used in textile substrates to create a sensorial environment, comfortable surroundings, or emotional entanglement. Different kinds of interiors or home textiles like wall hangings, bed sheets, curtains, carpets, and others are finished with natural fragrances to get rid of tiredness and weariness. Lavender, sandal, rose, champaka, lemongrass, palmarosa, etc. are mainly used for fragranced materials. Fragrances can be incorporated with the binder in the fabric for a long-lasting effect. Normally, the fragrance is evaluated by the sensorial evaluation technique by nail scratching on the surface of the fabric. Fragrance can be encapsulated into fabric by the microencapsulation process which produces an aroma as the finished fabric gets rubbed. Also, this encapsulation prevents the essential oil from moisture, oxygen, and light. Kumar et al. have reported a deodourising finish by applying lavender oil on cotton fabric using a nonionic binder [49]. It was found that the odour could last for 30 days when it got bound with a binder. Hipparagi et al. have reported aroma finishes on silk fabric [50]. A detailed comparison was made among the exhaust, pad-dry, and spraying methods to impart market available aromas. They found that the pad-dry method was the most efficient approach (with a minimum of 2% aroma content).

7.8 Wound healing textiles by natural materials

Health-related issues are promising areas of research for the survival of life. Different polymeric substances are used for various characteristics with special applications in the health sector. Generally, a wound means any kind of cut or injury in the tissues or internal systems of the body, which further can disrupt any functionality of human systems. Healing means the new generation of dermal or epidermal tissues. This can be done by a series of processes and can be affected by various parameters like body composition, age, previous disease record, any medicinal drug, and wound dimensions. World demand for wound healing products is increasing by 4.6 % from 2019 to 2024 and is expected to reach USD 24.8 billion [51]. Now, textiles are the emerging field in this application area. Various composite materials with porous, moisture-permeable properties can be incorporated for healing operations. Also, it can provide sufficient strength, cum extensibility, and flexural advancement. The wound dressing materials are composed of two layers, where the first layer provides adsorption and adaption as a scaffold for cell attachment and growth.

The second layer consists of a gel-forming agent, which helps to prevent cell adhesion. Cotton gauze dressings have some drawbacks, like microbial attack, bad adherence to the wound area, and improper permeability of air or gases. Nowadays, many branded dressing materials have properties like antimicrobial, biosynthetic, skin substitutive, non-adherent, super absorptive, etc. Textile-based wound dressing is important as it directly contacts the wound, promotes recovery, and prevents infections. Cotton, silk, linen, polyester, polypropylene, polyurethane, etc. are mainly used for wound dressing. Besides, chitosan, alginate, protein, etc. are also used. Hydrocolloids are incorporated with textile material to block the wound blood vessels. Gelatins, polysaccharides, and polyurethane gels are gel-forming agents for calcium ions. Calcium alginate gives a moist environment for better healing. Hydrogels are used for burning portion recovery. Dextran-dialdehyde, bovine-serum albumin, glycosaminoglycan, chitosan, and collagen are used for hydrogel formation. Cotton fabric can be modified by chitosan to interact with antibiotics for wound healing, and it depends on how much it gets modified. Gupta et al. have revealed chitosan as a wound-healing material [52]. Chitosan and chitin have properties like biocompatibility, hemostatic, and anti-inflectional activity that help the healing process efficiently [53].

Concluding remarks and future perspectives

This chapter represents a broad overview of sustainable textile finishing using biomolecules. The thought of making a sustainable fabric depends on what kind of biomaterials are used and the technology of application we adopt. Availability and extraction procedure for any selected natural materials and their application on the textile substrates are crucially handled by green technologists and textile researchers. Various factors can help to quantify the metabolites and compositions of an extract. The standard extraction process to obtain desired active ingredients is a vital and foremost task that varies from expert to expert, and this often causes differentiated end properties. The major tasks are to optimise integration techniques of a selected bio-extract into textiles and its add-on percent to get a particular functionality. The main drawbacks faced by finishing textiles with natural ingredients are the requirement for high chemical add-ons and poor fastness properties. These cause changes in the primary fabric properties along with the feel of the finished goods. From an overall standpoint, we can say that the implementation of green technology helps to reduce the pollution load and effluent treatment cost. The end products would be more user-friendly and toxicity-free. Apart from apparel usages, there are many application areas like defense, baggage sector, decorative or home furnishing and other outdoor use like making tents, protective clothes, etc. where textile substrates finished with green molecules can find enormous potential. The required functionalities can be imparted with the wise selection of different functional biomolecules.

References:

1. Senthil Kumar P, Gunasundari E (2018). Sustainable Wet Processing—An Alternative Source for Detoxifying Supply Chain in Textiles. In: Muthu, S. (eds) Detox Fashion. Textile Science and Clothing Technology. Springer, Singapore. https://doi.org/10.1007/978-981-10-4876-0_2
2. Saxena S, Raja ASM, Arputharaj A (2017). Challenges in Sustainable Wet Processing of Textiles. In: Muthu, S. (eds) Textiles and Clothing Sustainability. Textile Science and Clothing Technology. Springer, Singapore. https://doi.org/10.1007/978-981-10-2185-5_2
3. Joshi M, Ali SW, Purwar R, Rajendran S (2009), 'Ecofriendly antimicrobial finishing of textiles using bioactive agents based on natural products', *Indian J. Fibre Text. Res.*, 34 (3), 295–304.
4. Teli MD, Pandit P, Basak S (2018), 'Coconut shell extract imparting multifunction properties to ligno-cellulosic material', *J. Ind. Text.*, 47 (6), 1261–1290. https://doi.org/10.1177/1528083716686937.
5. El-Shafei A, Shaarawy S, Motawe FH, Refaei R (2018), 'Herbal extract as an ecofriendly antimicrobial finishing of cotton fabric', *Egypt. J. Chem.*, 61 (2), 317–327. https://doi.org/10.21608/EJCHEM.2018.2621.1209.
6. Saroj Y (2019), 'Assessment of antimicrobial activity of selected plant extracts for application on textiles', *Int. J. of Chem. Studies*, *7*(1), 33-36.
7. Das S, Das A, Bhavya T, Rama Nivashini S (2019), 'Molecular characterisation and antibacterial activity of Aloe barbadensis miller on textiles', *J. Text. Inst.*, 0 (0), 1–7. https://doi.org/10.1080/00405000.2019.1686268.
8. Ali SW, Purwar R, Joshi M, Rajendran S (2014), 'Antibacterial properties of Aloe vera gel-finished cotton fabric', *Cellulose*, 21 (3), 2063–2072. https://doi.org/10.1007/s10570-014-0175-9.
9. Teli MD, Pandit P (2018), 'A Novel Natural Source Sterculia foetida Fruit Shell Waste as Colorant and Ultraviolet Protection for Linen', *J. Nat. Fibers*, 15 (3), 337–343. https://doi.org/10.1080/15440478.2017.1328327.
10. Teli MD, Pandit P (2018), 'Application of Sterculia Foetida Fruit Shell Waste Biomolecules on Silk for Aesthetic and Wellness Properties', *Fibers Polym.*, 19 (1), 41–54. https://doi.org/10.1007/s12221-018-7315-4.
11. Prabhu KH, Teli MD (2014), 'Eco-dyeing using Tamarindus indica L. seed coat tannin as a natural mordant for textiles with antibacterial activity', *J. Saudi Chem. Soc.*, 18 (6), 864–872. https://doi.org/10.1016/j.jscs.2011.10.014.

12. Sheikh J, Bramhecha I (2018), 'Multifunctional modification of linen fabric using chitosan-based formulations', *Int. J. Biol. Macromol.*, 118, 896–902. https://doi.org/10.1016/j.ijbiomac.2018.06.150.

13. Saini S, Gupta A, Singh N, Sheikh J (2019), 'Functionalization of linen fabric using layer by layer treatment with chitosan and green tea extract', *J. Ind. Eng. Chem.*, *82*, 138-143. https://doi.org/10.1016/j.jiec.2019.10.005.

14. Joshi M, Ali SW, Rajendran S (2007), 'Antibacterial finishing of polyester/cotton blend fabrics using neem (Azadirachta indica): A natural bioactive agent', *J. Appl. Polym. Sci.*, 106 (2), 793–800. https://doi.org/10.1002/app.26323.

15. Vaideki K, Jayakumar S, Rajendran R, Thilagavathi G (2008), 'Investigation on the effect of RF air plasma and neem leaf extract treatment on the surface modification and antimicrobial activity of cotton fabric', *Appl. Surf. Sci.*, 254 (8), 2472–2478. https://doi.org/10.1016/j.apsusc.2007.09.088.

16. Murugesh Babu K, Ravindra KB (2015), 'Bioactive antimicrobial agents for finishing of textiles for health care products', *J. Text. Inst.*, 106 (7), 706–717. https://doi.org/10.1080/00405000.2014.936670.

17. Balamurugan GK, Babuskin S (2018), 'Antimicrobial coating of cotton twill tape with neem oil, eucalyptus oil and tulsi oil for medicinal application'. *Curr. Sci.*, 115 (4), 779–782. https://doi.org/10.18520/cs/v115/i4/779-782.

18. Ravindra KB, Murugesh Babu K (2016), 'Study of Antimicrobial Properties of Fabrics Treated with Ocimum Sanctum L (tulsi) Extract as a Natural Active Agent', *J. Nat. Fibers*, 13 (5), 619–627. https://doi.org/10.1080/15440478.2015.1093577.

19. Silva MG da, Barros MASD d, Almeida RTR de, Pilau EJ, Pinto E, Soares G, Santos JG (2018), 'Cleaner production of antimicrobial and anti-UV cotton materials through dyeing with eucalyptus leaves extract', *J. Clean. Prod.*, 199, 807–816. https://doi.org/10.1016/j.jclepro.2018.07.221.

20. Basak S, Ali SW (2016), 'Sustainable fire retardancy of textiles using bio-macromolecules', *Polym. Degrad. Stab.*, 133, 47–64. https://doi.org/10.1016/j.polymdegradstab.2016.07.019.

21. Horrocks AR (2011), 'Flame retardant challenges for textiles and fibres: New chemistry versus innovatory solutions', *Polym. Degrad. Stab.*, 96 (3), 377–392. https://doi.org/10.1016/j.polymdegradstab.2010.03.036.

22. Costes L, Aguedo M, Brison L, Brohez S, Richel A, Laoutid F (2018), 'Lignin fractionation as an efficient route for enhancing Polylactide thermal stability and flame retardancy', *Flame Retard. Therm. Stab. Mater.*, 1 (1), 14–24. https://doi.org/10.1515/flret-2018-0002.

23. Shukla A, Sharma V, Basak S, Ali SW (2019), 'Sodium lignin sulfonate: a bio-macromolecule for making fire retardant cotton fabric', *Cellulose*, 26 (13–14), 8191–8208. https://doi.org/10.1007/s10570-019-02668-7.

24. Basak S, Samanta KK, Chattopadhyay SK, Narkar RS, Mahangade R (2015), 'Flame retardant cellulosic textile using bannana pseudostem sap', *Int. J. Cloth. Sci. Technol.*, 27 (2), 247–261. https://doi.org/10.1108/IJCST-12-2013-0135.

25. Basak S, Samanta, KK, Saxena, S, Chattopadhyay, SK, Parmar MS (2017), 'Self -extinguishable cellulosic textile from spinacia oleracea', *Indian J. Fibre Text. Res.*, 42 (2), 215–222.

26. Basak S, Samanta KK, Chattopadhyay SK (2015), 'Fire retardant property of cotton fabric treated with herbal extract', *J. Text. Inst.*, 106 (12), 1338–1347. https://doi.org/10.1080/00405000.2014.995456.

27. Basak S, Ali SW (2017), 'Leveraging flame retardant efficacy of pomegranate rind extract, a novel biomolecule, on ligno-cellulosic materials', *Polym. Degrad. Stab.*, 144, 83–92. https://doi.org/10.1016/j.polymdegradstab.2017.07.025.

28. Basak S, Wazed Ali S (2019), 'Wastage pomegranate rind extracts (PRE): a one step green solution for bioactive and naturally dyed cotton substrate with special emphasis on its fire protection efficacy', *Cellulose*, 26 (5), 3601–3623. https://doi.org/10.1007/s10570-019-02327-x.

29. Teli MD, Pandit P (2018), 'Development of thermally stable and hygienic colored cotton fabric made by treatment with natural coconut shell extract', *J. of Ind. Text*, *48*(1), 87-118. https://doi.org/10.1177/1528083717725113.

30. Fang F, Zhang X, Meng Y, Gu Z, Bao C, Ding X, Li S, Chen X, Tian X (2015), 'Intumescent flame retardant coatings on cotton fabric of chitosan and ammonium polyphosphate via layer-by-layer assembly', *Surf. Coatings Technol.*, 262, 9–14. https://doi.org/10.1016/j.surfcoat.2014.11.011.

31. Carosio F, Di Blasio A, Cuttica F, Alongi J, Malucelli G (2014), 'Flame retardancy of polyester and polyester-cotton blends treated with caseins', *Ind. Eng. Chem. Res.*, 53 (10), 3917–3923. https://doi.org/10.1021/ie404089t.

32. Bosco F, Carletto RA, Alongi J, Marmo L, Di Blasio A, Malucelli G (2013), 'Thermal stability and flame resistance of cotton fabrics treated with whey proteins', *Carbohydr. Polym.*, 94 (1), 372–377. https://doi.org/10.1016/j.carbpol.2012.12.075.

33. Bosco F, Casale A, Mollea C, Terlizzi ME, Gribaudo G, Alongi J, Malucelli G (2015), 'DNA coatings on cotton fabrics: Effect of molecular size and pH on flame retardancy', *Surf. Coatings Technol.*, 272, 86–95. https://doi.org/10.1016/j.surfcoat.2015.04.019.

34. Alongi J, Carletto RA, Bosco F, Carosio F, Di Blasio A, Cuttica F, Antonucci V, Giordano M, Malucelli G (2014), 'Caseins and hydrophobins as novel green flame retardants for cotton fabrics', *Polym. Degrad. Stab.*, 99 (1), 111–117. https://doi.org/10.1016/j.polymdegradstab.2013.11.016.

35. Wang X, Lu C, Chen C (2014), 'Effect of chicken-feather protein-based flame retardant on flame retarding performance of cotton fabric', *J. Appl. Polym. Sci.*, 131 (15), 1–8. https://doi.org/10.1002/app.40584.

36. Ablett E, Whiteman DC, Boyle GM, Green AC, Parsons PG (2003), 'Induction of metallothionein in human skin by routine exposure to sunlight: Evidence for a systemic response and enhanced induction at certain body sites', *J. Invest. Dermatol.*, 120 (2), 318–324. https://doi.org/10.1046/j.1523-1747.2003.12025.x.

37. Teng C, Yu M (2003), 'Preparation and property of poly(ethylene terephthalate) fibers providing ultraviolet radiation protection', *J. Appl. Polym. Sci.*, 88 (5), 1180–1185. https://doi.org/10.1002/app.11773.

38. Kim S (2006), 'Dyeing characteristics and UV protection property of green tea dyed cotton fabrics', *Fibers Polym.*, 7 (3), 255–261. https://doi.org/10.1007/bf02875682.

39. Nattadon Rungruangkitkrail NT, Mongkholrattanasit R (2012), 'Uv Protection Properties of Wool Fabric Dyed With Eucalyptus Leaf Extract By the Padding Techniques'. *In RMUTP Int. Conf. Text. Fash. (Bangkok, 2012).*

40. Gies HP, Roy CR, Holmes G, McKinlay AF, Repacholi MH (2000), 'Ultraviolet radiation protection by clothing: Comparison of in vivo and in vitro measurements', *Radiat. Prot. Dosimetry*, 91 (1–3), 247–250. https://doi.org/10.1093/oxfordjournals.rpd.a033210.

41. Wilson CA, Gies PH, Niven BE, Mclennan A, Bevin NK (2008), 'The Relationship Between UV Transmittance and Color — Visual Description and Instrumental Measurement', *Text. Res. J.*, 78 (2), 128–137. https://doi.org/10.1177/0040517507081302.

42. Teli MD, Pandit P (2017), 'Novel Method of Ecofriendly Single Bath Dyeing and Functional Finishing of Wool Protein with Coconut Shell Extract Biomolecules', *ACS Sustain. Chem. Eng.*, 5 (9), 8323–8333. https://doi.org/10.1021/acssuschemeng.7b02078.

43. Pucciarini L, Ianni F, Petesse V, Pellati F, Brighenti V, Volpi C, Gargaro M, Natalini B, Clementi C, Sardella R (2019), 'Onion (Allium cepa L.) Skin: A rich resource of biomolecules for the sustainable production of colored biofunctional textiles', *Molecules*, 24 (3), 1–18. https://doi.org/10.3390/molecules24030634.

44. Iyer SN, Behary N, Nierstrasz V, Guan J, Chen G (2019), 'Study of photoluminescence property on cellulosic fabric using multifunctional biomaterials riboflavin and its derivative Flavin mononucleotide', *Sci. Rep.*, 9 (1), 1–16. https://doi.org/10.1038/s41598-019-45021-5.

45. Anitha R, Ramachandran T, Rajendran R, Mahalakshmi M (2011), 'Bio Physics Microencapsulation of lemon grass oil for mosquito repellent finishes in polyester textiles', *Elixir Bio Phys.*, 40, 5196–5200.

46. Banupriya J, Masheshwari V (2013), 'Effects of Mosquito Repellent finishes by Herbal Method on Textiles', *Int. J. Pharm. Life Sci.*, 4 (11), 3133–3134.

47. Sumithra M, Vasugi Raja N. (2012), 'Mosquito repellency finishes in blended denim fabrics', *Int. J. Pharm. Life Sci.*, 3 (4), 1614–1616.

48. Specos MMM, García JJ, Tornesello J, Marino P, Vecchia DM, Tesoriero DMV, Hermida LG (2010), 'Microencapsulated citronella oil for mosquito repellent finishing of cotton textiles', *Trans. R. Soc. Trop. Med. Hyg.*, 104 (10), 653–658. https://doi.org/10.1016/j.trstmh.2010.06.004.

49. Vasanth Kumar D, Boopathi N, Karthick N, Ramesh P (2012), 'Aesthetic Finishing for Home Textile Materials', *Int. J. Sci.*, 1 , 5–9. https://doi.org/10.5923/j.textile.20120103.01.

50. Hipparagi SA, Srinivasa T, Das B, Naik SV, Purushotham SP (2016), 'Studies on Application of Aroma Finish on Silk Fabric', *J. Inst. Eng. Ser. E*, 97 (2), 159–165. https://doi.org/10.1007/s40034-016-0082-8.

51. https://www.marketsandmarkets.com/Market-Reports/wound-care-market-371.html (accessed February, 2020).

52. Gupta B, Agarwal R, Alam MS (2010), 'Textile-based smart wound dressings', *Indian J. Fibre Text. Res.*, 35 (2), 174–187.

53. Moeini A, Pedram P, Makvandi P, Malinconico M, d'Ayala GG (2020), 'Wound healing and antimicrobial effect of active secondary metabolites in chitosan-based wound dressings: A review', *Carbohydr. Polym*, *233*, 115839. https://doi.org/10.1016/j.carbpol.2020.115839.

8

Replacement of harmful chemicals and Recycling/Reuse concepts for textile processing

V.D. Gotmare* S.S Kole

Textile Manufactures Department, V.J.T.I. Matunga, Mumbai-400 019, India

*Corresponding Author, Email: vdgotmare@tx.vjti.ac.in

Abstract: Textile processing industries have a strong environmental effect because most of the existing activities are unsustainable, and all stakeholders are looking at the solution to reduce the carbon footprint of textiles. To make textile processing greener, many eco-friendly steps have been taken. These consist of the replacement of harmful chemicals with their eco-friendly substitutes, use of greener pre-treatment chemicals, green dyes and textile auxiliaries along with greener solvents, optimised and efficient processing, bio-processing, recycling of textiles, water and chemicals. The chapter discusses the need of recycling and reuse of valuable textile materials and the effect of harmful chemicals on the environment and human beings, besides the replacement of PCB, HCHO, heavy metals and banned amines used in textile and garment processing. In order to make the processes more sustainable, the use of best available techniques (BAT) in the recycling of dyes and chemicals, such as the standing bath technique of dye bath replenishment, salt-free dyeing, utilisation of spent dyes, and supercritical CO_2 is also highlighted.

8.1 Introduction

The textile industry is considered as the most complicated and most extended manufacturing industry amongst the industrial chains, which consumes a high volume of natural resources starting from farming, and finishing up to garments which are dominated by small and medium enterprises. It is reported that the textile industry consumes around 2500 chemicals based on both natural and synthetic origin in the value addition in the textile products.

The textile industry is the backbone and has become an important element in building the nation's economy in the global market. Still, it is responsible for destroying the environment due to the use of toxic and harmful materials which includes pesticides during the cultivation of plants and toxic emissions during the production of synthetic fibres. There are some other processes where varieties of chemicals are used in the process of conversion of fibre to the finished fabric.

Various classes of toxic chemicals are used nowadays for enhancing the aesthetic and functional properties of textile materials. These chemicals create problems when exposed to the environment from different stages of wet processing.

There is a need to replace harmful dyes, auxiliaries and textile chemicals and use eco-friendly chemicals and process technology that is energy efficient, which produces less waste, and requires fewer resources such as chemicals and water. Lastly, it should be easy to operate. Technologists need to find a solution for recycling dyes, auxiliaries, and fibrous waste materials from the entire textile chain and produce an innovative solution to save the industries and the planet.

The recent development in the utilisation of "Green Materials" and "Green Processes" in the entire textile chain is discussed in this chapter. The ecological and economic restrictions that have become essential to the textile industry have led to its search for an alternative eco-friendly process that is sustainable are also outlined.

8.2 Harmful chemicals and their greener substitute in textile preparatory

Pretreatments of grey fabric are essential in order to make fabric free from all natural and added impurities so that it will be ready for value addition treatments. The pretreatment is required for almost all types of fabrics before the next process, such as dyeing, printing, or finishing. The pretreatment process also helps in removing hydrophobic impurities from the textile substrate like waxes, proteins from cotton, and vegetable impurities from wool. Textile industries were using harmful materials in sizing, desizing, scouring, bleaching and mercerization, posing a problem for textile workers and the environment. To develop innovative preparatory processes, it is required to understand the consumption of energy, labour costs, pollution control strategy and regulations and water requirements for environment-friendly textiles production [1,2].

8.2.1 Conventional sizing and desizing processes and their green substitute

Sizing is an important process in the warp yarn preparation for satisfactory performance in weaving operation. To protect the yarn from various stresses subjected during the weaving, many ingredients starting from size like starch,

PVA, CMC, acrylic-based up to antiseptic, defoamers and lubricants are used, which provides a uniform layer around the yarn. The selection of sizing agents is based on fabric type, eco-friendliness, desizing consideration and environmental considerations [3]. PVA creates ecological problems since it is a class of refractory biodegradable organic molecules. The use of exisiting chemicals and their alternatives are mentioned in the Table 8.1.

Generally, sizing is done with three agents:

- Polysaccharides, consisting of starch and its derivatives. This is the popularly used sizing agent. Starch can be obtained from various natural sources like corn and potatoes. Starch derivatives include dextrin, starch esters and starch ethers.
- Synthetic vinyl polymers such as polyvinyl alcohol (PV-OH), prepared from polyvinyl acetate, are used in many textile industries.
- Protein-based sizes derived from gelatin and casein. Animal proteins, such as casein or bone or skin glue, are mixed with softening additives such as glycerol and castor oil and used as sizing agents.

All types of grey substrates containing size material need to undergo the desizing process. The size paste is uniformly applied on the yarn to improve the weaveability of the yarn which sustains the various stress and strain during the high-speed weaving and facilitates better weaving efficiency. Still, for the subsequent process, the size has to be removed to facilitate further dyeing and finishing. Following are some of the methods of desizing depending upon the type of size used in the process:

- Acid desizing, in which dilute acids are used, and starch and other sizing materials are hydrolysed.
- Enzyme desizing is a biodegradation method, that converts starch and other materials into a soluble form that can be washed off during subsequent washes.
- Oxidative desizing in which oxidizing agents such as hydrogen peroxide or persulphates are is used.

Table 8.1 Existing sizing agents, desizing agents, other preservatives and their substitutes

Current Chemicals	Usages	Alternative Substitutes
Polyvinyl alcohol (PVA)	Yarn size	Potato starch or carboxy methyl cellulose (CMC)
Pentachlorophenol, formaldehyde	Size preservative	Sodium silicofluride
Acid Desizing	Desizing	Enzymatic desizing

8.2.1.1 Green sizing agents

The word green size refers to environmental protection from the preparation of size paste up to the application on warp yarns, and also, it should not cause water pollution when drained during desizing. The green sizing operation involves the use of biodegradable sizing material which helps in the reduction of water pollution. Currently, the materials like modified starch and acrylates are used for the formulation of green size, which enables ease of the desizing process because of their biodegradable nature [4].

The grafted starch can find a substitute for PVA in the preparation of sizing formulations for polyester /cotton blended on a large scale. The acrylic-based sizing agents are also used, in recent practice, in association with PVA for environmental protection [5].

Polyester sizing agent consisting of ester (-COO-) and a hydrophilic group is a new class of textile size. The presence of strong hydrophilic groups shows excellent adhesion to the polyester fibre with improved water solubility, which facilitates excellent desizing performance [6]. Inorganic nanometer materials and natural plant gum sizing agents have been developed by researchers to retain the strength of size paste after impregnation [7].

Atmospheric pressure plasma process based on He/O_2 recipes has been developed for evaluation of sizing property and desizing efficiency. In the study, it was found that the desizing efficiency of the plasma-based process is much superior compared to traditional sizing technology with the use of PVA [8].

8.2.1.2 Green desizing agents/process

In past, the acid desizing method was adopted before the suitable development of enzymes, and it was responsible for varieties of environmental problems.

The enzymes are available as alternative eco-friendly desizing agents; enzymes like amylases based on various sources are used to eliminate starch.

To remove starch-based size from the grey fabric, amylases are commonly used in wet processing. Dextrin, maltose and short-chain sugars are formed from the catalytic breakdown of starch by amylase. The desizing efficiency of this amylase is superior because it only acts on the starch. It works at optimum pH of 5.5 – 6.5 and a low temperature (30-60°C) [9]. The enzyme process offers benefits with no use of chemicals and complete biodegradability, which results in a reduction in effluent load with the elimination of harmful chemicals, such as mineral acid and others [10-11].

8.2.2 Conventional scouring and bleaching agents and their green substitutes

The partial or complete removal of the non-cellulosic parts from the native cotton along with added impurities of hydrophobic nature is done by a process called as scouring; traditionally, the use of various chemicals such as strong alkali like caustic soda and anionic detergent supported with rinsing operation is involved in the fulfilling the objective of scouring. The effluent from the scouring process contains a high volume of salts, acids, and alkali. The alkaline condition in the scouring process leads to loss of cellulose, resulting in a reduction in weight of fabric with considerable effect on fabric appearance. Additionally, the presence of such chemicals is responsible for increasing the level of TDS in wastewater since these harmful chemicals and auxiliaries show a high value of BOD to COD ratio [12-14].

Bleaching is a process designed for the production of white goods without damaging the aesthetic appearance of fabric after the process. The most common bleaching process involves the oxidation method, and typical oxidizing chemicals are sodium/calcium hypochlorite, hydrogen peroxide and sodium chlorite. The natural fibres, like cellulosic and protein fibres in which peroxide bleaching is commonly practiced in the industry because it is non-toxic, odourless and universally accepted as environment-friendly. H_2O_2 has lower pollution problems compared to those associated with chlorine-based bleaching agents. Chlorine-based bleaching agents are banned since they produce adsorbable organohalides (AOXs). During chlorine bleaching, chlorinated hydrocarbons are formed which increase the BOD and COD level of effluent and cause severe environmental problems [15]. The comparative analysis between the conventional pretreatment processes with enzyme has been studied, and it was found that use of enzyme shows better performance with respect to fabric handle and aesthetic sense than sodium hydroxide process where it creates additional effluent load with harsh fabric handle [16].

8.2.2.1 Bio-Scouring and eco-friendly bleaching as green alternatives

Enzymatic scouring (Bio-scouring)

Scouring refers to the process of removal of hydrophobic impurities from the cotton fabric. Pectinase and cellulase are the commonly used enzymes as far as the bio-scouring process is concerned. The pectinase removes the interaction between cuticles and the body of fibre by digesting the pectin, whereas cellulase abolishes cuticle by dividing the cellulose primary wall. The enzymatic process contributes low value of BOD/COD and it also requires a low working temperature as compared to alkaline scouring. It also results in the production of softer fabric than alkaline scouring. Enzymatic scouring is found to be more successful, and it does not affect the fabric or environment adversely. The key advantages of this process are that it consumes low energy and water, as it is conducted at mild pH and lower temperature compared to conventional scouring processes [17].

Enzymatic bleaching

The bleaching process is carried out to decolourise natural pigments and confer a pure white appearance to the cotton fabric [18-19]. The conventional bleaching process uses most of the oxidizing agents under alkaline conditions which create environmental problems; nowadays, H_2O_2 is extensively used the universal bleaching agent which is considered the most eco-friendly process. The alkaline bleaching process requires a huge quantity of water to wash the fabric. There is also a loss of cellulose due to radical reactions between bleaching agents and the fibre that may be responsible for decreasing the degree of polymerisation. Hence, replacing hydrogen peroxide with enzymatic bleaching allows for achieving quality product due to minimum loss of fibre with a considerable reduction in the water required for the removal of hydrogen peroxide. The enzymes such as amyloglucosidases, pectinase, and glucose oxidase are considered to be compatible with their active pH and temperature range for an effective bleaching process [20]. It is reported that the enhancement in the bleaching efficiency of cotton is achieved by using laccases enzyme, which triggers a reaction in low concentrations at low temperature with a shorter cycle time which helps to enhance fabric whiteness. [21,22]. The ultrasound-assisted laccase treatment for cotton bleaching shows an improved whiteness index without increasing the effluent load. It is found that with a low supply of ultrasound energy (7W), the bleaching performance of laccase on cotton fabrics is enhanced. Catalase enzyme is used to break down H_2O_2 bleaching liquor into water molecules and less reactive gaseous oxygen. The enzymatic process helps in cleaner contaminated water, lowers water consumption and assists in a reduction of time and energy as compared to conventional practices. The use of traditional materials and their possible substitutes in the preparatory process is shown in Table 8.2.

Table 8.2 List of conventional and alternative chemicals and auxiliaries in wet processing

Preparatory process	Objective of Process	Conventional Chemicals	Sustainable Chemicals
Desizing and Scouring	To remove size and the hydrophobic impurities from the grey fabric.	Alkaline and Acidic Compounds	Hydrogen peroxide and enzymes
		alkylphenol ethoxylates, TSP, NaOH	Fatty alcohol ethoxylates, Sodium carbonate
		Alkyl benzene sulphonates	Fatty alkyl sulphates, Polyglycolether
		NTA, EDTA	Zeolites (sodium aluminium Silicate)
Bleaching	To remove the coloured impurities from the grey fabric.	Reductive and Sulphur bleaches	Peracetic acid, potassium permanganate
		Chlorine-based compounds	Peroxide bleaches

8.2.3 Conventional mercerization and green alternatives

The success of the finishing process, especially in cotton, depends on the degree of mercerization. Mercerization treatments consist of the use of concentrated solutions of NaOH of about 20-22% at a low temperature (5-18°C). The demand for NaOH is high since NaOH is commonly used to accomplish other processes like desizing and scouring. It is needed to handle the concentrated NaOH; improper treatment of NaOH can lead to poor finishing and a decrease in the mechanical properties of the fabric. Also, the effluents of these operations are toxic and hazardous to the ecosystem [23].

It is reported that, six alternative mercerizing agents, namely liquid NH_3, NH_4OH, $(NH_4)_2C_2O_4$, CH_3CH_2OH, CH_3COOH, $(COOH)_2$ can be used for the mercerization of bleached fabric; the alternatives were found to be safe, and they improve the mercerization efficiency with the requirement of minimum energy. The pH of the alternative agents after the mercerizing process is considered as eco-friendly as compared to highly alkaline conditions used in the conventional method. The mentioned chemicals are reliable as these are cheaper with improved on dyeing and water imbibition properties of cotton/polyester blended fabric [24]. It is recommended to use alternative chemicals in the pre-wet processing to maintain sustainability in the process which is beneficial for both environment and consumer.

Ultrasonic-assisted mercerization method of the cotton fabric has been proposed and analysed using a statistical model based on response surface methodology [25]. It is reported that the process of mercerization through ultrasonic energy is superior and supports the eco-friendly mercerizing process with no use of water. The model also reports, that the optimum parameters to get maximum mercerization efficiency are at alkali concentration of mercerizing bath 17 % and a higher tenacity value at 23 %. The process can be used on a commercial scale as a more efficient process and alternative to the conventional mercerization process.

8.3 Green dyes and green processes in textile colouration

Currently, all synthetic dyes that have been used in the dyeing of textile goods mostly come under the class of soluble, disperse, and pigments dye. These dyes are not eco-friendly because of their non-degradable nature. To remove these toxic compounds from the aquatic environment, the industry is taking additional steps that are expensive. The dyeing of fabric via naturally derived dyes is considered as one of the best alternative techniques to protect the environment [26].

The active ingredient in the commercial dyes is typically ranging from 20 to 80%. Heavy metals like copper, cobalt, mercury, chromium and nickel are present in synthetic dyes that may be harmful to humans and the environment, especially with dyes containing metal-based impurities [27].

In the natural dyes, azo groups (-N=N-) are not available. In most of the synthetic dyes, the azo groups incorporated in benzenoids compounds. The structures of carcinogenic amines are given in Fig. 8.1. The dyes belonging to the azo group share approximately 70% of the total dye consumption in textile industries. A considerable amount of amine is released from less than 4.0% known azo dye structures. The azo groups in the presence of sodium dithionite form two amine groups under a given reductive condition, as shown below.

$$\text{A-N=N-B} \xrightarrow{Na_2S_2O_4} \text{A-NH}_2 + \text{B-NH}_2$$

Azo groups containing a small number of aromatic amines (listed in the list of banned amines) are considered to be potentially carcinogenic and harmful to humans. Twenty-two such aromatic amines are reported to be harmful and banned, according to EU Directive 2002/61/EC and the number is increasing day by day.

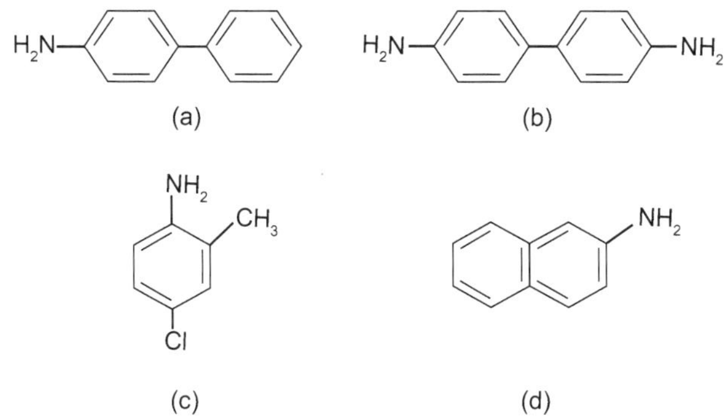

Figure 8.1 Chemical structure of four carcinogenic amines: (a) 4-aminodiphenyl; (b) benzidine; (c) 4-chloro-o-toluidine; and (d) 2-naphthylamine.

Textile colouration is the process involving the maximum effluent generation and maximum use of water. There are useful research and developments in this area to make the dyeing process greener and more sustainable [28-32]. Numerous auxiliary chemicals are added to the dyebath during the dyeing processes. To enhance the effects of other chemicals, some speciality chemicals have been developed. In other cases, the developed chemicals show side effects detrimental to the overall process. For example, to ensure proper penetration of chemicals wetting agents are often added for the preparation of dyeing steps, with few exceptions in the process of printing. The emissions are originating from dyeing and printing. The presence of metals and other auxiliary is responsible for originating aquatic toxicity. Apart from this, the basic chemicals, the presence of alkali, salts, reducing and oxidizing agents, and residual contaminants of dye on the fibre also add to damage to the aquatic environment.

The new technology should be designed in such a way that it should consume minimum water and energy. No use of harmful and restricted materials in the process is the key to achieving sustainability in wet processing. The dyestuff industry is continuously searching for new alternative colouring substances, particularly those which can be reduced to safer amino groups with reference to the guideline (Directive 2002/61/EC) due to restrictions on the use of azo dyes. While designing new technology, the emphasis should be given to the synthesis of dyes which will have a minimal impact on the environment with a reduction in the consumption of energy and raw materials and decreased waste. The use of biotechnology in industrial processes helps to maintain sustainability in the process with the use of biological systems which are more environment-friendly. Bio-dyes or natural dyes show a long-lasting effect and are environment-friendly with sufficiently fast colours. As a part of a bio-dye synthesis, the fungi or enzymes with precursors are utilised in the synthesis processes [33]. Some processes that can be used for making the conventional dyeing processes eco-friendly are as follows:

1. Pretreatment of cotton with poly(vinylamine chloride) for salt-free dyeing with reactive dyes [34]
2. Supercritical dyeing of PET with disperse dyes as an alternative for the complete elimination of water
3. Combining pretreatment and dyeing operations [35]

Table 8.3 Dyes, Chemicals and Auxiliaries and their ecofriendly alternative used in textile colouration.

Dyes, chemicals and auxiliaries	Ecofriendly alternatives
Benzidine-based dyestuffs and other banned amine releasing dyes	Mineral/pigment dyes, Natural dyes, Azo dyes not based on banned amines/not releasing banned amines
Dichromate used for oxidation of vat and sulphur dyes	Peroxide, air oxygen, metal-free agents
The acetic acid in the dyeing bath	Formic acid
Dispersants for dyes and chemicals	Water-based system
Copper sulphate used to treat direct dyes	Polymeric compounds
Dye powder in automatic injection	Liquid dyes
Sodium hydrosulphites stabilised	Sodium hydrosulphite
Formaldehyde and toxic metallic salts used as auxiliaries	Higher molecular weight polymeric compounds
Sodium sulphide	Glucose-based reducing agents
Dichlorobenzene/Trichlorobenzene	Butyl benzoate

8.4 Choice of natural dyes as green colourants

The use of bioresources-based colourants and non-toxic chemicals, in recent days, created a public awareness for health, eco-safety and eco-preservation, creating a significant revolution in textile research and development [38–42]. In western countries, the development of cleaner production strategies for value addition with a minimum cost and minimum damage to the aquatic environment to exploit high technical skills for the development of high-quality textile materials was observed [43]. The naturally obtained materials are nowadays given a top priority over the materials based on a synthetic source. In the past, painters used the dyes extracted from minerals, and plants for paintings. The natural dye in association with selected mordant helps in the cohesion. Natural dyes were also used in the pharmaceutical, food and cosmetic industries [36,37].

The extraction and application of natural dyes in the development of eco-textiles are interesting for creating a sustainable benchmark for establishing eco-friendly processes [44]. The current strict ecological legislation in European Union, USA and India has become a driving factor in restricting the use of synthetic chemicals in the textile value chain [45]. As a result, a green chemistry approach through the utility of eco-friendly and naturally occurring bio-colourants as a subsequent alternative to synthetic chemicals, has widespread applications in textile colouration and other biomedical aspects.

Based on the production sources, natural dyes can be classified into:

(a) Plant/Vegetable Origin - The natural dyes extracted from the plants created an opportunity to use in the textile process because of their non-toxic nature and equivalent performance with respect to synthetic dyes. The naturally occurring colour from various sources such as seeds, flowers, leaves, fruits, trunks, barks, and roots falls in this class of dyes and pigments. The natural dye's yield is quite low, and it has become a challenge in the textile industry to improve the same.

(b) Animal/Insect Origin - Red animal dyes are obtained from insect's dried bodies, namely, Laccifer lacca/Kerria lacca, Kermes, Cochineal and molluscs like carminic acid (cochineal), laccaic acid (Lac dye), kermesic acid (Kermes). In ancient times, Tyrian purple belonged to the current category known for dyeing purposes.

(c) Mineral Origin - Several pigments from metal oxides and inorganic metal salts belong to the current category. The red pigments originating from minerals are Red-Ochre, Cinnabar and Realgar.

The collaborations between scientists and designers focus on creating the foundation (green strategies) of many sustainable design projects that help to develop technological innovation and breakthroughs in the textile industry [46]. Many biological approaches are used in the production of microbiologically-assisted coloured compounds for textile applications [47,48]. Different grades of colours with different shades are produced from nature [49].

Latest research from at various organizations is being developed recently, many coloured products obtained by incorporating microbes are used as dyes for textiles [50]. Indigo, one of the oldest natural dyes, has been indigenously isolated with the naphthalene-degrading strain Pseudomonas sp. HOB1 producing a blue pigment [51]. Anthraquinone-based dyes have been extracted from a many of fungi such as *Curvularia, Trichoderma,*

Drechslera and Aspergillus. These types of techniques are helpful because they need inexpensive chemicals, and the entire process is performed at room temperature with neutral pH and avoids the use of non-eco-friendly strong acids and alkalis in the chemical synthesis [52]. Actinomycetes culture of Streptomyces coelicolor, capable of creating a red pigment that changes to blue with the culture ageing, has also been produced for dyeing silk and wool [53]. Also, Serratia marcescens and subspecies marcescens strains capable of creating innovative rose red pigment have been used for dyeing of protein fibres [54]. With the current nature and global focus of the eco-friendly system, there has been a great deal of interest in using herbal dyes and eco-friendly fibres [55].

8.5 Green textile auxiliary in dyeing, printing & finishing of textiles

The harmful and toxic chemicals in the form of surfactants, lubricants, solubilising agents, cleaning agents, and deformers are typically used in textile wet processing. The formaldehyde-based products, generally, are not eco-friendly as formaldehyde is toxic, carcinogenic, mutagenic and alarm safety concerns for customers and operators of textile wet processing. The effluents of textile industries comprising the above-mentioned unused chemicals pollute the ecosystem [56]. The principle of green chemistry to be adopted during manufacturing textile auxiliaries and solvents [57]:

(a) The textile auxiliaries to be designed to enhance the production
(b) Use renewable feedstock instead of non-renewable fossil fuels
(c) A safe method for auxiliaries and deemed synthesis
(d) Adoption of energy-efficient process
(e) Adoption of a single step or direct reaction to avoid the use of hazardous chemical
(f) Use of biodegradable product so that they should not accumulates and contaminates the environment.

8.5.1 Hazardous textile chemicals and auxiliaries and their replacement in dyeing

The production of sustainable textile chemicals and auxiliaries is essential for maintaining the human, process and environmental eco-system. A list of hazardous textile chemicals and auxiliaries utlised in the textile industry and their alternatives are given in Table 8.3 [58].

The production of green textile chemicals and related auxiliaries is important for maintaining sustainability in the process. The availability of a reducing agent based on sugar is also a green option for frequently used non-eco-friendly Na_2S reducing agents used in the reduction of dyes for their solution preparation for the colouration of cellulosic materials with vat and sulphur dyes. Newly introduced biodegradable detergents like fatty alcohol ethoxylate can be used against the objectionable use of non-eco-friendly detergents used in post dyeing process [59].

8.5.1.1 Ionic liquid as a green solvent in dyeing of textile

The unique property of the ionic liquid is gaining importance in the potential replacement of water in the dyeing process. The ionic liquids show zero emission and eco-toxicity due to superior dissolving power, non-volatility and low vapour pressure. A recent study proved a sustainable improvement in the dyeing process when ionic-liquid was used in natural fibre dyeing besides polyester colouration [60,61].

8.5.1.2 Hazardous textile chemical and auxiliaries and Eco-friendly alternative in textile printing

The multi-colouration of textile goods is also carried out with the use of printing technology apart from dyeing. Pigment printing technology is most popularly used and has a share of almost 50% in the printing of textiles. The other chemicals like fixing agents and binders are required to be added to the printing paste to fix pigment on the textile substrates. Therefore, no washing is required after the application of pigment printing, and this is one of the advantages of pigment printing. Water, emulsifier and thickening agent are the basic chemicals used a in typical printing paste recipe. The use of kerosene in the preparation of emulsion thickener causes major pollution, and researchers developed water-based polyacrylates copolymer as an eco-friendly substitute for kerosene [63]. In printing, the washing of residual print from the substrate is the main key parameter for ecological impact.

Recently the aqueous binders of polyurethane acrylate based on either glycerol ethoxylate-co-propoxylate or polyethylene glycol are developed with zero-volatile organic compounds for formulating printing paste for screen printing. The pigment printing based on this technology can be used for all types of fabrics and was reported as an innovative development of pigment printing [62].

8.5.1.3 *Eco-friendly alternatives chemicals for finishing of textiles*

Various finishing chemicals are used to improve the functional properties of textile substrates. The supporting chemicals, such as cross-linking resins, softeners, etc., are also used along with functional chemicals. To enhance the easy-care performance of the cotton fabric, formaldehyde-based resins are generally used, which form cross-links and reduce the formation of the crease. In 1992, formaldehyde was banned by Finland. Since 2004, as per the norm under group I 'carcinogenic to humans' by the International Agency for Research on Cancer (IARC), the release of formaldehyde is considered toxic with respect to an allergic reaction, skin rashes and eye irritation and conserved as a carcinogenic. The chemicals having perfluoro alkyl chains with eight or more fluorinated carbons for finishing of textiles to impart durable water, oil and stain repellency also come under harmful chemicals [64]. The fluorinated compounds containing long-chain perfluoro alkyl group create a problem as biodegradability is a concern since these chemicals are banned and reported globally under the toxic chemical category [65]. Some eco-friendly alternatives in textile finishing are listed in Table 8.4.

To protect from the growth of microorganisms, textile materials are finished with several antimicrobial chemicals which facilitate a hygienic and healthy lifestyle. The synthetic compounds based on triclosan (2, 4, 4'-trichloro-2'-hydroxydiphenyl ether) have been popularly used because of its high efficacy, but are limited as it changes to 2, 8-dichlorodibenzo-p-dioxin when get exposed to sunlight. Textiles finished with such a class of hazardous chemicals are harmful and responsible for increasing the intensity of occupational health hazards to the workers and end consumers. Any unfixed chemicals from the surface of textile during the washing cycle go into the effluent and disturb the ecosystem and finally affect the health of people. In textile wet processing, the finishing process involves the curing operation at an elevated temperature which leads to the release of harmful fumes and finally adds a new source to increase the level of carbon footprints [66].

Table 8.4 Non-ecofriendly chemical/auxiliaries and their alternatives in finishing.

Sr No	Green alternative	Non-eco friendly chemical/ auxiliaries	Application of wet processing
1	Plasma treatment	Chlorinated Shrink Proofing	Wool finishing
2	Polycarboxylic acid	Formaldehyde-based resin	Crease recovery chemicals
3	Combination of inorganic salt and phosphonates	Brominated diphenyl ether	Flame Retardant
4	C6 Fluorocarbon	C8 Fluorocarbon	Water repellent

Eco-friendly chemicals paving the way for a sustainable textile industry have been reported by introducing white biotechnology in the textile chemical industry [67].

There are many skin-related diseases and respiratory issues among users and workers due to the use of potentially toxic chemicals in textile factories. The demand for eco-friendly and alternative chemicals is now increasing to avoid health and environmental problems.

A novel method based on the i-CVD principle or initiated chemical vapour deposition has been considered an eco-friendly method. This technology facilitates the application of short fluorinated polymers on all classes of fibres, such as cotton, silk and wool, with a high value of fixation efficiency. These shorter polymers do not pose a significant threat like their longer counterpart and do not remain within the environment or the body. There is a need for newer green and biodegradable chemicals which will help to enhance the functionality of the textile material without significant loss of fibre, and at the same time, serve the environment.

8.6 Bio-materials and Bio-Processes in textile processing and value addition

Using materials of biological origin is an approach with excellent potential because these materials are usually biodegradable to reduce the pollution load safely. Pretreatments, dyeing and finishing of textile substrates using biotechnology-assisted products like enzymes and microbial dyes/pigments achieved varying degrees of success; however, they have gained great importance recently in textile industries.

The conventional peroxide removal technique is limited due to the consumption of high energy and water. Modified peroxidases reduce the hydrogen peroxide to water and oxygen without initiating side reactions with the dyestuffs or substrate [68].

The application of biological enzymes in industrial-scale desizing and scouring of textile fabrics enables eco-friendly processes in fibre treatment to get the desired level of product quality developed by using crude enzymes [69]. In recent years, the use of biopolymers made from renewable feed stalk has been explored. The polylactic acid (PLA) is a thermoplastic fibre and contains an aliphatic compound similar to polyester. Corn starch is the raw material used in the production of PLA fibre which finds application in the medical textile where biodegradability is required. The recycling of these fibres is easy since they can be biodegraded under certain conditions. PLA shows excellent biocompatibility and biodegradability; most of these fibres find application in tissue engineering for the development of favourable matrix materials.

The production of PLA fibre is completely free from toxic emissions, and it is reported that during production, it emits a lower amount of greenhouse gases. Almost 60% lesser greenhouse gases were emitted during the production of this fibre when compared to other synthetic fibre production [70].

Research and development in the area of biotechnology rapidly enhance a broad range of completely new bio-based materials and bio-polymeric substances used in the textile sectors. The Toray company created the world´s first functional PET fibre based on 100% biomass in 2011 and has now announced the next step, creating nylon fibre also based on 100% biomass [71]. The Bavarian AmSilk validation of a large-scale economic and eco-friendly production of recombinant spider silk proteins can be used as the next example of recent revolutionary bio-based fibre movements [72].

Sericin, a unique biomaterial, has proved itself as a highly potential biomaterial for various applications in medicine, pharmaceuticals, cosmetics, bio-sorbent and other areas. Sericin has a good capacity for use in biomedical and pharmaceutical areas. The fabric coated with sericin displayed the a significant bactericidal property. Therefore, sericin can be a useful component for the development of antibacterial textiles [73].

In the development of bacterial, filtration to reduce the number of free radicals and fungi, nylon and polyester fabrics coated with silk sericin can be used for such kind of applications. Sericin is a natural material that finds application in high-end areas and can be applied on all classes of fabrics using simple coating technique [74]. The use of enzymatic treatment for the cotton fabrics is more effective as compared to a single bath and continuous desizing, scouring and bleaching as per an effluent load concerned. Enzyme technology finds application in healthcare and hygienic textile products which covers safety aspects in textile industries [75].

The new class of products from renewable resources has been developed using a biotechnology approach. Also, it helps to create a road map for green alternatives at minimum energy consumption for making environmentally acceptable processes. Traditional textile processing consumes a large amount of energy, water, and many harmful substances, which are responsible for damaging the ecological balance, so there is a need to develop the technology which will help to substitute conventional textile chemical processes. Commercially, the use of enzymes in denim fading has created a benchmark to reduce the effluent load without loss of fibre quality [76].

8.7 Recycling, Reuse and Recovery of textile auxiliaries and chemicals and their environmental impact

Environmental pollution is a global issue that results in adverse effects on living beings. It is one of the major concern areas for the whole world. Desizing, scouring, and bleaching are the basic operations for pretreatment of cotton woven fabric before dyeing. Each process drains an excess amount of chemicals along with water in the effluent stream. The released chemicals by the global textile industry are continuously doing unimaginable harm to the environment. An attempt is made to investigate the opportunities to prevent pollution by recycling and reusing textile preparatory chemicals without the addition of any treatment.

8.7.1 Warp size recovery

Starch is the basic ingredient in the sizing of various commodity textile yarns. Polyvinyl alcohol (PVA) and maize starch are used in textile industries, often as a blend. The size is removed from the grey fabric in the desizing process. The effluent from the desizing bath contains high BOD/COD, which contributes to an increase in the effluent load to the plant's primary oxygenation treatment of water (POTW). Therefore, size recovery is very important from the environmental point of view. S. A. Pharmachem Pvt Ltd (India), in association with KOCH Membranes and DuPont (USA), has developed low capital-intensive ultrafiltration size recovery systems which are tailor-made for PVA recovery and has longer membrane life and are found useful to recover and reuse the size paste. The research contributes to the economic and ecological benefits of the textile industry in general, and the process was found globally viable [77].

Economically recoverable and reusable ELVANOL™ T Series is one of the unique molecular structure copolymers and an innovative textile sizing agent which is enormously chemical-stable. This is claimed to be chemically inert and stable to hydrolysis in presence of stresses and heat of a size-recovery system. This can be easily reused and recovered in any commonly available ultrafiltration system [78].

Environmental protection could be achieved by adopting state-of-the-art technology to minimise waste generation and recycling of size from effluent. By using the nanofiltration technique, synthetic size PVA or acrylic resin can be recovered for reuse. This recycling was successfully adopted for producing denim materials [79].

8.7.2 Desizing liquor recovery and reuse

In the two-stage enzymatic treatment, desizing liquors from the pretreatment of cotton fabrics can be used in the bleaching process which helps to decrease the energy and water consumption compared to conventional bleaching liquors. It is reported that this technique helps in lowering the consumption of ecologically harmful chemicals and process water [80].

The desizing effluent consisting of hydrolysed starch (if sizing is carried out using starch as a sizing agent) has big potential as the hydrolysed starch can be grafted and used as a textile finishing auxiliary; one application of such grafted starch is reported as an anti-crease agent and the results obtained were comparable with conventional finishing agents [81].

There is a need to develop strategies for sustainable recycling and reuse of process water in order to reduce water usage in textile wet processing. Approximately 93.1% hydrogen peroxide and 92.7% alkali from the scouring and bleaching process can be reused. Effluent from the bleaching process still contains 67.5% hydrogen peroxide and 55% alkali, which were reused to desize a new grey cotton fabric. Fabric properties such as whiteness and absorbency have been found to be better than the conventional enzyme-desized fabric [82].

8.7.3 Scouring and bleaching liquor recovery and reuse

To investigate the feasibility of reusing the process liquor for ecological benefits, a study on reusing of pretreatment liquor from scouring and bleaching was conducted. The reused residual chemicals can help for combining the process which will help to reduce the process cost; this can be confirmed by analysing and quantifying the concentrations of chemicals after and before the pretreatment. The performance of pretreatment can be assessed in terms of whiteness, yellowness, and absorbency of pretreated fabric, along with the effect in subsequent reactive dyeing and pigment printing in terms of colour difference, colour strength, and colour fastness. Reusing the pretreatment bath for the next subsequent process showed improvement with adequate results which can fulfil the product's requirement [83].

Wastewater from the textile process can be reused without any treatment which not only reduces the use of freshwater but also reduces the load of effluent. Weight loss of the scoured and bleached fabric was found to be 6.53% for freshwater and 6.65% for wastewater. The whiteness of fabrics bleached using freshwater and wastewater were 77.92 % and 76.68%, respectively.

The dyeing of these fabrics showed a colour difference of 0.62, which is acceptable. The bleached fabrics also showed excellent absorbency [84].

8.7.4 Mercerization liquor recovery

Mercerization is carried out by using 30 to 32% sodium hydroxide solution for treating cotton fabrics and certain blends. It is always a continuous process, and it is very easy to handle the waste from the mercerization bath. It is reported that about 98% caustic recovery is possible from the effluent drain. One of the alternative systems which work without caustic and also water in the mercerization process is the use of liquid ammonia. The ammonia gas used in the process was reused and showed the same mercerization effect; the highly alkaline wastewater stream, and the ammonia gas was recovered and reused [85]. $ZnCl_2$ can be used to increase the weight and dye uptake, which offers easy recovery of NaOH. Moreover, the process is eco-friendly and does not need acid neutralisation [86].

Recently, nanofiltration technique has been used for recycling alkali from the wastewater of mercerization process. It is reported that the nanofiltration gives about 80% recycling efficiency with reduced COD in mercerization, dyeing and printing processes [87].

8.8 Reuse of dyebath and auxiliaries

8.8.1 Dyebath reuse practices

Dyebath consists of dyes, chemicals and selected auxiliaries depending on the type of dyes and substrates to be dyed. The presence of heavy metals and harmful pollutants in the dye is highly hazardous because of their non-biodegradable nature and high bio-accumulation values with higher concentrations, thereby increasing toxic traces in the environment. Dyes and auxiliaries producers are now conscious of the environmental impact of colouring matter and surfactants along with the requirement for a better economy and ecology. Mercury-free dyestuff is also in practice. During the dyeing process, most of the chemicals used in textile wet processing are not fully consumed and could, therefore, contribute to pollution loads [88].

8.8.2 Dyebath replenishment techniques

A single treatment bath can be reused several times before discharge as an effluent. The reuse of used dyes and chemicals provides many economic and ecological benefits by being replenished for future use in the dyeing of another sample [88-90]. The 66.67% reduction in the process effluent was reported due to dyebath reuse through a replenishment program. It helps to reduce the cost of water and chemicals from the process, especially in dyebath with respect to direct dyes and acid dyes. It is possible to replenish due to the lesser number of chemicals and auxiliaries present and their ease of process reported by the investigator [91]. The method of reuse of dyebath helps to take care of the environmental implications of dyeing by regular use of chemicals, dyes and water [92,93].

A project study on the technical feasibility of batch dyebath reconstitution and reuse at a carpet mill was successfully carried out [94]. Environmental benefits are significant, even when the dyebath reuse is a practice in a non-optimal fashion. A reduction in pollutants and water was observed at around 25 to 50%. Environmental benefits were maximized when dilution of the dyebath through steam condensation and overflow cooling is kept to a minimum [90].

The influence of multiple reuses of residual dyebath on the colouration properties of wool fibre was investigated. The result showed that the qualities of samples dyed in reused dyebath were the same as those obtained from initial fabric dyeing and claimed that colour conformity was lower than 1 with good evenness and no deterioration in colour fastness despite multiple reuses of dyebath [94].

A study was conducted on the reuse of the spent dye bath in polyester dyeing. The result indicated that 5-7% saving on dye and 65-75% saving on auxiliaries was found by reuse of the spent dye bath, along with a substantial decrease in the discharge of pollutants. The dye bath reuse was attempted in mill trials with good success. No adverse effect was seen on the spinning performance of the fibres dyed in reused dyebath [95].

8.8.3 Indigo dye recovery and reuse

Indigo dye is commonly used in the dyeing of denim fabric. It is used to create a fancy effect in the denim fabrics, like a faded effect by means of stone washing. The dye coming out from this particular treatment is in a high level of concentration which creates environmental issues, and biologically, it is very tough to decompose Indigo dye since it has moderate chemical oxygen demand (COD) and solid particles with dark blue colour. It is essential to recover such a class of dye to maintain sustainability in textile processing. Various methods adopted in the recycling of Indigo dye have been practiced through adsorption, flocculation and membrane technology, out of which membrane technology is the more popularly used and commercially exploited technology in the industry. Membrane technology works on the principle of ultrafiltration technique, where the Indigo dye particles are filtered out from the drained dye bath, and the particle separation is governed by the membrane size. The recovered dye contains almost 30% of concentration and is effectively used in the dyeing process with minimum energy and other utility sources [96].

The recovery of indigo dyes was reported in one of the feasibility reports, where the recovery technique is based on membrane technology. About 80% reduction in COD and 99% colour removal were observed in the report. Concentrates containing 20 g/L of Indigo dye were eventually reused in new dyeing processes. The dyed fabrics showed similar characteristics to those obtained with industrial dyeing [97]. Textile dyes, particularly Indigo dye from the denim plant effluents, are successfully recovered and reused for the next process, as reported in another investigation [98].

8.8.4 Recycling and reuse of PET waste as textile auxiliaries

Polyethene terephthalate waste fibres were initially depolymerised in the presence of sodium sulphate as a catalyst, which is an ecofriendly chemical as compared to other heavy metal catalysts, using a glycolysis route. The process provides a good yield of pure monomer bis(2-hydroxyethylene terephthalate (BHET). In one of the investigators, purified BHET was converted to different fatty amide derivatives to obtain a quaternary ammonium salt that could be used as a softener in the textile finishing process. The chemicals used during depolymerisation and reuse of PET are cheaper and comparatively less hazardous to the environment; therefore, they offer advantages for the chemical processing of PET waste fibres [99].

8.9 Sustainability challenges of the textile dyeing and finishing

Various chemicals and water are used during colouration and finishing of textiles for the physico-chemical modifications of products. To improve and impart aesthetic and functional qualities (flame resistance, water repellency, wrinkle resistance, etc.), various classes of dyes and chemicals are applied during the finishing stages. Several technologies have been built from fibre to finished products in order to improve wastewater treatment, minimise electricity consumption, process cost and effluent loads [100].

The finished garment contains 5-15% of chemicals based on its weight. As far as technical textile products, like automotive textiles, are concerned, they also include dyed and finished textile substrates of around 20 to 25 % [101]. The presence of these harmful chemicals shows potential risks to human health, especially for products close to the skin. The presence of hazardous chemicals has consequences on material recovery due to a poor understanding of the chemical's interaction found in feedstocks with process chemicals used in recycling treatment. The removal of harmful chemicals from pre and post-treatment needs proper measures to be taken [102].

The dyeing process involves the application of colour to the textile substrates. Colourants are substances that can contain pigments or dyes. Today, mostly synthetic dyes are used which are generally synthesised from petrochemicals and their byproducts; at the same time, some natural dyes are also used which are extracted from plant and animal-based resources [103]. The yield of natural dyes is very poor, the number of colours available is less, and it provides moderate fastness properties; on the other side, synthetic dyes are found to be superior to natural ones and widely accepted in the industry provided they are from eco-friendly category. Natural biopolymers are abundantly available in nature and are widely used in textiles. Various natural materials such as chitosan, neem, aloe vera, etc., are used for the production of antimicrobial textile products. Curcumin was used in the form of nanoemulsion for imparting the antimicrobial activity to the textile material in association with β-cyclodextrin and BCTA (1,1,4,4-Butanetetracarboxylic acid). The curcumin showed enhancement in the antimicrobial activity, as reported in a recent investigation [104].

Conventionally there are three modes of dyeing operation, i.e., exhaust, semi-continuous and continuous process. Exhaust is the most common dyeing method in which dyeing is carried out in a batch mode, and the substrate is dyed in a dye bath. It is suitable for 10kg to 1000kg lot sizes. During the application of dye on fabric, the dye concentration, dyeing temperature, additives, solvent, electrolytes and mechanical stirring need to be controlled. In a continuous process, the fabric is padded with a dyeing solution/dispersion, dried and further fixation is carried out using suitable methods like baking, steaming, curing, etc. The method of fixation and interaction between fibre and dye molecules can affect the release of dyes in the effluent, colour fastness and dye removal at the recycling stage. The sustainability of the dyeing process is assessed by the ability to discharge low amounts of dyes in the effluent; less discharge is achieved by introducing chemical interaction between the dye molecules and fibre. Table 8.5 summarises the different dye classes applied to textile with respect to the type of fibre, type of interaction between dye and fibre and hazardous substances involved.

Table 8.5 List of dyes and supporting chemicals used in textile dyeing

Class of dye	Fibre type	Hazardous substances	Fixation degree in %	Interaction between fibre and dye
Acid	Wool, Silk	Banned Aryl amines due to carcinogenic effects (from azo dyes)	85-98	Ionic bonding, van der Waals, hydrogen bonding,
Acid Subclass: Metal complex (Replaced banned azo dyes)	Wool, Silk	Toxic metals from heavy metal content in dyestuff	82-98	hydrophobic bonding, ionic bonding, Hydrogen bonding, van der Waals forces,
Basic	Silk, Acrylic	Carcinogenic dyestuff, complexing agents (quaternary ammonium compounds)	95-100	Ionic bonding, Hydrogen bonding,
Direct (Azo-based direct dyes banned)	Cellulose	Salts, aftertreatment with water, toxic cationic agents	64-96	Van der waals forces, Hydrogen bonding
Mordant	Wool, Silk	Chromium VI, some banned	95-98	Dye fixative
Reactive	Cellulose, Wool, Nylon	Salt emissions, unfixed dyestuff in effluent require treatment, metal complexes	50-97	Covalent bonding, Hydrogen bonding (hydrolysed reactive dye)
Sulphur	Cellulose	Sodium hydrosulphite	60-95	Van der Waals forces, Hydrogen bonding
Vat	Cellulose	Alkali,Sodium hydrosulphite	75-95	Van der Waals forces, Hydrogen bonding
Disperse	Nylon, Polyester	Allergenic and carcinogenic dyestuff, chlorinated solvents	88-100	Hydrophobic bonding, Van der Waals forces
Pigment	Polyester, Cellulose, Nylon, Wool, Silk, Acrylic	Chlorinated or aromatic, aliphatic solvents, residues (binders, VOC, etc.)	100	Require additional substrate compound for attachment

It is reported that the above-listed chemicals affect the environment, which is categorised as 1 to 5 numbers from lowest to highest polluting category. Cotton dyeing is primarily done by reactive dyeing. It has been found to emit greater quantities of effluent compared to dyeing of other fibres, which are correlated with high cleaning costs. Apart from that, reactive dyes need large volumes of fresh water to remove the unfixed dyes, which accounts for up to 40-50% of the overall cost of the dyeing process [105]. Therefore, sulphur, vat and metal complex dyes are considered to have lower impacts on the environment [106]. Various dyes and their impacts, as reported in Table 8.6, indicate the interaction of various fibres and their dye class requirement, and polluting agents and their impact.

Table 8.6 Various dyes and their pollution impact

Fibre	Dye class	Type of pollution from dye	Pollution category	Other chemical and finishing effects
Cotton	Direct	Cationic fixing agent Salt Copper salts Unfixed dye (5-30%)	1 3 5	Huge amounts of chemicals are required during production. Must be treated with chemicals to absorb dyes (pre-treatment and post-treatment).
	Reactive	Alkali, Salt Unfixed dye (10-40%)	1 3	
	Vat	Oxidising agent, Alkali, Reducing agents	1 2	
	Sulphur	Alkali, Oxidising agent Reducing agents	1 2	
Wool	Mordant (Chrome) (banned)	Heavy metal salts Organic acid	2 5	Preparation steps often involve harsh chemical treatments.
	Acid; Metal complex (single-stage application; dye and mordant combined prior to dyeing)	Unfixed dye (5-20%) Organic acid	2 3	
Polyester	Disperse	Reducing agents Organic acid Carriers Unfixed dye	2 5 3	Heavy metal catalysts (antimony) used in production, carcinogenic. Dyeing requires, high temperatures

Contd...

Contd...

Fibre	Dye class	Type of pollution from dye	Pollution category	Other chemical and finishing effects
Nylon	Disperse	Reducing agents	2	Production emits nitrous oxide gas.
		Organic acid	5	
		Carriers		
	Reactive	Salt, Alkali	1	
		Unfixed dye (10-40%)	3	

Textile finishing is carried out using either chemical or mechanical methods. In some chemical finishes, actual reactions are performed with the association of cross-linking agents, which imparts the functional properties and enhances the physical characteristics of the textiles [107].

Some typical finishes used for finishing of textiles are as follows:

- Easy-care finish
- Flameproof or flame retardant finish
- Water Repellent finish
- Moisture management finish
- UV-protective finish
- Antimicrobial finish
- Stain release finish
- Antistatic finish

A variety of chemicals, such as formaldehyde-based, halogenated compounds and other metal-based derivatives, are used for physico-chemical modification of textile products. However, these chemicals create serious problems for the environment. There are some alternative chemicals for available to impart the functional properties to the textiles; the list of such chemicals is given in Table 8.7.

Table 8.7 List of alternative chemicals for the textile finishing

Finishing Chemical/ Solvent	Function	Alternative Chemicals (under development, commercially available)
Finishes based on Formaldehyde, other short-chain aldehydes	Easy-care, wrinkle-free	Butane teracarboxylic acid (BTCA) Polycarboxylic acids (PCA), citric acid (CA) and CA/xylitol
Halogenated flame retardants	Flame retardant	Patented atmospheric plasma/UV laser technology on viscose/flax and cellulosic
		Phosphorus, or Nitrogen (melamines) based, and inorganic flame retardants on cotton, polyester, and polycotton blends

Contd...

Contd...

Finishing Chemical/ Solvent	Function	Alternative Chemicals (under development, commercially available)
Triclosan and triclocarbon Nanosilver	Antimicrobial fabric treatments	Zinc oxide nanoparticles applied to polypropylene, Nylon 6, polyester Nano chitosan particles applied to 100% cotton, viscose, polyester fabrics Siloxane sulfopropylbetaine (SSPB) covalently bonded to cotton Sol-gel based surface coatings
Per- and poly-fluorinated compounds	Durable water repellency (DWR)	Dendritic/hyperbranched chemistry Silicon-based, hydrocarbons, wax-based repellents Sol-gel (Si-based) surface coatings Plasma treatment for water repellent finishing
Chlorinated cleaning solvents, alkali detergents	Dry cleaning, spot cleaning, scouring	Enzyme-based textile processing

8.10 Best Available Techniques (BAT) in textile processing

8.10.1 Supercritical carbon dioxide for dyeing of textiles

The textile industry consumes a huge amount of water in both pre and post-processing sections, and finally, it increases the effluent load which is now a challenging task. The use of supercritical carbon dioxide offers waterless dyeing in the dyeing of polyester fabric and helps to maintain sustainability as per effluent load concerns. The penetration of dye becomes easy because of the low value of surface tension in the sCO2. Furthermore, there is no need for drying the fabric after dyeing, and this offers a reduction in the energy that is required for drying in the conventional dyeing techniques [108].

8.10.2 Plasma technology for finishing of textiles

The physico-chemical modification of the textile substrate can be possible with the use of plasma techniques, which are environment-friendly with no use of chemicals and water, without modifying the bulk properties of different materials. For the surface modification of textiles, the plasma technique expected to grow because of its versatility in the textile processing area [109]. The low-temperature atmospheric plasma has been used for the functionalisation of cotton fibre. The plasma-treated cotton fibre showed an excellent hand and aesthetic sense [110].

8.10.3 Diamond finishing technology

This is a newly emerging technique used for surface modification of both apparel and industrial-grade fabrics. The technology developed is completely eco-friendly as there is no use of water, chemicals and energy. This is a proven technology to get a consistent quality product without sacrificing the aesthetic and mechanical properties of denim materials.

8.11 Green garment processing

8.11.1 Harmful materials in conventional garment processing

As per Market Line data of 2017, the global apparel market was hit to more than $1,652 billion in year 2020, a rise of above 31.8 % sine 2015 onward. During this year the apparel industry grew by 11.1% as announced by world trade organization (WTO) [111-113]. Without a doubt, this number will increase with every passing year. Nowadays, clothing has become more than mere basic needs. People buy more clothing than they actually wear in their lifetime. The life cycle of clothing has a significant effect on the environment.

Hazardous textile chemicals such as alkylphenol compound (Nonyl Phenol) are commonly used in the textile industry in dyeing and cleaning process. Since 2005, NPs have been heavily regulated in Europe, and there has been an EU-wide ban imposed on its primary application. Phthalates are a group of chemicals utilised in some of the dye's formulations and as plasticisers. Brominated and chlorinated compounds have been listed as priority hazardous substances. Azo dyes, which on breakdown release aromatic amines as cancer-causing compounds, are also on the banned list of EUs. The organotin compound, per-fluorinated chemicals, chlorobenzene, short-chain chlorinated paraffin and heavy metals such as chromium (VI), mercury, lead and cadmium utilised in some textile clothing processes are restricted in some applications in the EU since the last decade [114].

Consumers today are more aware of the need to protect the environment, and companies can seek to advertise their products or services with eco-labels. Clothing is an important part of our lives, and green apparel has begun to gain more and more interest in the fashion and clothing industry [115].

Environmental issues turn out to be crucial aspects of the selection of consumer products all over the globe. The edified user is demanding 'green' clothes and is willing to pay a higher cost. Throughout the production of textiles, there are two main aspects to be considered, such as limiting harmful products in textiles under stipulated limits and minimising air and water pollution [116,117].

During the whole life cycle of clothes, each stage (from cloth production to end of life) leads to the quantity of waste disposed of and ultimately affects the environment. The rising demand, altering trends, and the growing uproar on sustainability have directed the garment manufacturing industry to invent eco-friendly techniques. Garment processing is not only a washing or laundry unit but comprises curing, dyeing, drying, washes, sandblasting, spraying and many others value addition processes. [118].

Dow Corning Corporation has developed a new granulated silicone technology-based softener finish that processes denim and denim garments. This eliminates separate washing requirements and reduces water consumption by 30%-50%, leading to energy, time and ultimately cost savings. This led to a relatively positive ecological effect without martyrs in garment qualities [119,120].

The environmental performance of textile firms from Bangladesh improves by cleaning up the production process by saving resources, particularly energy and water. This is done merely by making low or zero cost-effective changes like water metering, enhanced boiler efficiency, reusing cooling water and insulation of pipe [121]. A leading brand in Asia has been supporting its environment-friendly viscose staple fibre. It has developed backward integration of waste into pulp [122].

Bleaching technologies such as ozone and laser technology, likewise high exhaustion dyes that decrease effluent load, are also in practice as a green process. The ozone use for fading and bleaching effects produces fewer loads on effluent.

Atlantic Care Chemicals has developed sustainable washing techniques for zero discharge of hazardous chemicals. This can be achieved by using innovative techniques, environmental xenobiotics using microorganisms; green chemistry which focuses on minimizing hazardous chemicals; and enzymes-based chemistry by replacing stoichiometric chemical additives. This results in enhancing the performance of the finished quality, less effect on the ecosystem, decreased consumption of natural resources and a reduction in time and economics. They had a breakthrough accomplishment of substituting potassium permanganate with modified natural clay. Sustainable laundry techniques application is to accomplish effective use of energy, water, time and chemicals.

SF Dyes has developed green procedures for garment colouration such as Eco Dip, Eco tint / Eco coat and salt-free dyeing. Eco Dip process using an environment-friendly reducing agent and even excluding post-washout results in reduced water utilisation by more than 80%, along with a decrease in energy and more production of coloured garments. Eco tint / Eco coat processes are based on application by spray-cure technology and zero sulphides. The salt-free dyeing process can be used for all dye classes with no salt addition, no post-wash-out and effectiveness in low liquor ratio machines along with less effluent load (low TDS) and no machine staining [123].

BASF has developed an electrochemical dyeing process for the production of blue denim which consists of pre-reduced indigo dye. This technique helps to replace the chemical used as a reducing agent based on hydrosulfite source. The process is not successful in the market because of high capital investment, but this technology reduces the load on the effluent. It offers ecological and economical situations in the dye house.

In the finishing operation, in order to get a vintage look, especially for denim, the garments are mostly treated using chemical or physical methods. The conventional processes based on chlorine bleach (sodium hypochlorite) and potassium permanganate bleach are not used on the commercial-scale [124].

Levi Strauss and Co. have developed a screened chemistry system that evaluates the human health and environmental impact of chemicals used in the garment finishing process to manufacture their products. Gammon chemicals, in collaboration with INVISTA, owner of the LYCRA brand, have developed a range of knit denim garments using a fabric made with Lycra Hybridization technology and finished with Gammon's screened chemical formulation [125]. Achieving new cleaner looks, oxygen, peroxide and ozone bleach are picked up in the finishing of a garment because of lower environmental impact.

The application of stone wash treatment is considered more effective than the chemical methods which are more labour-intensive. The sandblasting method is also banned due to excessive exposure to silica and the firing of sand under high pressure. Several brands such as H&M, Levi's, Gucci and Versace have prohibited sandblasting and are pushing towards greener alternatives. The novel sodium hydroxide treatment on the denim surface after dyeing creates improved results than sandblasting technique. The technique offers the same results and maintains fabric strength, shortens the processing time, and reduces the concentration of chemicals, which is also beneficial to the environment.

The US-based denim company Levi Strauss has now proposed a "waterless" idea to achieve major water savings in denim finishing. One element of the concept is the use of ozone bleaching and fading. This technology is invented by DyStar Textilfarben GmbH for quality control of the Indigo-dyed fabric. The water-free denim finishing technique by Japan Rags is "Japan-RAGS ZERO", which is an ozone-based finishing process without the use of any water.

The ozonization of process water after denim finishing in Southern Germany is invented by German Jeans maker Joker Jeans for eco-awareness amongst the denim finisher (laundries). Freshtex, Heilbronn/Germany, a denim corporation, has revealed a company-wide recycling program this year. The Italian Martelli Company, one of the world's most creative denim finishers, also introduced in 2011 a new ecologically preferable finishing concept that supports a specific production phase. The used products are removed after "having done the job" mostly by washing off into the wastewater.

An additional environmental-related theme in the textile industry is "Recycling". The recycling and reuse of wastewater, from fibre production to finishing, are getting studied. Turkey has developed the production of denim from worn-out denim which shows 20-30% cost savings compared to virgin denim production [126].

The development of single-step dyeing and fading treatment as eco-friendly garments washing is considered a future sustainable process in denim processing. Unconventional yet eco-friendly methods to produce fading effect on denim fabric dyed with Indigo dye have been studied using three techniques such as fading by UV radiation of sunlight, calcium hydrate and extra fabric stitching. In the study, it was observed that treated fabric shows improved fading efficiency up to 84.58% with superior handle and aesthetic sense [127].

The latest innovation is the dyeing of denim with eco-friendly concentrated sulphur dyes. The sulphur dyes assist in increasing the affinity for cotton. After application, the dye is oxidized and remains on the surface of denim without any rinsing treatment. This process is more efficient than the conventional dyeing with Indigo, where it saves about 87% of cotton waste, 92% of water and 30% of energy with no additional load on the environment. The most trusted brand Levi Strauss has introduced one sustainable process in denim production that reduces 28% of water used in finishing the garment.

Recycling and reusing technology has become the slogan of many apparel brands to demonstrate commitment and meet the demands of consumers. Big brands launched denim made from plastic bottles. The plastic bottles are split into pieces and then recycled into polyester fibres. Such fibres are blended with cotton to produce denim fabrics. The post-recycled rug by Swedish label

Nudie jeans rugs offers zero water, washing by laser technique with no use of harmful chemicals; the technology promotes green process in the complete value chain. H&M has launched denim made of donated and recycled clothing. This denim collection uses 20% recycled cotton, which is the topmost limit that can be utilised without compromising the quality of the final product [128].

The reduction and removal of waste in the textile industry will reduce the cost of raw materials and manufacturing. However, production with zero waste has so far been too idealistic. Direct Pattern on Loom (DPOL) brings a new approach to design with virtually no waste in the process of pattern making and cutting, and it can also produce ready-to-stitch garment components. This novel technique can increase fabric efficiency by 15% to 20% and reduce the lead time by 50%. DPOL is the perfect mixture of garment engineering, fashion designing and textile technology. The fabrics for making garments, according to this concept, are specifically designed for a particular purpose. Any type of embellishment shall be designed when the fabric is produced. It can be utilised to design both woven and knitted high fashion garments [129].

Processing and garment production was discussed in a briefing at the European Parliament recently. A significant amount of energy is involved in the garment process, from sewing, gluing, and welding up to seam taping operation. 20% of the total waste is generated during the cutting of left-over patterns and considered in the overall garment waste. Many steps have been taken to minimise the environmental impact in the processing and manufacturing phases. Minimum consumption of chemicals, replacement of harmful chemicals with safer alternatives, controlling the concentration of dye and maintaining the machinery and resources are some of those steps. Knitting operation is considered more effective for the production of garments where there is no need for cutting and sewing. Some industries are trying innovative dyeing processes, like using supercritical CO2 as the dyeing medium instead of water (e.g., Dutch company DyeCoo), while others are testing with different cuts, computer-controlled patterning tools to utilise more fabric with fewer cuts, garments with fewer or no seams, gluing or bonding as a substitute of stitching [130].

The textile and apparel industry is waking up to changing times and making attempts to keep the environment safe. Today, companies are encouraging consumers to wear recycled garments and move towards more sustainable items. To reduce the volume of waste, the recent technology for the production of garments is concentrating on designing the process which will consume less water and energy with less or no chemicals.

Concluding remarks and future perspectives

The search for solutions to waste minimisation and prevention, recovery and recycling is of paramount importance for the complex textile industry as a major contributor to landfill, water and environmental issues. With the growing public awareness and legal pressure in recent years, attempts are being made in all areas of textile wet processing to reuse and recycle dyes, chemicals, water and auxiliaries. Nonetheless, for a variety of processes and even for the chemicals involved, recovery mechanisms remain largely unexplored. Textile researchers are facing challenges in the proper separation and purification of chemicals, the recovery of chemicals in adequate quantities and the adoption of an economic balance. Green processing technology for material and garment processing is a need of the hour, and more and more efforts are required for creating sustainable green processes and products for textile value addition. Innovative technologies, such as water-free dyeing, enzymatic methods and green fabric manufacturing, need open-mindedness to be embraced in an industrial setting; however, they have tremendous potential to solve environmental issues.

The textile industry is one of the largest manufacturing industries which consumes a high volume of water, chemicals and other natural resources and harms the environment. Today the use of modified, greener materials and processes in textiles has led to a drastic reduction in environmental problems. There is great potential to develop sustainability in the entire textile value chain, from the production of fibre to finished goods with the help of sustainable materials and practices. The use of bio-materials in the textile is also helpful for the development of a new class of sustainable materials for technical applications. Nowadays, the application of bio-nanotechnology in textiles plays a vital role in the development of smart and sustainable materials for various applications. Modern tools like life cycle analysis are helpful for the assessment of all kinds of textile materials with respect to their sustainability. In recent years, the industrial revolution 4.0 has gained importance for a reduction in the process cycle which helps to maintain and monitor the process. There is need to have a collaborative research in the area of textile science in conjunction with biotechnology, material science, environmental engineering, pharmaceutics and medical science to surge up the research and development activity in the field of sustainable textile materials and processes.

References

1. Rao JV (2001) Development of grey preparatory processes of cotton materials. *Indian J Fibre Text Res* 26:78-92.

2. Emission Scenario Document on Textile Finishing Industry (2004) Environment Directorate Organisation for Economic Co-operation and Development.

3. http://www.bdtextileinfo.blogspot.xn.com.tooa/Accessed on September 26, 2021.

4. Qin ZJ (2006) Development and application of green sizing agent. *Progress in Textile Science & Technology* 4:5-6.

5. Qu CX (2005) Application analyses on environment protection sizing agent. *Cotton Textile Technology* 33(5):257-260.

6. Yang ZQ (2008) Green environment protection textile size. *J. Text. Sci.* 1:5-6.

7. Zhang C, Yu S-M, Gao X-L (2007) Research progress of green textile size at home and abroad. *Progress in Textile Science & Technology* 6:15-16.

8. Sun S, Yu H, Williams T, Hicks RF, Qiu Y (2013) Ecofriendly sizing technology of cotton yarn with He/ O2 atmospheric pressure plasma treatment and green size recipes. *Tex. Res. J* 80(30):2177-2190. https://doi.org/10.1177/0040517513490061.

9. Cavaco-Paulo A, Gubitz G (2003) Textile processing with enzyme. *Woodhead Publishing, Cambridge*, pp 22-36.

10. Xiao H, Zang W (2009) Current Situation of Environment Protection sizing Agents. *J. Sustain. Dev.* (3):172-175. https://doi.org/10.5539/jsd.v2n3p172.

11. Sheikh J, Bramhecha I (2019) In *The Impact and Prospects of Green Chemistry for Textile Technology* (Islam S and Butola BS Eds.), The Textile Institute Book series, Woodhead publication, Cambridge pp135-160.

12. Battan B, Dhiman SS, Ahlawat S, Mahajan R, Sharma J (2012) Application thermostable xylanase of bacillus pumilus in textile processing. *Indian J microbial* 52: 222-229. https://doi.org/10.1007/s12088-011-0118-1.

13. Wahid MS (2010) Introducing Bio-scouring: A cost effective and ecofriendly process for the generation next. *Textile Today. Accessed on September 28 2021.*

14. Menezes E, Choudhari M (2011) Pre-treatment of textiles prior to dyeing. Peter Hauser ISBN: InTech, pp-1-22.

15. Koushic S, Uddain K (2011) Comparative Analysis Between Conventional Pretreatment and Bio-Preparation. *Int. J. Eng. Sci. Technol.*11:14-19.

16. Siddiquee AB, Mahbubul MD (2014) Comparative study of conventional and enzymatic pretreatment (scouring & bleaching) of cotton knitted fabric. *Int. J. Eng. Sci. Technol.* 3 (1): 37-43.

17. Cavaco-Paulo A (1998) Processing Textile Fibres with Enzymes: An Overview. *ACS Symposium Series* 687, 180-189.

18. Ardon O, Kerem Z, Hadar Y (1996) Enhancement of laccase activity in liquid cultures of the ligninolytic fungus Pleurotusostreatus by cotton stalk extract. *J. Biotechnol.* 51: 201-207. https://doi.org/10.1016/S0168-1656(96)01597-0.

19. Hedin PA, Jenkis JN, Parrot WL (1992) Evaluation of flavonoids in Gossypium arboretum cottons as potential source of resistance to tobacco budworm. *J. Chem Ecol.* 18: 105-114. https://doi.org/10.1007/BF00993746

20. Tzanov T, Basto C, Gubitz GM, Cavaco-Paulo A (2003) Laccases to Improve the Whiteness in a Conventional Bleaching of Cotton. *Macromol. Mater. Eng* 288: 807-810. https://doi.org/10.1002/mame.200300100.

21. Pereira L, Bastos C, Tzanov T, Cavaco-Paulo A, Gubitz GM (2005) Environmentally friendly bleaching of cotton using laccases. *Environ. Chem. Lett* 3: 66-69. https://doi.org/10.1007/s10311-005-0004-3.

22. Basto C, Tzanov T, Cavaco-Paulo A (2007) Combined ultrasound-laccase assisted bleaching of cotton. *Ultrason. Sonochem.*14: 350-354 4. https://doi.org/10.1016/j.ultsonch.2006.07.006.

23. Boryo DE, Bello KA, Ibrahim AQ, Gin NS, Ezeribe AI, Wasiu KA (2014) Effects of Alternative Mercerizing Agents on Some Mechanical Properties of Cotton / Polyester Blend Fabric. *Acad. J. Interdiscip. Stud.* 3(5): 91-103. https://doi.org/10.5901/ajis.2014.v3n5p91.

24. Boryo DE, Bello KA, Ibrahim AQ, Ibrahim AQ, Mashat GUM, Okakwu AA (2017) Improvement on the dyeing and water of imbibitions properties of cotton/polyester blend fabric by alternative mercerizing agents. *J. Eng. Res.*17 (1):43-66.

25. Khajavi R, Berendjchi A, Mohammad B, MoghaddamM, Akhani M (2015) Ultrasound-assisted mercerizing process of cotton fabric a numerical model based on response surface methodology (RSM). *Fibres Polym.* 16(6), 1281-1288. https://doi.org/doi 10.1007/s12221-015-1281-5

26. Sengupta S, Sing BR (2003) Natural green dyes for textile industry, the Massachuetts toxic use Reduction Institute. University of Massachusetts. Lowell Technical Report No.57.

27. Water Quality Guidelines and Pollutant Fact Sheets, Business for Social Responsibility, San Francisco, CA, USA, (2002), (http://www.bsr.org/CSRResources/Environment/WQG_Pollutant_FactSheet.pdf/Accessed on 31 October 2022.)

28. Choudhury AK (2017) Green Chemistry and Textile Industry. *J. Textile Eng. Fashion Technol.* 2(3):351-361. https://doi.org/10.15406/jteft.2017.02.00056.

29. Chequer FMD, de Oliveira GAR, Ferraz ERA, Cardoso JC, Zanoni MVB, de Oliveira DP (2013). Textile Dyes: Dyeing Process and Environmental Impact. In (Ed.), *Eco-Friendly Textile Dyeing and Finishing.* IntechOpen. https://doi.org/10.5772/53659 59.

30. Kale RD, Kane PB (2017) Colour removal using nanoparticles. *Cloth. Text. Res. J* 2(1):4. https://doi10.1186/s40689-016-0015-4.

31. Kale RD, Kane PB (2019) Colour removal of phthalocyanine based reactive dye by nanoparticles. *Groundw. Sustain. Dev.*8: 309–318. https://doi.org/10.1016/j.gsd.2018.11.007.

32. Kale RD, Kane PB (2018) Synthesis of PVP stabilized bimetallic nanoparticles for removal of azo based reactive dye from aqueous solution. *Sustain. Chem. Pharm.* 10: 153-162. https://doi.org/10.1016/j.scp.2018.11.002.

33. Jarosz-Wilkołazka A (2015) Dyeing of Textile Fabrics with Bio-dyes. *Fibres Text. East. Eur.* 1(109): 120-126.

34. Ma W, Zhang S, Tang B, Yang J (2015) Pretreatment of cotton with poly (vinylamine chloride) for salt free dyeing with reactive dyes. *Color. Technol.* 121:193-197. https://doi.org/10.1111/j.1478-4408.2005.tb00272.x.

35. Harane R, Mehra N, Tayade P, Adivarekar RV (2014) A facile energy and water-conserving process for cotton dyeing. *International Journal of Energy and Environmental Engineering* 5 (2): 96. https://doi 10.1007/s40095-014-0096-2.

36. Shahid M., Mohammad F (2013) Recent advancements in natural dyes applications: a review. *J. Clean. Prod* 53: 310–331. https://doi.org/10.1016/j.jclepro.2013.03.031.

37. Melo MJ. (2009) *Handbook of Natural Colorants. Wiley Series,* pp 150-160.

38. Shahid M, Ahmad A, Yusuf M, Khan MI, Khan SA, Manzoor N (2012) Dyeing, fastness and antimicrobial properties of woolen yarns dyed with gallnut (Quercus infectoria Oliv.) extract. *Dyes Pigm.* 95(1):53–61. https://doi.org/10.1016/j.dyepig.2012.03.029.

39. Sarkar AK (2004) An evaluation of UV protection imparted by cotton fabrics dyed with natural colorants. *BMC Dermatology* 4 (1):1. https://doi 10.1186/1471-5945-4-15.

40. Khan MA, Srivastava M (2005). Extraction of natural dyes from myrobalan, gallnut and pomegranate, and their application on wool. *Colourage 52*(12): 53–60.

41. Yusuf M, Mohmmad F, Shabbir M (2016) Eco-friendly and effective dyeing of wool with anthraquinone colorants extracted from Rubia cordifolia roots; optimization colorimetric and fastness assay. *J. King Saud Univ. Sci* 29(2) 137-144. https://doi.org/10.1016/j.jksus.2016.06.005.

42. Shabbir M, Mohammad I (2016) Application of terminalia chebula natural dye on wool fibre-evaluation of color and fastness properties. *Text. Cloth. Sust.* 2(1) 1–9. https://doi 10.1186/s40689-016-0011-8.

43. Bechtold T, Mahmud-Ali A, Mussak R (2007) Anthocyanin dyes extracted from grape pomace for the purpose of textile dyeing. *J. Sci of Food Agric.* 87: 2589-2595. https://10.1002/jsfa.3013.

44. Gulrajani ML (2001) Present status of natural dyes. *Indian J Fibre Text Res* 26 (2): 191-201.

45. German Legislation on Azo Dyes, http://www.greenriv ertech.com.tw/stand-method-pdf/rohs/azo.pdf. Accessed 22 Feb 2016.

46. Myers W (2012) Bio Design. Thames & Hudson Publishing London. pp 112-162.

47. Lu Y, Wang L, Xue Y, Zhang C, Xing XH (2009) Production of violet pigment by a newly isolated psychrotrophic bacterium from a glacier in Xinjiang. *Biochem Engineering J.* 43: 135-141.https://doi.org/10.1016/j.bej.2008.09.009.

48. Chua JM (2015) London Designer Dyes Silk Scarves with Living Soil Bacteria. *Eco-Textiles*/Accessed on 16 October 2021.

49. Zhao H (2013) Synthetic biology tools and applications. *Elsevier/Academic Press, Boston.*

50. Polak J, Jarosz-Wilkolazka A (2012) Fungal laccases as green catalysts for dye synthesis. *Process Biochem.* 47:1295-1307. https://doi.org/10.1016/j.procbio.2012.05.006.

51. Pathak H, Madamwar D (2010) Biosynthesis of indigo dye by newly isolated naphthalene-degrading strain pseudomonas sp. HOb1 and its application in dyeing cotton fabric. *Appl. Biochem.Biotech.* 160 (6):1616-1626. https://doi10.1007/s12010-009-8638-4.

52. Fouillaud M (2016) Anthraquinones and Derivatives from Marine-Derived Fungi: Structural Diversity and Selected Biological Activities. *Mar Drugs* 14 (4):64. https://doi10.3390/md14040064.

53. Nerurkar M (2012) Dyeing of natural fibres with a red pigment produced by Streptomyces coelicolor', *Journal of Textile Association* 12(2): 16-26.

54. Nerurkar M, Vaidyanathan J, Adivarekara RV, Bhathena LZ (2013) Use of a Natural Dye From Serratia Marcescens Subspecies Marcescens in Dyeing of Textile Fabrics. *Octa. Jour. Env. Res* 1(2): 129-135.

55. Annapoorani SG, Divya S (2015) Extraction Of Eco-Friendly Natural Dyes Obtained From Aloe Vera, Green Chirayta and Indian Whitehead Leaves And Their Application On Cotton Fabric. *International Journal of Interdisciplinary Research and Innovations* 3(3): 35-38.

56. Gulzar T (2018) Current Scenarios of sustainable chemistry in Textile Applications. In *The impact and prospects of Green Chemistry for textile technology* (Islam S and Butola BS Eds.), The Textile Institute book Series, Woodhead Publication.

57. Anasta PT, Warner JC (2000) Green Chemistry: Theory and Practice', Oxford University Press.

58. Shenai VA (2001) Non-ecofriendly textile chemicals and their probable substitutes, *Indian J Fibre Text Res* (26): 50-54.

59. Markandeya S, Sukle P, Mohan D (2018) Toxicity of Disperse Dyes and removal from waste water using various adsorbents: A Review. *Res. J. Environ. Toxicol.*11 (2): 72-89. https://doi. 10.3923/rjet.2017.72.89.

60. Zhang Y, Bakshi BR, Deovessie ES (2008) Lifecycle assessment of an ionic liquid versus molecular solvent and their application. *Environ. Sci Technol.* 42(5): 1724-1730. https://doi.org/10.1021/es0713983.

61. Bianchini R, Cevasco G, Cheappe C, Pomelli CS, Douton MJR, (2015) Ionic liquid can significantly improve textile dyeing: an innovative application assuring economical and environmental benefits. *ACS Sustain. Chem. Eng.* 3(9): 2303-2308. https://doi.org/10.1021/acssuschemeng.5b00578.

62. El-Molla MM, Schneider R (2006) Development of eco-friendly binders for pigment printing of all types of textile fabrics. Dyes Pigm. 71: 130–137. https://doi.org/10.1016/j.dyepig.2005.06.017

63. Intelligence ethical fashion.com/article/ (2015)10-toxic. chemicals to avoid in your product used for dyeing/Accessed on 19 October 2021.

64. Saxena S, Raja ASM, Arputaraj A (2017) Challenges in sustainable wet processing of textiles In *Textiles and Clothing Sustainability* (Muthu S. *Eds*), Textile Science and Clothing Technology. Springer, Singapore. https://doi.org/10.1007/978-981-10-2185-5_2.

65. Kissa E (2001) Fluorinated surfactants and repellents. Surfactant science series. CRC Press pp 110-140.

66. Buck RC (2011) Perfluoroalkyl and polyfluoroalkyl substances in the environment: terminology, classification, and origins. *Integr. Environ. Assess. Manag.* 7(4): 513–541. https://doi.10.1002/ieam.258.

67. Textile Today news-letter, Assessed on 30 September 2018.

68. Schonberger H, Schäfer T (2003) Best Available Techniques in Textile Industry. *Federal Environmental Agency in the Textile series*. pp 162-212.

69. Das S, Singh S, Sharma D, Anowar M, Kaur R, Kaurandeep S (2017) Bio scouring and Desizing of Textile Fabrics Using crude Enzyme Produced by White Rot Fungus (Basidiomycetes) Isolated from Rotten Wood. *Int. J. Chem. Environ. Biol. Sci.* 5(1): 8-12.

70. Choudhury AKR (2017) Green chemistry and textile industry. *J. Text. Eng. Fash. Technol* 2(3): 351-361. https://doi.10.15406/jteft.2017.02.00056.

71. Future Materials (2012). www.toray.com, Accessed on October 2, 2021.

72. Future materials. (2013). www.amsilk.com.Accessed on October 3, 2021.

73. Rajput SK, Singh MK (2015) Sericin-A unique biomaterials. *J. Polym. Text. Eng.* 2 (3):29-35. https://10.9790/019X-0232935.

74. Sarovart S, Sudatis B, Meesilpa P, Grady BP, Rathanawan M (2003) The use of Sericin as an antioxidant and antimicrobial for polluted air treatment, *Rev. Adv. Mater Sci.* 5, 193-198.

75. Vighneswaran C, Subramaniananth M, Kandhavadiv P (2014) Bioprocessing of Textiles: Fundamentals for Applications and Research Prospective, *Woodhead Publishing in India, Delhi.*

76. Kiro D, Mojsov (2014) Trends in Bio-Processing of Textiles: A Review. *Advanced Technologies* 3(4): 135-138.

77. Gotmare VD, Patil VD, Gajul S, Jadhao AR (2013) Innovative techniques of size recovery & their economical and ecological aspects. *Colourage* 36(4):12-18.

78. Robinson GD (1993) ELVANOL-Polyvinyl Alcohol Warp Size Recovery. *Kuraray Co., Ltd.*-Accessed on 25 October 2021.

79. Sule AD, Bhardhan MK (2001) Recycling of textile waste for environment protection-An Overview of some practical case in the textile industry. *Indian J. Fibre Text. Res.* (26): 223-232.

80. Opwis Klaus, Echard Schollmeyer (2000) Advance textiles. *Wiley online library.*

81. El-Sheikh MA(2016) Novel photo-grafting of acrylamide onto carboxymethyl starch. 1. Utilization of CMS-g-PAAm in easy care finishing of cotton fabrics. *Carbohydr. Polym.* 152: 105-118.

82. Harane RS, Adivarekar RV (2017) Sustainable processes for pre-treatment of cotton fabric. *Cloth. Text. Res. J.* 2 (2). https://doi10.1186/s40689-016-0012-7.

83. Govindan N (2018) Reusing of pre-treatment liquor bath in textile wet processing. *J. Text. Eng. Fash. Technol* 4(6): 378-384. https://doi. 10.15406/jteft.2018.04.00169.

84. Shaid A, Osman MS, Hannan MA, Rahman Bhuiyan MA (2013) Direct Reusing of Textile Wastewater in Scouring-Bleaching of Cotton Goods Devoid of Any Treatment. *Int. J. Eng. Res. Dev* 5(8): 45-54.

85. Agler B (1995) Pollution Prevention in the Textile Industry, Training Manual. *Environmental Pollution Prevention Project. U.S. Agency for International Development.*

86. Karim MM, Dasa AK, Lee SH (2006) Treatment of colored effluent of the textile industry in Bangladesh using zinc chloride treated indigenous activated carbons. *Anal Chim Acta* 576: 37-42. https://doi.org/10.1016/j.aca.2006.01.079.

87. Zhang Y, Shao S, Yu W, Yang F, Xu X (2014) Study on Recycling Alkali from the Wastewater of Textile Mercerization Process by Nanofiltration, *Proceedings of International Conference on Environment systems Science and Engineering*, IERI Procedia Vol 9, Elsevier, 71-76.

88. Bergenthal JF, Tawa AJ (1984) Investigation of Textile Dye bath Reconstitution and Reuse, Volume I, Technical report, EPA-600/2-84-100. *Research Triangle Park, North Carolina, US Environmental Protection Agency.*

89. Bergenthal JF (1985) Textile dyebath reconstitution and reuse', Proceedings of Water Reuse Symposium III. *American Water Works Association Research Foundation*, TMP.

90. Bergenthal JF (1984) Wastewater Recycle and Reuse: Potential for Indirect Discharge Textile Finishing Mills, Vol.I, Technical Report. EPA-600/ 2-84-070., *Research Triangle Park, North Carolina*, US Environmental Protection Agency.

91. Cook FL (1983). Direct dye bath reuse: the future is now. *Textile World* 133(9):144–7.

92. Carr WW, Cook FL (1980) Savings in dye bath reuse depend on variations in impurity concentrations. *Text. Chem. Color.* 12(5): 33–37.

93. Pollution Prevention Tips for Wet-Processing Textile Mills Georgia Pollution Prevention Assistance Division', Georgia Department of Natural Resources, (http:// www.p2ad.org/ Assets/Documents/ma_textiletips.htm), *Replacement of harmful chemicals and Recycling/Reuse concepts*, 221/Accessed on 3 October 2021.

94. Tincher WC, Cook FL, Carr WW, Olson LH, Bergenthal (1985) Reconstitution and Reuse of Dyebaths in Carpet Dyeing. *Journal of the Society of Dyers and Colorists* 12: 51-58.

95. Shenai VA (2003) Effluent treatment and / or reuse of spent dye baths. *Colourage*, 39-40.

96. Wambuguh D, Chainelli RR (2008) Indigo dye waste recovery from blue denim textile effluent: a by-product synergy approach, *New J. Chem.* (32):2189-2194.

97. Buscio V, Crespi M, Gutiérrez-Bouzán C (2015) Sustainable dyeing of denim using indigo dye recovered with polyvinylidene difluoride ultrafiltration membranes. *J. Clean. Prod.* 91: 201-207. https://doi: 10.1016/j.jclepro.2014.12.016.

98. Amaral MCS, Neta LSF, Souza M, Cerqueira N, Carvalho RB (2014) Evaluation of operational parameters from *a microfiltration system for indigo blue dye recovery from textile dye effluent. Desalin. Water Treat* 52(1): 257-266. https://doi.org/10.1080 /19443994.2013.793618.

99. Sukla SR, Harad AM, Jawale LS (2007) Recycling of waste PET in to useful textile auxiliaries. *Waste manage.* (28): 51-56.

100. INVISTA. (2017) The Recyclability of Type 6,6 Nylon. Accessed on 2 November 2022.

101. Ammayappan L, Seiko J, Raj AA (2016) Sustainable Production Processes in Textile Dyeing. In *Environmental Footprints and Eco-design of Products and Processes* (Muthu SS *Ed.*), *Springer*. pp 232-248.

102. Schmidt A, Watson D, Roos S, Askham S, Poulsen PB (2016) Gaining benefits from discarded textiles - LCA of different treatment pathways. *Nordic Council of Ministers*. Accessed on 28 October 2021.

103. Stein Fibers Ltd (2018) Environment, [Online] Available :https://www.steinfibers.com/environment/Accessed on August 2018.

104. Kole SS, Gotmare VD, Athawale RB (2019) Novel approach for development of eco-friendly antimicrobial textile material for health care application. *J. Text. Inst.* 110*: 254-266.*

105. Hussain T, Wahab A (2018) A critical review of the current water conservation practices in textile wet processing. *J. Clean. Prod.* 198: 806-819. https://doi.org/10.1016/j.jclepro.2018.07.051

106. ACS Webinars (2017) Sustainability Challenges of the Textiles, *Dyeing and Finishing Industries: Opportunities for Innovation - Dr. Richard Blackburn.* ACS Green Chemistry. Accessed on 28 October 2021.

107. https://www.acs.org/content/dam/acsorg/events/popular-chemsitry/Slides/2017-03-30-*textile-chem-slides.pdf.* Accessed on 29 October 2021.

108. Bach E, Cleve E, Schollmeyer E (2002) Past, present and future of supercritical fluid dyeing technology an overview. *Prog. Color. Color. Coat* 32: 88:94.

109. Höcker, Hartwig (2002) Plasma treatment of textile fibers. Pure and applied chemistry, *The Textile Institute Book Series*. pp 265-277.

110. Gotmare VD., Samanta KK, Patil VD (2014) Surface modification of cotton textile using low temperature plasma. *Int. J. Bioresour. Sci.* 2(1): 37-45.

111. Data Source: Market Line, (2017), Accessed on 10 October 2021.

112. www.greenpeace.org/eatasia, Accessed on 10 October 2021

113. https://shenglufashion.com/tag/wtom, Accessed on 10 October 2021

114. www.greenpeace.org/eatasia, Accessed on 12 October 2021.

115. Eryuruk SH (2012) Greening of the Textiles and Clothing Industry. *Fibres Text. East. Eur.* 95 (6): 22-27.

116. Paul R, Thampi J, Jayesh M (1996) Eco-friendly textile processing - a global Challenge, *Textile Dyer and Printer* 29 (16): 17-20.

117. Parthiban M, Shanmugasundaram OL (2005) Eco-friendly textiles – A comprehensive overview (Review). *Journal of the Textile Association* 66(3): 121-125.

118. Ali A, Hossain D, Shahid MA (2018) Development of Eco-friendly Garments Washing for Localized Fading Effect on Garments: A Future Sustainable Process for Single Step Dyeing Fading Effect. *Adv. Res. Test. Eng* 3(1),10-22.

119. Nair P (2008) New softener for green processing of denim. *ATA Journal* 19 (1): 46-47.

120. Phong M (2008) More options for green processing. *ATA Journal* 19 (2): 48-49.

121. Srivastava SK (2007) Green supply-chain management-a state-of-the-art literature Review. *Int. J. Manage.* 9(1): 53-80.

122. Wang Fangquing (2013) Green Garments: sustainable practices in AsiaJust-stylehttps://www.just-style.com/management-briefing/sustainable-practices-in-asia_id118586.aspx, Accessed on 16 October 2021

123. The SDC EC organised seminar on Clean and Green Garment Processing (2016) http://colour.sdc.org.uk/2016/06/clean-and-green-garment-processing, Accessed on 16 October 2021.

124. Bikash J, Bishnu PD, Khandual A, Sahu S, Behera L (2015) Ecofriendly Processing of Textiles. *Materials Today: Proceedings* 2: 1776 – 1791.

125. https;//www.greenscreenchemicals.org, Accessed on 17 October 2021.

126. Wolfgang (2011) Sustainable denim Eco-Labeling And Environmentally Friendly Denim Production" denims and jeans. Sustainable-denim-ways environmentally-friendly-denim-production/3563/https://www.denimsandjeans.com/environment/ Accessed on17 October 2021.

127. Wilson A, Uncapher J, McManigal J, Lovins L, Brownin H, Cureton M (1998) Green development: Integrating ecology and real estate (Vol. 9).John Wiley & Sons. pp 235-247.

128. Blue goes green new approach to make denim eco-friendly. *Fibre2Fashion.com*, Accessed on 18 October 2021.

129. Clothing technologies that keep it green. *Fibre2Fashion.com,* Accessed on 20 October 2021.

130. EPRS |Environmental impact of textile and clothes industry (2019), 'European Parliamentary Research Service Author: Nikolina **Šajn** Members' *Research Service PE*-633.143.

9

Innovation in textile auxiliaries for sustainable processing

C. N. Sivaramakrishnan
Chemical technologist, Mumbai, India
Email: cns8024@gmail.com

Abstract: Innovative auxiliaries play an important part in the manufacturing of textiles products. The knowledge of chemical processes, polymer chemistry and comprehension of complicated biochemical processes culminated in what we see in a processor's minds a drastic change. Advance chemical innovation has played a crucial role in maintaining textile chemicals rising in line with increasingly changing health, protection and environmental regulations. A speciality in this category is textile wet processing which can help to lower textile chemical manufacturing cost. The supply of the chemical in bulk or semi-bulk containers is a solution to reduce the cost of a compound. By this system, the drums and drum clearance costs can be eliminated. That quality requirements, followed by rising cost pressure, are keeping textile processors and the textile industry as a whole in a challenging role.

There is an enormous demand to provide low-cost textile auxiliaries. Continuing the ten-year-old old pattern of the textile industry's fastest development, textile chemical producers have now come up with innovative ideas to make textile chemicals accessible in concentrated conditions by lowering their costs. Special auxiliaries for low-temperature bleaching of cotton, spin finishing, salt-free dyeing of cellulose with reactive dyes, single bath dyeing of cotton and polyester and functional finishing are discussed. Moreover, the upcoming chemistries like sugar-based surfactants, acrylic-based synthetic thickeners and silicone copolymers are also discussed. All these technologies are taking action to attain more sustainable textile processing.

9.1 Introduction

The fibres are the prime raw materials used in the textile industry that is converted into yarns, then knitted, woven and turned into fabrics. These are further processed to get bleached, dyed, printed or finished fabrics. During this conversion, various auxiliary chemicals are used at different stages for different purposes, generally called as textile auxiliaries. Within this family, a significant number of them are surfactants or surface-active agents.

The textile industry and chemical industry have been linked together since the beginning of the industrial revolution, and without chemical industry, there would not have been a modern textile industry. The international economy, accompanied by a shift in many areas, is currently undergoing a profound change. With an ever-increasing threat to the environment, the textile processing is under severe pressure to comply with various regulations set by the local pollution control bodies. Chemists, scientists and technologists are called upon to produce innovative products putting their theoretical knowledge and practical shop floor experience. Undoubtedly, knowledge of chemistry is regarded as fundamental, and thus, a clear understanding of various other disciplines related to applied science, macromolecular chemistry, dye chemistry, surfactant chemistry, colloid chemistry and physical chemistry is required. There are more than one hundred thousand chemicals in practice, but that constitutes just an insignificant portion of those known. More than 20 million chemicals are registered with the Chemical Abstract Service (CAS), and innovative ones come at a much faster speed than before. It is all the more challenging to distinguish one from the other. Over 2000, different textile auxiliaries and speciality chemicals are used during various stages of processing to meet the growing requirements [1,2].

Given the wide variety of advanced textiles being produced and launched today, there is a growing understanding of the value of textiles as building block resources for the production of new goods. The textile industry is witnessing spectacular growth with the advent of more sustainable fibres making the processing operations more complex. Textile processing plants have meanwhile seen significant changes in simplification and automation. The benefits of nanotechnology are made use of to upgrade many functional finishes.

9.2 Invention and Innovation

The invention may be well-defined as the formation of a method or a product for the first time, else ways, and innovation is an improvement to an existing product or process. Institutions, which is often a neglected area, are the place where innovations happen. Technology and innovations go hand in hand. Innovation in textile processing is entering a new zone with the increasing importance of sustainability, whether it means developing bio-based raw materials instead of those based on fossil fuels or developing completely new products that are more sustainable innovation is also critical to another key area - the sustainability agenda – finding ways to use less energy and use renewable energy. The focus, therefore, is on "knowledge-based" textile chemicals that have inherent functionalities, often referred to as "functional chemicals with the only goal of claiming modern body-friendly abilities. They are also meant to redefine the role of modern textile chemicals by expanding their capabilities to meet the modern living style requirements. Functional finishing is one of the recent advancements in textile finishing that is used as a technical bridge for achieving the latest in the field of textiles.

Ideal sustainable auxiliary chemical

- To provide a similar function and or perform better than the existing product.
- Have minimal environmental footprint (water – air – energy)
- They are preferably manufactured from renewable raw materials with no adverse impact on humans.
- No adverse impact on food supply or water

Factors contributing to innovation

- Ecology
- Legislation
- Cost
- New substrate and new process
- Changing demographics
- Fashion and media
- Productivity constraint
- Sustainability

9.3 Changing scenario

In early times, the textile business was in fabric. The mills produced fabrics and sold them in retail. From fibre to customer, it has many stages of production, and it is not uncommon for each stage to be delinked from the one beyond the next one. For example, the yarn manufacturer rarely has any contact or interest in the end product manufacturer. The processor has little to do with the fibre manufacturer. This creates many isolated pockets and causes a massive divide in understanding. But now things are changing. Textile processing is accused as one of the chief contributors to environmental pollution. The production of toxic dyes and chemicals used in the wet processing of textile cause serious health-related issues. The growing awareness has resulted in the innovation of products across various segments of processing, resulting in the reduction of water-energy consumption and carbon footprints.

9.4 Rating sustainability

Each step of processing uses different chemicals using different techniques which can result in various sustainability issues. The risk evaluations tackle a range of concerns relevant to social, economic and ecological sustainability. The risk may be regarded as a likelihood and impact (consequence).

Likelihood	Impact (Consequences of occurrence)
H: Highly likely	H: High impact
M: Might happen	M: Moderate impact
L: Less likely	L: Limited impact

Table 9.1 Pollution categories for various chemicals used in the textile industry

Chemicals utilised in textile processing	Pollution characteristics	Pollution category
Acids, alkali, salts, oxidizing agents	Inorganic pollutants, comparatively inoffensive	1
Natural oils, waxes, fats, surfactants, sizing agents, reducing agents, organic acids.	Easy biodegradables; with a reasonable - high BOD5	2
Dyes –OBAs, and impurities of polymeric nature, silicones, synthetic polymeric resins.	Hard to biodegrade	3
PVA, surfactants, and mineral oils	Difficult to biodegrade moderate BOD5	4
Formaldehyde, resins, salts of heavy metals, cationic surfactants complexing agents and coloured compounds or accelerators	Cannot be removed by conventional biological treatment, low BOD5	5

9.5 Textile processing – Ecology and RSLs

As the world awakens to the damaging effects of many chemicals that were synthesised by humans, ecology and pollution have become one of the main focus issues. Since the late 1980s and even more rapidly in the 1990s, requirements related to environmental protection, consumer health and safety have brought in more laws and regulations than ever before. All recent regulations have forced the processor to seriously ponder the production process, the chemicals being used, dye manufacturers to look at the dyes that are being made and auxiliary manufacturers to contemplate a change in the chemicals being produced and ensure that there are no left-over harmful substances. The government and third-party accreditation bodies have released restricted substances lists [RSLs] that relate development ecology to human ecology. These lists provide opportunities to encourage the usage of safer chemical inputs and set goals for the monitoring of the sustainable output of textile goods. It is vital to recognise the pollutants generated and their harmful effects on the environment. Comprehensive approaches are essential for the study of all leading groups of environmental contaminants of significance, such as herbicides, perfluorinated substances, pesticides, endocrine disruptors, dioxins and several others [3,4].

9.6 New fibres / Complex blends

Fibres can be natural and synthetic, and they can be spun into yarns. The yarns can be converted into fabric through weaving, knitting, or other methods. Fabric care and performance are directly related to fibre behaviour and properties. Learning regarding fibres and their properties can boost a dyer knowledge of fabrics. Majorly four natural fibres and twenty-three man-made fibres are commercially available, which include synthetic fibres and natural fibres derived from plants and animals. The plant fibres are generally composed of cellulose (e.g. cotton and linen). The animal fibres are composed of protein (e.g. silk and wool). The man-made fibres are constructed under two classes semisynthetic and synthetic.

Blends hold a special place in the world textile market, as the number of possible blend compositions and blend levels is exceedingly broad. Blend represents a variety of options, and most are the deliberate combinations of various physical and chemical fibrous polymers. Blended fabrics can deliver the best of all worlds by mixing materials to accentuate their strengths and restrain their disadvantages. The method of blending produces thousands of different fabrics. Many times, fibres are combined for aesthetics and economic reasons; the most significant explanation is to reach a variety of physical properties in the blends that cannot be achieved with the individual components. The most evident and perhaps the most effective method is the blending of cotton/polyester. Polyester provides abrasion resistance, tensile strength and dimensional stability, whereas cotton offers water absorption, reduced pilling and comfort. It is very important to understand the different properties of fibre blends and how these properties affect their dyeing behaviour during colouration. The fibre that makes up at least 50% of the blend will have a more significant effect on the properties of the fabric.

Identification of the fibres and the regulation of the variables are the root of effective colour management. Blends generate more complexity than single-fibre, and synthetic fibres are more durable than natural fibres. Yarns and fabrics created show a variety of beneficial properties that cannot otherwise be accomplished by a single textile element. Based on the chosen substrate, various forms of dyeing and application methods are possible. Consequently, the blending of two or more separate forms of fibres is essential and important. Industrial and technical criteria for blended fibres present tremendous challenges for dyers to produce reproducible performance and high-quality under manufacturing conditions. With the number of available substrates and a multitude of variables, the dyeing technologists need to remember to render complicated dyeing blends.

Three approaches can produce mixed fibre systems:

1. Different types of fibres in the staple form are combined at the processing stage of the yarn during the spinning process.
2. Different types of fibres are spun separately, and the resulting yarns are twisted together to create a blended yarn.
3. Different types of fibres are spun separately and mixed only at the weaving level (union fabrics), where one or two fibre yarns are used as weft and the others as warp.

Single fibre dyeing is a lot easier and simple as an operation compared to dyeing of blended fibres and union fabrics. Given these drawbacks, dyeing tends to happen as near as possible to the end of the finishing cycle. Besides, this allows the dyer to meet the demands of the industry without the need to store vast quantities of material already dyed in yarn form in all usable shades.

Fabrics made of mixed fibres may be dyed to give effects such as:

Union or solid dyeing: In this instance, all the components are dyed to provide solids shades

Reserve or resist dyeing: where at least one of the components is essentially undyed, i.e., almost white.

Cross dyeing: This can be represented as purposely producing colour contrasting fibres. The contract can be significant variations in brightness, hue, and, in certain situations, depth only recognised as shadow effects or tone in tone.

Some of the approaches to the economical dyeing of blends is to be able to achieve the desired colour result in a cumulative dyeing cycle times less than the amount of the dyeing times required for the subsequent dyeing of the blend elements.

This must be attained without compromising the operating flexibility or quality within the dye house. Processes in which different mixing elements are dyed individually can be stated as conservative. The techniques are said to be rapid where the specific steps of the conventional procedures are run together or omitted (with corresponding time savings). Fast dyeing processes have the highest commercial importance and demand the most creativity be invented. All the big dye firms have their own sets of such operations.

Cross Staining: It is one of the unexpected possibilities while dyeing blends, where one or more of the elements of the blends are coloured by the dye bath elements to create a stain rather than dye subsequent in poor fastness.

9.6.1 Commonly used blends

Polyester /cellulose: A substantial portion of polyester production is utilised to produce this combination. Polyester-cellulose hybrids are being used for all types of apparel and home textiles. Cellulose is typically a cotton component, although viscose staple fibres and sometimes linen is often used. The suggested mixture ratios are 50: 50, 80: 20, and 67:33 polyester to cellulose. Microfibre polyester which is 60 to 100 times finer than a human hair is also utilised. They are mixed with natural fibres and are utilised for active sportswear, outdoor pursuits, and clothing. They are also used for carpets, knitwear, sportswear and underwear. The disperse dyes are used to dye the polyester in the polyester-cellulose blends, whereas the cellulose can be dyed with direct, vat and reactive dyes. During polyester-cellulose blend dyeing, the cellulose fibres get stained by disperse dye, which can be easily removed by washing. Apart from that, the dyes used for cellulosic dyeing cannot stain the polyester or can only slightly stain the polyester. In the one-step or one-bath process, specific auxiliaries like acid donors and pH sliding agents are used that lower the pH when the temperature increases. Therefore, it is likely to fix the reactive dyes in alkaline mediums and then by increasing the temperature, the optimal dyeing conditions (pH 5 - 6) for disperse dyes can be attained. Otherwise, it is desirable to use alkali-stable disperse dyestuffs at pH 8-10, which often prevent oligomer issues. In a continuous process, the applications of dyes on the fabric are usually done in one bath. Subsequently, the fabric is dried, and the disperse dye is fixed to the polyester portion by a thermosol procedure. Then, the second dye is formulated based on the standard procedure of each class, using pad-batch, pad-jig or pad-steam methods.

Thus, the complexities involved in the dyeing due to many variables present a daunting task in choosing the right auxiliary for the desired operation [5].

9.7 Innovation in textile auxiliaries

Natural resources are depleting, forcing scientists and chemists to develop innovative ways to replace petrochemical raw materials. The concept of green chemistry is getting a foothold and is being implemented with a broader vision.

Surfactants and polymers have a special place in the processing of textile fabrics. Over 80% of the products used for textile processing are either based on surfactants and polymers or used as one of the components in the formulation for an application. They are synergistic and unique in their ways and perform excellent properties.

9.8 Oleochemicals

Surfactants are smart components that support our lives in many ways, even if much of the time, they are opaque to us. Their versatility allows them the gateway to a variety of textile processing operations.

The cluster of surfactants and oleochemicals is a clear definition of a value-chain cluster. Oleochemicals are derived from animal fats and vegetable oils through enzymatic and chemical reactions. Some of them (alcohols and glycerol) are also derived from petrochemical feedstocks. The major process for transforming animal and vegetable oils and fats into oleochemicals is hydrolysis, the separation of natural triglycerides into crude blended fatty acids and crude glycerine under the control of pressure, temperature and water. Also, fatty acids are the main oleochemicals produced from vegetable oils and animal fats. The fatty acid is a carboxylic acid with a long aliphatic chain that is either unsaturated or saturated. Many naturally coming fatty acids have an unbranched chain with an even number of carbon atoms [6].

9.9 Novel surfactants

Diffusion accelerator for Polyester fibres

Polyester fibre is made by condensation, involving an acid with two carboxylic acid groups (Purified terephthalic acid PTA) and alcohol with two hydroxyl groups (monoethylene glycol MEG) at high temperature under vacuum.

Benzene - 1,4 dicarboxylic acid

Ethane - 1,2-diol

Figure 9.1 The acid with two -COOH groups and alcohol with two -OH groups

The tensile strength of polyester fibres is relatively high, which makes them resistant to stretching, shrinkage and crease, thus can embrace their shape well. As a best 'wash-and-wear' fibre, polyester is washable and easy to care; therefore, it is widely used for apparel, home furnishings and other textile products. The durability of blended fabrics will be increased when polyester is blended with natural fibres [7].

Auxiliaries are not an integral part of the dyeing and printing process; however, they can enhance the colouration properties of dye during dyeing and printing. Polyester is hydrophobic and thermoplastic in nature. A highly compact and crystalline structure is a deterrent for trouble-free dyeing of polyester with disperse dyes. The fibres have a high glass transition temperature, requiring a high temperature of up to 130°C for dyeing. Organic carriers swell polyester fibre at a boil, thereby facilitating the diffusion of disperse dyes into polyester. Carriers have limitations as it leads to poor light fastness. Another drawback is carrier spots in the event of a breakdown of emulsion at high temperatures. Novel surfactant blends based on a mixture of fatty acid condensates/polyion compounds and solvents dramatically enhance the dye pick up of disperse dyes on polyester fibre by virtue of its high diffusion property.

9.10 Auxiliaries for low-temperature bleaching

The pretreatment of fabrics is the primary step of the textile wet processing unit. This pretreatment process removes the impurities present in the fabric and makes the fabric absorbent and white, which is the prerequisite for dyeing, printing and finishing of the fabric.

Special bleach activators have been produced for bleaching at low temperatures. Pentaacetyl glucose (PAG) is a common and economic intermediate for cotton bleaching at a low temperature and has non-toxic, biocompatible, and sustainable properties. For the pretreatment of cotton fabric, the PAG is utilised as a bleach activator for H_2O_2 bleaching. The bleaching results of the H_2O_2/PAG bleaching system in terms of whiteness index (WI), bursting strength and H_2O_2 decomposition rate are comparable with conventional hydrogen peroxide bleaching. H_2O_2 decomposition rate increased significantly between 55 to 60°C with little damage to the strength, thus showing good promise. Potassium triple salt can also trigger the decomposition of hydrogen peroxide, allowing bleaching to be carried out at 50°C [8–11].

9.11 Cleavable surfactants

Surfactants are critical raw materials in the development of textile auxiliaries. Surfactants are used right from the pretreatment stage to the finishing stage. Surfactants can reduce the surface tension of liquids at low concentrations. The degree to which these molecules achieve this depends on the balance of their hydrophobic and hydrophilic areas. Once the surfactant is appropriately balanced, causing a force at the liquid-air interface, which reduces the surface tension of surface energy. There is an important relationship between surface pressure and surface tension. The surface pressure at the concentration of a solution is equal to the surface tension of the pure solvent minus that of the surfactant solution at the same concentration. Mostly, surfactants are used in emulsion polymerisation, where surfactants regulate the molecular weight and its distribution through micelle formation. Nowadays, cleavable surfactants showed sparked interest in various emulsion-based applications because of their ability to turn off the surfactant character according to requirements. Cleavable surfactants have attractive potential from both environmental and economic points of view. The surfactant is capable of being reformed again after decomposition, which also opposes the thermodynamic breaking.

Sodium vinyl sulphonate (SVS) is a typical example of a cleavable surfactant. With the advancement of SVS, other cleavable surfactants have been developed, having decomposition, tension effects, and micellization properties. All have been utilised to cleave the polar head from the non-polar tail. The cleavable surfactants improve efficiency by reducing the amount and extent of aqueous waste stream contamination. They can be used productively during emulsion polymerisation of the acrylic binders used for pigment printing.

Surfactants are also known for their performances at low critical micelles concentration (CMC). Many surface-related properties either maximise or minimise around the CMC. These foundations are traditionally based on the CMC determined at equilibrium state by static surface tension (SST) measurements. Since surfactants face severe conditions of non-equilibrium state (dynamic processes) in any industrial applications, the dynamic surface tension (DST) measured at a non-equilibrium state may be preferred over SST equilibrium state data. Various types of surfactants can be used to alleviate the wetting problems associated with aqueous-based applications. The effectiveness of acetylenic diols-based surfactants in providing good defoaming and surface wetting characteristics stems from the unique structure of the molecule. Most conventional surfactants, in contrast, have a long chain hydrophilic or anionic species, all of which have disadvantages as high-performance wetting agents due to their inherent problem of foaming.

The backbone is a 10-carbon chain with a carbon-carbon triple bond in the middle and two adjacent hydroxyl groups. The combination of these groups yields a region of electron density, making the molecule polar. There is also an asymmetrical, highly branched alkyl group on each side of this region, supplying the molecule with two hydrophobic regions. Overall, the molecule has a hydrophobic-hydrophilic-hydrophobic structure, making it an excellent wetting agent or surface tension reducer. Furthermore, by careful tinkering of the hydrophilic – hydrophobic ratio, these acetylenic diols can be designed as anti-foaming and/or defoaming agents. These high-performance low foam acetylenic glycol surfactants (wetting-cum-antifoaming) are used in a wide variety of applications [12–16].

9.12 Polymeric surfactants

During the past three decades, safety concerns in all areas of surfactant applications have guided the replacement of solvent-based formulations with aqueous systems. The process has gained further momentum owing to limits on the release of volatile compounds in the ecosystem. In areas like cotton cultivation, concentrated emulsions have displaced the emulsifiable concentrates. In the field of polymer emulsion, new colloidal classes have been established. The area of polymeric surfactants has been of great interest in recent years. Such polymeric surfactants can covalently bind to the dispersed phase and, as such, have a significant benefit over typical surfactants that are only physically adsorbed and can be separated from the interface by a change in the phase or shear, resulting in the instability of emulsion. These surfactants become an integral part of the finished product due to the binding of polymeric surfactants to the dispersed phase, which prevents the release of surfactants in the effluent, thus reducing the environmental impact of the intermediate product and commercial formulations.

The repulsive barriers of the polymer chain prevent coalescence and agglomeration as compared to monomeric surfactants both in aqueous and non-aqueous suspension, thus making it an excellent choice for emulsion polymerisation, where surfactants control the molecular weight distribution through micelle formation. The polymeric word suggests a large number of repeating molecular units with high molecular weight. Many polymeric surfactants have only a few repeating units and the oligomeric phrase. Polymeric surfactants offer an excellent opportunity regarding functionality, diversity, and flexibility. This advanced polymer chemistry allowed synthetic polymer researchers to develop novel, and well-defined amphiphilic block copolymers, which show exciting surfactant behaviour. It is especially valid considering the recent development in controlled radical polymerisation chemistry. One of the salient features of polymeric surfactants is their ability to impart excellent particle stability. The multiple anchoring points of polymeric surfactants facilitate excellent interaction between the surfactant and the substrate. Combined with steric stabilisation, this results in highly robust and stable finish product formulations. Polymeric surfactants can stabilise very high load suspensions and still maintain the viscosity of the product to an acceptable level. A class of polymeric surfactants that has multiple end applications are ethylene oxide–propylene copolymers, acrylic styrene copolymers, methacrylic acid copolymers and alkyd PEG resin derivatives. Polyacrylic acid of low molecular weight (<15000), sulfonated styrene-maleic anhydride copolymers and condensates of naphthalene sulphonate with formaldehyde are also involved in the family of polymeric surfactants. Amongst the natural class, alginates–pectin and protein-based products figure in the list of polymeric surfactants. Polymeric surfactants exhibit low surfactant characteristics in terms of the formation of micelles and reduction of the surface tension reduction in an aqueous solution. Polymeric surfactants outstand in minimum two separate forms, a combination of super-low CMC: The right mixture of conventional surfactant and polymer will contribute to a significant reduction in CMC. The question of whether you like a decreased CMC is another thing; it is not the primary consideration in certain situations. Nevertheless, if a low CMC is required the suitable polymer could attract the surfactant to itself, leading to an initial "virtual micellization" and, hence, a low critical concentration of micelles.

Some polymeric surfactant innovations are the domain of hydrophobic and hydrophilic deblock copolymers, where the hydrophobic block is always hydrophobic. Hydrophilic deblocks, in which the less hydrophilic block can be modified to become hydrophobic by adjusting the condition of a solution such as a temperature, pH of a solution, or concentration of electrolytes, have recently been granted attention. For example, to polymerise, vinyl monomers such as styrene or methacrylate, an initiator based on alkyl halide in combination with a transition metal catalyst, generally, Cu has been used. It is assumed that the propagating intermediate is radical-based; hence such chemistry is especially resistant to functional monomers and to be conducted in protic solvents with water.

It is anticipated that these tailor-made surfactants can be used in several specific fields, with the preparation of novel colloidal nanostructures, advanced latex emulsifiers and stabilisers. In the area of film-forming polymers, the water-resistance of polymeric surfactants is far superior compared to conventional surfactants [17].

9.13 Sugar surfactants

The major driving factor for the polyacrylonitrile fibre market is its high demand in blending applications. This fibre is mixed with other fibres such as wool and cotton, to enhance the physical and chemical characteristics of the final product. Polyacrylonitrile fibre is also used to make knitted fabric such as socks and sweaters. The fibre possesses aesthetic appeal by its ability to accept bright colours and subdued, muted tones.

The big obstacle for producers and researchers is to develop more environmentally friendly products that conform to the stringent government regulations on volatile organic compounds (VOC), while at the same time formulating less flammable and hazardous processes to reduce costs and improve safety for workers. Because of their beneficial properties (in the absence of adverse effects on the ecosystem, as in the usage of synthetic surface-active agents), natural surfactants are gaining recognition year after year.

Given its biodegradable and toxicological properties, the eco-friendly sugar-based cationic surfactant is an advanced and novel auxiliary for polyacrylonitrile dyeing. Sugar surfactants are next-generation surfactants. Built from sugar or starch in combination with fatty acids, these surfactants have a renewable carbon index of up to 95%. The degree to which these molecules can perform depends on the balance of their hydrophobic and hydrophilic regions. Retarding agents provide safe control of the pickup speed and facilitates level dyeing. Thus, levelling agents and retarding agents can be termed as synonymous when dyeing polyacrylonitrile fibres, especially in pale and medium shades. Retarding agents also level out the rate of pickup of the dye and promote migration of the dyestuffs on the fibre. Here, a distinction has to be made between retarding agents and migrating agents, depending on the predominance of the retarding or migration-promotion action of the cationic levelling agent. Thus, the affinity of retarding agents is higher than that of migrating agents.

The retarders will go on the acrylic fibre during the dyeing process in the same manner as a cationic dyestuff occupying the dye sites, or dye receptors, on the fibre molecule. Retarders remain on the acrylic fibre as long as the saturation limit of fibre is not exceeded. They can be replaced on the fibre by cationic compounds of higher affinity. Cationic dyes of lower affinity can be prevented from going on to the fibre that is blocked by retarders. The cleavage products produced by the chemical dissociation of the retarders have a low affinity, enabling them to migrate back into the liquor. This can prevent the blocking of the dyestuffs. Migrating agents, whose affinity is lower than that of the retarders, exercise only a very low retarding action on cationic dyestuffs.

Sugar surfactants and levelling agents used in dyeing are distinctive in some ways. Primarily, both demonstrate the principles of levelling, and secondly, the sorption properties, reaction kinetics during and after dyeing, evaluating dyeing rate constant, and calculation of diffusion coefficient which are very important parameters for dyeing [18,19].

9.14 Sugar derivatives

The usage of sodium hydrosulphite and sodium sulphide to develop a water-soluble dye form is raising environmental issues and concerns. There are alternate methods; either the dye is sold in a reproduced form, or different reducing agents are utilised. There are sugar-based systems; glucose can be useful either alone or in conjunction with hydroxyacetone or sodium dithionite.

The importance of sulphur dyeing of cellulosic fibres, mainly cotton, is being economically realised in the dyeing industry. Sulphur dyeing involves the use of multiple auxiliaries, all of which have detrimental environmental consequences. The most destructive of these is the sodium sulphide reducing agent, which is needed to reduce the dye molecules to a water-soluble form to allow adsorption and diffusion into the fibre. Efforts have been made to substitute sodium sulfide utilised with several eco-friendly reduction sugars in the sulfur dyeing process. The redox potential of different pentose and hexose monosaccharides and reduction of disaccharides is equivalent to that of conventional reducing agents. Cotton was consequently dyed with the most effective Sulphur dye in the world, C.I Sulfur Black 1, the usage of reducing sugars under alkaline conditions, and the use of commercial sulphide reducing agents relative to dyeing obtained. Reducing sugars provides equal, and in many instances better, outcomes in colour intensity and wash fastness with regards to commercially available sulphide-based reducing agents, which also account for the overwhelming majority of Sulphur dyeing process that is of considerable environmental concern. Reducing sugars in Sulphur dyeing will, therefore, offer an alternative to sodium hydrogen sulphide and sodium polysulphide, which is biodegradable, non-toxic, and cost-effective [20, 21].

9.15 Sugar acrylates

Both sequestering agents and chelating agents perform the same function in a system, i.e., making a metal ion by creating a stable metal ion complex. This helps to prevent these metal ions from participating in ongoing chemical reactions or interacting with specific chemical reactions. Thus, such components are very important. Although both these compounds do the same thing; however, there are differences between them. A chelating agent can attach at a time to a single metal ion, while a sequestering agent can attach at a time to a few metal ions. The sequestering value of sugar acrylates is as high as phosphonates or amino polycarboxylates. They are biodegradable, efficient components used in the pretreatment of the cellulosics, such as desizing, scouring, bleaching and mercerizing, and as conditioners for the dye bath. Salient features of sugar acrylates are their excellent chelation values both in acidic and alkaline mediums and at varying temperature ranges. Sugar acrylates have no detrimental effects on metal-containing dyestuffs and broadly are non-foaming products [22].

9.16 Ester quats-based surfactants and softeners

Because of their inherent weaknesses, the product based on pure fatty acid amine condensates is being used less and less in everyday practice. Paraffin is widely known for its usage for smoothening chemicals. Paraffins with a carbon chain length of C_{24} to C_{32} are quite impressive. Paraffins with a shorter carbon chain can quickly be vaporized, while those with longer carbon chains will only transform under pressure to functional emulsions.

Softeners also provide an array of other benefits such as softness and fluffiness, decreasing static electricity build-up, facilitating ironing, eliminating ironing of permanent-press fabrics, repelling water and oily stains, reducing soiling, improving hand, reducing drying time, reducing wrinkling, improving flammability resistance, imparting a pleasant smell, decreasing pilling on man-made fibres, decreasing absorbency, and acting as a bacteriostatic and so on. Nevertheless, the performance of all types of fabric softeners may not be commensurate all. It requires a combination of two or more products to achieve desired properties, although some of them may appear contradictory.

Ester quats constitute an important class of surface-active – cationic surfactant - quaternary ammonium compounds synthesised by reacting long-chain (C_{12} - C_{22} or C_{16} - C_{18}) fatty acids with a tertiary amine in the presence of an acid catalyst and subsequently quaternised with different molar ratios of dimethyl sulphate to obtain a viscous quaternary ammonium compound. For the triethanolamine ester quat (TEAQ), shorter chain lengths may be present. For the manufacturing of ester quats typical distribution of the carbon chain length of the natural fats and oils is used as starting materials. Ester quats containing some degree of unsaturation having an iodine value range between 20 to 30 exhibit good softening properties.

9.17 Methyl ester route

The processing of ester quats through the methyl ester route requires transesterification of palm-based triethanolamine (>99%) and methyl ester in the presence of a basic catalyst that produces methanol as a by-product. Likewise, dimethyl sulphate is quartered to the ester amines. The methyl ester route is favoured over the fatty acid route because milder-lower temperature and vacuum are needed for the process conditions. The reaction time is 1 hr vs 4 hr by the fatty acid route.

In the early 1980s, the ester quat surfactants were introduced in the French market, and after the early 1990s in the European market, when questions were posed in Europe regarding the environmental quality of the typical cationic surfactants used hitherto (Di-hardened tallow di-methyl ammonium chloride, or DHTDMAC). Ester quats are structurally identical to the previously used unesterified dialkyl quats, except that ester links have been added to bind the alkyl chains to the molecule's head-group, rendering them more prone to biodegradation and hydrolysis. The integration of ester linkages into the aliphatic chains greatly enhances the kinetics of cationic surfactants' biodegradation. These quats are solids with varying melting points and decompose upon heating. Lower molecular weight quats are soluble in water.

An example is tetramethylammonium chloride, while the solubility decreases in polar solvents and increases in non-polar solvents with an increase in molecular weights of quaternary ammonium compounds. Fabric conditioners generally comprise of three esterquat groups, DEEDMAC (diethyl oxy ester dimethylammonium chloride), TEAQ (triethanolamine quat), and HEQ ((Z)-2-hydroxy-3-[(1-oxo-9- octad ecenyl) oxy propyl trimethyl ammonium chloride). They combine a good environmental profile, especially when it comes to ready and ultimate biodegradability (OECD criteria), with the structural features needed for an effective fabric conditioner. Certain TEA ester quats are composed of mono-, di-, and tri-ester quats with residual mono-, di-, and tri-ester amines. This detailed chemistry results in emulsions that contain several types of emulsion structures, some of which do not effectively contribute to softening performance upon dilution in water during the rinse cycle of a fabric washing process because of their high solubility.

Although softeners have been used for decades, there is still a lack of agreement on both the active deposition on cotton fabrics and softening process. The surfactants used in conditioners are self-assembled into vesicles of micron size. The softening effect is usually thought to be related to the development of a lubricating surfactant layer on the fibres. Research studies have suggested that frictional fabric characteristics are correlated with smoothness and softness. In contrast, others have postulated that the friction between the human skin and fabrics plays a more dominant role. The softening effect will eventually come from the decrease of H-bonding between the attached cotton fibres and the water molecules. As the world-leading component for fabric softeners, the future demand for ester quats is in textiles. Palm-based ester quats are utilised as an active component in fabric softeners to substitute formulations of tallow-based ester quats [23–25].

9.18 Silicone surfactants

The functional polysiloxanes have a solid in the field of finishing by an introduction of silicone chemistry for textile finishing. Silicones are based on Si-O-polymer structure. On the one hand, the oil viscosity differs; on the other hand, besides the amino-functional polysiloxanes, ethoxylated, propoxylated silicones and those with epoxy functionality have also played an important role. In textile softening agents, amino-functional polysiloxanes have become the utmost important item group.

The siloxane surfactants are composed of a methylated siloxane hydrophobe coupled to one or more polar groups. This surfactants class finds a number of applications in areas where certain surfactant forms are reasonably unsuccessful. In recent years significant advancement has been made regarding their unusual rapid wetting of hydrophobic substrates and aqueous phase behaviour. Siloxane surfactants are distinct from conventional hydrocarbon surfactants.

Organo-modified silicones are a very versatile class of polymers. Depending on the degree of modification and the overall silicone oil content, they can be employed as substrate wetting agents, flow and levelling agents and slip agents. In general, silicone-based additives reduce the surface tension of a formulation. Silicones are very surface-active polymers that provide low surface tension, and they always try to orientate themselves on the air/liquid interface. This makes them ideal raw materials for active interfacial additives.

1. They are found to be surface-active in non-aqueous media
2. They are capable of reducing the surface tension of 20 dynes/cm compared to approximately 30 dynes/cm for typical hydrocarbon surfactants and
3. They are high molecular weight fluids. This holds for polyoxyethylene groups shorter than around EO17. One consequence of this group of surfactants is that this usually does not display either a gel point or a kraft point for aqueous lamellar phase dispersions. Another feature of siloxane surfactants that contributes to their utility is the range of available synthetic routes and the subsequent variety of probable structures. The methyl groups define the surface-active character of the methylated siloxane component of the molecule – the siloxane backbone acts mainly as a flexible framework to which the methyl groups are added. The methyl saturated surface has a surface energy of around 20 dynes/cm. This is the lowest tension obtainable with siloxane surfactants. In comparison, methylene-saturated surfactants' surface energy is ≥ 32 dynes/cm. Most of the hydrocarbon surfactants pack loosely at the interface of air and liquid. The methylene group dominates the surface energy of such a surface; thus, hydrocarbon surfactants usually attain surface tensions ≥ 30 dynes/cm. Therefore, one of the fundamental differences between the siloxane surfactants and hydrocarbon surfactants is the varied surface energy of - CH_3 and - CH_2.

In many common surfactant properties, the siloxane surfactants are identical to hydrocarbon surfactants.

1. They display a break in the surface tension and log activity (concentration) that are typically symbolic of the initiation of self-aggression, such as the formation of micelles.
2. Lowering surface tension and critical aggregation concentration differ, in the same way, with molecular structure; larger hydrophobic groups result in lower surface tension and smaller essential concentrations of aggregation.
3. They display identical aggregation activity in aqueous solution, creating the same aggregation types, and following the same molecular structure trends.
4. Siloxane surfactants containing polyoxyethylene and mixed polyoxyethylene polypropylene groups also display comparable solubility in reverse temperature and cloud points.

Some of the copolymers of the siloxane polyether can act as a defoamer above the cloud points; however, the presence of a cloud point for certain siloxane surfactants should be treated with caution as other siloxane surfactants produce cloudy dispersions in the dilute concentration range which are contrary to the presence of a cloud point as is generally known. Siloxane surfactants are used as emulsifiers, wetting agents, antistatic agents, lubricants and mould release agents in a broad range of applications.

In several different aqueous and non-aqueous applications, the potential of siloxane surfactants to improve wetting plays a significant role. One standard method of characterizing wetting that is beneficial in textiles is the Drave's wetting test in which the time is taken by weighted cotton skein to wet (or sink) is measured. Cotton has a relatively high hydrophilicity. It is observed that the time to wet relies on the length of the EO groups and the size of a siloxane hydrophobe; the quickest wetting surfactants with the smaller EO groups and the shortest siloxane groups are shown.

Silicone surfactants (polyethers and copolymers of silicone) are beneficial for uses where their silicone character and high surface activity give performance benefits. They have high efficiency for microemulsion and nanostructured applications. The unique wetting properties of the trisiloxane surfactants contributing to a deeper understanding of the function of surfactant aggregation and diffusion render it a good option for several applications [26–28].

9.19 Dye transfer Inhibiting polymers

An area that needs particular attention during processing is the washing-off process. Washing off is the final step after the dyeing or printing process. Different fibres have different affinities to absorb synthetic or natural dyes and must be tailor-made accordingly. Many times, even if no washing-off agent is employed, the dyed and printed goods exhibit good fastness properties. Washing with water alone makes the fabric bright. Water-soluble dyes, mainly, reactive dye molecules, contain one or two reactive groups, and in conventional dyeing, over 35 to 40% of the dye gets hydrolysed. To achieve high colourfastness, these hydrolysed or unfixed dyes need to be removed completely. This involves a series of hot and cold washes resulting in vast consumption of water and energy. The washing-off stage plays a crucial role in maintaining shade reproducibility for several reactive dyes, and pH control is also needed. Efficiently removal of hydrolysed reactive dye takes place at four different stages. The washing-off operation involves both physical and chemical processes and calls for a proper understanding of dye chemistry and surfactant applications. During this stage, chemical reaction seldom occurs, and the process is more of a mass transfer and diffusion. Each of the available rinsing systems follows a general trend. Novel washing-off formulations based on PVP copolymers under its anti-redepositing properties can make it possible to obtain dyeing and prints with outstanding wet fastness properties on cellulosic fabrics. The washing-off of disperse and reactive dye mixtures printed materials present a rather complex two-fold problem. In the first place, the efficiency of fixation is inevitably lower than when single-fibre fabrics are printed, and therefore, the amount of unfixed dye to be removed is enormous. Secondly, the ideal conditions for the reactive dye-cellulose system (high washing temperature) are the worst possible for the disperse dye–polyester system.

Further, reduction clearing cannot be used to remove unfixed disperse dye because the reactive dyes within the cellulosic fibres would be attacked. A lengthy, washing process, therefore, must be adopted, with the use of selected surfactant mixtures. After thorough cold rinsing, one wash in 5 g/L (gpl) of a non-ionic–anionic surfactant blend at 50–60 °C is followed by the use of the same agent at a higher temperature until no other dye is removed. Rinsing, soaping and washing-off processes can be optimised to a large extent to improve efficiency and maintain high wet fastness levels of both reactive and disperse dyed fabrics. Some dye transfer inhibiting polymers are presented in Figs. 9.2–9.5.

Dye transfer inhibiting polymers (DTI) when added during the washing-off stage significantly helps in controlling unfixed dyes. DTI reduces operation time along with water and energy consumption. Polymers based on polyvinyl pyrrolidine N oxide and polyvinyl pyrrolidine betaine are the most effective and efficient for controlling the staining of unfixed/hydrolysed dyes. PVP polymers perform because of their outstanding protective colloidal and suspending action and can be made available in several viscosity grades and different molecular weights [29, 30].

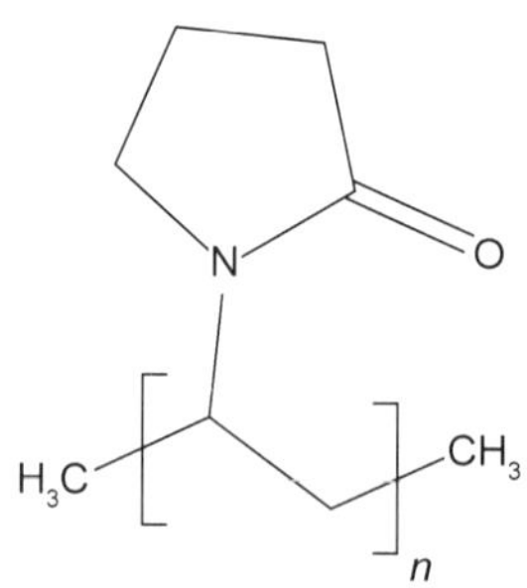

Figure 9.2 Polyvinylpyrrolidone (PVP)

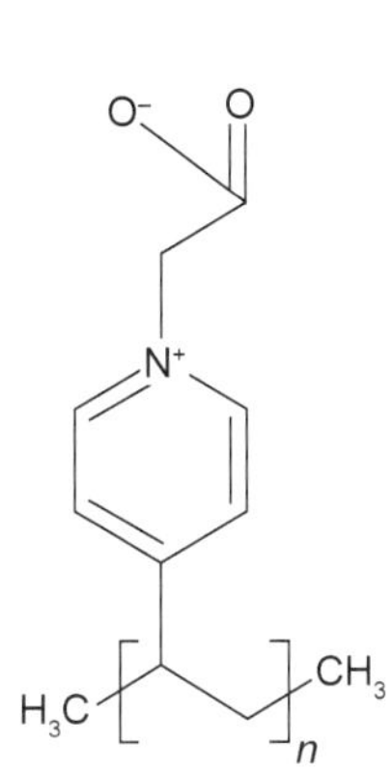

Figure 9.3 Polyvinylpyridine betaine (Polybetanine)

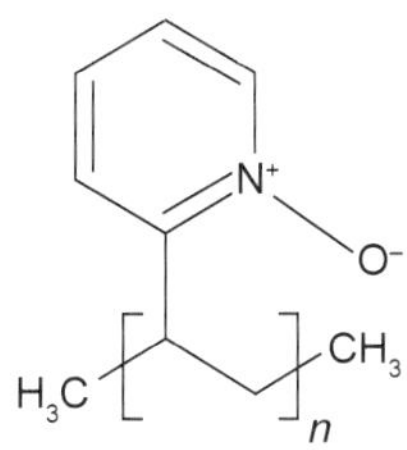

Figure 9.4 Polyvinylpyridine N-oxide (PVNO)

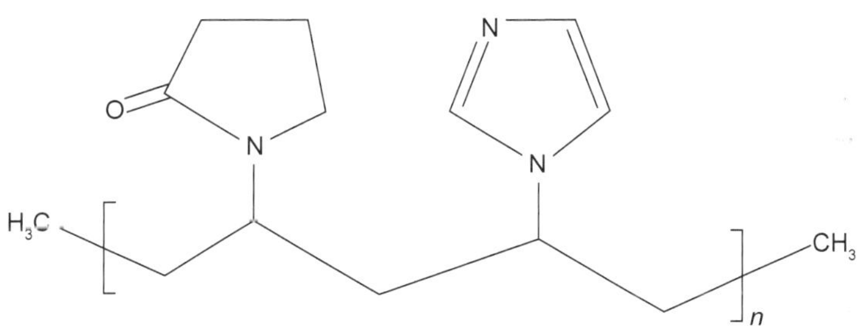

Figure 9.5 Polyvinylpyrrolidone-Vinylimidazole (PVP VI)

9.20 Acrylic polymers

Acrylic polymers and copolymers find multiple uses from sizing to finishing. Copolymerisation of an alkyl acrylate (ethyl or n-butyl or methyl methacrylate) with a small proportion of a monomer containing a reactive substituent gives film-forming polymers that can be insolubilised again by covalent cross-linking. The reactive groups may be provided by methacrylic acid, acrylic acid, itaconic acid or acrylate esters containing aminoethyl or glycidyl groups serving the same purpose. The amide groups may further be mentholated by reaction with formaldehyde. Water-soluble acrylic polymers can be categorised into two sections: Thermosetting and Thermoplastic. They are primarily used as thickening agents in sizing and textile finishing. Thermosetting and thermoplastic resins can be manufactured in different molecular weights. Polyacrylic acids swell and thicken under alkaline conditions due to conversion to their sodium or ammonium salts. Polyacrylic acid and its derivatives are used in detergents, adhesives and ion-exchange resins. Copolymers of acrylic acid are used in detergent and washing formulations. They also find applications as a thickening, dispersing and suspending agents. Acrylic acid is also the principal component of superabsorbent polymers. Acrylic polymers with a higher proportion of carboxylic acids such as methacrylic acid/ethyl acrylate (75:25) are useful as soil release finishing agents for textile fabrics.

Owing to the inherently superficial nature of indigo dyeing, the dyeing of cellulosic yarns and fabrics produces a fabric susceptible to significant and constant washing down or colour loss through prolonged usage. Indigo-dyed cellulosic fabrics have poor fastness to crocking, especially at full depths of shade. Customer acceptance and preference, mainly in denim fabrics such as jeans and overalls, have been extremely promising to the so-called washed-down look for more than two decades. The denim fabrics are also pre-washed by fabric many times to achieve the "used" look that is so often searched after. In certain instances, one or more chlorine bleaches are used to obtain an improved appearance. Today, the trend of consumer choice and style has turned toward denim that is more robust for washing, either on commercial washing machines or at home. In jeans, not only is more wash-fastness required but also introduced it as a high-fashion fabric for use in dresses, slacks, suiting, etc. Acrylic binders and aliphatic polyurethanes are used to improve the crock fastness. Indigo-dyed fabrics after treated with the polyurethane finishing agents improve the rating by 1 unit in the greyscale rating 1-5 for dry crocking fastness. Polyacrylic dispersions are self-cross-linking polymers giving the soft, transparent and flexible film, thereby improving the crock fastness of Indigo-dyed fabrics [31].

9.21 Special polymers for anti-pilling finishing

To prevent the pilling on fabric surfaces, many chemical finishing methods are available. The polymeric coating on the fabric that can bind the fibres into the fabric prevents the loose fibres from forming the initial 'fuzz'. Usually, these finishes contain friction-reducing lubricants to reduce abrasion damage. Liquid acrylic copolymers reduce pilling and seam slippage of treated fabrics, and in addition, give a flexible handle. These polymers can be modified to produce flexible and durable films with strong adhesion to fibre surfaces. Polymers based on silicic acid dispersions reduce the formation of pills on the fabrics. This polymer suppresses the ability of fibres to slacken and thereby reduces the pilling of fibres. Amide polymers also work very all as anti-pilling agents. These surface protection polymers comprise of a polymer with at least one amide monomer unit, where the amide monomers are free of amide linkages; the amide functionality aids in attracting the polymer to the surface. Once associated with the surface, the copolymer forms a protective barrier. Rubber-based latex compounds (styrene-butadiene emulsion polymer) and non-ionic polyurethane can also be used as effective anti-pilling agents. Bio-polishing is an enzymatic process designed to improve fabric quality. Besides giving a superior handle and lustre to cellulose they can be effectively used as anti-pilling agents.

A non-chemical method of pilling reduction enables the pills to drop from the fabric as soon as they are formed; this can be achieved by reducing the fibre strength. In synthetic fibres, the anti-pilling properties can be imparted by changing the polymer structure before extrusion. One yarn can be produced using standard polyester staple fibres and modified polyester fibres with lower strength. Therefore, the fabric created using the low-strength fibres shows significantly less pilling, which confirms the effect of low-strength fibres on pilling. The utilisation of these low-strength polyester fibres in cotton/polyester blend fabrics improves the anti-pilling property.

9.22 Inherently low-pilling polyester

Most of the attempts are directed at modifying the polymer, especially lowering the average molecular weight. This can be accomplished by incorporating cross-linking agents or modifying the polymer structure using permanent branches. The addition of 0.01 to 2% by weight of diphenyl silane diol is done prior to ester interchange. This results in low-pilling polyester. However, the breaking extension is too high to permit processability with cellulosic fibres, and the breaking strength is insufficient to weave staple fibre yarn. As seen above, conventional methods for improving pill resistance generally involve weakening of fabric so that the loosened fibres break away from the fabric rather than forming a pill. Hence care must be taken to attain a proper balance between the pill resistance and the fabric strength.

9.23 Polyamine condensates

Many cationic polymers can be applied to cellulosic fabrics to increase the fastness properties due to the action of electrostatic forces between the dyes and cationic sites in the polymer. Within this class, DADMAC has a special place. This monomer can be copolymerised with allylamine to enhance the fastness properties of cotton. Reactive dyes being anionic, get attracted to the cationic groups of poly-DADMAC, and the reaction between the reactive groups in the dye and the amino-functional nucleophiles takes place through a nucleophilic substitution mechanism.

The resin dye-fixatives with the reactive groups that can react with cellulose or dyes are classified as reactive resin dye-fixatives. The most widely used groups are the condensation products derived from epicholohydrin, reactive polyurethane or reactive organosiloxane units and the double bond of carbon-carbon. Within this class, reaction products of diethyelenetetraamine and dicyanamide with epichlorohydrin give outstanding results in cotton dyed with fibre reactive dyes [32].

9.24 Speciality waxes

Polyethylene and paraffin waxes are used in textile processing for five main applications.

1. Waterproofing
2. Enhancing the processability of fibre, yarns and sewing threads by modifying their frictional properties
3. Adding and improving the value of finishing auxiliary such as resins, silicones and softeners.
4. Enhancing fabric properties such as tear strength, flex resistance, sewability, yarn knittability and feel of the finished textiles.
5. In warp sizing, where the use of paraffin waxes improves the processing of yarns. It decreases fibre-to-fibre and fibre-to-metal friction during weaving operation, thus eliminating the processing problems throughout the weaving process and maximizing production efficiency.

On the other hand, polyethylene waxes can be modified to have hydrophilic characteristics by air oxidation in the melt at high pressure. In the presence of alkali, the emulsification offers more stable products with better quality. They display high lubricity, which is not fast to dry-cleaning; however, it is firm in extreme thermal and pH conditions in standard textile processing [33].

9.25 Polyester resins

Soil can be any material that in any way adversely affects a textile substrate's required demand profile during its usage or washing.

It can be:

- Water-insoluble inorganic (cement)
- Water-soluble organic/inorganic soil
- Water-soluble organic, polar soil (fatty acids in the form of sweat, proteins-egg yellow).
- Water-soluble organic, non-polar type soil (pigments)

Polyethylene glycol polyester/copolyester has the unique property of depositing a uniform film on the fabric during the washing cycle. Aromatic polyesters are water-insoluble. A portion of ethylene glycol is removed and replaced with a high molecular weight polyethylene glycol (PEG). The long-chain PEG renders water solubility to the product. The hydrophobic portion absorbs on the hydrophobic surface, making the exposed surface more hydrophilic, thereby reducing the interactions between the soil the fabric. Polyester resins work as an excellent anti-soil finish for polyester, expected to minimise these interactions between the soil and the textile material [34].

9.26 Fluorine-free alternatives in repellent finishing

Flurochemicals pose threat to human, environment and health because of the release of perfluorooctanoic acid (PFOA) and perfluorooctanesulfonate (PFOS). PFOA and PFOS are bioaccumulative, toxic and bio-persistent fluoro-organic compounds. Significant work has been done to develop alternatives for fluorine-free alternatives.

Some alternatives to fluorocarbon chemistry are:

1. Paraffin wax
2. Stearic acid /melamine derivatives
3. Silicone oil
4. Dendrimers

Table 9.2 Comparative results of non-fluro and fluro chemicals

Properties	C_8 Chemistry	C_6 Chemistry	Alternatives
Water repellency	Excellent	Excellent	Very good
Oil repellency	Good	Good	Nil
Dry cleaning	Good	Fair	Nil
Fabric handle	Moderate	Soft	Soft
Curing conditions	High	High	Low

Table 9.3 Oil and water repellency of different fluorocarbons

Substituent	Oil repellency	Water repellency
CF_3	0	50
C_2F_5	60	70
C_3F_7	90	70
C_5F_{11}	100	70
C_7F_{15}	120	70
C_9F_{19}	130	80

Oil repellency: 150 is the best; Water repellency spray test: 100 is the best

In finishing, fluorocarbon compounds can be utilised alone or in combination with other chemical agents where multi-functional properties are required. These are used in mixtures with durable-press finishes to render cotton repellent simultaneously as well as easy-care and crease-resistant. Following are the finishing auxiliaries used with fluorochemicals:

1. Crosslinker: DMDHEU or other resins can be used to give durability to the textile finish.
2. Extenders: The wax or aliphatic-based water repellent could be utilised to improve the quality and decrease the quantity of the fluorochemicals.
3. Non-rewetting agent: The isopropyl alcohol is applied along with fluorochemical to support substrate wetting. Such fugitive wetting agents evaporate during drying or curing. The conventional wetting agents must not be utilised because they interfere with the key elements affecting both oil and water repellency [35, 36]. 1,2,3,4-Butane tetracarboxylic acid (BTCA) represents an environmentally safe alternative for formaldehyde-based DMDHEU (Dimethyloldihydroxyethyleneurea) for imparting easy-care properties to cotton and its blends. However, there are several drawbacks encountered with BTCA finishing, like excessive fabric tendering and yellowing. Incorporating polyvinyl alcohol or CMC (Carboxy Methyl cellulose) either alone or in combination with chitosan significantly improves the crease recovery angle, tear strength and abrasion resistance [37].

9.27 Coloured fibres

Interesting developments are taking place in the field, with the major thrust coming from the manipulation of dopes to give a range of products. Speciality chemical is injected into the dopes to get the desired effects. Doped or solution-dyed yarns are developed via adding a masterbatch colourant to the polymer melt during spinning or extrusion. This results in a one-step process in which filaments and fibres flowing out of the spinnerets are entirely impregnated with pigment.

Advantages

- Environment-friendly since no chemical or water is utilised in the dyeing.
- Both regenerated and synthetic fibres can be dope dyed.
- Dope-dyed yarns are highly resistant to UV rays and less susceptible to shade change during prolonged storage.
- Unlike conventional dyeing, dope-dyed yarns do not vary from batch to batch.
- Dope-dyed yarns show good fastness properties against repeated washing cycles.
- Dope-dyed yarns have the same energy costs as natural yarns.

9.28 Cationic and anionic dyeable fibres

By introducing anionic components in the dopes, nearly all polymer varieties, including cellulosic, can be made cationic-dyeable, which is close to the acrylic fibre manufacturing in which an anionic charged co-monomer becomes an integral part of the fibre backbone. The fibres can be made anionic-dyeable by adding certain cationic groups to the fibre. The workability and processing parameters determine the chemical molecules' selection, and some are used for regenerated cellulosics, whereas some others are used for synthetic hydrophobic fibres. Here, the crucial role is played by quaternary ammonium and nitrogen compound chemistry, and accomplishments have originated on quaternary polymeric molecules that do not impair the spinnability of these polymers. Apart from that, chitosan as a naturally occurring chemical has arisen as an intermediate, although the stability and dosages are being studied and optimised [38].

9.29 Salt-free/high fixation dyeing using reactive dyes

Reactive dyes comprise a reactive group, either an activated double bond or a heterocycle, which creates a chemical bond with the OH group of cotton when added in an alkaline dye bath containing cotton fibres. However, improper dye usage, a considerable number of electrolytes, and a high quantity of untreated wastewater released also make the dyeing process extremely unsustainable and inefficient. Each year more than 80,000 tons of reactive dyes are manufactured and consumed, causing the overall amount of pollution induced by their usage to be quantified. Recently, with the increasing popularity of reactive dyes for cotton dyeing, the environmental concerns connected with their uses have been gaining consideration. Subsequently, most reactive dyes have only a mild affinity for cellulose; therefore, a considerable number of electrolytes such as Na_2SO_4 or NaCl (40-100gpl) are usually needed for exhaustion. The dye bath exhaustion, in many cases, does not exceed more than 70% despite using vast quantities of electrolytes. This results in a high volume of effluents discharged, having high total dissolved solids and hydrolyzed dye at the end of the pipe treatment. Therefore, a dyeing process that results in high dye affinity can be of countless advantages to minimizing the environmental issues. There are several approaches to improve dye bath exhaustion.

Techniques also advanced whereby these techniques make pigment dyeable substrates usable for daily processing. The charging type cationizer develops positive changes on the polymer that aids in the virtual 'sucking' of the reactive dye components to get them near to the fibre without the assistance of electrolytes such as salt, only to be reacted in an alkaline medium for consequent fixation. Cationic agents based on 3-chloro-2-hydroxypropyl trimethylammonium chloride create cationic sites on cotton and cellulosic materials in the alkaline medium. The cationisation process involves an etherification reaction between the hydroxyl group of cellulosic fibre with the epoxide group of the cationising agent. Cellulosic fabric and cationising agent do not react with each other without the addition of alkali, like caustic soda. Some of the popular cationising agents are CHPTAC (3-chloro-2-hydroxypropyl trimethyl ammonium chloride), PDADMAC (Polydiallyldimethylammonium chloride), and Polyhexamethylene biguanide (PHMB). Textile material treated with these cationising agents does not require any electrolytes and produces a fabric with good fastness properties [39–43].

9.30 Auxiliaries for single bath dyeing of polyester/cotton

The conventional method of dyeing polyester/cotton is a two bath method where dyeing steps are independent of each other, the selection of dyes is unrestricted, and optimum fastness properties are achieved. The issue is that the overall duration of the dyeing process is 9-10 hr. A single bath two-step process can be a solution to these challenges. Hydrolysable esters such as gama-butyrolactone and ethyl lactate are used to slide the initial alkaline pH to the final acidic pH, thereby allowing the fixation of reactive dyes on the cellulose initially, followed by fixation of disperse dyes on the polyester component, as the temperature of the bath is raised to130°C. The hydrolysis behaviour and fixation of reactive dyes work well.

The hydrolysable esters gamma-butyrolactone and ethyl lactate or diethyl tartrate are employed individually or as a 1:1 mixture for the pH slide from an initially alkaline pH to an acidic pH with the controlled rise in the dye bath temperature. The cotton component is dyed first with reactive dyes, followed by dyeing of polyester with disperse dyes. Gamma-butyrolactone (BLO) hydrolyses at high temperature to release 4-Hydroxy butyric acid. In an unbuffered system, it results in a linear drop in pH with a rise in temperature. Ethyl lactate (EL) also hydrolyses with a rise in temperature, releasing lactic acid and ethyl alcohol. The rate of hydrolysis of the esters BLO and EL is dependent on temperature, pH as well as a time of reaction. Thus, when the rate of temperature rise between 60°C and 100°C is kept at 1°C/min, it is found that esters hydrolyse rapidly, lowering the pH drastically. For the reactive dyes present in a dye bath, alkaline pH is essential for its fixation, for which the bath should remain considerably alkaline up to 90°C and, after that, shift towards acidic pH, finally attaining slightly acidic pH for the disperse dyeing of polyester. The rate of temperature rise is, therefore, to be kept low at 0.5°C/min till the temperature reaches 100°C

The addition of 0.4 g/L soda ash and 0.6 g/L diethanolamine along with 2 g/L sodium silicate in the final bath is found to be desirable for efficient dyeing of the reactive and disperse dyestuffs. Thus, pH values at 90°C are 7.7 and 7.5, whereas those at 130°C are 5.8 and 5.2 for BLO and EL additions, respectively. The addition of 5 g/L of hydrolysable esters in the ratio of 1:1 gives good results. Alkali stable disperse dyes give good reproducible dyeing.

9.31 Ionic liquids

Textile processing is a water-intensive industry. Generally, processing 1kg of textile material is estimated to consume 200 litres of water. The invention of ionic liquids that show valuable and unique properties will generate tremendous untapped opportunities for industrial applications to improve the operational productivities of a great deal of chemical manufacturing, including textiles processing.

The non-aqueous process can either reduce the volume of water utilised in textile processing or remove it entirely. Reducing the amount of water used in textile processing offers both environmental and cost-saving benefits. Some studies showed that ionic liquids have the great potential to be utilised in some textile processes in place of water. Ionic liquid's outstanding low pressure makes it easy to handle compared to conventional organic solvents. Ionic liquids are salts with a melting point lower than 100°C and are sometimes temperature stable far beyond 200°C, making them an ideal solvent for the disperse dyeing of polyester. Ionic liquids show slight vapour pressure, rendering them to manage efficiently in comparison to conventional organic solvents. Additionally, ionic liquids demonstrate high dielectric constants, consequently displaying outstanding solvent power for keratin, cellulose, etc. Ionic liquids selected for cellulose chemistry are generally based on 1- allyl -3 methyl imidazolium salts (Figs. 9.6–9.9).

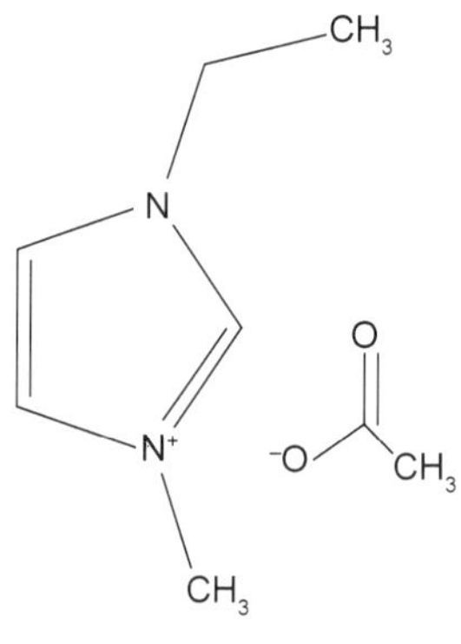

Figure 9.6 1-Ethyl-3-methylimidazolium acetate (CAS no: 143314-17-4)

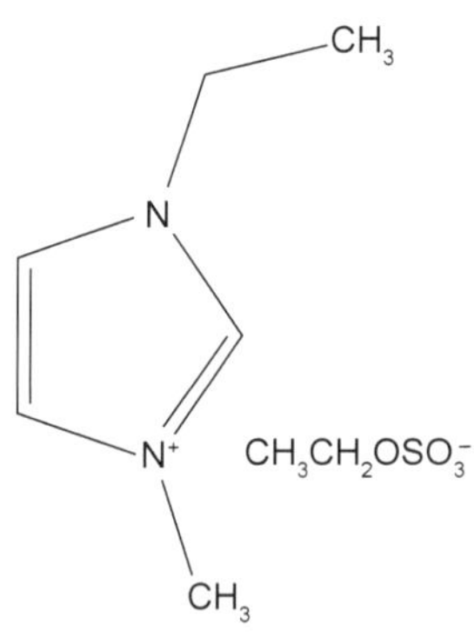

Figure 9.7 1-Ethyl-3-methylimidazolium ethyl sulfate (CAS no: 342573 – 75- 5)

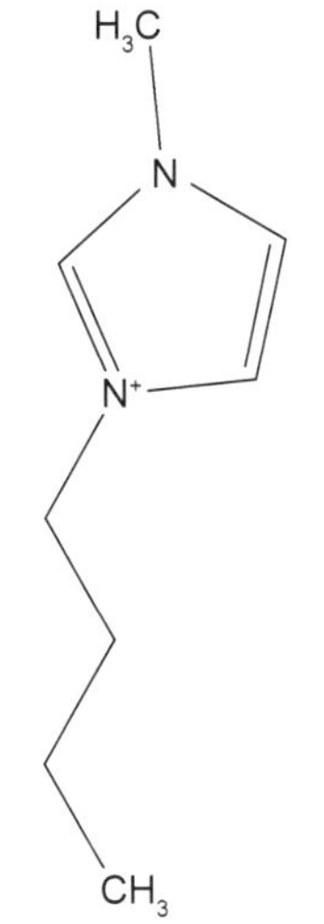

Figure 9.8 1-Butyl-3-methylimidazolium chloride (CAS no: 79917-90-1)

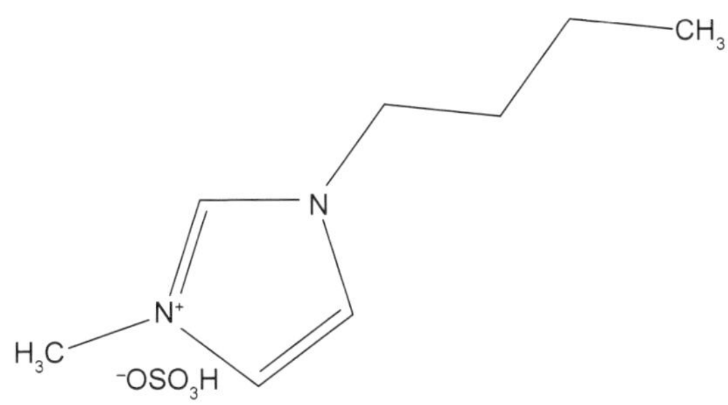

Figure 9.9 1-Butyl-3-methylimidazolium hydrogen sulfate (CAS no: 202297 – 13 -2) [38]

9.32 Auxiliaries for digital printing

The digital revolution has influenced every part of life today. Textile printing has often embraced modern technology to suit its requirement, powered above all by the need for quicker and cheaper sampling and printing. Originally, developed for printing on paper, digital printing is, therefore, a new technology for textile printing. Digital printing describes a collection of techniques that can be used to move an image onto the desired surface (or substrate) in digital form. Because various target surfaces have specific properties, it is not feasible to apply any printing technique to every defined substratum. Of the many non-impact digital technologies, three have emerged as leaders: laser printing, thermal transfer printing, and, more recently, digital printing. However, there are several requirements that any digital printing technology must meet to make a severe impact on traditional printing.

- Give equal or better results to screen printing
- Give reproducible printing
- Should be easy to manage with reliable technology
- Low investment.

Digital printing meets all of the above requirements. A major driver for this shift is the lengthy, polluting, capital-intensive nature of conventional printing. Changing fashion trends and new market requirements have also contributed to the drive for digital printing.

Different formulations and colourants are used in digital printing. The screen printing process uses viscous paste, while inkjet printing involves very thin ink formulations. The average particle size of the inks must be small relative to the orifices of the nozzle to allow smooth passage of the ink flow. The choice of the auxiliary system is vital for the stability of ink formulations.

The inkjet printing ink can be divided into two types:

1. Water-based inks
2. Solvent-based inks

9.32.1 Water-based inks

A lot of good work has been done over the years in the area of water-based inks. Water-based inks are appealing from a sustainability point of view compared to solvent-based ones. Water is inexpensive and readily accessible, and the solvents dependent on petroleum are unlikely to increase in demand. At no risk of contamination, water will be vented directly into the environment. However, inks dependent on the water are generally safe and non-toxic to use.

Inkjet is a dot-matrix printing technique in which ink droplets are jetted directly from a small nozzle to a designated location on a media for producing an image. The magnitude of the smallest size decides the most exquisite detail in the inkjet process that can be repeated. Dots as tiny as 3 Pico-liters (diameter of about 18 microns) can be produced at the current level of inkjet technology. Inkjet printing is the only contactless printing technique and is the perfect system of printing. There are two types of inkjet inks. One is ink based on the dye, and the other is ink based on pigment. Some inkjet printing inks require several component optimisations of formulations, making the design a difficult job. Those inks, depending on the printer, they are designed for, must follow particular requirements in terms of surface tension, viscosity, storage stability, good fastness and conductivity. However, when printed on fabric, most dye-based inkjet printing inks suffer from low wet-fastness properties. Different techniques can be employed to improve the water fastness properties. For better jet-ability, pigment inks should be of low viscosity, as much of the print head for aqueous ink operates at a low viscosity. The high functional polymeric dispersing agents are utilised not only to avoid precipitation in a solution of low viscosity but also for jetting well through the print head.

9.32.2 Pigments for digital printing

Inks for digital printing can be both in pigment and dye-based formulations that combine pigment dispersion, polymer and ink-formulation technologies. Conventional classes such as acid, disperse, reactive and pigment as used in textile printing, can be formulated to give excellent printing results in digital printing, having all-round fastness properties and print reliability. These nanoparticle-based inks are based on alkyl phenol-free emulsifiers. The digital printing process using pigments as inks has many advantages over traditional pigment printing. The main reason is its risk of nozzle clogging caused by pigment particles. The variability in K/S of printed fabrics may be attributed to both the chemical composition of the pigment and the pigment dispersion stage.

Innovation in the field of polymer chemistry enables the low-viscosity ink to eliminate many chemicals which were otherwise used before. The new ink formulation involves multi-functional auxiliaries based on surfactants and polymeric dispersant combinations to address problems encountered during printing. These high-tech inks have proper consistency to pass through the heads at very high speed, dry directly on the textile, and leave a pleasant and soft handle. Pigment printing only requires a heat fixation step for post-treatment and therefore provides an eco-friendly process with a reduction in water consumption and waste. Advanced pigment ink system, where the binder is in the pigment ink, gives a one-step print process with a good wash and lightfastness with no requirement for pre-coating.

9.33 Dispersants

Dispersing agent disperses the pigment particles in ink and holds them in a stable state at the dispersing stage. Surfactants improve the affinity of pigment particles and ensure rapid wetting. To break down the agglomerates, adequate force must be applied to counteract the effects that hold the agglomerates together: particle-particle attrition, physical impact, and shear transmitted through an intermediate layer of fluid. The surfactant systems are known to dissolve insoluble or partly soluble dyes by integrating them into surfactant micelles. There are three main types of forces that determine the stability of pigment dispersion: The Van Der Waals' forces of attraction, the electrostatic forces of repulsion and steric stabilisation. Ethylene glycol added as a humectant in the print formulation to control the viscosity and the surface tension prevents the inks from drying and, at the same time, helps to penetrate the substrate quickly.

Having too many items in an ink formulation will lead to multiple problems, and it is difficult to detect and solve in the event of any problem during production. With pigment inks, the correct choice of pigment binder plays an important role. Binder is a film-forming material; it encloses the pigment or the dye and adheres to the substrate. A pigment print's washing, rubbing, and the binder film's fastness dictates dry-cleaning fastness, and the quality of the print depends on the print quality. Generally, binders are composed of polymer dispersions in water. It can be copolymers and polymers which are developed via polymerising monomers in the presence of catalysts or initiators. Acrylics are usually the best versatile binders, with good dry/wet strength and colourfastness. They also deliver significant durability and a wide variety of fabric hand properties [44].

Concluding remarks and future perspectives

One technology can bring a novel invention, the so-called 'spin-off' technologies, into existence. Not only that, new science offer rise to novel technologies, but the reverse is valid too: novel technology brings modern science. Knowledge will control tomorrow's world and not the tangible or physical assets, but intangible assets will be the priority. Most of the advances and innovations in the area of textile auxiliaries are fundamentally incremental improvements in existing technologies. In the future, combined with technology, the conventional growth factors like labour, land and capital will become less critical.

References

1. Sivaramakrishnan CN (2009), Enhancing aesthetics of textile fibers, *Anthology of Speciality Chemicals for Textiles*, Colour Publications Pvt. Limited, Mumbai.
2. https://www.ecotextile.com/2018102223813/dyes-chemicals-news/revealed-hidden-source-of-hazardous-textile-pollutants.html (Accessed Sept. 2020).

3. Sivaramakrishnan CN (2012), Sustainability in wet Textile Processing, *Textiles & The Environment*, Colour Publications Pvt. Limited, Mumbai.

4. Sivaramakrishnan CN (2012), Textile Processing Industry: An overview, *Textiles & The Environment*, Colour Publications Pvt. Limited, Mumbai.

5. Sivaramakrishnan CN (2018), *Dyeing, Textile Doctor: Comprehensive Solutions for Processing,* Colour Publications Pvt. Limited, Mumbai.

6. Marangoni AG, Ghazani SM (2012). *Trends in Interesterification of Fats and Oils*, ILSI NA, Washington DC.

7. https://www.essentialchemicalindustry.org/polymers/polyesters.html (Accessed Sept. 2020).

8. Liu K, Zhang X, Yan K (2017). Low-temperature bleaching of cotton knitting fabric with H2O2/PAG system. *Cellulose* 24, 1555-1561.

9. Tavčer PF (2012). Low-temperature bleaching of cotton induced by glucose oxidase enzymes and hydrogen peroxide activators. *Biocatal. Biotransformation* 30, 20-26.

10. Reinhardt G, Miranda C, Martin M (2013). Low temperature bleach catalysts for improved tea stain removal. *HPC Today* 8, 36-42.

11. Opwis K, Knittel D, Schollmeyer E, Hoferichter P, Cordes A (2008). Simultaneous application of glucose oxidases and peroxidases in bleaching processes. *Eng. Life Sci.*, 8, 175-178.

12. Farn RJ (Ed.) (2008), *Chemistry and technology of surfactants*, John Wiley & Sons, New York.

13. https://biokhimact.com.ua/images/catalogs/Surfactants.pdf (Accessed Sept. 2020).

14. Sivaramakrishnan CN (2013), The use of surfactants in the finishing of technical textiles In Advances in the Dyeing and Finishing of Technical Textiles, Woodhead Publishing: Bradford, pp. 199-235,

15. Azarmi R, Ashjaran A (2015). Type and application of some common surfactants, *J. Chem. Pharm. Res.*, 7, 632-640.

16. Laurent JB, Buzzaccarini F, Clerck K, Demeyere H, Labeque R, Lodewick R, Van Langenhove L (2007), Laundry cleaning of textiles, *Handbook for cleaning/ decontamination of surfaces*. Elsevier, Netherlands.

17. Sivaramakrishnan CN (2009). Polymeric surfactants, *Anthology of Speciality Chemicals for Textiles*, Colour Publications Pvt. Limited, Mumbai.

18. Esmaeilian N, Malek RMA, Arami M, Mazaheri FM, Dabir B (2018). Environmentally friendly sugar-based cationic surfactant as a new auxiliary in polyacrylonitrile dyeing. *Colloids Surf. A: Physicochem. Eng. Asp.* 552, 103-109.

19. Mańko D, Zdziennicka A (2015). Sugar-based surfactants as alternative to synthetic ones, *Annales UMCS Chemia* 70, 161-168.

20. Teklehaimanot M, Amsalu T (2008). Hydrlosis of sugar for dyeing of cotton fabric with sulphur black. *J. Text. Sci. Eng.* 8, 384 https://doi:10.4172/2165-8064.1000384.

21. Zouhaier R, Sofiène D, Faouzi S (2014). The use of glucose as ecological reducing agent for sulphur dyes: Optimization of experimental conditions. *Eur. Sci. J.* 10, 436-446.

22. Saraf NM, Sabale AG (2010). Complex Problems Can Have Sugary Solutions. *Colourage* 57, 86.

23. Mishra S (2007). Ester quats: the novel class of cationic fabric softeners. *J. Oleo Sci.* 56, 269-276.

24. Oikonomou EK, Christov N, Cristobal G, Bourgaux C, Heux L, Boucenna I, Berret, JF (2018). Design of eco-friendly fabric softeners: Structure, rheology and interaction with cellulose nanocrystals. *J. Colloid Interface Sci.*, 525, 206-215.

25. Akram M, Zaman W, Hussain Z, Salman M (2010). Synthesis of tallow based esterquat. *J. Sci. Res.* 40, 31-36.

26. Sivaramakrishnan CN (2009), Silicone Surfactants, *Anthology of Speciality Chemicals for Textiles*, Colour Publications Pvt. Limited, Mumbai.

27. https://www.momentive.com/en-us/categories/antifoams/sagtex-phd (Accessed Sept. 2020).

28. https://www.tissueadditives.com/product/tissue-additives/en/products/product-search/function/pages/product-details.aspx?productId=13279&category=1070 (Accessed Sept. 2020).

29. https://www.ashland.com/industries/energy/batteries/pvp-k-series (Accessed Sept. 2020).

30. Rathinamoorthy R (2019). Performance Analysis of Pyridine N-oxide as Dye Transfer Inhibitor in Household Laundry. *Fibers Polym.* 20, 1218-1225.

31. Sivaramakrishnan CN (2018), Printing, *Textile Doctor: Comprehensive Solutions for Processing*, Colour Publications Pvt. Limited, Mumbai.

32. Yu Y, Zhang Y (2009). Review of study on resin dye-fixatives on cotton fabrics. *Mod. Appl. Sci.* 3(10), 9-16.

33. http://www.aknet.biz.pl/psz/09_08_eng.pdf (Accessed Sept. 2020).

34. Lark JC (1981), U.S. Patent No. 4,268,645, Washington, DC: U.S. Patent and Trademark Office.

35. Sivaramakrishnan CN (2018), Insights in to Fluro Carbon finishes, *Textile Doctor: Comprehensive Solutions for Processing*, Colour Publications Pvt. Limited, Mumbai.

36. http://www.ecetoc.org/publication/tr-103-toxicity-of-possible-impurities-and-by-products-in-fluorocarbon-products/ (Accessed Sept. 2020).

37. Aksoy SA, Genc E (2015). Functionalization of cotton fabrics by esterification cross-linking with 1, 2, 3, 4-butanetetracarboxylic acid (BTCA). *Cellul. Chem. Technol.* 49, 405-413.

38. Mehra A, Sivaramakrishnan CN (2005), Some Frontiers in specialty chemicals for processing. 13th Asian Textile Conference, Geelong Victoria Australia.

39. Logan D (2001), Synthesis of Water-Soluble Quaternary Ammonium Chitosan Derivatives and Their Potential Applications as Biocides, LSU Historical Dissertations and Theses.

40. Neenaz, S. (2015), Comparison of Different Cationizing Agents on Cotton, Faculty of North Carolina State University Graduation Thesis.

41. Farrell MJ (2017). Cationic cotton Prepared with Hydrophobic Alkyl Chlorohydrin Quats: A New Fiber with New Properties, Sustainable Cotton Dyeing. *AATCC 2017—2017 AATCC international conference proceedings*, 97-125.

42. Singha K, Maity S, Singha M (2012). The salt-free dyeing on cotton: An approach to effluent free mechanism; can chitosan be a potential option. *Int. J. Text. Sci.* 1, 69-77.

43. https://www.dow.com/en-us/pdp.quat-188-cationic-reagent-65-active.64284z.html (Accessed Sept. 2020)

44. Sivaramakrishnan CN (2018), Inks for Digital Printing, *Textile Doctor: Comprehensive Solutions for Processing*, Colour Publications Pvt. Limited, Mumbai.

10

Sustainable chemical processing of denim

Aravin Prince Periyasamy*, Jiri Militky, Mohanapriya Venkataraman
Department of Material Engineering, Technical University of Liberec, Studentska 1402/2, 46117, Liberec, Czech Republic.
*Corresponding Author, Email: aravinprince@gmail.com

Abstract: World over, the culture of denim is widely spreading and blooming due to its modern trends and fast adoption of emerging fashion. The washing of denim has grown to a significant industrial scale. This sector has a huge chemical footprint that alarms ecological concerns. Without modifying the look and feel of their favourite jeans, brands as well as consumers prefer to switch to more sustainable options. There are opportunities for saving water, energy, time and chemicals throughout the process chain to produce the desired designs and look of denim. The concerning issues regarding sustainable and coordinated denim wet processing with special emphasis on the usage of colouring matters, chemicals and auxiliaries, water and energy are analysed in this chapter. Additionally, an alternative sustainable process of denim wet processing has been discussed thoroughly. Further, the environmental impacts of conventional and sustainable denim dry finishing technologies are discussed in this chapter.

10.1 Introduction

The existence of life-supporting conditions is the vital reason for the planet Earth to be exceptionally good. The enduring the human actions refurbished the environment due to the biosphere. The term 'environment' or "surroundings" is defined as the complex of physical, chemical, and biotic factors (such as climate, soil, and living things) that act upon an organism or an ecological community and ultimately determine its form and survival. The biosphere is the zone of air, land and water where organisms exist which is narrowed when compared to the planet Earth. It is commonly known as the global sum of all ecosystems and consists of several layers, including the atmosphere, the lithosphere and the hydrosphere made up of different biomes (ecological systems) where survival is possible without any artificial equipment. To dwell as human beings on Earth is accomplished by agriculture and industrial revolutions, and further inadvertent and deliberate discharge influences the environment and biodiversity by the recent sophisticated world of synthetic and artificial materials. The world made remarkable progress in child mortality rate due to medical development reassured in global progress turned the birth rate 2 to 3 times greater than the death rate in 2050, implying the increase in global population to around 9.8 billion than the existing level of 7.96 billion [1]. Human population stress has pushed the Earth into major environmental devastation leading to affect land habitation and resources like water, minerals, fossil fuels and food. Deforestation has extended exponentially leading to desertification, soil erosion, fewer crops, flooding, increased greenhouse gases in the atmosphere and a host of problems for indigenous people, subsequently increasing global warming and climatic change. Deforestation, hunting, urbanisation and agricultural land reformation have paved the way for endangering a lot of species and rare animals. The earth's biodiversity is in grave danger due to habitat destruction which is the major cause of biodiversity loss. The industrial revolution resurrected several industries, including the denim industry. In order to achieve the desired anticipated finishing, it is essential to undergo varied chemical processes to remove impurities and perform colouration by denim industry. Due to toxicity and other health hazards, chemical processing is replaced by sustainable colouration techniques. Emissions from the dyeing process are an exemplary source of anthropogenic pollution. Global textile dye utilisation has already reached 700,000 metric tonnes approximately on the overall consumption of dyes, whereas 50% of it comprises of azo dyes [2]. Chemical reagents of various compositions, from the beginning simple organic and inorganic compounds to various polymers and complex synthesised organic products [3]. Henceforth researchers investigate detailed analysis to design a sustainable dyeing technique compensating conventional dyeing process.This chapter is devoted to discussing the electrochemical reduction, enzymatic processing, plasma, ultrasonic and laser-assisted processing and advanced oxidizing process to enrich the precise properties by significantly reducing the impact on the environment.

10.2 Denim manufacturing

Currently, denim has created the anomalies of work wear look to fashion commodity. Globally denim has accepted by the young generations as a trendsetting garment. Consumers presently prefer to be more casual, relaxed yet, sophisticated jeans manufactured sustainably. Both in developed and developing countries the market growth for sustainable textiles maximises year by year. Cotton Inc confirms the huge market potential of denim. The denim has the market value of $65 billion, expected to increase by 8% in coming years [4–9] purchasing, using and maintaining a product (here it is jeans. However, the expected increase may be strongly affected due to global pandemic (COVID-19). Warps uses Coarse yarns of 6's-20's Ne than weft. Warp was treated with the indigo dyes whereas weft was left undyed. Fig. 10.1 shows the overall manufacturing sequence of denim. Nevertheless, raw material for textile is acquired either naturally or artificially. Henceforth, it is essential to implement the first stage of categorisation to cultivate or produce the fibres [10,11].

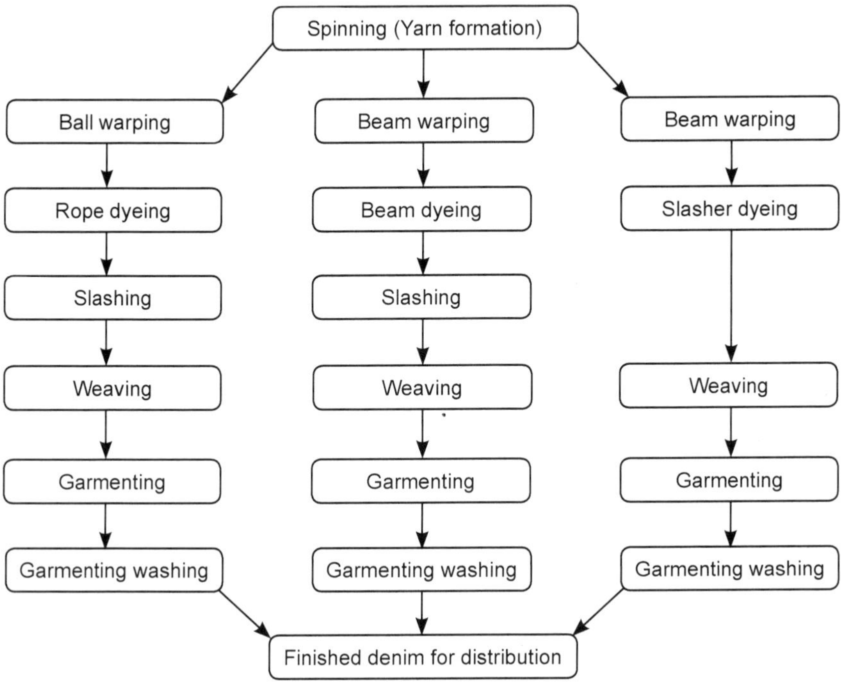

Figure 10.1 Process sequence for denim manufacturing from fibre to finished garment (adopted and modified from [5]).

10.2.1 Yarn manufacturing

The foremost aim to open the materials without damage is the first operation at the blowroom line. The final features of the yarn are determined by carding operations in the spinning process. The most significant is the number of draw frame passages to produce open-end denim yarn and ring denim yarn [12,13].

10.2.2 Warping

Generally, a creel of single-end packages of many yarns is transferred by warping to form a parallel sheet wound to form a beam or a section beam of yarn. Weaver's beam is the warp beam installed on a weaving machine.

10.2.3 Sizing

The strength of the yarn is increased by sizing through binding the fibres to one another using various size formulations. Yarn encapsulated with a protective coating is the main objective of sizing warp yarns as it diminishes the yarn abrasion that occurs during the operation of weaving. It also lessens the hairiness of yarn by avoiding the adjacent yarns interwoven with one other at the weaving machine. During the process of weaving, the size protective coating prevents the dye of indigo from rubbing. The yarns are accumulated together in a long-chain beamer involving the multiplication of warp sheets in sizing operations to construct one weaver's beam. Usually, a sizing percentage of 8-12 is applied [14]. Various chemicals are involved in the sizing process, which creates different forms of pollution, they are

- All kinds of starch increase the BOD and COD
- Wax, oil and other organic ingredients increase the BOD
- Synthetic adhesives

Ingredients weigh about 12.5-15 kg in 100 litres of water, with the size add-on being 8-11%. The bath contains the existing size formulations along with some ingredients which could be detached during the process of rinsing, generating pollution. It is essential to remove all these materials from the warp sheet during further processing. When the fabric is treated for the removal of size, it generates wastewater in which most of the additives have high BOD and COD. In denim dyeing, many pollutants such as dyes and chemicals are associated.

10.2.4 Dyeing

In tropical zones, the Indigo plant is seen predominantly where its leaf is used to extract the Indigo. Natural Indigo dye has a history of practical applications and lasts for longer without fading for many centuries. Even though, secrecy is maintained in cultivating and extracting of paint from this culture in the countries like India, China and Japan since it is being a peculiar and undisclosed art of famous craftsman which is delivered only from fathers to sons secretly. Egyptians used to dye the cloths of mummies in the beginning. For the past thousand years, the blue dye utilised in cloth has been Indigo.

Synthesis of Indigo

During the industrial revolution, demand for Indigo increased dramatically. The production of bulk quantity was not possible since extracting it naturally was costly for the flourishing garment industry. Hence, researches were conducted to produce the dye in synthetic methods. Indigo's chemical structure was formulated by Adolf von Baeyer in 1883, treating omega-bromoacetanilide with an alkali (a substance that is high in pH) to produce oxindole. Based on this research, synthesis pathway was introduced by K.Heumann lately [15]. The foremost synthetic dye commercially was produced because of their work within 14 years. The Nobel prize was awarded to Baeyer in 1905 for his great discovery. Even though this synthesis was not enough to manufacture a large amount of Indigo for applications in the industry, it can be utilised well in small-scale industries. The Baeyer-Drewson reaction was implemented to produce Indigo in this experiment involving the reaction of *o*-nitrobenzaldehyde with acetone under highly basic conditions. The final product is formed by the reactants undergoing a series of reactions shown in Fig. 10.2.

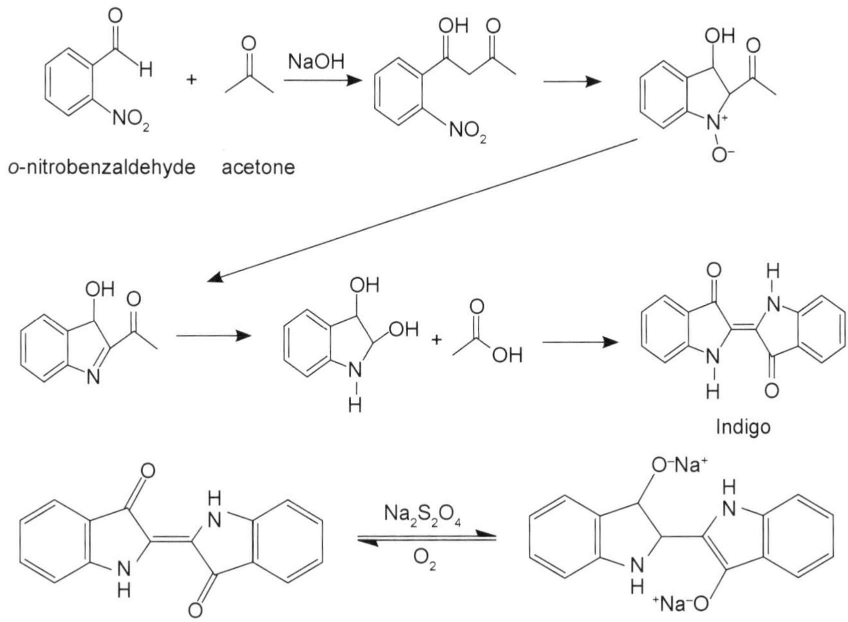

Figure 10.2 Synthesis of artificial Indigo and reduction and oxidation reaction of Indigo dyes.

Indigo is a water-insoluble dye; unfortunately, fabrics cannot be introduced by indigo directly by simply immersing in an aqueous solution of dye. Vat dyeing used to introduce Indigo into the fabric involves the reduction of the Indigo dye in the presence of alkali, which can be applied to the fabric easily. The reaction involved in converting to leuco form is given in Fig. 10.2. Leuco compound, through hydrogen bonding, gets adhered to the fabric when the fabric is immersed in this solution. Leuco compound is oxidised into the dye by exposing the fabric to air.

The depth of penetration is less in ring dyeing (Fig. 10.3a); approximately 70 to 80 % of yarn is not dyed in this category. The penetration of the indigo into the fibre's surface is at most 40%, whereas 60% of the overall cross-section of the yarn is left undyed (Fig. 10.3b) belongs to the deeper ring dyeing effect. In general, for denim getting fixed on warp sheets, not all the dyes are utilised. Generally, 100 % fixation of dyes is not possible, since there are many parameters that influence the dye fixation directly, such as dyeing method, the structure of dye molecules and type of fibres. Therefore, unfixed dyes are released into water streams resulting in turbidity and disturbing the ecosystems, in addition to its toxicity, carcinogenicity or mutagenicity [16,17], suspended solids, metal ions, etc. Most dyes are synthetic compounds with aromatic molecular structures and are non-biodegradable. The oxidative destruction via homogenous oxidation processes with hydrogen peroxide (simple chemical oxidation with H_2O_2 or advanced oxidation processes (AOPs) are required. From dye to dye, the unfixed dye percentage varies apart from the fact that dyes also add an enormous quantity of total dissolved solids (TDS). The normal process is not enough to reduce the TDS; instead, expensive processes like reverse osmosis or evaporation are required.

Figure 10.3 Computer simulated cross-sections, Thinner ring (a); Deeper ring (b).

10.2.5 Post-weaving processes

Pre-wetting

In order to soften the denim fabric, it is dipped in a solution including 1-2g/L detergent for 2-5 min to avoid the wash streaks.

Desizing

Removal of size from denim fabric is the main aim of desizing. It can be achieved by using various chemicals such as detergents, Na_2CO_3, HCl, enzymes, or some oxidative chemicals. Generally, except for the enzyme method, the remaining methods involve fabric damage; therefore, enzyme-based desizing is used by the majority of industries.

10.3 Why sustainable denim

10.3.1 Denim industries and their water pollution

Water resources are polluted through any chemical, physical or biological change in the quality of water, subsequently creating a harmful effect on any living thing that drinks or uses it or lives in it. When humans drink polluted water, it often has serious effects on their health. Water pollution can also make water unsuitable for the desired use, animal health, industry, pisciculture and agriculture or recreation. The degree of pollution effectiveness varies from land to the marine [18]. Fig. 10.4 describes the numerous levels of effluent liberation from public sewers, in and outland surface water and marine coastal area. It is essential to ensure the quality standards of the river and marine water for all sets of streams that are not yet laid in spite of their importance. Table 10.2 illustrates the raw water quality used as a drinking water source and for bathing as advised by WHO and ISO, and limiting the wastewater process and domestic sewage [5,19].

In the process of denim production, processing of products per kilogram release 40-65 liters of wastewater accompanied by various fibres utilised for colourations [20]. 20% of water pollution globally is caused by textile processing. Prominent NGO (Greenpeace International) says, it is predominant in the most populated countries like China, India and Bangladesh [21]; more than half of India's $1.25 billion worth of textile exports to the U.S. came from the southern city of Tirupur. While the business has brought economic benefits, its environmental and social costs are many. Downstream of Tirupur and its more than 300 textile factories, the Noyyal River has become foamy and discoloured. Pollution from this industry is blamed for causing illness among local people and sapping the productivity of nearby farms. Tirupur is not an isolated case. According to the World Bank, 20% of water pollution globally is caused by textile processing. The entire process of manufacturing involves toxic chemicals, which create an extensive diversity of wastewater dangerous to the environment. The key source of textile pollution is mainly comprised of dyes, dyeing additives and other chemicals. Evidently, dyes may contain heavy metals that are highly toxic [22,23]. These contaminants cause marine toxicity due to the presence of varied toxic elements such as salts, surfactants, ionic metals and their complexes, formaldehyde, toxic organic chemicals, biocides and toxic anions, detergents, emulsifiers and dispersants [24].

The wastewater disposed from the textile industry is highly polluted and dangerous precisely when other chemicals are combined with it and released without treatment. Textile industry effluents predominantly affect the agricultural sector and further affect day-to-day activities due to untreated pollutants. All the residues are mixed up in groundwater streams, polluting them and making them non-consumable. During the process of denim manufacturing, the effluents have the characteristics of high pH, BOD, COD, high concentrations of TDS, suspended solids (SS), chlorides, sulphates and phenols, which are listed in Fig. 10.5; the required pollution treatment is summarised in Tables 10.3 and 10.4. Since the chemicals and dyes are not easily biodegradable, it causes severe health complications [3,24]. During the treatment, sludge produced comprises of chromium and other heavy metals which are highly toxic. Therefore, it is mandatory to treat and dispose of the sludge in a secure landfill.

Figure 10.4 Denim effluent contaminates of in nearby river at Perundurai, Tamilnadu, India (Picture captured in April 2019).

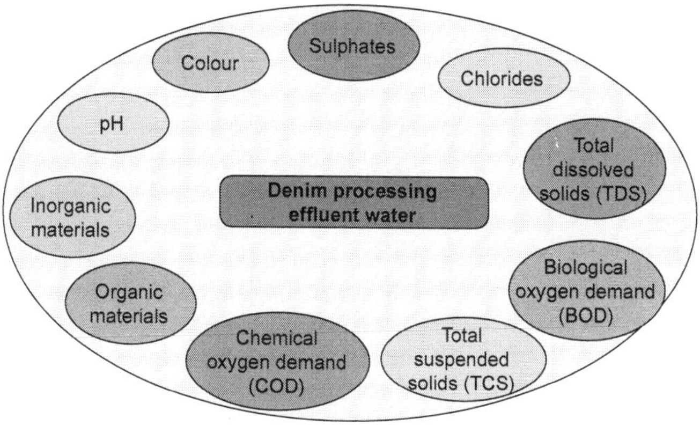

Figure 10.5 Characteristic of denim processing effluents

Table 10.1 Various level of effluent discharge from public sewers, in and outland surface water and marine coastal area

Parameter	Standards			
	Inland surface water	Public sewers	Irrigation	Marine
pH	5.5 to 9.0	5.5 to 9.0	5.5 to 9.0	5.5 to 9.0
Suspended solids (SS) (mg/L)	90	500	190	90
TDS (mg/L)	2200	-	2200	-
BOD (mg/L)	31	352	102	102
COD (mg/L)	248	-	-	250.0
Sulphates (mg/L)	1050	1050	1050	-
Chlorides (mg/L)	950	950	950	-
Oils & grease (mg/L)	5	5	5	5
Lead (as pb) (mg/L)	0.1	1.0	-	2.0
Total Chromium	2.0	-	2.0	2.0

Table 10.2 Indian standards (IS: 15000 -1983), water quality standards for drinking water

Characteristics	Desirable Limit	Permissible limit
pH value	6.5 – 8.5	No relaxation
Odour	Un objectionable	-
Colour (hazen unit)	10	-
Taste	Agreeable	-
Turbidity (max. ntu)	5	10
Total dissolved solids (max. ppm)	500	-
Total hardness (as $CaCO_3$) (max. ppm)	300	600
Chloride (as Cl) (max. ppm)	250	1000
Residual free chlorine (min. ppm)	0.2	0.5
Total Coli form organisms (max. MPN/100 mL)	10	10
Pesticides	Nil	Nil
Radioactive materials		
(a) a - emitters, mC/mL, maximum	10^{-8}	-
(b) b - emitters, mC/mL, maximum	10^{-7}	-

Table 10.3 Water pollution during the colouration of denim

Process	Pollutants and its characteristics	Pollution treatment
Sizing	Starch/PVA, BOD, SS, softener, oils/fats.	Polyelectrolyte, cationic adsorbent coagulants, biological aerobic, UV radiation, nano filtration, reverse osmosis, evaporation.
Dyeing	Unfixed dyes, salts, NaOH, H_2O_2, $Na_2S_2O_3$, TDS, TSS and turbidity.	

Table 10.4 Water pollution during the denim washing/ finishing process

Process	Pollutants and its impacts	Effluent treatments for pollutants
Desizing	Starch/PVA, BOD, SS, enzymes, softener, oils/fats.	
Prewash	Enzyme, surfactants, detergents.	
Bleaching	Chlorines, NaOH, H_2O_2, silicates, suspended solids, fatty alcohol, $KMNO_4$, AOX, detergent.	Polyelectrolyte, cationic adsorbent coagulants, biological aerobic, UV radiation, nano filtration, reverse osmosis, evaporation.
Antichlor wash	Sodium/ disodium thiosulfate.	
Stone/Enzyme wash	Stones, enzymes, acids, pH, BOD issues.	
Softener	Fatty acids, BOD issues.	

The effect of water quality on denim chemical processing

Denim processing requires water, and it is affected due to the additions of chemicals and dyes, resulting in a change in water properties. The effect of water with various constituents are:

- Turbidity
- Suspended solids
- Dissolved solids
- pH values
- Metals
- Hardness
- Oil and other impurities

Turbidity

Stain formation, dye precipitation, coagulation, uneven dyeing, specky dyeing and patchy dyeing are results of suspended solids causing turbidity which causes several problems in the package dyeing machine where these pores are closed by the solids and affecting the uniformity circulation of dye liquor.

Suspended solids

Fabrics are affected by similar faults due to suspended solids as that of turbidity.

Dissolved solids

Dyer faces severe problems due to dissolved solids and the nature of salts. The properties of dyes constituting solids affect exhaustion, rate of dyeing, even dyeing, level dyeing and fastness property.

pH Value

The textile processing activities have a remarkable impact on the pH of groundwater in the nearby area. The enzyme activity is inhibited by the extreme acidic or alkaline pH. Therefore, it is essential to maintain appropriate pH levels for various processes. Likewise, alkaline pH in acid dyeing of silk/wool will have an impact on dyeing properties and fibres, making the acidic pH mandatory for dyeing.

Metals

Staining is caused by the metals like manganese, aluminum, iron, copper, and other heavy metals, causing low dye exhaustion, colour stains, precipitation, corrosion of tanks, pipes, tone variations and high effluent load. Metal soaps are formed by combining the hydroxide of iron and manganese and are highly objectionable with fatty acids. Heavy metal affects the natural colour of silk, e.g., Ferrous ions give a greenish tone and chrome ions give an orangish tone.

Salts

Staining and corrosion are caused by the discharge of sulphates, sulphites, sulphides, chlorides and nitrites in the water. H_2SO_4 used in solubilised vat dyeing along with nitrates prevents corrosion. Excess nitrates cause stains by establishing compounds of the amino group.

Hardness

Generally, the hardness of water is increased by the calcium and magnesium ions as salts, sulphates, carbonates and bicarbonates resulting in patchy dyeing, specky dyeing, poor exhaustion of dyes, precipitation of dyes causing colour stains, tone variations, etc. Soap, major factor, gets precipitated in hard water resulting in inappropriate soaping, emulsification and saponification.

Oils and greasy contaminants

Water bodies are affected by the oils, grease and fatty materials coming through effluent discharges. They also spoil the garments bringing forth the stains and obstructing dye exhaustion and level dyeing since fabric absorbency develops unevenly in their presence.

10.3.2 Denim industries and their air pollution

Energy from the burning of fossil fuels is used to produce and transport denim clothing which is the primary reasons for the air pollution. In turn, it will create greenhouse gasses (GHGs) that will make way for climate change. Volatile organic compound (VOC) is created as the fumes spread all over the place [25].

Dust

Cotton produces dust owing to the spinning process, storage and transportation, in addition to the functions of machines during the production [6]. Throughout denim processing, various forms of dust are produced, such as coal dust, ash, sawdust and grain dust [8]. In order to avoid such limitations of air, there is a necessity to develop a proper methodology for managing dust particles [26].

10.3.3 Raw materials and environmental concerns

Cotton is perhaps the essential natural fibre derived from plants that produce a fruit called boll. According to the Cotton Incorporation survey, around ~12 % of cotton is used to produce denim garments [27]. Cotton has very good comfort properties such as moisture absorption and no static energy, and it provides a better feeling to the human skin. With respect to environmental concerns for landfills and it disposal, cotton is a readily decomposable fibre, and there are no severe issues like those caused by synthetic fibres such as polyester and polyamide. On the other hand, cotton has environmental issues during its production. It pollutes the land and water due to excess amount of fertilizer and pesticide utilisation. Recently, some organisations have been creating awareness to reduce the environmental impact by reducing fertilizer and pesticide usage.

Organic cotton can be grown in a way that utilises less water and uses no fertilizers and pesticides, which lessen the environmental impact. Conventional cotton consumes huge insecticides and pesticides. Based on the WHO survey, humans are highly affected by pesticide poisoning. In addition, it causes more than 20000 deaths per year. Chemicals from conventional cotton may mix with water streams during the rain and affect drinking water, and the residual pesticide can be found in animal meat, birds and even breast milk. For the past two decades, there has been a big movement toward organic cotton. As of now, nearly 3.7 million hectares of land is utilised to cultivate organic cotton, and more than 75 countries are involved in the production of organic cotton.

Lyocell

Lyocell an a eco-friendly fibre made from wood pulp with an amine oxide solvent system. Lyocell is introduced as a substitute for viscose [28–33]. Lyocell is made by using solvent spinning technology which initiates the many new generation cellulosic fibres [34,35]. It is made from the managed forest and does not require to feed chemicals and pesticides [31,36–39]. Generally, lyocell has better properties like good wet strength, lustre, etc.; therefore, it can be used in denim industries as an alternative to cotton [17,40]

Polyester

Polyester mostly refers to "polyethylene terephthalate" which has a significant variety of ester groups in the backbone structure. Generally, it is a cheap fibre compared to natural fibres; therefore, it has become ubiquitous in textiles, including denim. In addition, it provides properties similar to natural fibres except moisture absorption, so it is a popular choice for the textile sector. Polyester is a synthetic fibre manufactured from petroleum products. It is made from carbon-intensive non-renewable resources. Every year more than 70 million barrels of crude oil is used to produce polyester fibres/films and resins [41]. Environmental concerns with polyester, it is not biodegradable, and remains in the ecosystem for a long time. On the other hand, polyester garments (including denim) are the majorly responsible for the microplastic generation in the ocean. Some studies show that 2000 microfibres go to the ocean for washing of one polyester garment [42,43]. Apart from the microplastic issue, it is not a decomposable fibre, with some studies showing that takes it more than 200 years.

10.4 Sustainability in denim processing

10.4.1 Environmentally friendly dyeing

The whole colouring, printing and finishing process involves more than 8000 chemicals [44]. The garment industry produces 80 billion garments per day, most of which are coloured [45]. In the case of dyeing, where reactive, vat, and disperse dyes are involved in the colouration process, a lot of chemicals and auxiliaries such as salt, alkali, and acid are required [30,46–49]. This is another cause of pollution load; in addition, most of them are toxic and hazardous. Recently some alternatives for alkali/salt-free [48,50] reactive dyeing on cotton are developed. Table 10.5 provides some information on the alternatives for these toxic chemicals.

Table 10.5 Environmentally friendly alternative chemicals and processes for textile colourations.

Dyes	Traditional	Alternatives	References
Sulphur dyes	Na_2S	$C_6H_{12}O_6$, $HSCH_2CH_2OH$	[51,52]
Vat dyes	$Na_2S_2O_4$, NaOH	Electrochemical method	[53–57]
Vat and sulphur dyes	$K_2Cr_2O_7$	H_2O_2, $[Na^+]_2 \cdot [B_2O_4(OH)_4]^{2-}$	[52,57,58]
Sulphur, vat dyes	–	Solubilized dyes	–
Hydrotropic agents	CH_4N_2O	$NaN(CN)_2$	[59]
Neutralizing agents	CH_3COOH	HCOOH	[59]

10.4.1.1 Environmentally friendly reducing agents

$Na_2S_2O_4$ is used in the conventional reduction of indigo dyes. However, if the amount of $Na_2S_2O_4$ is more than the stoichiometric amount, the possibilities for by-products [sulphite (SO_3^{2-}) and sulphate (SO_4^{2-})] formation are very high. The formation of by-products is shown in Scheme 1.

$$S_2O_4^{2-} + O_2 \longrightarrow SO_3^{2-} + SO_4^{2-} + H_2O$$

Scheme 10.1 By-product formation from $Na_2S_2O_4$ during Indigo dyeing

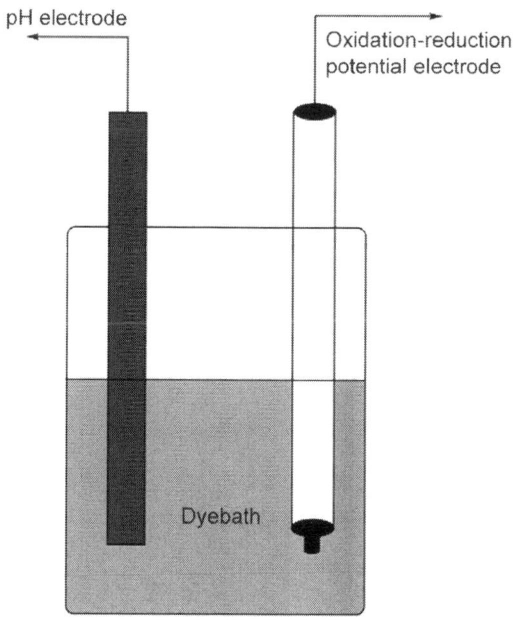

Figure 10.6 Measurement setup for the redox potential

Reduction potential is one of the important properties to measure when we use the non-conventional reducing agent. It can be measured with the oxidation-reduction potential platinum electrode. Because of regard to the Ag/AgCl electrode, that further operates as a working electrode. The redox potential evaluation can be conducted under the KCl electrolyte, that is attached to the pH meter. The simple setup of the instrument is shown in Fig. 10.6 [60,61]. The presence of plants for the long contact time scenario significantly impacted the removal efficiencies of nutrients for chemical oxygen demand (COD).

Hydroxyacetone

Due to environmental impacts from conventional reducing agents, it is important to develop green reducing agents like hydroxyacetone. It provides a reduction potential of up to 810 mV *vs.* Ag/AgCl/ 3 M KCl. The main advantage of this reducing agent is that it can reduce the drastic amount of COD in the dyeing waste. In addition, there are no residual sulphites, sulphide, hydroxide, or strong alkali in the wastewater discharged from the dyeing process. Also, it is suitable for the pad-steam method and needs further research to use in the continuous dyeing method. As it is expensive, this reducing agent could not be commercialised.

Fe^{2+} complexes

Since ancient times Fe^{2+} has been used for the reduction of vat dyes under the "copperas method" along with $FeSO_4$ and $Ca(OH)_2$. On the other hand, the solubility of $Fe(OH)_2$ is not enough under alkaline conditions resulting in sedimentation. Overall, Fe^{2+} has lower redox potential as compared to a conventional reducing agent. The chemical reaction is presented in Scheme 2.

$$FeSO_4 + 2NaOH \longrightarrow Fe(OH)_2 + Na_2SO_4$$

$$2Fe(OH)_2 + 2H_2O \longrightarrow 2Fe(OH)_3 + H_2$$

Scheme 10.2 Redox potential reaction of Fe^{2+} based compounds

Glucose

Glucose is one form of carbohydrate and is used to reduce Indigo [62] and sulphur dyes [57]. It is a non-toxic, biodegradable reducing agent. However, glucose is mostly used to reduce sulphur dyes (i.e., black colour). The glucose-based reduction has very good results in closed form (continuous dyeing) and poor results in open form (jigger machines). The oxidation of the aldehyde group of glucose forms the carboxylic acid (Scheme 3). Using a similar process, Indigo could be reduced to leuco form. However, to get the best results, a higher temperature with a strong alkaline bath is required, which results in negative redox potential suitable for a reduction of Indigo dyes. Similar attempts have been made on hexose, fructose, galactose, lactose and maltose [57].

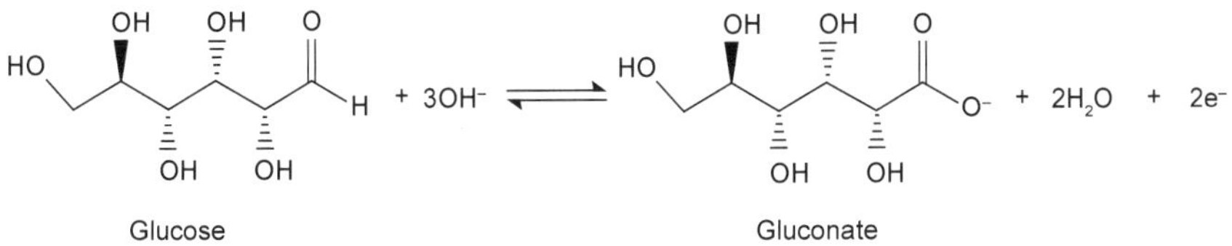

Scheme 10.3 Oxidation reaction of aldehyde group of glucose

Temperature plays a vital role in the reduction of Indigo; in the minimum temperature, there is no reduction of indigo dyes. In the first 10 min, the redox potential for all dyebath is around 650 to 700 mV. Among the different reducing sugars, fructose gives the highest negative redox potential under the same conditions [57]. The highest absorbance of dyebath was found with fructose; it is directly correlated with the amount of leuco form of dye molecule or highest negative redox potential. In all cases, the absorbance is time-dependent [57]. One of the studies compared the redox potentials of glucose and sodium dithionite. Overall, the redox potential depends on the temperature, reducing time and dyeing time. The lower the temperature, higher the negative redox potential (Fig. 10.7-a), and there is no significant difference in redox potential on the reducing time (Fig. 10.7-b). However, during the dyeing time, there is a slight decrease in redox potential with respect to glucose. In all cases, the conventional reducing agent produces more negative redox potential than glucose (Fig. 10.7-c). Apart from that, there are some attempts made with fruit sugars, such as banana [63], ripped banana [63], and bokbunja [64], for the reduction of dyes.

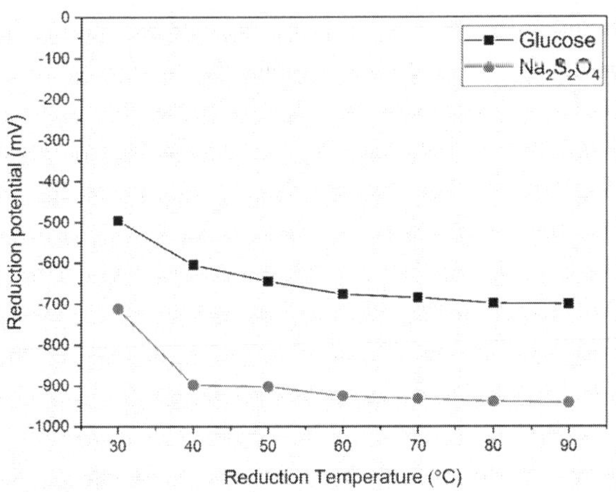

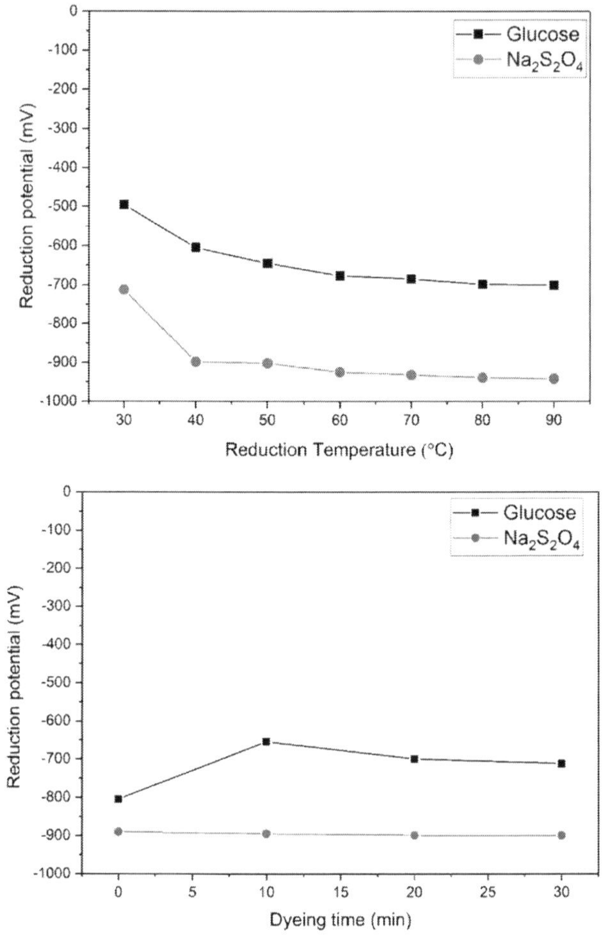

Figure 10.7 Effect of redox potential on reduction temperature (a); effect of redox potential on reduction time (b) and effect of redox potential on dyeing time (c).

10.4.1.2 Electrochemical reduction of vat dyes

As discussed in previous sections, a conventional reducing agent has severe environmental and health risks. In addition, it is very difficult to process those effluents since they contain the most toxic contents (sulphite and sulphate from reducing agents and high COD from organic reducing agents). Last three decades, there have been more attempts for the reduction process of vat (Indigo) and sulphur dyes without the addition of organic and inorganic chemicals; the electrochemical reduction technique is the best choice since it reduces the utilisation of chemicals in the reduction process. However, still, the electrochemical reduction technique is at the laboratory level and not yet commercialised. Electrochemical reduction involved two methods for the dyeing of Indigo dyes. The dyes are directly reduced in the direct electrochemical dyeing method; however, it requires a small quantity of conventional reducing agents to reduce the small quantity of dyes to start the chemical reaction, and once the reaction starts, the electrochemical reaction can continue as well as sustained. Usually, this technique is most suitable with sulphur dyes, particularly sulphur black. On the other hand, in indirect electrochemical dyeing, the organic dyes cannot reduce with the help of an electrode. First, the dyes are reduced with a conventional reducing agent like the conventional method. The reducing agent can initiate the dyes to reduce in the cathode surface to get a complete reduction of dyes. This system is simply known as a reversible redox system; this cycle can repeat to make maximum dyes convert to leuco form.

10.4.2 Enzymatic desizing

Hydrolysis, the most common reaction, along with other reactions such as oxidation, reduction, coagulation, and decomposition, are several reactions involved in enzymatic processes. Denim processing widely utilises enzymes due to their specificity of action and biodegradability. Lock-key theory explains the specific action of an enzyme with a single substrate. There are many enzymes that are used in the textile/ denim industry; Table 10.6 shows the main enzymes with the reactions they catalyse. Desizing is the process that can remove the starch and other ingredients added during the sizing process. Generally, size is applied to the denim warp sheets to increase the weaving efficiency by improving the wear and tear properties. In other words, the size ingredients provide temporary strength to the denim warp sheets. For cellulosic fibres, starch-based materials are used as the main ingredient. On the other hand, fibres like polyester require synthetic polymers. Conventional desizing requires strong acids, which may cause fabric damage due to the formation of hydrocellulose. In addition, it creates a higher TDS and results in the acidic nature of wastewater. The enzyme-based desizing process overcomes the limitations and drawbacks of traditional desizing. Generally, the alpha-amylase enzyme is utilised for the enzymatic desizing process. The α-amylase can help to remove the insoluble starches from the denim fabric by a simple hydrolysis process to make insoluble starch into soluble products. The optimum conditions for the enzymatic desizing on cotton are shown in Table 10.7.

Table 10.6 Classification of main industrial enzymes and their examples and reaction

Enzyme class	Types of reactions catalyzed	Enzymes
Hydrolases	Hydrolysis of molecules and degradation in some cases.	Cellulase, Protease, Amylase, Lipase
Lyases	Nonhydrolytic cleavage of degradation of the molecule.	Fumarase
Transferases	Transfer a group from one molecule to another.	Transaminase
Oxireductases	Oxidation or reduction of molecules.	Laccases
Isomerases	Conversion of one isomer to another.	Glucosephoshate, Isomerase
Ligases	Joining of two molecules with ATP.	Glutamine Synthetase

Table 10.7 Simple recipe for desizing of denim with enzyme

Chemicals	Quantity (% owf)	Purpose
Amylase	1-4%	Breaks down the longer starch molecular chin into water-soluble smaller chain.
Surfactant	0.1-1.5%	Helps to wet the fabric.
Fixing agent	0.2-2.5%	Helps to fix the dyes (reduce more wash-off of dyes).
Temperature & Time	60 to 65 °C & 1 hr	–
pH	6-8	–

10.4.3 Sustainable abrasion (bio-stone/bio-abrasion)

Over the past few years, denim washing methods have improved. In denim garment, effective wash-down effects (i.e., vintage look) has accomplished. There is a necessity to use environment-friendly process due to ecological issues associated with conventional abrasion techniques in which enzymatic abrasion is one among them. Cellulases, group of enzymes utilised for abrasion of denim fabrics is useful in creating the stonewashed look of denim. As discussed earlier, the cellulase is sensitive to pH and temperature and the optimum conditions for cellulase enzyme. The biological processing of catalysing specific chemical reaction cellulose of high molecular weight protein naturally is known as "bio-catalyst". In general, the explicit bonds in cellulose are hydrolysed by cellulases. The 1,4-β- glucosidic bonds of the cellulose molecule are under catalytic action performed by enzyme. The hydrolysis of other bond is initiated and moved to another place due to the enzyme which is only a catalyst and is not used up by the reaction. Due to the repeated action of enzymes, a minor quantity of enzyme is enough to diminish a superior quality of substrate. Various fungal and bacterial species produce cellulase naturally to allow the microorganism to live from cellulose comprising material, such as wood and plants. Mixtures of several cellulase types produced by microorganisms act on cellulose in different ways. The cellulase can be divided into several classes; they are,

- Endo-glucanases (EGs) attack only soluble cellulose.
- Cellobiohydrolases (CBHs) attack the crystalline form of cellulose to convert into water-soluble by making them as two glucose units.
- Cellobioases attack the cellobiose.
- β-glucanases attack larger chains to convert smaller chains.

The obnoxious movement of the stones on the fabric surface helps to fade the blue denim to provide appearance of worn-out effect in the traditional stone washing process. Abrasion accelerates the cellulose in bio-stone finish. The Indigo dye is eased on the denim used by the cellulase in this process. Catalytic hydrolysis of 1,4-β glycoside bonds of cellulose molecules is caused by the cellulase [65,66]. For bio-washing, both acid and neutral cellulase are used; however, the acid cellulase is more aggressive and rapidly attacks 1,4 - β-glycoside bonds of cellulose. Hence, it requires less process time and it generates wide range of abrasion effects on denim by modifying the parameters. However, it increases the degree of back-staining as well as fabric damage.

Advantages of neutral cellulase

- Enzymes are economical as compared to the conventional washing.
- Enzymatic treatment is environmentally friendly.
- Less damage to the materials as compared to conventional pumice stone.
- Enzymes can be recycled.

Back staining

During the process of washing, the solution turns dark blue when indigo is released into the wash liquor. Redeposition of this Indigo takes place on the white parts of denim fabric to enhance the lighter blue shade resulting in a dull look with less contrast on the front side of the jeans known to be the back staining. The white weft yarns and the white pockets are mostly affected by back-staining which depends on the pH and/or the type of enzymes used [67,68]. The back-staining hinge mainly on the type of enzyme used. Nevertheless, additional features such as Indigo yarn dyeing conditions, mechanical effect, liquor ratio, temperature, pH, process time, and stones ratio are also inducing the final aspects [67–70].

Combined process

In some cases, the combinations of cellulase enzyme with the pumice stones are used to develop the specific vintage look on the denim. Usually, such recipe is recommended only for the heavy vintage look finishes enzyme alone cannot produce. However, this recipe contains 50 % of pumice stones which is required for very huge quantity. Maryan et al. [71] studied the discolouration of denim with a combination of nano-bio treatment. In this process, first the Indigo-dyed denim was treated with enzymes, and later it was treated with nanoclay (montmorillonite). As a result, a very good vintage effect, as compared to the conventional finishing techniques, was observed. On other hand, there are many possibilities to use other clay materials such as illite, bentonite, chlorite and kaolinite. Perhaps, montmorillonite is the biodegradable one.

10.4.4 Bio-polishing

Bio-polishing is the process for cellulosic fibres to enhance the physical appearance of the fabric. Generally, the cellulosic fibres like cotton and lyocell have microfibres due to the mechanical action during the production as well as the presence of short fibre. These microfibres are usually called as fuzz. Fuzz is acceptable in some of the applications; however, it reduces the quality of the fabric and makes a duller effect. In addition, these fibres form the pilling balls in lyocell fibre. Therefore, bio-polishing is the process which removes these microfibres from the fabric and significantly reduces the pilling formation. This process enhances the smoothness and brightness of the fabric appearance. In bio-polishing process, cellulase enzymes are used which specifically attack the 1,4-β-glucoside linkage of the cellulosic fibres. The molecule size of cellulase is very big and cannot penetrate inside the structure of cellulosic fibre to attack; therefore, it attacks the surface of fabric. Bio-polishing is one of the important processes for denim made with lyocell fibres. Lyocell fibres are regenerated cellulosic fibres which are regenerated using the organic solvent called NMMO [72]. Fibrillation on the lyocell fibres leads to pilling and affects the denim fabric's appearance [73,74]. Generally, fibrillation can be controlled by several methods such as alkali treatment [74–78], enzymes and dyeing [79]. The enzymatic treatment makes lyocell denim fabric with a more remarkable appearance. Nisha et al. [80] studied the bio-polishing on cotton fabrics; from this study, it is confirmed that the enzymatic treatment removes the protruding fibres on the surface of fabric, since the untreated fabric contains a higher amount of protruding fibres. On other hand the immobilised enzymes make removal efficiency than conventional enzymes. In addition, it reduces the weight loss as well as the strength loss of the treated fabric.

Table 10.8 Various enzymatic bio-polishing methods

Enzyme	Work carried	Reference
Cellulase	Optimised bio-polishing on cotton/ jute blended fabrics	[81]
	Enzyme activity on dyed cotton fabric	[82]
	Enzyme activity on Bamboo cotton blended fabric	[83]
	Enzyme processing on natural fibres	[84]
	Bio-polishing on cellulosic fibres	[85]
	Reduce the pilling of Polyester viscose fabric	[86]
	Combined desizing and bio-polishing of cotton fabric	[87]

10.4.5 Sustainable bleaching

When the stone or enzyme washing gives the desired pattern to match the customer design, colour depth of denim is tuned by bleaching process. The desired shade is the main objective of the bleaching on denim. In the past era, sodium hypochlorite (NaOCl) has been utilised as a bleaching agent. Due to its dreadful, poisonous nature it discharges chlorine and hypochlorous acid spoiling the surroundings and also endangering the living organisms by acidification. This causes lung problems, such as acute respiratory distress syndrome (ARDS), owing to aspiration, which often pays way to death. As soon as the bleaching process is completed, it is essential to remove the residual hypochlorite from the denim fabric. This process is commonly called as anti-chlorine treatment. In denim, sodium-reducing agents such as metabisulphite or thiosulphate are used for the anti-chlorine process. As it generates a pungent SO_2 which generates unpleasant-smelling gas, it spoils the water and ecosystem. The process of hypochlorite bleaching is very dreadful due to chlorite which is harmful; further, the neutralisation process releases high amount of salts.

Nowadays, people concentrate more on environment and eco-friendly processes which are turning to be the trend. Additionally, government agencies act by putting more restrictions on controlling the quality of effluent. Under these situations, it is necessary for the processors to update the alternative with an eco-friendly methodology. Considering the issues, excessive efforts are taken in denim processing. In order to reduce the chemicals, time, energy and water consumption in denim bleaching industry, various environment-friendly approaches such as ozone, electrochemical, plasma, laser and advanced oxidation process (AOP) methods have been familiarised in recent times. The various discolouration methods studied by numerous investigation groups are summarised in Table 10.9.

Table 10.9 Various sustainable discolouration methods for denim processing

Discolouration techniques	Advantages	Disadvantages	References
Oxidizing-bleaching treatment (Ozone)	• Strong discolouration effect • Environmentally friendly • Simple application • Efficient process	• Strength loss • Possible of yellowing on washdown surface • Yellowing effect • Required high investment cost	[88–91]
Bio-washing	• Simple process • Environmentally friendly • Low risk in denim damage (i.e. except Acid cellulase) • Possibility of enzyme recycle and reuse.	• Possibility of back-staining • Possibility of health hazards	[92–96]
Electrochemical decolouration	• High energy efficient • Required less time • Environmentally friendly	• Not available beyond laboratory level	[52,53, 97–99]
Plasma-assisted decolouration	• Faster discolourations • Energy efficient • Does not produce any solid/ wastewater. • Totally dry operation • Not a simple process	• Special skill required • Technical challenge • Possibility of yellowing • Difficult to adjust parameters. • Less possibilities of repeatability.	[19, 23, 91, 100, 101]

Contd...

Contd...

Discolouration techniques	Advantages	Disadvantages	References
Laser-assisted decolouration	• Faster discolourations • Energy efficient • Does not produce any solid/ wastewater. • Totally dry operation • Repeatability of designs	• Special skill required • Technical challenge • Possibility of yellowing • Difficult to adjust parameters. • Strength loss	[102–106]
Advanced oxidation process (AOP)	• No strength loss • Environmentally friendly • Corrosion free • Yellowing free • No neutralisation required, in addition, it does not require various chemicals namely stabilisers, acids and alkalis.	• Recently introduced. • Many experiments must be done. • Not available in commercially.	[107–109]

10.4.5.1 Ozone fading

Ozone (O_3) is a colourless gas and it is formed with the help of three oxygen atoms. Ozone is one of the highly reactant chemicals, and it kills the living cells if the gas is inhaled, and therefore, care should be taken before its application to the denim bleaching. The natural ozone is formed in the upper atmosphere due to the reaction of sunlight with oxygen gas. For the industrial purpose, ozone is created artificially with the help of UV light. Naturally, the ozone is formed in the stratosphere atmosphere when the high energetic solar radiation hits a molecule of oxygen.

Ozone (O_3) fading on denim is one of the alternative sustainable processes. Chemically it is a strong oxidizing agent, and therefore it bleaches (i.e., worn-out effects) faster than other oxidizing agents. Indigo dyes on the surface of denim get isatin and anthranilic acid (i.e., oxidation by-products), resulting in fading or tending to yellow when it is treated with ozone (Fig. 10.8). The main drawback of ozone processing, the running cost is higher than the conventional hypochlorite bleaching. The advantages of ozone fading are [8],

- Faster bleaching process
- Require less rinsing process
- Saves water as well as reduce the wastewater load

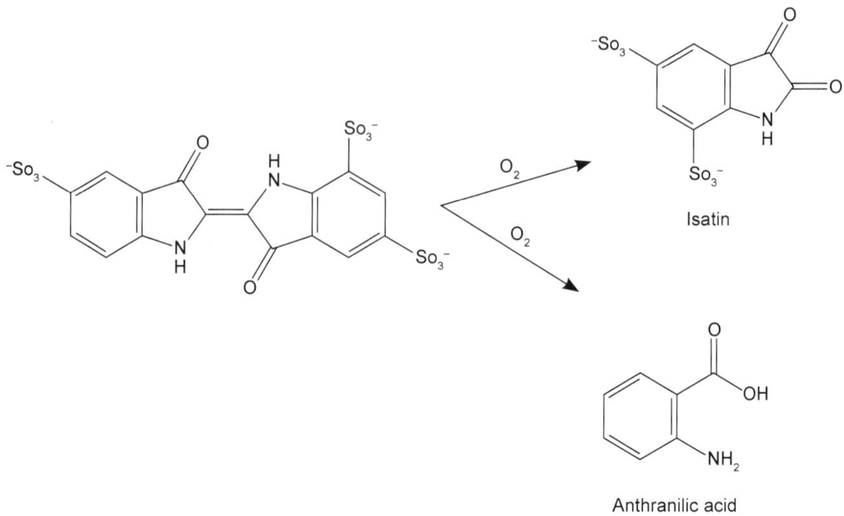

Figure 10.8 Schematic mechanism of denim bleaching with ozone.

Generally, there are many factors which influence the fading rate, they are,

- Diffusion of dyes and its position in the fibre
- Moisture of fibre
- Relative humidity during the process
- Ozone concentration
- Impurities present in the fibre and dye

Ozone fading or bleaching is an interesting field in research due to environmental potential, it has been studied widely (Table 10.9). In another study, ozone treatment on Indigo was enhanced by carboxylate carbon nanotubes [110]. The authors [110] described the method to improve the ozonation efficiency by using carbon nanotubes functionalised with carboxyl groups (CNTs-COOH) as catalysts. It was observed that ozonation in presence of CNTs- COOH provides a strong influence on the decolourisation which is increased as compared to the conventional process. Authors [110] tried to compare various catalysts and found that carboxylic functional groups have a potential catalyst for the ozonation treatment. In the molecular orbital theory, location and orbital energy are important for the reactivity of attacking compounds, in this case Indigo dyes. In Indigo dyes, the hydroxyl radical of double bond could be attacked during this process.

10.4.5.2 Enzymatic bleaching

Compared to ligninolytic peroxidases (>1V), laccases possess comparatively low redox potentials (0.385-0.820 V versus normal hydrogen electrode) [10]. However, later, laccases oxidizing the series of substrates maximises through mechanism involving a mediator. In textile industry, the usage of laccase has maximised in process of bleaching, specifically where it is exceptional. As mentioned in section 10.4.5, lacasse-based bleaching is a sustainable alternative to conventional NaOCl bleaching. It has the advantages of being inexpensive and taking place at room temperature. However, it attacks cotton and reduces its strength since being an oxidizing agent, which is not anticipated for light-ounce denim. Additionally, the laccase usage cannot be extended to Lycra-containing denim. Generally, the laccase is "redox" type of enzyme which have a molecular oxygen as an electron acceptor. Lacasse gets oxidised under aqueous medium and attacks the mediator which converts into free radicals. Their free radicals further attack Indigo to convert it into anthranlic acid and isatin (Table 10.10).

Advantages of laccase-based bleaching of denim

Laccase works faster than the conventional bleaching methods. Its addition, it works faster than acid cellulase with the lowest quantity of enzyme feed. The main benefits of lacasse-based bleaching are:

- Lowest possible strength loss (i.e. if the parameters are optimised).
- Faster discolouration
- Less water consumption
- A safe and eco-friendly alternative for conventional chlorite bleaching.

Table 10.10 Various research on enzymatic bleaching on fabric

Type of enzyme	Work carried out	Reference
Laccase	Laccase and peroxide combined bleaching	[111]
	Ultrasonic assisted laccase bleaching	[112,113]
	Whiteness improvement over conventional bleaching	[114]
	Improvement of whiteness on combining bleaching with ultrasonic and laccase	[115,116]
Glucose-oxidase	Bleaching of cotton with glucose	[117–119]
	Combined desizing, bleaching	[120]
	Improved whiteness on bleaching	[121]

Marie et al. [122] carried out the enzymatic (laccase) bleaching on denim fabric under violuric acid mediator. In the colour measurement, higher reflection values were obtained for the laccase bleached denim which confirms the efficiency of enzymes.

Cold enzyme DeniLite®

The popular enzyme company Novozymes developed the new enzymes "DeniLite®"[123]. The specialty of this enzyme is that it can work at room temperature. It is developed for the bleaching process of textiles including denim. This enzyme is based on peroxidase which does not require oxygen (usually from water, air and fabric surface) to carry out the bleaching process like laccase enzyme. Laccase works slightly slower than peroxidase due to the available oxygen content during the process. It required 10 mins to reach 90% of the entire bleaching process.

10.4.5.3 Water jet fading

In order to pattern or enrich the surface finish, texture, durability, and other characteristics of denim garments, waterjet or hydrojet treatment has been established. In general, the exposure of one or both the surfaces of the garment through hydrojet nozzles is involved in hydrojet treatment. The grade of colour washout, clarity of patterns, and softness of the resulting fabric are associated with the category of dye used in the fabric and the quantity with mode of fluid influences energy applied to the fabric. Predominantly worthy results are achieved with blue Indigo-dyed denim. Due to its chemical-free involvement, it is pollution-free. The technique of water recycling system is used as an economical and environmentally-friendly denim processing. A faded effect without blurring, loss of fabric strength or durability, or excessive warp shrinkage is created in the stripped areas while colour washout of dye takes place [124].

10.4.5.4 Laser fading

The laser generates light energy which is possible to control by the intensity and power [125–127]. This is one of the alternative sustainable dry finishing of denim fabric to create stone wash/sandblasting effects [105]. This process can be easily controlled by a computer which enables to simplify the process; however, it requires special skill to operate. The patterns like lines, dots, pictures and text can be fed into the computer to create similar effects on denim fabric surface. Laser fading is achieved through the thermal degradation of Indigo. Sometimes the pigments or dyes can be sprayed prior to the laser fading to achieve some special effects. Laser fading involves threes steps:

- Computer generated (CAD) pattern or design for fading
- Controlling of laser
- Fading process

For laser fading, solid and gas types of lasers were used since solid-based is a very narrow waveband (i.e. 1 μm), whereas 10 μm for gas-based lasers [102]. There are many parameters (wavelength, power density, duration of exposure and width of the laser beam [128]) that influence the final effects [106]. The primary one is the duration of laser exposure. However, altering these parameters ensures to create different effects. This process has many advantages,

- Water-free denim fading process
- Environmentally friendly process
- Lower process cost
- Easy to transform the design to fabric.

Prabhuraj [89] studied the fading effect of denim by using the laser with carbon dioxide. The colour difference (dE*) as well as colour strength (K/S) values before and after laser and carbon treatments were also measured. The results observed that the brightness of treated denim fabric had been changed according to the laser power used. Higher laser power has lowered the brightness of denim fabric, and it is due to the degradation of dyes. Also, the tone of the fabric surface was dependent on the supplied laser power. Also, they observed a tone change from blue to greenish-yellow due to the oxidation of cotton fabrics; however, there were no results associated with the amount of oxidation and their impact on the fabric's physical and chemical properties. Chi-wai Kan [129] studied the fading of denim garment by laser treatment, and it was compared with stone washing. The laser fading process can save time and water. Since the conventional process (stone washing) involves seven steps whereas the laser involves six steps, on the contrary, the conventional process requires three rinsing processes and laser involves only two rinsing processes. In terms of time and temperature, stone wash required 45 to 60 mins at 55-60 °C. However, the laser fading on denim can be achieved within 3 mins at room temperature. Therefore, this process saves time, energy and reduces the generation of wastewater, resulting in a sustainable process.

10.4.5.5 Photocatalytic discolouration of denim (AOP)

Due to the cost-effective process, photocatalytic discolouration is one of the popular study areas in the field of wastewater treatments. This method ensures various dyestuff to degrade during the wastewater treatment, and this method belongs to the advanced oxidation process (AOP). Izadyar Ebrahimi et al. [130] studied the photocatalytic discolouration of denim using AOP with H_2O_2/UV. First, the denim fabric was impregnated with H_2O_2 (30%) solution for 5 mins; later, the fabric was rinsed and kept under UV radiation for different durations. Finally, the denim fabric was washed and dried. From

FTIR analysis, it was confirmed that there were OH and OOH radicals due to the absorbed UV radiation by hydrogen peroxide and it initiated the photolysis reactions, which led to the degradation of Indigo into anthranilic acid and isatin. Meanwhile, the sample with treatment shows that higher reflectance is nothing but lower colour strength values. They concluded that this process is very promising since it is more effective and sustainable.

10.4.5.6 Plasma associated with denim processing

Plasma-associated processing is one of the promising technologies to modify the surface characteristic of polymers and textiles [131–133], low-temperature plasma (LTP) is one among them. The final effects of the plasma-associated process depend on the nature of gas, duration of process, pressure during the process and discharge power [131,134–136]. Generally, the final core property of final result depends on the type and nature of gas used during the plasma processing [137–139]. Ghoranneviss et al. [140] compared the fading effects of denim fabric by using Ar and O_2 in the LTP process with different exposure times; this study shows that Ar-treated samples show lower K/S values with the same frequency and same duration.

10.4.6 Developments in enzyme application on denim processing

As discussed, enzymes are green alternatives to non-sustainable conventional processes. It is a bioprocess that is not only a cleaner process; it saves huge energy and time which is indirectly proportional to the carbon credit due to the denim industry. However, there are some drawbacks of the enzyme application in the denim industry; therefore, the importance of research in this field has tremendously increased. So, the recent research is focused on the enzymes with superior activity which includes surpassing temperature and pH conditions. It can be framed into two categories where most of the industry-oriented research could take place.

- Immobilisation of enzyme
- Combined enzymatic process

10.4.6.1 Immobilisation of enzymes for denim processing

Immobilisation is a process in the protein engineering to enhance or increase the resistance to the enzyme activity towards varying temperatures or pH. Usually, it is done by the insoluble material (i.e. calcium alginate) [141–147].

This immobilisation is one of the promising processes which provide a variety of applications in various industries. Immobilisation of enzymes has many advantages such as:

- Continuous process
- Rapid enzyme separation
- Possible stabilisation
- Durability
- Recovery and repeated use [80]

Bio-washing

Bio-washing is one of the alternative processes to conventional stone-washing. Bio-washing has several advantages such as improved garment appearance, softness, etc. [148]. Similarly, the disadvantage of bio-washing is that it causes severe strength loss of the fabric. It is due to the uncontrolled hydrolytic reaction on the fibres by cellulase enzyme. Also, it has the possibility to diffuse into the fibres resulting the strength loss [149–153]. Yuanyuan Yu et al. [154] attempted to analyse the immobilised cellulase on denim washing and its impact on weight loss and decolouration efficiency. Generally, CIE L value is used to describe the decolouration efficiency of denim fabric. Based on the CIE LAB colour space, higher L values are nothing but more whiteness, since the immobilised cellulase shows a similar CIE L value as compared to native cellulase when the concentration is 6 % (on weight of fabric). The decolouration is increased by increasing the concentration of both native and immobilised cellulase. Results confirm that the immobilised cellulase has similar decolouration efficiency with reduced weight loss. For both the enzyme, the weight loss is linearly dependent on the concentration; however, immobilised cellulase has lower weight loss as compared to native cellulase. It reveals that there is less damage to the fabric since the immobilised enzyme has a higher molecular weight which does not allow to carry the hydrolytic reaction in diffused state and it happens only on the surface of the denim fabric. Back-staining is another severe issue associated with denim washing. Immobilised cellulase-washed denim shows higher CIE L values which are nothing but lower K/S values. On the contrary to the back-staining measurement, the K/S values increased with increasing cellulase concentration. This trend was found in both native and immobilised enzymes. Finally, this study concludes that the back-staining amount of Indigo dyes in the washed bath is higher in the case of native cellulase as compared with immobilised cellulase.

10.4.6.2 Combined process

Wang et al. [112] introduced the ultrasonic (US) energy for enzymatic desizing. The results improved with the introduction of US energy with the same period. However, there was no significant improvement with the treatment of more than 30 mins. They reported that first 10 mins treatment with US energy provided the desizing efficiency of approximately 73 %, and in 20 mins it reached 80 % with no significant improvement afterward. The advantage of these techniques is that they provide better efficiency within a short period. The similar results were obtained in research works [155 -157]. In another work, Nima et al. [158] introduced the fermentative desizing of cotton, and results found that the fermentative desizing is economically cheaper than the enzymatic desizing; in addition, it utilises fewer resources.

Maryan et al. [159] studied the combination of amylase/ cellulase/ laccase for the bio-washing of denim. Combining amylase/ cellulase/ laccase enzymatic treatment on denim reduces the water, energy and other resources. The use of amylase/ cellulase/ laccase in one bath ensures desizing and bio-washing at the same time, and the result showed that there was no significant colour change compared to the conventional process. The denim subjected to combined enzymatic treatment withstands 10000 abrasion cycles with respect to the back-staining.

Concluding remarks and future perspectives

This chapter summarised the sustainable chemical processing of denim and its environmental impacts. In the denim processing, most of the processes create numerous amounts of pollutants which are hazardous and toxic in nature. Numerous chemicals are involved in the process of denim dyeing other than the dyes mentioned in section 10.2. Generally, the BOD/COD increases tremendously due to the various chemicals involved during the wet processing. It is hard to remove it in effluent treatment. It is therefore essential to recover these chemicals for reuse in order to reduce the dependence on effluent treatment. Consequently, reuse pays a method to lessen the load of wastewater.

There is a strong market for the dyeing and finishing of denim garments in society using eco-friendly methods. Thus, it is worth setting up an innovative manufacturing methodology. In this area, plasma technology, laser treatment and supercritical fluids, which are environmentally friendly and with which the surface properties of inert materials can be easily modified, have varied benefits. In the coming years, it could be anticipated that the ecological challenges due to the denim processing will be resolved by using physical processes. So, it is necessary to discover solutions to resolve the problems in processing. As of now, the research activities are going on toward finding sustainable alternatives. However, some of them have been commercialised and the rest are being researched in laboratories. We believe that such processes could be commercialised in the future. Nevertheless, the above reason is secondary; the prime reason is that the consumer is ready to buy a sustainable product even if it is expensive. In addition, awareness needs to be created among the consumers about their buying behaviours of not buying the new denim unless if they really need it. The alternative sustainable process of denim is tabulated in Table 10.11.

Table 10.11 Conventional and alternative sustainable denim processing [5,8]

Conventional process	Environmental issues	Alternative process	Advantages	Limitation
Sandblasting	Potential Silicosis, Lung cancer, Life threat.	Ice blast, Eco blast, Light stoner by Jeanologia (Spain) [160].	Environmentally friendly, Harmless process.	High machinery cost, Required qualified employers, Finished effects not good as sandblasting.
PP Spray & PP Brush	Environmental impacts.	Light PP Spray by Jeanologia (Spain), Laser fading.	Less pollution load, Natural effect.	The laser is not suitable for all fabric (Polyester may melt during laser treatment), proper safety precaution is required.
Hand scrap, Grinding (Large area damage)	Flying lint and dyes particles, Potential health hazardous.			
3D finishing	Environmental impacts.	Zero- formaldehyde chemicals, or reduced formaldehyde content chemicals (DMDHEU).	Reduction of environmental impacts and consumer safety	High cost for formaldehyde content, Low durability on effects
Bleaching (hypochlorite)	AOX issues, toxic in nature	Ozone bleaching, Laser fading	Enhance productivity. Energy efficiency. Water consumption. Reduction of pollution load.	Poor bleach effect, High machinery cost.
Sulphur Dyeing	Change colour of water, High BOD, High COD, Environmental pollution.	Reduction by Glucose [161,162].	Easily mineralized by bacteria. Less pollution loads. Odour less environment	Not yet commercialised due to less efficiency
		Pre-reduced dyes	Less effluent load Easy process	High cost Maintaining the proper process conditions

Contd...

Contd...

Conventional process	Environmental issues	Alternative process	Advantages	Limitation
Indigo dyeing		Electrochemical reduction [163] Pre-reduced indigo dyes (From BASF, Dystar)	Reduction of using strong oxidizing/ reducing agent In future, it might be BAT.	Required special machinery Poor fastness properties High machinery cost
Sodium hydrosulphite		Catalytic hydrogenation process (by Dystar and BASF) [164].	Reduces the need for hydrosulphite Produce deeper colour	High chemical cost

Acknowledgement

I (Aravin Prince Periyasamy) would like to thank Ministry of Education, Youth and Sports of the Czech Republic and the European Union - European Structural and Investment Funds in the frames of Operational Programme Research, Development and Education - project Hybrid Materials for Hierarchical Structures (HyHi, Reg. No. CZ.02.1.01/0.0/0.0/16_019/0000843).

References

1. ONU, World population to hit 9.8 billion by 2050, (2017). http://www.un.org/apps/news/story.asp?NewsID=57028#.WeSCiFtL_IU (accessed August 15, 2019).

2. Bauer C, Jacques P, Kalt A (2001), 'Photooxidation of an azo dye induced by visible light incident on the surface of TiO2', *J. Photochem. Photobiol. A Chem.* 140,87–92. doi:http://dx.doi.org/10.1016/S1010-6030(01)00391-4.

3. Choudhury AKC (2013), 'Green chemistry and the textile industry', *Text. Prog.* 45,3–143. doi:10.1080/00405167.2013.807601.

4. Periyasamy AP, Militky J (2017), 'Denim and consumers' phase of life cycle'. doi:10.1016/B978-0-08-102043-2.00010-1.

5. Periyasamy AP, Ramamoorthy SK, Lavate SS, (2018) 'Eco-friendly Denim Processing', in: Martínez LMT, Kharissova OV, Kharisov BI (Eds.), Handb. Ecomater., Springer International Publishing, Cham, 2018: pp. 1–21. doi:10.1007/978-3-319-48281-1_102-1.

6. Periyasamy AP, Wiener J, Militky J (2017), 'Life-cycle assessment of denim', doi:10.1016/B978-0-08-102043-2.00004-6 830110.

7. Periyasamy AP (2019), 'Life Cycle assessment of denim: A review', *Colourage* 66 (4).

8. Periyasamy AP, Militky J (2017), 'Denim processing and health hazards'. 161-196. doi:10.1016/B978-0-08-102043-2.00007-1.

9. Periyasamy AP, Duraisamy G (2018), 'Carbon Footprint on Denim Manufacturing', in: L.M.T. Martínez, O.V. Kharissova, B.I. Kharisov (Eds.), Handb. Ecomater., Springer International Publishing, Cham,: pp. 1–18. doi:10.1007/978-3-319-48281-1_112-1.

10. McLoughlin J, Hayes S, Paul R (2015), 'Cotton fibre for denim manufacture', in: R. Paul (Ed.), Denim Manuf. Finish. Appl., Woodhead Publishing Limited, U.K: pp. 15–36. doi:10.1016/B978-0-85709-843-6.00002-0.

11. Patra AK, Pattanayak AK (2015), 'Novel varieties of denim fabrics', in: Roshan Paul (Ed.), Denim Manuf. Finish. Appl., Woodhead Publishing Limited, U.K: pp. 483–506. doi:10.1016/B978-0-85709-843-6.00016-0.

12. Rahman M, Nur G (2014), 'Recent Innovations in Yarn Technology: A Review', *Int. J. Sci. Res. Publ.* 4, 1–7.

13. Haleem N, Wang X (2015), 'Recent research and developments on yarn hairiness', *Text. Res. J.* 85, 211–224. doi:10.1177/0040517514538692.

14. Achwal W Ba (1990), 'Environmental aspects of textile chemical processing: part-I', *Colourage* 37, 40–44.

15. The Nobel Prize in Chemistry (1905). https://www.nobelprize.org/prizes/chemistry/1905/summary/.

16. Zaharia C, Suteu D, Muresan A, Muresan R, Popescu A (2009), 'Textile Wastewater Treatment by Homogeneous Oxidation with Hydrogen Peroxide', *Environ. Eng. Manag. J.* 8,1359–1369.

17. Periyasamy AP, Dhurai B, Thangamani K (2011), 'Salt free dyeing: A new method of dyeing of Lyocell fabrics with reactive dyes', *Autex Res. J.* 11, 14–17.

18. Khan S, Malik A (2014), 'Environmental and Health Effects of Textile Industry Wastewater', in: A. Malik, E. Grohmann, R. Akhtar (Eds.), Environ. Deterior. Hum. Heal., Springer Netherlands, Dordrecht : pp. 55–71. doi:10.1007/978-94-007-7890-0_4.

19. Periyasamy AP, Venkatesan H (2018), 'Eco-Materials in Textile Finishing, in: L.M.T'. Martínez, O.V. Kharissova, B.I. Kharisov (Eds.), Handb. Ecomater., Springer International Publishing, Cham : pp. 1–22. doi:10.1007/978-3-319-48281-1_55-1.

20. Manu B, Chaudhari S (2002), 'Anaerobic decolorisation of simulated textile wastewater containing azo dyes', *Bioresour. Technol.* 82, 225–231. doi:10.1016/S0960-8524(01)00190-0.

21. Scott A (2015), 'Cutting Out Textile Pollution', *Chem. Eng. News.* 93,18–19. doi:10.1021/cen-09341-bus1.

22. Periyasamy AP, Ramamoorthy SK, Rwawiire S, Zhao S (2018), 'Sustainable Wastewater Treatment Methods for Textile Industry', in: S.S. Muthu (Ed.), Sustain. Innov. Appar. Prod., Springer Singapore, Singapore, 2018: pp. 21–87. doi:10.1007/978-981-10-8591-8_2.

23. Periyasamy AP, Rwahwire S, Zhao Y (2018), 'Environmentally Friendly Textile Processing', in: L.M.T. Martínez, O.V. Kharissova, B.I. Kharisov (Eds.), Handb. Ecomater'., Springer International Publishing, Cham: pp. 1–38. doi:10.1007/978-3-319-48281-1_176-1.

24. Karthik T, Gopalakrishnan D (2015), 'Roadmap to Sustainable Textiles and Clothing', in: S.S. Muthu (Ed.), Roadmap to Sustain. Text. Cloth. Environ. Soc. Asp. Text. Cloth. Supply Chain, Springer Singapore, Singapore: pp. 153–188. doi:10.1007/978-981-287-164-0.

25. Spengler JD, Sexton K (1983), 'Indoor air pollution: a public health perspective', *Science* (80). 221, 9–17. http://science.sciencemag.org/content/221/4605/9.abstract.

26. Lai PS, Christiani DC (2013), 'Long-term respiratory health effects in textile workers', *Curr. Opin. Pulm. Med.* 19 ,152–157. doi:10.1097/MCP.0b013e32835cee9a.

27. Kininmonth M (2012), 'Greening of the denim supply chain', *Lenz. Beri.* 90, 8–15.

28. Periyasamy AP, Eklahare SR (2013), 'Effect of alkali on lyocell fabric', *Text. Asia* 44,20-22 .

29. Periyasamy AP, Dhurai B (2011), 'Salt free dyeing: A new method of dyeing of Lyocell fabrics with reactive dyes', *Pakistan Text. J.* 60, 40–43.

30. Periyasamy AP (2022), 'Natural dyeing of cellulose fibers using syzygium cumini fruit extracts and a bio-mordant: A step toward sustainable dyeing'. *Sust. Mat. and Technologies* 33, e00472.

31. Periyasamy AP, Dhurai B, Thangamani B (2011), 'A study on fibrillation properties of Lyocell fiber', *Colourage* 58, 45–48.

32. Periyasamy AP, Mehta P (2013), 'Lyocell fibers for nonwovens', *Tech. Textilien.* 56(2), E70-E71.

33. Periyasamy AP, Mehta P (2013), 'Lyocell fibers for nonwovens', *Chem. Fibers Int.* 63, 104-105.

34. Firgo H, Eibl M, Eichinger D, (1995) 'Lyocell - An ecological alternative', *Lenz. Beri.* 47–50.

35. Rosenau T, Potthast A, Sixta H, Kosma P (2001), 'The chemistry of side reactions and byproduct formation in the system NMMO/cellulose (Lyocell process)', *Prog. Polym. Sci.* 26,1763–1837. doi:https://doi.org/10.1016/S0079-6700(01)00023-5.

36. Bhambure SD, Lavate SS, Periyasamy AP (2013), 'Textiles in footwear', *Asian Text. J.* 22, 54-59.

37. Periyasamy AP (2012), 'Effect of alkali pretreatment and dyeing on fibrillation properties of lyocell fiber', in: RMUTP Int. Conf. Text. Fash. 2012, Rajamangala University of Technology, Thanyaburi, Thailand, 2012: pp. 1–11. http://decade1.ird.rmutp.ac.th/wp-content/uploads/2012/09/Section-II-3.pdf.

38. Manian AP, Ruef H, Bechtold T (2007), 'Spun-dyed lyocell', *Dye. Pigment* 74 (2007) 519–524. doi:http://dx.doi.org/10.1016/j.dyepig.2006.03.015.

39. Periyasamy AP, Dhurai B (2011), 'Salt free dyeing: A new method of dyeing of Lyocell fabrics with reactive dyes', *Pakistan Text. J.* 60.

40. Periyasamy AP (2016), 'Effect of PVAmHCl Pre-treatment on the Properties of Modal Fabric Dyed with Reactive Dyes: An Approach for Salt Free Dyeing', *J. Text. Sci. Eng.* 6: 262 ,1–9. doi:10.4172/2165-8064.1000262.

41. What is polyester. https://www.thehouseofpillows.eu/polyester-production-blog/ (accessed August 15, 2019).

42. Belzagui F, Crespi M, Álvarez A, Gutiérrez-Bouzán C, Vilaseca M (2019), 'Microplastics' emissions: Microfibers' detachment from textile garments', *Environ. Pollut.* 248, 1028–1035. doi:10.1016/j.envpol.2019.02.059.

43. Álvarez G, Barros A, Velando A (2018), 'The use of European shag pellets as indicators of microplastic fibers in the marine environment', *Mar. Pollut. Bull.* 137, 444–448. doi:10.1016/j.marpolbul.2018.10.050.

44. Kumar PS, Narayan AS, Dutta A (2017), 'Textiles and Clothing Sustainability', in: S.S. Muthu (Ed.), Text. Cloth. Sustain. Text. Sci. Cloth. Technol., Springer Singapore, Singapore: pp. 57–96. doi:10.1007/978-981-10-2188-6.

45. Environmental Impact. https://truecostmovie.com/learn-more/environmental-impact/ (accessed August 8, 2019).

46. Ma W, Meng M, Yan S, Zhang S (2016), 'Salt-free reactive dyeing of betaine-modified cationic cotton fabrics with enhanced dye fixation', *Chinese J. Chem. Eng.* 24 ,175–179. doi:10.1016/j.cjche.2015.07.008.

47. Wang L, Ma W, Zhang S, Teng X, Yang J (2009), 'Preparation of cationic cotton with two-bath pad-bake process and its application in salt-free dyeing', *Carbohydr. Polym.* 78., 602–608. doi:10.1016/j.carbpol.2009.05.022.

48. Zhang Y, Zhang W (2015), 'Clean dyeing of cotton fiber using a novel nicotinic acid quaternary triazine cationic reactive dye: Salt-free, alkali-free, and non-toxic by-product', *Clean Technol. Environ. Policy* 17, 563–569. doi:10.1007/s10098-014-0821-9.

49. Periyasamy AP, Dhurai B (2011), 'Salt free dyeing', *Asian Dyer* 8, 47–50.

50. Dong X, Gu Z, Hang C, Ke G, Jiang L, He, (2019) 'Study on the salt-free low-alkaline reactive cotton dyeing in high concentration of ethanol in volume', *J. Clean. Prod.* 226, 316–323. doi:https://doi.org/10.1016/j.jclepro.2019.04.006.

51. Kulandainathan MA, Patil K, Muthukumaran A, Chavan RB (2007), 'Review of the process development aspects of electrochemical dyeing: Its impact and commercial applications', *Color. Technol.* 123,143–151. doi:10.1111/j.1478-4408.2007.00082.x.

52. Božič M, Kokol V (2008), 'Ecological alternatives to the reduction and oxidation processes in dyeing with vat and sulphur dyes', *Dye. Pigment* 76, 299–309. doi:10.1016/j.dyepig.2006.05.041.

53. Kulandainathan MA, Muthukumaran A, Patil K, Chavan RB (2007), 'Potentiostatic studies on indirect electrochemical reduction of vat dyes', *Dye. Pigment* 73, 47–54. doi:10.1016/j.dyepig.2005.10.007.

54. Roessler A, Crettenand D, Dossenbach O, Marte W, Rys P (2002), 'Direct electrochemical reduction of indigo', *Electrochim. Acta.* 47,1989–1995. doi:10.1016/S0013-4686(02)00028-2.

55. Chavan RB (2015), 'Indigo dye and reduction techniques, in: R.B.T.-D. Paul (Ed.)', Denim Manuf. Finish. Appl., Woodhead Publishing: pp. 37–67. doi:10.1016/B978-0-85709-843-6.00003-2.

56. Roessler A, Jin X (2003), 'State of the art technologies and new electrochemical methods for the reduction of vat dyes', *Dye. Pigment* 59, 223–235. doi:10.1016/S0143-7208(03)00108-6.

57. Saikhao L, Setthayanond J, Karpkird T, Bechtold T, Suwanruji P (2018), 'Green reducing agents for indigo dyeing on cotton fabrics', *J. Clean. Prod.* 197,106–113. doi:10.1016/j.jclepro.2018.06.199.

58. Chakraborty JN, Sulphur dyes (2011), in: T. Clark, I. Dyeing (Eds.), Handb. Text. Ind. Dye. Princ. Process. Types Dye., Woodhead Publishing: pp. 466–485. doi:10.1533/9780857093974.2.466.

59. Gulzar T, Farooq T, Kiran S, Ahmad I, Hameed A, (2018) 'Green chemistry in the wet processing of textiles', in: Shahid-ul-Islam, B.S.B.T.-T.I. Butola, P. of Green Chemistry for Textile Technology (Eds.), Impact Prospect. Green Chem. Text. Technol., Woodhead Publishing, 2018: pp. 1–20. doi:10.1016/b978-0-08-102491-1.00001-0.

60. Scholz M (2019), 'Dye Wastewater Treatment by Vertical-Flow Constructed Wetlands', in: M.B.T.-S.W.T. Scholz (Ed.), Sustain. Water Treat., Elsevier, 2019: pp. 191–213. doi:10.1016/B978-0-12-816246-0.00008-2.

61. Jono R, Tateyama Y, Yamashita K (2015), 'A method to calculate redox potentials relative to the normal hydrogen electrode in nonaqueous solution by using density functional theory-based molecular dynamics', *Phys. Chem. Chem. Phys.* 17,27103–27108. doi:10.1039/C5CP05029D.

62. Shin Y, Son K, Il Yoo D (2016), 'Indigo dyeing onto ramie fabric via microbial reduction: Reducing power evaluation of some bacterial strains isolated from fermented indigo vat', *Fibers Polym.* 17,1000–1006. doi:10.1007/s12221-016-6353-7.

63. Hossain MD, Khan MMR, Uddin MZ (2017), 'Fastness Properties and Color Analysis of Natural Indigo Dye and Compatibility Study of Different Natural Reducing Agents', *J. Polym. Environ.* 25, 1219–1230. doi:10.1007/s10924-016-0900-6.

64. Shin Y, Choi M, Yoo D. Il (2014), 'Eco-friendly indigo reduction using bokbunja (Rubus coreanus Miq.) sludge', *Fash. Text.* 1,6. doi:10.1186/s40691-014-0006-5.

65. Berrin JG, Herpoel-Gimbert I, Ferreira NL, Margeot A, Heiss-Blanquet S (2014), 'Use of Cellulases from Trichoderma reesei in the Twenty-First Century Part II: Optimization of Cellulolytic Cocktails for Saccharification of Lignocellulosic Feedstocks', in: V.K. Gupta, M. Schmoll, A. Herrera-Estrella, R.S. Upadhyay, I. Druzhinina, M.G.B.T.-B. Tuohy, B. of Trichoderma (Eds.), Biotechnol. Biol. Trichoderma, Elsevier, Amsterdam, : pp. 263–280. doi:10.1016/B978-0-444-59576-8.00018-7.

66. Puranen T, Alapuranen M, Vehmaanperä J (2014), 'Trichoderma Enzymes for Textile Industries, in: V.K. Gupta, M. Schmoll, A. Herrera-Estrella, R.S. Upadhyay, I. Druzhinina, M.G.B.T.-B. Tuohy', B. of Trichoderma (Eds.), Biotechnol. Biol. Trichoderma, Elsevier, Amsterdam,: pp. 351–362. doi:10.1016/B978-0-444-59576-8.00026-6.

67. Sinitsyn AP, Gusakov AV, Grishutin SG, Sinitsyna OA, Ankudimova NV(2001), 'Application of microassays for investigation of cellulase abrasive activity and backstaining', *J. Biotechnol.* 89,233–238. doi:https://doi.org/10.1016/S0168-1656(01)00306-6.

68. Cavaco-Paulo A, Campos R, Morgado J (1999), 'Use of enzymes in textile processing - indigo backstaining during cellulase washing', in: A.R.B.T.-E. '98 Horrocks (Ed.), Woodhead Publ. Ser. Text., Woodhead Publishing,: pp. 191–194. doi:https://doi.org/10.1533/9780857092991.3.191.

69. Cavaco-Paulo A, Morgado J, Almeida L, Kilburn D (1998), 'Indigo Backstaining During Cellulase Washing', *Text. Res. J.* 68,398–401. doi:10.1177/004051759806800602.

70. De Souza Moreira LR, Sciuto D Lo , Filho E.X.F. (2016), 'An Overview of Cellulose-Degrading Enzymes and Their Applications in Textile Industry', in: V.K.B.T.-N. Gupta, F.D. in Microbial Biotechnology, Bioengineering (Eds.), New Futur. Dev. Microb. Biotechnol. Bioeng. Microb. Cell. Syst. Prop. Appl., Elsevier, Amsterdam, pp. 165–175. doi:10.1016/B978-0-444-63507-5.00014-9.

71. Maryan AS, Montazer M, Damerchely R (2015), 'Discoloration of denim garment with color free effluent using montmorillonite based nano clay and enzymes: Nano bio-treatment on denim garment', *J. Clean. Prod.* 91,208–215. doi:10.1016/j.jclepro.2014.12.014.

72. Meister G, Wechsler M (1998), 'Biodegradation of N-methylmorpholine-N-oxide', *Biodegradation* 9, 91–102. doi:10.1023/A:1008264908921.

73. Kreze T, Stana-Kleinschek K, Ribitsch V (2001), 'The sorption behaviour of cellulose fibres', *Lenz. Beri.* 28–33.

74. Carrillo F, Colom X, Suñol JJ, Saurina J (2004), 'Structural FTIR analysis and thermal characterisation of lyocell and viscose-type fibres', *Eur. Polym. J.* 40,2229–2234. doi:10.1016/j.eurpolymj.2004.05.003.

75. Xu Y, Lu Z, Tang R (2007), 'Structure and thermal properties of bamboo viscose, Tencel and conventional viscose fiber', *J. Therm. Anal. Calorim.* 89,197–201. doi:10.1007/s10973-005-7539-1.

76. Borbly V (2008), 'Lyocell, the new generation of regenerated cellulose', *Acta Polytech. Hungarica.* 5, 11–18.

77. Biganska O, Navard P (2009), 'Morphology of cellulose objects regenerated from cellulose-N-methylmorpholine N-oxide-water solutions', *Cellulose* 16,179–188. doi:10.1007/s10570-008-9256-y.

78. Lonnberg B (2001), '2 - Industrial cellulose, in: C. Woodings (Ed.), Regen. Cellul. Fibres', Woodhead Publishing Limited,: pp. 22–36. doi:http://doi.org/10.1533/9781855737587.22.

79. Periyasamy AP (2013), 'Non-formaldehyde crease resistance finishing on tencel with poly maleic acid', *Int. Dye.* 198, 37–39.

80. Sankarraj N, Nallathambi G (2018), 'Enzymatic biopolishing of cotton fabric with free/immobilized cellulase', *Carbohydr. Polym.* 191,95–102. doi:10.1016/j.carbpol.2018.02.067.

81. Sreenath HK, Shah AB, Yang VW, Gharia MM, Jeffries TW (1996), 'Enzymatic polishing of jute/cotton blended fabrics', *J. Ferment. Bioeng.* 81,18–20. doi:https://doi.org/10.1016/0922-338X(96)83113-8.

82. Yamada M, Amano Y, Horikawa E, Nozaki K, Kanda T (2005), 'Mode of Action of Cellulases on Dyed Cotton with a Reactive Dye', *Biosci. Biotechnol. Biochem.* 69, 45–50. doi:10.1271/bbb.69.45.

83. El-Khatib HS, Badr AA, Diyab WA, Atia RM (2019), 'Part I: Enzymatic treatment of Bamboo, Bamboo/Cotton knitted fabric using brewer's yeast suspension', *Alexandria Eng. J.* 58, 819-825. doi:https://doi.org/10.1016/j.aej.2019.05.017.

84. Shahid M, Mohammad F, Chen G, Tang RC, Xing T (2016), 'Enzymatic processing of natural fibres: White biotechnology for sustainable development', *Green Chem.* 18,2256–2281. doi:10.1039/c6gc00201c.

85. Wadham M (1994), 'Bio-polishing of cellulosic fabrics', *J. Soc. Dye. Colour.* 110,367–368. doi:10.1111/j.1478-4408.1994.tb01599.x.

86. Montazer M, Mazaheri F, Khosravian S, Azimi M, M Bameni Moghadam, Sadeghi AH, Application of resins and crosslinking agents on fiber blend fabric to reduce pilling performance, optimized by response surface methodology, *J. Vinyl Addit. Technol.* 17 (2011) 213–221. doi:10.1002/vnl.20274.

87. Esfandiari A, Firouzi-Pouyaei E, Aghaei-Meibodi P (2014), 'Effect of enzymatic and mechanical treatment on combined desizing and bio-polishing of cotton fabrics', *J. Text. Inst.* 105,1193–1202. doi:10.1080/00405000.2014.880222.

88. Choudhury AKR (2017), 'Environmental impacts of denim washing - Muthu, Subramanian Senthilkannan - Sustainability in Denim', in: Text. Inst. B. Ser., Woodhead Publishing : pp. 49–81. doi:https://doi.org/10.1016/B978-0-08-102043-2.00003-4.

89. Venkatraman PD, Liauw CM (2019), 'Use of a carbon dioxide laser for environmentally beneficial generation of distressed/faded effects on indigo dyed denim fabric: Evaluation of colour change, fibre morphology', degradation and textile properties, *Opt. Laser Technol.* 111, 701–713. doi:10.1016/j.optlastec.2018.09.004.

90. Choudhury AKR (2017), 'Finishing of denim fabrics, Roy Choudhury (Ed.)', Princ. Text. Finish., Elsevier, 2017: pp. 383–415. doi:10.1016/B978-0-08-100646-7.00012-6.

91. Kan C, Cheung H, Chan Q (2016), 'A study of plasma-induced ozone treatment on the colour fading of dyed cotton', *J. Clean. Prod.* 112,3514–3524. doi:https://doi.org/10.1016/j.jclepro.2015.10.100.

92. Focher B, Marzetti A, Cattaneo M, Beltrame PL, Carniti P (1981), 'Effects of structural features of cotton cellulose on enzymatic hydrolysis', *J. Appl. Polym. Sci.* 26,1989–1999. doi:10.1002/app.1981.070260622.

93. Aly AS, Moustafa AB, Hebeish A (2004), 'Bio-technological treatment of cellulosic textiles', *J. Clean. Prod.* 12,697–705. doi:https://doi.org/10.1016/S0959-6526(03)00074-X.

94. Rousselle MA, Bertoniere NR, Howley PS, Wilton JR, (2002), 'Effect of Whole Cellulase on the Supramolecular Structure of Cotton Cellulose', *Text. Res. J.* 72,963–972. doi:10.1177/004051750207201106.

95. Sheikh J, Bramhecha I (2019), 'Enzymes for green chemical processing of cotton', in: Shahid-ul-Islam, Butola, P. of Green Chemistry for Textile Technology (Eds.), Impact Prospect. Green Chem. Text. Technol., Woodhead Publishing,: pp. 135–160. doi:10.1016/b978-0-08-102491-1.00006-x.

96. Cavaco-Paulo A (1998), 'Mechanism of cellulase action in textile processes', *Carbohydr. Polym.* 37,273–277. doi:https://doi.org/10.1016/S0144-8617(98)00070-8.

97. Sarkar AK (2015), 'Dyeing technologies for denim garments', in: R.B.T.-D. Paul (Ed.), Denim Manuf. Finish. Appl., Woodhead Publishing: pp. 271–285. doi:10.1016/B978-0-85709-843-6.00009-3.

98. Bechtold T, Turcanu A (2009), 'Electrochemical reduction in vat dyeing: greener chemistry replaces traditional processes', *J. Clean. Prod.* 17,1669–1679. doi:https://doi.org/10.1016/j.jclepro.2009.08.004.

99. Roessler A, Crettenand D (2004), 'Direct electrochemical reduction of vat dyes in a fixed bed of graphite granules', *Dye. Pigment* 63,29–37. doi:10.1016/j.dyepig.2004.01.005.

100. Shahidi S, Ghoranneviss M, Moazzenchi B (2014), 'New Advances in Plasma Technology for Textile', *J. Fusion Energy* 33, 97–102. doi:10.1007/s10894-013-9657-2.

101. Jelil RA (2015), 'A review of low-temperature plasma treatment of textile materials', *J. Mater. Sci.* 50, 5913–5943. doi:10.1007/s10853-015-9152-4.

102. Nayak R, Padhye R, (2016) 'The use of laser in garment manufacturing: an overview', *Fash. Text.* 3, 5. doi:10.1186/s40691-016-0057-x.

103. Hung ON, Chan CK, Kan CW, Yuen CWM, Song LJ, (2014) 'Artificial neural network approach for predicting colour properties of laser-treated denim fabrics', *Fibers Polym.* 15,1330–1336. doi:10.1007/s12221-014-1330-5.

104. Yuan G, Chen Z, Luzzi D,(2018) 'Application of Laser Technology in Fashion Industry BT - Contemporary Case Studies on Fashion Production, Marketing and Operations', in: P.-S. Chow, C.-H. Chiu, A. C. Y. Yip, A. K. Y. Tang (Eds.), Springer Singapore, Singapore : pp. 43–56. doi:10.1007/978-981-10-7007-5_3.

105. Kan CW, (2014) 'Colour fading effect of indigo-dyed cotton denim fabric by CO2 laser', *Fibers Polym.* 15,426–429. doi:10.1007/s12221-014-0426-2.

106. Kan CW, Yuen CWM, Cheng CW, (2010) 'Technical study of the effect of CO2 laser surface engraving on the colour properties of denim fabric', *Color. Technol.* 126 (2010) 365–371. doi:10.1111/j.1478-4408.2010.00270.x.

107. Aleboyeh A, Moussa Y, Aleboyeh H,(2005) 'The effect of operational parameters on UV/H2O2 decolourisation of Acid Blue 74', *Dye. Pigment* 66,129–134. doi:https://doi.org/10.1016/j.dyepig.2004.09.008.

108. Zuorro A, Lavecchia R, (2014) 'Evaluation of UV/H2O2 advanced oxidation process (AOP) for the degradation of diazo dye Reactive Green 19 in aqueous solution', *Desalin. Water Treat.* 52 ,1571–1577. doi:10.1080/19443994.2013.787553.

109. Al-Momani F, Touraud E, Degorce-Dumas JR, Roussy J, Thomas O, (2002), 'Biodegradability enhancement of textile dyes and textile wastewater by VUV photolysis', *J. Photochem. Photobiol. A Chem.* 153,191–197. doi:https://doi.org/10.1016/S1010-6030(02)00298-8.

110. Qu R, Xu B, Meng L, Wang L, Wang Z, (2015), 'Ozonation of indigo enhanced by carboxylated carbon nanotubes: Performance optimization, degradation products, reaction mechanism and toxicity evaluation', *Water Res.* 68, 316–327. doi:10.1016/j.watres.2014.10.017.

111. Tzanov T, Andreaus J, Guebitz G, Cavaco-Paulo A, (2003), 'Protein interactions in enzymatic processes in textiles, Electron'. *J. Biotechnol.* 6,147–155. doi:10.2225/vol6-issue3-fulltext-8.

112. Wang WM, Yu B, Zhong CJ, (2012), 'Use of ultrasonic energy in the enzymatic desizing of cotton fabric', *J. Clean. Prod.* 33,179–182. doi:10.1016/j.jclepro.2012.04.010.

113. Abou-Okeil A, El-Shafie A, El Zawahry MM, 'Ecofriendly laccase-hydrogen peroxide/ultrasound-assisted bleaching of linen fabrics and its influence on dyeing efficiency', *Ultrason. Sonochem.* 17,383–390. doi:10.1016/j.ultsonch.2009.08.007.

114. Tzanov T, Basto C, Gübitz GM, Cavaco-Paulo A, 'Laccases to improve the whiteness in a conventional: Bleaching of cotton', *Macromol. Mater. Eng.* 288,807–810. doi:10.1002/mame.200300100.

115. Basto C, Tzanov T, Cavaco-Paulo A,(2007) 'Combined ultrasound-laccase assisted bleaching of cotton', *Ultrason. Sonochem.* 14, 350–354. doi:https://doi.org/10.1016/j.ultsonch.2006.07.006.

116. Gonçalves I, Herrero-Yniesta V, Perales Arce I, Escrigas Castañeda M, Cavaco-Paulo A, Silva C, (2014), 'Ultrasonic pilot-scale reactor for enzymatic bleaching of cotton fabrics', *Ultrason. Sonochem.* 21,1535–1543. doi:10.1016/j.ultsonch.2014.02.009.

117. Mojsov K, (2019), 'Enzymatic desizing, bioscouring and enzymatic bleaching of cotton fabric with glucose oxidase', *J. Text. Inst.* 110,1032–1041. doi:10.1080/00405000.2018.1535240.

118. Tzanov T, Calafell M, Guebitz GM, Cavaco-Paulo A, (2001) ' Bio-preparation of cotton fabrics', *Enzyme Microb. Technol.* 29, 357–362. doi:10.1016/S0141-0229(01)00388-X.

119. Tavčer PF, (2012), 'Low-temperature bleaching of cotton induced by glucose oxidase enzymes and hydrogen peroxide activators', *Biocatal. Biotransformation.* 30, 20–26. doi:10.3109/10242422.2012.644437.

120. Hebeish A, Ramadan MA, Hashem M, Sadek B, Abdel-Hady M, (2013), 'New Development for Combined Bioscouring and Bleaching of Cotton-Based Fabrics', *Res. J. Text. Appar.* 17, 94–103. doi:10.1108/RJTA-17-01-2013-B010.

121. Davulcu A, Eren HA, Avinc O, Erişmiş B, (2014) 'Ultrasound assisted biobleaching of cotton', *Cellulose* 21, 2973–2981. doi:10.1007/s10570-014-0273-8.

122. M.M. Iracheta-Cárdenas, M.A. Rocha-Peña, L.J. Galán-Wong, K. Arévalo-Niño, O.E. Tovar-Herrera (2016), 'A Pycnoporus sanguineus laccase for denim bleaching and its comparison with an enzymatic commercial formulation', *J. Environ. Manage.* 177, 93–100. doi:10.1016/j.jenvman.2016.04.008.

123. Novozymes' new technology offers cold bleaching to denim manufactures, (2014). https://www.novozymes.com/en/news/news-archive/2014/11/novozymes-new-technology-offers-cold-bleaching-to-denim-manufactures (accessed August 24, 2019).

124. Samanta KK, Basak S, Chattopadhyay SK, (2017), 'Environmentally friendly denim processing using water-free technologies', in: S.S.B.T.-S. in D. Muthu (Ed.), Sustain. Denim, Woodhead Publishing, pp. 319–348. doi:10.1016/B978-0-08-102043-2.00012-5.

125. Patel R, Baisch G, (2012), 'Single Pass Laser Cutting of Polymers', *J. Laser Appl.* 5,13–16. doi:10.2351/1.4745319.

126. READY JF, (2014), 'Properties of Lasers', in: K. Thyagarajan, A. Ghatak (Eds.), Eff. High-Power Laser Radiat., Springer US, Boston, MA,: pp. 1–31. doi:10.1016/b978-0-12-583950-1.50006-4.

127. Riedl MJ (2009), 'Basic Optics', in: K. Thyagarajan, A. Ghatak (Eds.), Opt. Des. Fundam. Infrared Syst., Springer US, Boston, MA,: pp. 19–35. doi:10.1117/3.412729.ch2.

128. Štěpánková M, Wiener J, Rusinová K, (2011), 'Decolourization of vat dyes on cotton fabric with infrared laser light', *Cellulose* 18, 469–478. doi:10.1007/s10570-011-9494-2.

129. Kan CW (2014), 'CO2 laser treatment as a clean process for treating denim fabric', *J. Clean. Prod.* 66, 624–631. doi:10.1016/j.jclepro.2013.11.054.

130. Ebrahimi I, Parvinzadeh Gashti M, Sarafpour M (2018), 'Photocatalytic discoloration of denim using advanced oxidation process with H2O2/UV', *J. Photochem. Photobiol. A Chem.* 360,278–288. doi:10.1016/j.jphotochem.2018.04.053.

131. Yip J, Chan K, Sin KM, Lau KS, (2002) 'Low temperature plasma-treated nylon fabrics', *J. Mater. Process. Technol.* 123, 5–12. doi:10.1016/S0924-0136(02)00024-9.

132. Kravets LI, Dmitriev SN, Sleptsov VV, Elinson VM (2002), 'Production of asymmetric track membranes with a high permeability and separation selectivity', *Desal.* 144,27–34. doi:10.1016/S0011-9164(02)00284-9.

133. Pu FR, Williams RL, Markkula TK, Hunt JA, (2002), 'Effects of plasma treated PET and PTFE on expression of adhesion molecules by human endothelial cells in vitro', *Biomaterials* 23, 2411–2428. doi:10.1016/S0142-9612(01)00377-5.

134. Nageswaran G, Jothi L, Jagannathan S, (2019) 'plasma Assisted Polymer Modifications', in: S. Thomas, M. Mozetič, U. Cvelbar, P. Špatenka, P.B.T.-N.-T.P.T. for P.M. K.M. (Eds.), Non-Thermal Plasma Technol. Polym. Mater., Elsevier, 2019: pp. 95–127. doi:10.1016/b978-0-12-813152-7.00004-4.

135. Ražić SE, Čunko R, Bautista L, Bukošek V, (2017), 'Plasma effect on the chemical structure of cellulose fabric for modification of some functional properties', *Procedia Eng.* 200,333–340. doi:https://doi.org/10.1016/j.proeng.2017.07.047.

136. Shahidi S, Ghoranneviss M, Wiener J, (2015), 'Improving synthetic and natural dyeability of polyester fabrics by dielectric barrier discharge', *J. Plast. Film Sheeting.* 31, 286–308. doi:10.1177/8756087914567013.

137. Kazanskii NL, Kolpakov VA (2009), 'Effect of bulk modification of polymers in a directional low-temperature plasma flow', *Tech. Phys.* 54,1284–1289. doi:10.1134/S1063784209090060.

138. Yeh YS, Yasuda H, (1990) 'Modification of Surface State of Polymers by Glow Discharge Plasmas', in: F. Garbassi, E. Occhiello (Eds.), High Energy Density Technol. Mater. Sci., Springer Netherlands, Dordrecht,: pp. 33–45. doi:10.1007/978-94-009-0499-6_3.

139. Chi-wai K, Kwong C, Chun-wah MY (2004), 'The possibility of low-temperature plasma treated wool fabric for industrial use', *Autex Res. J.* 1 (2004) 37–44.

140. Ghoranneviss M, Shahidi S, Moazzenchi B, Anvari A, A. Rashidi, H. Hosseini, (2007) 'Comparison between decolorization of denim fabrics with Oxygen and Argon glow discharge', *Surf. Coatings Technol.* 201, 4926–4930. doi:10.1016/j.surfcoat.2006.07.162.

141. Ali Khan A, Alzohairy MA, (2010) 'Recent Advances and Applications of Immobilized Enzyme Technologies: A Review', *Res. J. Biol. Sci.* 5, 565–575. doi:10.3923/rjbsci.2010.565.575.

142. Yim TJ, Kim D-Y, Karajanagi SS, Lu TM, Kane R, Dordicka JS, (2003) 'Si Nanocolumns as Novel Nanostructured Supports for Enzyme Immobilization', *J. Nanosci. Nanotechnol.* 3,479–482. doi:10.1166/jnn.2003.240.

143. Katchalski-Katzir E (1993) ,'Immobilized enzymes — learning from past successes and failures', *Trends Biotechnol.* 11, 471–478. doi:https://doi.org/10.1016/0167-7799(93)90080-S.

144. Brena B, González-Pombo P, Batista-Viera F, (2013), 'Immobilization of enzymes: A literature survey', in: J.M. Guisan (Ed.), Methods Mol. Biol., Humana Press, Totowa: pp. 15–31. doi:10.1007/978-1-62703-550-7_2.

145. Ottolina G, Carrea G, Riva S, Sartore L, Veronese FM, (1992) 'Effect of the enzyme form on the activity, stability and enantioselectivity of lipoprotein lipase in toluene', *Biotechnol. Lett.* 14, 947–952. doi:10.1007/BF01020635.

146. Tischer W, Wedekind F, (1999) 'Immobilized Enzymes: Methods and Applications, in: W.-D. Fessner, A. Archelas, D.C. Demirjian, R. Furstoss, H. Griengl, K.-E. Jaeger, E. Morís-Varas, R. Öhrlein, M.T. Reetz, J.-L. Reymond, M. Schmidt, S. Servi, P.C. Shah, W. Tischer, F. Wedekind (Eds.), Springer Berlin Heidelberg, Berlin, 1999: pp. 95–126. doi:10.1007/3-540-68116-7_4.

147. Sheldon RA (2007), 'Enzyme immobilization: The quest for optimum performance', *Adv. Synth. Catal.* 349,1289–1307. doi:10.1002/adsc.200700082.

148. Belghith H, Ellouz-Chaabouni S, Gargouri A, (2001) 'Biostoning of denims by Penicillium occitanis (Pol6) cellulases', *J. Biotechnol.* 89, 257–262. doi:10.1016/S0168-1656(01)00309-1.

149. Gübitz GM, Cavaco-Paulo A, (2001), 'Biotechnology in the textile industry - Perspectives for the new millennium', *J. Biotechnol.* 89, 89–90. doi:10.1016/S0168-1656(01)00317-0.

150. Lenting HBM, Warmoeskerken MMCG (2001), 'Guidelines to come to minimized tensile strength loss upon cellulase application', *J. Biotechnol.* 89, 227–232. doi:10.1016/S0168-1656(01)00301-7.

151. Ibrahim NA, El-Badry K, Eid bm, Hassan TM, (2011), 'A new approach for biofinishing of cellulose-containing fabrics using acid cellulases', *Carbohydr. Polym.* 83,116–121. doi:10.1016/j.carbpol.2010.07.025.

152. Yassin MA, Gad AAM, Ghanem AF, Abdel Rehim MH (2019), 'Green synthesis of cellulose nanofibers using immobilized cellulase', *Carbohydr. Polym.* 205, 255–260. doi:10.1016/j.carbpol.2018.10.040.

153. Khoshnevisan K, Vakhshiteh F, Barkhi M, Baharifar H, Poor-Akbar E, Zari N, Stamatis H, Bordbar AK, (2017), 'Immobilization of cellulase enzyme onto magnetic nanoparticles: Applications and recent advances', *Mol. Catal.* 442, 66–73. doi:10.1016/j.mcat.2017.09.006.

154. Yu y, Yuan J, Wang Q, Fan X, Ni X, Wang P, Cui L, (2013), 'Cellulase immobilization onto the reversibly soluble methacrylate copolymer for denim washing', *Carbohydr. Polym.* 95, 675–680. doi:10.1016/j.carbpol.2013.03.043.

155. Hao L, Wang R, Fang K, Liu J, (2013) 'Ultrasonic effect on the desizing efficiency of α-amylase on starch-sized cotton fabrics', *Carbohydr. Polym.* 96,474–480. doi:https://doi.org/10.1016/j.carbpol.2013.04.003.

156. Şahinbaşkan BY, Kahraman MV (2011), 'Desizing of untreated cotton fabric with the conventional and ultrasonic bath procedures by immobilized and native α-amylase', *Starch - Stärke.* 63,154–159. doi:10.1002/star.201000109.

157. Benli H, Bahtiyari MI, (2015) 'Use of ultrasound in biopreparation and natural dyeing of cotton fabric in a single bath', *Cellulose* 22, 867–877. doi:10.1007/s10570-014-0494-x.

158. Chand N, Sajedi RH, Nateri AS, Khajeh K, Rassa M, (2014), 'Fermentative desizing of cotton fabric using an $α$-amylase-producing Bacillus strain: Optimization of simultaneous enzyme production and desizing', *Process Biochem.* 49,1884–1888. doi:https://doi.org/10.1016/j.procbio.2014.07.007.

159. Maryan AS, Montazer M, (2013), 'A cleaner production of denim garment using one step treatment with amylase/cellulase/laccase', *J. Clean. Prod.* 57,320–326. doi:10.1016/j.jclepro.2013.05.041.

160. Sustainable Frontiers in Denim Finishing, Appar. Resour. (2016). https://www.apparelresources.com/business-news/sustainability/sustainable-frontiers-in-denim-finishing/ (accessed July 10, 2017).

161. Bechtold T, Berktold F, A. Turcanu, 'The redox behaviour of CI Sulphur Black 1 - a'basis for improved understanding of sulphur dyeing', *Color. Technol.* 116 (2000) 215–221. doi:10.1111/j.1478-4408.2000.tb00041.x.

162. Teli MD, Paul R, Sachin LM, (2001) 'Ecofriendly processing of sulphur and vat dyes -an overview'., *Indian J. Fiber Text. Res.* 26,101–107.

163. Roessler A, Dossenbach O, Rys P, Marte W, (2002) 'Direct electrochemical reduction of indigo: Process optimization and scale-up in a flow cell', *J. Appl. Electrochem.* 32, 647–651. doi:10.1023/A:1020198116170.

164. Schrott W, Paul R, (2015) 'Environmental impacts of denim manufacture', in: Denim, Elsevier : pp. 563–580. doi:10.1016/B978-0-85709-843-6.00020-2.

11

An approach on saving water and energy for a sustainable textile production

Sanjay Kumar Bhikari Charan Panda, Samrat Mukhopadhyay*
Department of Textile and Fibre Engineering
Indian Institute of Technology (I.I.T.) Delhi, India
*Corresponding author, Email: samrat@textile.iitd.ac.in

Abstract: Rapid population growth, the rise in affluence, and demanding technologies are few known culprits that bring the ecosystem at risk, which poses a direct or indirect threat to biotic and abiotic environments. The textile industry is well known as a water-intensive industry. The water consumption is 100-180 L per kilogram of textiles produced, and the textile industries discharge around 80% of the total wastewater. Almost 25% of the total energy consumed in textile wet processing is lost in the effluent in textile colouration. Such unwanted overuse of resources leads to an eye on a continuous focus of researchers in this area of research engaged in the textile industry. This chapter covers the current practices and investigations on water and energy conservation in textiles. A brief discussion is also included on innovative technologies like plasma application, sonication, waterless processing, developments in dyes and auxiliaries, promising energy conservation methods for heat recovery, and economical use of resources for sustainability.

11.1 Introduction

We must consider water as an essential resource due to its limited availability in a pure form. Sufficient availability of food, access to buy it in a nutritious way, and stability are primary requirements for human beings. Food utilisation cannot be fulfilled with a proper diet unless there is an ample volume of clean water. Getting sufficient clean water is a challenge for many countries. Day by day, the cost of living is getting higher in terms of energy, water, and other expenses. Experts believe that the main driving forces for such a high price of electricity and pure water are the rapid growth of population, rise in affluence, and the demanding technology for sophistication [3]. The water consumed for a product during manufacturing and its use, along with evapotranspiration, is called blue water. The blue water footprint is estimated to be 24% of the global utilisation in India. 70% of the blue water footprint is primarily consumed in agriculture, while the industry and households utilise the rest. The significant water consumers among industries are power plants, engineering, pulp industries, textile mills, steel plants, sugar manufacturers, etc. Though the textile sector comes fourth in the list of water consumption, the water demand for unit production is the highest among all industries [45]. A large portion of the water used in the textile industry is in its wet processing unit. There are different stages like preparation, dyeing, printing, and finishing in this process. 100-180 L of water for 1 kg of textiles consumed during wet processing. Around 80% of the total wastewater generated is by the textile industry [66]. Higher water consumption leads to higher energy consumption too. China is considered one of the world's largest exporters of textiles and clothing. Its average energy efficiency growth rate from 2000 to 2012 was only 10.4% and was affected by negative energy structure [99]. High water and energy consumption is the bottleneck in the textile manufacturing process. Henceforth, a thorough review is necessary on water and energy conservation methods in textiles. It is vital to find out the plausible scopes of research for a sustainable future and mesmerise the textile trade business.

11.2 Water consumption in wet processing

In textiles, there are two types of processing; one is dry processing, and the other is very important in terms of water and energy usage, called wet processing. From fibre to grey fabric manufacturing comes under dry processing, where water usage is significantly less. The primary use of water is in wet processing during the conversion of the grey fabric to make it suitable for end-use. In wet processing, the fabric is processed through different intermediate processes like desizing, scouring, mercerizing, bleaching, dyeing, printing, and finishing [63]. All these processes need a massive volume of blue water. In textile, 72% of the total water is spent in wet processing. The rest of the water is for firefighting, sanitary use, steam production, cooling, and other tasks, as depicted in Fig. 11.1. Water consumption for firefighting purposes is approximately 1% of the total requirement, but a large volume of water needs to be stored for an emergency. Stage-wise water consumption in wet processing is shown in Fig. 11.2, which indicates that 30% of the water is used for pretreatment (desizing, scouring, bleaching, and mercerizing), colouration, and finishing. The printing and finishing process consumes less water, but these processes generate effluents with high concentrations. A large portion of the water is consumed in the washing-off operations, accounting for 70%. The intermediate washing removes the unwanted residual chemicals, dyes, or pigments from each specific process [46]. From a water and energy conservation point of view, few measures are necessary during the process and production, which are not capital intensive and essential. At a time for analyzing such a large volume of water consumption, the textile industry's state of the art must be evident [75]. There is a vast scope in water recycling, while the reuse and recycling of wastewater require a few remedial measures to avoid risk [79]. A processor needs to find the gap in establishing water and energy conservation methods.

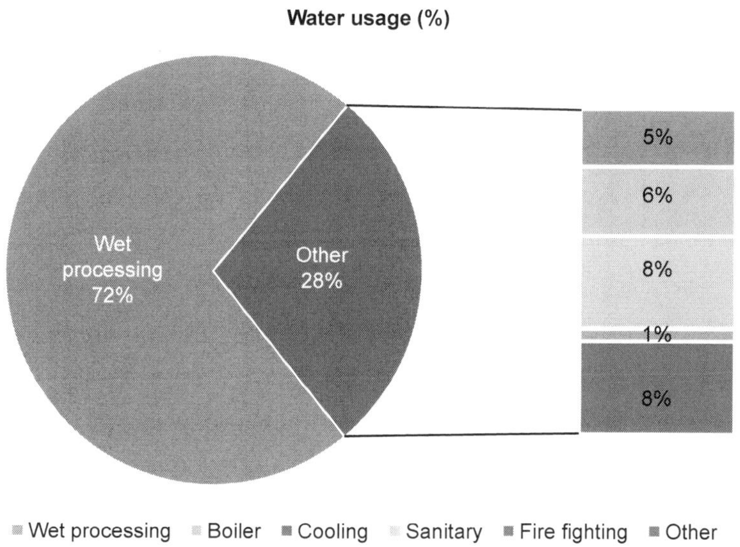

Figure 11.1 Water usage for different purposes in textile industry

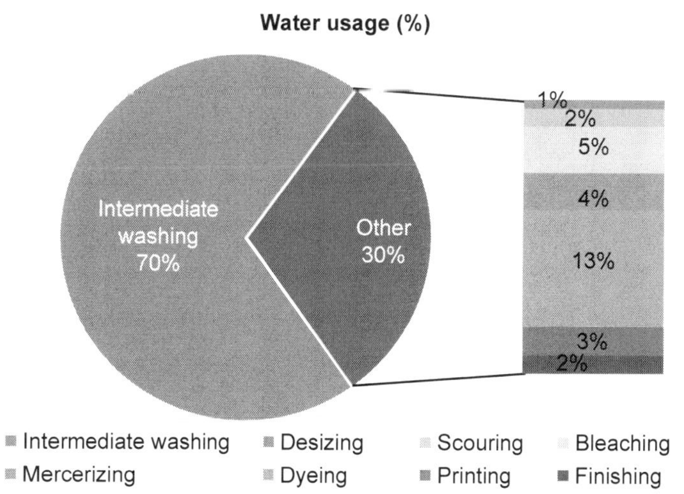

Figure 11.2 Water usage for various processes in textile industry

11.3 Energy consumption in textile

Textile machine makers are continuously researching to manufacture more efficient robotic machines. The developed new machinery requires less workforce but consumes much more energy in terms of power. Fig. 11.3 depicts the electricity consumption for different sections in a textile organisation. The highest power-consuming machine, the ring frame, consumes 28% of the total power consumption.

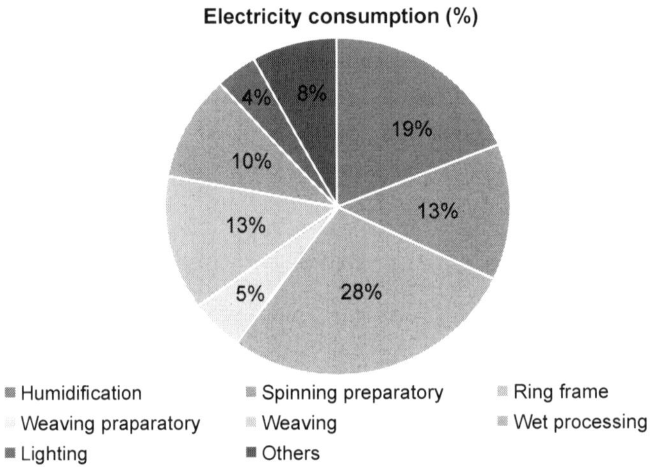

Figure 11.3 Electricity usage for different units in textile industry

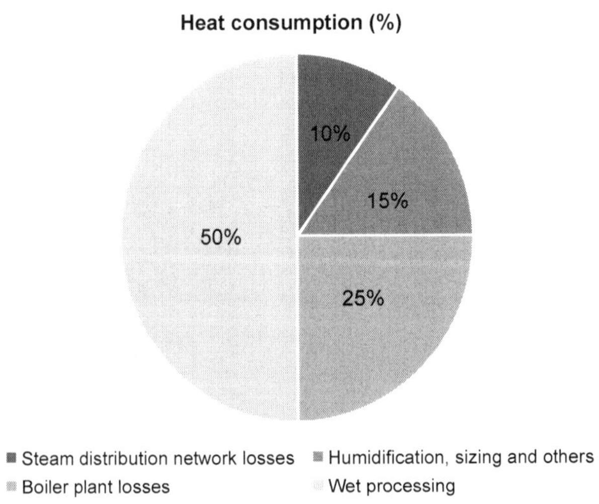

Figure 11.4 Distribution of thermal energy in different stages

The heat consumption distribution is shown in Fig. 11.4. More than 50% of the heat energy is getting lost during the generation and distribution of steam. Manufacturers are spending up to 17% of capital on energy. It is estimated that 80% of the total energy used in textile plants accounts for heating [30].

11.4 Water and energy management techniques

Textile producers' focus is always on controlling the high water and energy consumption. A thorough investigation of the current water and energy conservation practices and future scopes for a sustainable textile manufacturing process would be necessary. The studied research and development practices in water and energy conservation are classified as,

(a) Developments in machinery
(b) Developments in the process/method
(c) Technologically assisted innovative processes
(d) Use of 3R's (Reduce, Reuse, Recycle)

11.4.1 Developments in machinery

The profit margin for products is relatively low, especially in spinning and weaving, with high competition in the market. Even at a marginal price, selling textile products is a worry for textile manufacturers [92]. The processing unit's high inventory and operational cost is also a significant challenge for textile processors to exist in business [39]. The manufacturing cost needs to be reduced with higher productivity and lower operating costs [37]. The attempts on water and energy conservation methods are crucial for eliminating the high operational cost. Advances in engineering are an integral part of this process, which brought many innovations in textile machinery manufacturing.

11.4.1.1 Low liquor dyeing machine

Textile involves mainly two types of processes, i.e., batch process and continuous process. Both processes have their own merits and limitations. In a batch process, water and energy consumption is too high. Different technologies are being used to reduce such high consumption of water. The THEN airflow machine is one of the latest examples of low liquor operational machines for batch processing. The material to liquor ratio (MLR) in conventional winch machines is 1:30 to 1:50. The THEN aerodynamic system has reduced the MLR substantially to 1:2 for synthetics and 1:3 for natural fibres. The THEN research and development department has made it possible to claim approx. 40% reduction in energy compared to old hydraulic jet dyeing machines with a 25% reduction in process time. The newly developed THEN-AIRFLOW SYNERGY is a solution to ecology [29].

11.4.1.2 Economical washing systems

Washing off process to remove either residual chemicals or dyes from the fabric needs a considerable quantity of water and heat energy. Benninger, a leading manufacturer of continuous range machines in textile, has brought many new features to save water and electricity. A continuous washing system, BEN-WASH has claimed a 50% saving in energy and water compared to the batch process. TRIKOFLEX, an efficient washing system, increases its washing effect by up to 40% with counter-flow washing technology [9]. UNIWASH system manufactured by Arioli, designed with unique features for the highest washing efficiency with liquor recirculation up to 100000 Lh^{-1} [4].

11.4.1.3 Waterless processes

NOVA dry cleaning range marketed by Santex completely replaced water by using organic solvents for the washing-off process. Nova claimed a reasonable payback period for investors with features like reducing second quality fabrics, low energy consumption, minimal water usage in the cooling system, a dry process that avoids additional drying costs, etc. [77]. Zheng et al. elaborated on the design and implementation of a vessel capable of working in supercritical carbon dioxide fluid for package dyeing. This process completely replaces water and reduces energy usage with no generation of effluent [100].

11.4.1.4 Heat recovery systems

Heat recovery for its reuse is an integral part of energy management. As a lot of water is used in the textile process, the generation of wastewater is also massive. Out of such wastewater, a significant portion is a hot effluent. Recovery of the heat from the hot effluent is eco-friendly and economical in reuse. Heat recovery can be obtained in many ways, like the heated effluent from synthetic dyeing, reuse of cooling water, condensate recovery, heat exchanging from effluent, heat recovery from the air, etc.[25]. A heat exchanger is used in hot air drying equipment where the fresh air is drawn with the contact of exhaust outlet air to absorb the heat before its reuse, which will require less energy to heat it to the required drying temperature [32]. Drying is one of the most energy-intensive processes, and heat recovery is crucial for energy saving [58]. Waste heat recovery systems can modify existing equipment with an investment of 6 month payback period [70]. According to a case study in Bangladesh, investment in terms of the payback period of 6 months to 3 years maximum with considerable energy saving by installations on heat recovery from the generation of flue gas, using economizer in boilers, heat recovery from effluents, steam condensate recovery, waste heat recovery from stenter exhaust, and blowdown heat recovery [71].

11.4.2 Developments in the process methods

Weaving-preparatory involves a process called sizing. Sizing is an essential process that enhances the weavability during fabric weaving. Starch, polyvinyl alcohol (PVA), and modified starch are mainly used as sizing agents. However, removing these sizing materials after fabric manufacturing is a water and energy-consuming task [53]. The water consumption for sizing and removal of sizing ingredients during desizing is 30% of the total water consumption with the generation of an equal volume of effluent. PVA is an attractive sizing agent for sizers due to its outstanding adhesion property and effortless removal during the washing process. However, PVA is not biodegradable. Soy protein can replace PVA as a biodegradable sizing agent [12]. A novel technique at low temperature for yarn reinforcing can eliminate the sizing process with substantial water and energy-saving [42]. In conventional pretreatment processes, there is a use of strong alkalis. Removal of the residual alkali with other auxiliaries consumes considerable water and energy. The replacement of such harsh chemicals is possible by using biocatalysts of specific nature known as enzymes [41]. Enzymes are proteins used under mild conditions with less water and energy. Enzymatic techniques are considered a substitute for conventional processes in cotton fabrics' pretreatment with an opportunity to save power and water. Biotechnology is also beneficial in the pretreatment of cotton as it generates less volume of biodegradable effluent [87]. A rapid enzymatic single bath treatment combines pretreatment (desizing, scouring, bleaching) and dyeing with special category enzymes. This treatment method uses only 33.33% of the water of a conventional process [60]. The enzymatic process saves water and energy in different stages. In the scouring process, pectinases replace a strong alkali like sodium hydroxide. An enzymatic scouring process can save 20-50% of the water required in conventional alkaline scouring. Also, bio-preparation results in a 20-40% reduction in biochemical oxygen demand (BOD) and chemical oxygen demand (COD) with a reduction in process time [35]. The scouring and bleaching process using an enzyme with peracetic acid is carried out at a temperature of 60 °C, while the average process temperature is 90 – 110 °C. Different attempts are taken to use enzymatic, low water, and energy-consuming approach [69]. A solvent dyeing method used for dyeing polyesters using liquid paraffin completely replaces water. The process cycles can be followed similarly to polyester dyeing in a water bath with the benefit of low oligomer release and no variation in dyeing-related properties [96]. The reactive dyeing process consumes a lot of water, mainly for the removal of hydrolysed dyes. Removal of hydrolysed dyes from fabric requires too much water and a high level of thermal energy. A treatment with laccase/lipopeptide decolourises the rinsing bath of reactive dyeing. It reduces water consumption and uses eco-friendly technology [54].

11.4.3 Technologically assisted innovative processes

Several emerging technologies are being implemented in textiles like ozone processing, ultrasonic processing, microwave processing, supercritical carbon dioxide processing, and plasma processing. Very few are in a real application in bulk, while most are either in the laboratory or pilot stages [40].

11.4.3.1 Process using trioxygen (O_3)

Hydrogen peroxide (H_2O_2) is popularly used as a bleaching agent. However, the application of hydrogen peroxide requires a considerable volume of water and thermal energy. H_2O_2 bleaching is operated at a boiling temperature and 10.0- 10.5 pH. Ozone (O_3) is having potential for oxidizing impurities (organic/inorganic). Ozone can be used as an alternative to H_2O_2 in the bleaching of textiles. Also, ozone bleaching can be carried out at a lower temperature and requires no water. Ozone bleaching is carried out at 23-25 °C in neutral pH for a shorter time than conventional peroxide bleaching [65]. Artworks need to protect from an environment with ozone to avoid fading. The property of ozone to decolourise dyes and pigments was investigated long back in 1983 [83]. Fading of silk dyed with Chinese plant dyes was discussed by Ye et al. [98]. Similarly, fading of traditional Japanese colourants on silk and paper was found by Paul et al.[94, 95]. Fading spoils the aesthetic look of dyed or printed materials and is unwanted, but intentional fading on denim garments is a new fashion. Denim garments are processed through various treatments like bleaching, enzyme wash, acid wash, stone wash, and micro sanding to get different fashion looks. [20]. As these processes are partly or more not ecofriendly, few sustainable processes developed like a water jet, laser, and ozone fading [47]. At the most, ozone bleaching for denim is sustainable as it conserves water and energy [8].

11.4.3.2 Process using acoustic waves

The audible frequency of human sound is 20 Hz-20 kHz. Ultrasonic waves are generated above 20 kHz. Ultrasound energy can be obtained using a maximum frequency of 5 MHz and 500 MHz for gasses and liquids, respectively [86]. An ultrasonic-assisted enzymatic desizing on phosphate modified starch-sized cotton fabric was reduced 50% of the processing time with increased efficiency. The cavitation energy was generated at 53 kHz from 180 W power [93]. Ultrasonic washing was effective for soiled polyester fabric at a lower liquor bath [33]. The effect of cavitation using ultrasound energy can enhance the washing effect, dye diffusion, which accounts for a 20-30% reduction in dyes and chemical consumption, a 20% reduction in water, and 30% energy saving by lowering the process temperature [76].

11.4.3.3 Process using short electromagnetic waves

The application of microwaves started at the beginning of World War II in radar equipment. The primary use of microwaves is in security systems, mainly for communication. Though the low-frequency range of microwaves has not been officially set, it is somewhere about 1 GHz, and the upper limit is around 300 GHz [85]. A conventional dyeing process requires a lot of time and water consumption of 10-20 times the weight of dyed goods. A microwave (below 2450 MHz) assisted dyeing method used a low bath ratio of 1:2 at a lower temperature [38]. The pretreatment of cotton fabric takes 2.5-3 h to complete. A microwave-assisted pretreatment, which combines the intermediate processes (desizing-scouring-bleaching), can be processed in 5 min to achieve the same pretreatment results [36].

11.4.3.4 Process using the fourth state of matter

Plasma was introduced by Lewi Tonk and Irving Langmuir in 1929. It is the fourth state of matter, including exciting species like ions, atoms, molecules, fragments of molecules, and electrically neutral. Plasma is used for textile surface modification in different ways [55]. Atmospheric pressure plasma-assisted desizing of polyester-cotton blended fabric, sized with polyvinyl alcohol (PVA), is possible with a low water consumption rate and at a lower temperature than the conventional process to save energy [18]. Atmospheric pressure glow plasma treated polyester increases its dyeability and can be dyed at 35 °C lower than the conventional dyeing temperature [24]. Polyamide 6, 6 was dyed at a lower temperature with improved dyeability when the surfaces were modified using dielectric barrier discharge (DBD) air plasma [59]. Plasma was applied to silk to alter its surface so that it could be dyed in supercritical carbon dioxide. This process eliminates the usage of water [97].

11.4.4 Use of 3R's (Reduce, Reuse, Recycle)

Reduce, reuse, and recycle are the most critical strategies in US EPA integrated waste management hierarchy. This use of 3R's can be applied universally to any waste management. The first strategy is to reduce the quantity and toxicity of waste during its generation. Reusing the materials several times will help in reducing waste. Recycling the articles will take care of the conservation of resources [68]. These strategies will reduce water and energy consumption in the textile industry and lead to sustainable manufacturing.

11.4.4.1 Process using low liquor

There are several techniques adopted in textile industries to conserve water and energy. Low liquor application is one such technique in which different methods are discussed. In general, most of the energy is consumed in the textile to dry the fabrics. The time for drying depends on the amount of liquor present in the material. Researchers have made efforts to reduce the wet pick-up before drying. In the conventional padding system, water is extracted using padding mangles where the pick-up usually is 60-100%. The water-of-imbibition for cotton fabric is 30%. One cannot get the wet pick-up below 30% by the padding method. Roberto fibre-filled rollers replaced the ordinary rubber roller for improving wet pick-up below 50%. This technique increases the processing speed by 66%, with a 20% reduction in power cost. In Kleinewefers' Hydrofuga system, the fabric can be dehydrated using a nonwoven blanket sandwiched squeezing system. This system works better for synthetic materials. Vacuum extractor developed by Textile Vacuum Extractor Co claimed a wet pick up of 40-50% was achieved using a suction roller extractor called. Low liquor application techniques are also available using a pneumatic ejector manufactured by Pletec, Brugmann, and Hikosaka. Such air-jet ejector techniques bring down the wet pick-up below 10-20% for specialized fabrics and 40-50% for cotton blends. Also, other low-add-on systems are available like the nip-padding system, kiss roller system, wicking system, loop transfer system, pad transfer system, and spray techniques [72].

11.4.4.2 Process using minimum liquor

A method using minimum liquor is widely accepted as foam technology. Foam is a mixture of gas and liquid, where a large proportion of liquid, about 50-95%, is replaced by gas. Foam application is not a new technology, but to apply a chemical or dye in the form of foam on the substrate uniformly is a challenge. The realization of the fabric with foam finishing is environmentally and economically attractive as less water used in foam leads to a faster drying process. Foam density, blow ratio, and foam stability are essential parameters for its application. The application of durable press finish and softeners by foam saves 26-39% of energy due to higher production speed [64]. Serwar et al. mentioned a stable foam application of zero formaldehyde-based dihydroxy ethylene urea product on cotton lycra denim fabric as a cost-efficient process for saving chemicals, water, and energy [78].

11.4.4.3 Process using management techniques

The Best Available Techniques (BAT) are quite helpful in reducing the consumption of water and energy and avoiding waste generation. System measures with good housekeeping are generally part of the best management practices. Employee training, maintenance of equipment, optimal use of chemicals and dyes, storage of materials and proper handling, upgrading raw material knowledge, water management, and solid waste management are required.

Optimisation of scheduling in the process, adjustment and escaping of unnecessary procedures in pretreatment, optimisation of overflow rinsing with auto controls, the discovery of substituting process to overflow rinsing, arresting leaks and spills on an immediate base, reuse of rinsing baths, taking measures before using fresh water to avoid any rejection or reprocessing, application of counter-current flow washing in a continuous process, separation of water from cooling system for reuse, and reuse of low contaminated rinsing baths are part of best water management practices.

Optimising systems generating heat, controlling process temperature, appropriate maintenance of the heating equipment, suitable insulation heat distribution arrangements, installing HRP (heat recovery plants) for conserving energy, preventing steam and air leaks, using electronically controlled motors, considering co-generation where feasible, and minimizing water content before drying by using low application techniques are the measures required for using rational energy [80]. Additional management techniques, such as water auditing, analysis of water cost, obtaining vendor technical know-how, and on-site education by water conservation teams, can benefit the textile industry[14]. There are many best available techniques to apply in textile for sustainable manufacturing. According to the European Commission (EC) directive, the European Union (EU) countries have followed the legal arrangements with general approaches to prevent industrial pollution. These EC directives are published by Integrated Pollution Prevention and Control (IPPC) to protect the ecosystem with the help of Best Available Techniques (BAT) [28]. BAT reference documents (BREF) published by IPPC are essential tools for manufacturers to conserve energy and water. Kocabas et al. mentioned a 29.5% reduction in water and a 9% reduction in power by adapting BREF in a textile mill in Turkey. The net savings on electricity and water were expected to be around 22 times the investment's cost [48]. A similar application of BREF in a Turkish textile plant resulted in a reduction of water cost by 49%, with a 28% saving in chemicals [62]. Using BREF guidelines with a newly designed BAT procedure, 50% of the freshwater would be saved with due concern on reduced environmental impact is proposed by Li Rosi et al. [74].

The analytical hierarchy process (AHP) is widespread in management control systems. This tool was applied in the dyeing and printing industry for water conservation purposes [88]. Wastewater minimization must be addressed by identifying the need for minimizing, evaluation, and system balancing using mathematical analysis. Almost 70% of freshwater reduction is claimed by Dokwala et al. for a starch industry and obtained a decrease of 64% in the pilot stage using networking and piping. This methodology can be applied in other water-intensive sectors, like textile industries[21]. The water demand for textile processing might be controlled using a process-level evaluation tool to reduce freshwater withdrawal, water consumption, and water for dilution with wastewater before releasing to a river. This evaluation tool was implemented in a polyester processing plant and found helpful [13]. Another tool for water conservation in water-intensive industries is industrial water conservation potential analysis (IWCPA). Strategic product planning and forecasting are essential functions of the IWCPA[31]. Cao et al. designed a network that uses a method for regenerating and reusing water to reduce water consumption[10]. In Australia, an integrated water management strategy (IWMS) is used to conserve water in two major water-intensive industries [1].

In the Chinese textile industry, proposed algorithms for optimising production schedules are being implemented. Compared to manual scheduling, software-based scheduling saves 20-30% of freshwater while saving 10-15% production time. Such a tool looks like a good technique with low investment for water and energy conservation [43]. Another study shows an 18-20% reduction in water consumption utilising MATLAB-based database management coupled with a genetic algorithm. [102]. A single nonlinear optimisation program developed mathematically can be applied to batch and continuous processes. This technique is used in the USA for continuous application and in Denmark for batch operation where the effluent is recycled. Internal reuse of the wastewater is also included in such a program. This method is found economical and promising for water and energy saving for multiple contaminants[23].

11.4.4.4 3R's techniques

The reuse of wastewater generated by the textile industry is cost-effective, eco-friendly, and necessary due to the scarcity of water. From 2003 to

2012, the European Union (EU) funded numerous projects in Europe, including FOTOTEX, ADOPBIO, PROWATER, BATTLE, PURIFAST, and AQUAFIT4USE. Development in water recycling and reuse penetration methodologies must be adopted rigorously by all nations in the textile business [90]. Different techniques are available to minimise water consumption. Application of cleaner technology that is effective in monitoring the quality and quantity of product, implementation of first-time-right dyeing, reusing wastewater with or without process as committed, recycling RO concentrate, fixing boiler water, using level dyeing machines, optimising or eliminating rinsing were the techniques used to save 41-69% of water in a woolen mill [61]. Various processes like coagulation and flocculation, foam floatation, ozonation, electrolysis, Fenton oxidation, fungi/H_2O_2, photocatalysis, biodegradation with activated sludge, sequential aerobic and anaerobic process, fixed bed, filtration, and sorption are used to treat textile effluents [91].

The water consumption in a batch process is so high. Reuse of the auxiliary and dye bath has been demonstrated by Cook and Tinche for polyester and nylon dyeing. The output from a cost-benefit analysis claims such a process with substantial savings in chemicals, water, and energy and capital investment recovery in less than one year payback period[17]. A closed-loop recycle system is implemented in a textile plant. It is most suitable for larger plants with different fabrics to be processed where water usage can be minimised by reusing the effluents with minor treatments [89]. Sometimes microfiltration is a solution to reuse a few streams of wastewater but may not be suitable for some manufacturers due to its high cost. In contrast, zeolite can be used as an alternative [26].

Shaid Abu et al. proposed that by segregating the different types of effluents generated from the source, the collection of low contaminated wastewater can be directly used for scouring and bleaching the cotton fabric without harming fabric quality. Without any treatment, reusing wastewater within the system reduces freshwater consumption without incurring additional costs [82]. A similar method was adopted for the wool and textile blend process to save up to 75% water by reusing the segregated reusable effluents [6]. Wastewater generated from reactive dyeing can be reused several times after decolourising with UV-assisted peroxide treatment. Though it has some limitations, it cannot be used in all shades for dyeing but can be used for specific colours to save water and salt [73].

A system is designed for recycling laundry wastewater by sedimentation and filtration using sand and gravels, where the investment cost can be recovered in less than a year [2]. In a continuous process, too much water is used for washing. A methodology developed on a water source diagram to reuse wastewater was demonstrated in a textile unit. This case study shows a 64% reduction in water consumption [5]. A case study shows a 55% water saving by implementing ultra-filtration and RO in a textile cluster that aims for water conservation of ground and surface water. This additional installation to the CETP plant also reduces the health risk by restricting the contaminated water released to the cultivation lands [34]. Ultra-filtration of silk degummed water separates sericin; however, the permeate is not suitable for reuse due to its high COD value. Incorporating RO for the UF makes it ideal for reuse[27]. Textile effluents are treated through physicochemical, biological, and tertiary treatment like pressure sand filtration before discharge to surface water. Due to stringent pollution acts on wastewater, and the quality of the release, it is recommended to reuse such wastewater with additional treatments like ultrafiltration accompanied by reverse osmosis (RO). Adopting such an advanced treatment process on tertiary wastewater is suitable for reuse as boiler feedwater. A case study on the above method from 7 textile sectors shows a 30-35% reduction in water consumption [15, 22]. Under membrane technology, there are many applications of ultra-filtration, nanofiltration, reverse osmosis, forward osmosis, and membrane bioreactor for textile wastewater. However, the different techniques are suitable for wastewater with different characteristics [67]. Wastewater is generated from reactive dyeing of cotton, disperse dyeing of polyester, and both reactive, disperse of blends considered for reuse after membrane filtration. The output of nanofiltration is most suitable for rinsing as it contains monovalent salts, whereas RO permeates are ideal for complete reuse. Such a membrane system can be installed with a payback period of fewer than three years[84]. RO generates 40-50% concentrate, and such concentrates are not biodegradable even after advanced treatments. When combined with a soda-lime process, treating RO concentrates with persulphate oxidation is very useful for reusing in the finishing process [101]. Recovery of caustic soda and water for reuse incorporating ultra-filtration and RO used in the paper factory is suitable for treating textile-mercerizing effluent [44]. The biological method combined with ceramic subfilters is an impressive technology used on a pilot scale to recycle effluent for its reuse[52].

Further biological treatments with membrane technology are an alternative for wastewater reclamation from textile effluent and its application in the textile processing sector [51]. A system developed for wastewater treatment on membrane and electrodialysis principle. Nanofiltration, combined with bipolar electrodialysis, is a promising tool for 100% wastewater utilization and zero liquid discharge [49]. A membrane bioreactor with nanofiltration is required for wastewater reclamation and water reuse. However, a case study shows only ultrafiltration of washing liquor can reduce 87.5% water consumption by reusing it in washing-off treatments itself [81]. A specially designed membrane bioreactor consisting of ultra-filtration assembly is used to make the textile effluent reusable with an acceptable toxicity level [7]. An electrochemical method, chemical coagulation, and ion exchange combined process can improve the secondary effluent to make it suitable for reuse in the dyeing process. An optimal ratio of 2 is recommended for the cationic to anionic resin in the ion exchange process for a reusable water output [50]. Ultrafiltration and RO make water with a quality suit for boiler feedwater [57].

Reusing reactive dyeing wastewater after ozonation catalyzed by Ag-Fe_2O_3 carbon aerogel in wash-off steps is a promising method for reducing water consumption [11]. Dispersed-air floatation of reactive dyes using cetyltrimethylammonium bromide can reduce the dye concentration to less than 1mg/L for single dyes. For trichromatic dyes, less than 300 ADMI units can be obtained to reuse water. The residual coloured effluent can be further modified to bring down the colour intensity by ultrasonic irradiation [19]. The reuse of wastewater recycled from textile effluent is not suitable for dyeing due to the massive contamination of colourants. Many adsorption techniques tried to make such tertiary effluents colourless. Bentonite-based industrial waste from oil clarification can be used as an adsorbent both in raw and thermally treated form for reusing the water in cotton fabric dyeing [56]. Ozonation for the degradation of dyes is already discussed. However, its application in effluent treatment is very suitable for water reuse in dyeing. Similar results can be obtained using elctroflocculation. These treatments may need to accompany a sand filtration unit for their application in bulk[16].

Concluding remarks and future perspectives

To conclude the above study of various techniques, the best way of water and energy conservation is to apply the management techniques where the investment is not so high. Adopting a few methods and minor modifications in manufacturing is quite attractive for a substantial saving on manufacturing costs. Low liquor and minimum liquor application techniques are adoptable and need up-gradation to enhance efficiency further. Reuse and recycle methods are essential and, most of the time, found to be capital intensive. Direct reuse of wastewater techniques must be promoted on industrial scales, which may require low investment. Heat recovery systems are an integral part of energy saving. Modification in the process by using smart dyes and auxiliaries needs continuous research. It would be suggested that as no single technology can fulfill all the requirements for water and energy conservations, a thorough audit of the system can guide for a suitable up-gradation in the manufacturing plant. Technological up-gradation may need a considerable capital investment but is necessary for sustainable growth.

References

1. Agana BA, Reeve D, Orbell JD (2013), 'An Approach to Industrial Water Conservation - A Case Study Involving Two Large Manufacturing Companies Based in Australia', *J. Environ. Manage.*, 114, 445–460. https://doi:10.1016/j.jenvman.2012.10.047.
2. Ahmad J, El-dessouky H (2008), 'Design of a modified low cost treatment system for the recycling and reuse of laundry waste water', *Resour. Conserv. Recycl.*, 52, 973–978. https://doi:10.1016/j.resconrec.2008.03.001.
3. Alder J, Barling D, Dugan P, Herren HR, Josupeit H, Lang T (2012), 'Avoiding Future Famines : Strengthening the Ecological Foundation of Food Security through Sustainable Food Systems', *A UNEP Synth. Report., UNEP*.
4. https://arioligroup.com/wp-content/uploads/2019/05/UNIWASH_ENG.pdf (accessed 30 September 2019).
5. De Souza AAU, Melo AR, Fernando LPP, De Souza SMAGU (2010), 'The modified water source diagram method applied to reuse of textile industry continuous washing water', *Resour. Conserv. Recycl.*, 54, 1405–1411. https://doi:10.1016/j.resconrec.2010.06.001.
6. Baban A, Yediler A, Ciliz NK (2010), 'Integrated water management and CP implementation for wool and textile blend processes', *Clean - Soil, Air, Water*, 38, 84–90. https://doi:10.1002/clen.200900102.
7. Ben F, Feki F, Han J, Isoda H, Sayadi S (2015), 'Treatment of textile wastewater by submerged membrane bioreactor : In vitro bioassays for the assessment of stress response elicited by raw and reclaimed wastewater', *J. Environ. Manage.*, 160, 184–192. https://doi:10.1016/j.jenvman.2015.06.008.
8. Ben Hmida S, Ladhari N (2016), 'Study of Parameters Affecting Dry and Wet Ozone Bleaching of Denim Fabric', *Ozone Sci. Eng.* 38, 175–180. https://doi:10.1080/01919512.2015.1113380
9. https://benningergroup.com/fileadmin/user_upload/downloads/textile_finishing/only_en/washing_en.pdf (accessed 30 September 2019).
10. Cao D, Feng X, Duan X (2004), 'Design of Water Network with Internal Mains for Multi-contaminant Wastewater Regeneration Recycle', *Chem. Eng. Res. Des.*, 82, 1331–1336. https://doi:10.1205/cerd.82.10.1331.46732.
11. Hu E, Shang S, Tao X, Jiang S, Chiu KL (2018), 'Minimizing freshwater consumption in the wash-offstep in textile reactive dyeing by catalytic ozonation with carbon aerogel hosted bimetallic catalyst', *Polymers (Basel).*, 10, 1–18. https://doi:10.3390/polym10020193.
12. Chen L, Reddy N, Yang Y (2013), 'Soy proteins as environmentally friendly sizing

agents to replace poly(vinyl alcohol)', *Environ. Sci. Pollut. Res.,* 20, 6085–6095. https://doi:10.1007/s11356-013-1601-5.

13. Chen L, Wang L, Wu X, Ding X (2017), 'A process-level water conservation and pollution control performance evaluation tool of cleaner production technology in textile industry', *J. Clean. Prod.,* 143, 1137–1143. https://doi:10.1016/j.jclepro.2016.12.006.
14. Chougule MB, Sonaje NP (2012), 'Novel Techniques of Water Recycling in Textile Wet Processing through Best Management Practices (BMP ' S)', *Int. J. Appl. Sci. Adv. Technol.,* 1, 29–33.
15. Ciardelli G, Corsi L, Marcucci M (2000), Membrane separation for wastewater reuse in the textile industry', *Resour. Conserv. Recycl.,* 31, 189–197.
16. Ciardelli G, Ranieri N (2001), 'The treatment and reuse of wastewater in the textile industry by means of ozonation and electroflocculation', *Water Res.,* 35, 567–572. https://doi:10.1016/S0043-1354(00)00286-4.
17. Cook FL, Tincher WC (1979), Dyebath and Auxiliary Bath Reuse for Energy and Mass Conservation In: *Energy Conservation in Textile and Polymer Processing.* American Chemical Society.,Washington, DC: pp. 201-241.
18. Cornelius C, McCord M, Bourham M, Hauser P (2018), 'Desizing of PVA Sized PET/Cotton Fabrics with Atmospheric Pressure Plasma', *Cellulose,* 25, 869–881. https://doi:10.1007/s10570-017-1586-1.
19. Dafnopatidou EK, Gallios GP, Tsatsaroni EG, Lazaridis NK (2007), 'Reactive dyestuffs removal from aqueous solutions by flotation, possibility of water reuse, and dyestuff degradation', *Ind. Eng. Chem. Res.,* 46, 2125–2132. https://doi:10.1021/ie060993v.
20. Dakuri A, Hiranmayee J, Farheen MN (2013), 'Technology of Industrial Denim Washing: Review', *Int. J. Ind. Eng. Technol.,* 3, 25–34.
21. Dakwala M, Mohanty B, Bhargava R (2009), 'A process integration approach to industrial water conservation: a case study for an Indian starch industry', *J. Clean. Prod.,* 17, 1654–1662. https://doi:10.1016/j.jclepro.2009.08.007
22. Dhodapkar RS, Pophali GR, Nandy T, Devotta S (2007), 'Exploitation results of seven RO plants for recovery and reuse of treated effluents in textile industries', *Desalination,* 217, 291–300. https://doi:10.1016/j.desal.2007.02.015.
23. Dunn RF, Wenzel H (2001), 'Process integration design method for water conservation and wastewater reduction in industry', *Clean. Pro.d Process.,* 3, 319–329. https://doi:10.1007/s10098-001-0122-y.
24. Elabid AEA, Zhang J, Shi J, Guo Y, Ding K, Zhang J (2016), 'Improving the low temperature dyeability of polyethylene terephthalate fabric with dispersive dyes by atmospheric pressure plasma discharge', *Appl. Surf. Sci.,* 375, 26–34. https://doi:10.1016/j.apsusc.2015.12.015.
25. Elahee K (2010), 'Heat recovery in the textile dyeing and finishing industry: Lessons from developing economies', *J. Energy South. Africa.,* 21, 9–15. https://doi:10.17159/2413-3051/2010/v21i3a3255.

26. Erdumlu N, Ozipek B, Yilmaz G, Topatan Z (2012), 'Reuse of effluent water obtained in different textile finishing processes', *Autex Res. J.*, 12, 23–28. https://doi:10.2478/v10304-012-0005-9.

27. Fabiani C, Pizzichini M, Spadoni M, Zeddita G (1996), 'Treatment of waste water from silk degumming processes for protein recovery and water reuse', *Desalination*, 105, 1–9. https://doi:10.1016/0011-9164(96)00050-1.

28. EC-European Commission (2008), 'Directive 2008/1/EC of the European Parliament and of the Council of 15 January 2008 Concerning Integrated Pollution Prevention and Control', *Off. J. Eur. Union*, 24, 8–29.

29. https://www.fongs.eu/assets/templates/fongs/img/pdf/THEN/THEN_SYNERGY/THEN_SYNERGY_Eng.pdf (accessed 30 September 2019).

30. Jayant S, Lynn P, Stephane de la R du C, David F (2005), 'Assessment of Energy Use and Energy Savings Potential in Selected Industrial Sectors in India', *Energy Anal. Dep. Environ. Energy Technol. Div. Lawrence Berkeley Natl. Lab.*

31. Gao L, Zongguo W, Bin D, Chao Z, Jining C (2008), 'An Analysis of Industrial Water Conservation Potential and Selection of Key Technologies based on the IWCPA Model', *Resour. , Conserv. Recycl.*, 52, 1141–1152. https://doi:10.1016/j.resconrec.2008.06.005.

32. Gordon JS (1984), U.S. Patent No. US4532721A, Washington, DC: U.S. Patent and Trademark Office.

33. Gotoh K, Harayama K (2013), 'Application of ultrasound to textiles washing in aqueous solutions', *Ultrason. Sonochem.*, 20, 747–753. https://doi:10.1016/j.ultsonch.2012.10.001.

34. Gupta BG, Agrawal KM, Biswas JK (2018), 'Conservation of Ground Water at Maheshtala Bleaching and Dyeing Cluster, a Populated Area in West Bengal, India by Implementing Ultra filtration and Reverse Osmosis Based Effluent Treatment Plant—A Case Study', *J. Inst. Eng. Ser. A*, 99, 705–718. https://doi:10.1007/s40030-018-0314-7.

35. Hasanbeigi A, Price L (2015), 'A Technical Review of Emerging Technologies for Energy and Water Efficiency and Pollution Reduction in the Textile Industry', *J. Clean. Prod.*, 95, 30–44. https://doi:10.1016/j.jclepro.2015.02.079.

36. Hashem M, Taleb MA, El-Shall FN, Haggag K (2014), 'New Prospects in Pretreatment of Cotton fabrics using Microwave Heating', *Carbohydr. Polym.*, 103, 385–391. https://doi:10.1016/j.carbpol.2013.11.064.

37. Hill M (1987), 'Towards a Vanishing Middle : Competition in the World Garment Industry', *World Dev.*, 15, 1269–1284.

38. Hong SM (2002), U.S. Patent No. US6381995B1. Washington, DC: U.S. Patent and Trademark Office.

39. Hoque A, Clarke A (2013), 'Greening of industries in Bangladesh : pollution prevention practices', *J. Clean. Prod.*, 51, 47–56. https://doi:10.1016/j.jclepro.2012.09.008.

40. Hussain T, Wahab A (2018), 'A critical review of the current water conservation practices in textile wet processing', *J. Clean. Prod.*, 198, 806–819. https://doi.

org/10.1016/j.jclepro.2018.07.051

41. Illanes A (2008), Introduction. In *Enzyme biocatalysis: Principles and Applications* (ed A Illanes), Editorial Springer-Verlag New York Inc., United States: pp. 1–30.

42. Jiang Q, Chen S, Deng X, Feng Y, Reddy N, Zhu Q, Liu W, Qiu Y (2019), 'A sustainable low temperature yarn reinforcing process to reduce water and energy consumptions and pollution in the textile industry', *J. Clean. Prod.,* 210, 646–652. https://doi:10.1016/j.jclepro.2018.11.034.

43. Jiang W, Yuan Z, Bi J, Sun L (2010), 'Conserving water by optimizing production schedules in the dyeing industry', *J. Clean. Prod.,* 18, 1696–1702. https://doi:10.1016/j.jclepro.2010.07.004.

44. Johakimu JK, Andrew J, Sithole BB, Syphus E (2016), 'Preliminary techno-economic assessment of recovering water and caustic soda from alkaline bleach plant ef fl uent', *J. Clean. Prod.,* 139, 914–921. https://doi:10.1016/j.jclepro.2016.08.088.

45. Joseph N, Ryu D, Malano HM, George B, Sudheer KP, Anshuman (2019), 'Estimation of industrial water demand in India using census-based statistical data', *Resour. Conserv. Recycl.,* 149, 31–44. https://doi:10.1016/j.resconrec.2019.05.036.

46. Karthik T, Rathinamoorthy R (2015), Recycling and reuse of textile effluent sludge. In *Environmental Footprints and Eco-Design of Products and Processes* (ed Subramanian Senthilkannan Muthu), Springer Science+Business Media, Singapore: pp. 213–258.

47. Khalil E (2015), 'Sustainable and Ecological Finishing Technology for Denim Jeans', *AASCIT Commun.* 2. https://doi:10.6084/M9.FIGSHARE.1480400.

48. Kocabas AM, Yukseler H, Dilek FB, Yetis U (2009), 'Adoption of European Union's IPPC Directive to a textile mill: Analysis of water and energy consumption', *J. Environ. Manage.,* 91, 102–113. https://doi:10.1016/j.jenvman.2009.07.012.

49. Lin J,Ye W, Huang J, Ricard B, Baltaru MC, Greydanus B, Balta S, Shen J, Vlad M, Sotto A, Luis P,Van der Bruggen B (2015), 'Toward Resource Recovery from Textile Wastewater: Dye Extraction, Water and Base/Acid Regeneration Using a Hybrid NF-BMED Process', *ACS Sustain. Chem. Eng.* 3, 1993–2001. https://doi:10.1021/acssuschemeng.5b00234.

50. Lin SH, Chen ML (1997), 'Treatment of textile wastewater by-chemical methods for reuse', *Water Res.,* 31, 868–876. https://doi:10.1016/S0043-1354(96)00318-1.

51. Lu X, Liu L, Liu R, Chen J (2010), 'Textile wastewater reuse as an alternative water source for dyeing and finishing processes : A case study', *DES,* 258, 229–232. https://doi:10.1016/j.desal.2010.04.002.

52. Lu X, Liu L, Yang B, Chen J (2009), 'Reuse of printing and dyeing wastewater in processess assessed by pilot-scale test using combined biological process and sub-filter technology', *J. Clean. Prod.,* 17, 111–114. https://doi:10.1016/j.jclepro.2008.03.003.

53. Ma P, Wang X, Xu W, Cao G (2009), 'Application of Corona Discharge on Desizing of Polyvinyl Alcohol on Cotton Fabrics', *J. Appl. Polym. Sci.,* 114, 2887–2892. https://doi:10.1002/app.

54. Mao YH, Guan Y, Song Q shuang, Zheng QK, Wang XX (2015), 'Application of laccase and lipopeptide in rinsing procedure of cotton fabric dyed with reactive dye', *Fibers Polym.*, 16, 2202–2214. https://doi:10.1007/s12221-015-4876-y.

55. Massimo P (2010), Introduction to plasma and plasama technology In: *Plasma Technology for Hyperfunctional Surfaces*. WILEY-VCH Verlag GmbH & Co. KGaA, Weinheim: pp 3-32.

56. Mittersteiner M, Schmitz F, Barcellos IO (2017), 'Reuse of dye-colored water post-treated with industrial waste: Its adsorption kinetics and evaluation of method efficiency in cotton fabric dyeing', *J. Water Process Eng.*, 17, 181–187. https://doi:10.1016/j.jwpe.2017.04.004.

57. Nandy T, Manekar P, Dhodapkar R, Pophali G, Devotta S (2007), 'Water conservation through implementation of ultrafiltration and reverse osmosis system with recourse to recycling of effluent in textile industry-A case study', *Resour. Conserv. Recycl.*, 51, 64–77. https://doi:10.1016/j.resconrec.2006.08.004.

58. Oğulata RT (2004), 'Utilization of waste-heat recovery in textile drying', *Appl. Energy.*, 79, 41–49. https://doi:10.1016/j.apenergy.2003.12.002

59. Oliveira FR, Zille A, Souto AP (2014), 'Dyeing mechanism and optimization of polyamide 6,6 functionalized with double barrier discharge (DBD) plasma in air', *Appl. Surf. Sci.*, 293, 177–186. https://doi:10.1016/j.apsusc.2013.12.126.

60. Öner E, Sahinbaskan BY. 2011 A new process of combined pretreatment and dyeing: REST. *J. Clean. Prod.* 19, 1668–1675. https://doi:10.1016/j.jclepro.2011.05.008.

61. Ozturk E, Cinperi NC (2018), 'Water Efficiency and Wastewater Reduction in an Integrated Woolen Textile Mill', *J. Clean. Prod.*, 201, 686–696. https://doi:10.1016/j.jclepro.2018.08.021.

62. Ozturk E, Koseoglu H, Karaboyaci M, Yigit NO, Yetis U, Kitis M (2016), 'Minimization of water and chemical use in a cotton/polyester fabric dyeing textile mill', *J. Clean. Prod.*, 130, 92–102. https://doi:10.1016/j.jclepro.2016.01.080.

63. Paul SA, Chavan SK, Khambe SD (2012), 'Studies on Characterization of Textile Industrial Waste Water in Solapur City', *Int. J. Chem. Sci.*, 10, 635–642.

64. Elbadawi AM, Pearson JS (2003), 'FOAM TECHNOLOGY IN TEXTILE FINISHING', *Text. Prog.* 33, 1–31. http://dx.doi.org/10.1080/00405160308688960.

65. Perincek SD, Duran K, Korlu AE, Bahtiyari IM (2007), 'An Investigation in the Use of Ozone Gas in the Bleaching of Cotton Fabrics', *Ozone Sci. Eng.*, 29, 325–333. https://doi:10.1080/01919510701509578.

66. Petek J, Glavič P (1996), 'An integral approach to waste minimization in process industries', *Resour. Conserv. Recycl.*, 17(3), 169–188. https://doi.org/10.1016/0921-3449(96)01151-2

67. Petrinić I, Bajraktari N, Hélix-Nielsen C (2015), Membrane technologies for water treatment and reuse in the textile industry In: *Advances in Membrane Technologies for Water Treatment*. Woodhead Publishing, Cambridge, UK: pp. 537-550.

68. Pichtel J (2005), Recycling Solid Wastes. In *Waste management practices* (ed J

Pichtel), CRC Press Taylor & Francis Group., Boca Raton: pp. 4–6.

69. Preša P, Forte Tavčer P (2009), 'Low Water and Energy Saving Process for Cotton Pretreatment', *Text. Res. J.,* 79, 76–88. https://doi:10.1177/0040517508092019.

70. Pulat E, Etemoglu AB, Can M (2009), 'Waste-heat recovery potential in Turkish textile industry: Case study for city of Bursa', *Renew. Sustain. Energy Rev.,* 13, 663–672. https://doi:10.1016/j.rser.2007.10.002.

71. Rakib MI, Saidur R, Mohamad EN, Afifi AM (2017), 'Waste-heat utilization – The sustainable technologies to minimize energy consumption in Bangladesh textile sector', *J. Clean. Prod.,* 142, 1867–1876. https://doi:10.1016/j.jclepro.2016.11.098.

72. Van der Walt GHJ, Van Rensburg NJJ (1986), 'low-liquor dyeing and finishing', *Text. Prog.* 14, 1–50. http://dx.doi.org/10.1080/00405168608688900.

73. Rosa JM, Fileti AMF, Tambourgi EB, Santana JCC (2015), 'Dyeing of cotton with reactive dyestuffs: The continuous reuse of textile wastewater effluent treated by Ultraviolet / Hydrogen peroxide homogeneous photocatalysis', *J. Clean. Prod.,* 90, 60–65. https://doi:10.1016/j.jclepro.2014.11.043.

74. Rosi OL, Casarci M, Mattioli D, De Florio L (2007), 'Best available technique for water reuse in textile SMEs (BATTLE LIFE Project)', *Desalination,* 206, 614–619. https://doi:10.1016/j.desal.2006.06.010.

75. Rott U, Minke R (1999), Overview of wastewater treatment and recycling in the textile processing industry. In *Water Science and Technology*, International Association on Water Quality: pp. 137–144.

76. Vouters M, Rumeau P, Tierce P, Costes S (2004), 'Ultrasounds: An industrial solution to optimise costs, environmental requests and quality for textile finishing', *Ultrason. Sonochem.*, 11, 33–38.

77. https://mareintex.com.ar/wp-content/uploads/sperottorimar/NOVA.pdf (accessed 30 September 2019).

78. Sarwar N, Mohsin M, Bhatti AA, Ahmmad SW, Husaain A (2017), 'Development of water and energy efficient environment friendly easy care finishing by foam coating on stretch denim fabric', *J. Clean. Prod.,* 154, 159–166. https://doi:10.1016/j.jclepro.2017.03.171.

79. Schäfer AI, Beder S (2006), 'Relevance of the Precautionary Principle in Water Recycling', *Desalination,* 187, 241–252. https://doi:10.1016/j.desal.2005.04.083.

80. Schönberger, H., & Schäfer, T. (2003), 'Best Available Techniques in Textile Industry,. Retrieved from http://www.umweltbundesamt.de/sites/default/files/medien/publikation/long/2274.pdf (accessed 30 September 2019).

81. Schoeberl P, Brik M, Braun R, Fuchs W (2004), 'Treatment and recycling of textile wastewater case study and development of a recycling concept', *Desalination* 171, 173–183. https://doi:10.1016/j.desal.2004.02.105.

82. Shaid A, Osman S, Hannan A, Bhuiyan MAR (2013), 'Direct Reusing of Textile Wastewater in Scouring-Bleaching of Cotton Goods Devoid of Any Treatment', *Int.*

J. Eng. Res. Dev., 5, 45–54.

83. Shaver CL, Cass GR, Druzik JR (1983), 'Ozone and the Deterioration of Works of Art', *Environ. Sci. Technol.*, 17, 748–752. https://doi:10.1021/es00118a011.

84. Sójka-Ledakowicz J, Koprowski T, Machnowski W, Knudsen HH (1998), 'Membrane filtration of textile dyehouse wastewater for technological water reuse', *Desalination*, 119, 1–9. https://doi:10.1016/S0011-9164(98)00078-2.

85. Adam SF (1969), Introduction. In *Microwavewave theory and applications*, NJ, Prentice-Hall.,USA: pp. 1–3.

86. Mason TJ, Lorimer JP (2002), General Principles, In: In *Applied Sonochemistry: The Uses of Power Ultrasound in Chemistry and Processing* (eds TJ Mason, JP Lorimer), Wiley-VCH Verlag GmbH.,Weinheim: pp. 25-74.

87. Mojsov K (2016), 'Bioscouring and bleaching process of cotton fabrics – an opportunity of saving water and energy', *J. Text. Inst.*, 107, 905–911. https:doi:10.1080/00405000.2015.1070603.

88. Tong O, Shao S, Zhang Y, Chen Y, Liu SL, Zhang SS (2012), 'An AHP-Based water-Conservation and waste-Reduction indicator system for cleaner production of textile-Printing industry in China and technique integration', *Clean Technol. Environ. Policy*, 14, 857–868. https://doi:10.1007/s10098-012-0453-x.

89. Turnbull RH, Groves GR, Buckley CA (1979), 'Closed Looped Recycle Systems for Textile Effluents', *J. Water Pollut. Control Fed.*, 51, 499–517.

90. Vajnhandl S, Valh JV (2014), 'The status of water reuse in European textile sector', *J. Environ. Manage.*, 141, 29–35. https://doi:10.1016/j.jenvman.2014.03.014.

91. Vandevivere PC, Bianchi R, Verstraete W (1998), 'Treatment and Reuse of Wastewater from the Textile Wet-Processing Industry : Review of Emerging Technologies', *J. Chem. T echnol. Biotechnol.*, 72, 289–302.

92. Verma S (2002), Export comprtitiveness of Indian textile and garment industry, ICRIER, India, New Delhi, 94,52.

93. Wang WM, Yu B, Zhong CJ (2012), 'Use of ultrasonic energy in the enzymatic desizing of cotton fabric', *J. Clean. Prod.*, 33, 179–182. https://doi:10.1016/j.jclepro.2012.04.010.

94. Whitmore PM, Cass GR (1988), 'The ozone fading of traditional Japanese colorants', *Stud. Conserv.*, 33, 29–40. https://doi:10.1179/sic.1988.33.1.29

95. Whitmore PM, Cass GR, Druzik JR (1987), 'The ozone fading of traditional natural organic colorants on paper', *J. Am. Inst. Conserv.*, 26, 45–58. https://doi:10.1179/019713687806027906.

96. Xu S, Chen J, Wang B, Yang Y (2016), 'An environmentally responsible polyester dyeing technology using liquid paraf fin', *J. Clean. Prod.*, 112, 987–994. https://doi:10.1016/j.jclepro.2015.08.114.

97. Yan J, Zheng LJ (2011), 'Study on Dyeing of the Plasma Modification Silk Fabric in Supercritical Carbon Dioxide', *Adv. Mater. Res.*, 175–176, 661–666. https://

doi:10.4028/www.scientific.net/amr.175-176.661.

98. Ye Y, Salmon LG, Cass GR (2000), 'The ozone fading of traditional chinese plant dyes', *J. Am. Inst. Conserv.*, 39, 245–257. https://doi:10.1179/019713600806082685.

99. Zhao H, Lin B (2019), 'Assessing the energy productivity of China ' s textile industry under carbon emission constraints', *J. Clean. Prod.*, 228, 197–207. https://doi:10.1016/j.jclepro.2019.03.327.

100. Zheng H, Zhang J, Yan J, Zheng L (2016), 'An industrial scale multiple supercritical carbon dioxide apparatus and its eco-friendly dyeing production', *J. CO2 Util.* 16, 272–281. https://doi:10.1016/j.jcou.2016.08.002.

101. Zheng L, Wang X, Wang X (2015), 'Reuse of reverse osmosis concentrate in textile and dyeing industry by combined process of persulfate oxidation and lime-soda softening', *J. Clean. Prod.*, 108, 525–533. https://doi:10.1016/j.jclepro.2015.09.027.

102. Zhou L, Xu K, Cheng X, Xu Y, Jia Q (2017), 'Study on Optimizing Production Scheduling for Water-saving in Textile Dyeing Industry', *J. Clean. Prod.*, 141, 721–727. https://doi:10.1016/j.jclepro.2016.09.047

12

Effluent management in textile chemical processing

Girendra Pal Singh[1*], Abdul Aziz Ansari[2]

[1] Govt Arts Girls College Kota, [2] Northern India Textile Research Association, Ghaziabad

*Corresponding Author, Email: singhgirendra2000@gmail.com

Abstract: The textile industry generates all types of pollutants, viz. water, air, soil and noise. Since textile wet processing is a water-intensive process, it generates an equally large volume of liquid effluent. The effluent is mostly intensively coloured and highly alkaline due to the presence of unspent dyes, chemicals and auxiliaries. If untreated effluent is discharged into the environment, it pollutes the water bodies and soil.

This chapter deals with the management of liquid waste generated during the wet processing of textiles. Water consumption in wet processing, sources of effluent generation, characteristics of various process effluents and their adverse effects on the environment have been discussed. The general approach to the problem of pollution abatement, including a reduction in waste volume and waste load with simultaneous recovery and reuse of water and chemicals, has been explained. Standard limits from pollution control board and their purpose has been described. The conventional treatment technologies used to treat textile effluents have been discussed. Various disposal methods for the textile effluent and problems faced by the industry have also been discussed.

12.1 Introduction

The textile industry is the largest after agriculture in terms of employment generation on the industrial map of India. A large variety of cotton textiles, synthetic textiles and cotton blended with synthetic fibres are produced by the industry. The manufacturing of textiles includes many processes such as spinning, weaving, knitting and wet processing. The textile wet processing houses consume a large quantity of fresh water and discharge an equal volume of liquid effluents [1-2]. The presence of various organic and inorganic chemicals in the dissolved, colloidal and suspended forms make the textile effluent very complex.

These effluents are highly variable, both concerning quantity and characteristics. They are mostly intensively coloured and highly alkaline due to the presence of unconsumed dyes, auxiliaries and chemicals. A great deal of organic matter in the effluent comes from the fibres, particularly natural fibres.

The textile effluents are discharged into sewers, usually after conventional physico-chemical and biological treatments. These industries are located mostly in large cities where sewerage systems are well equipped. A large number of industries are situated in cities and towns where infrastructural facilities are underdeveloped for the collection, transport and disposal of effluents.

The effluent generated by the industry is subjected to strict scrutiny by the local pollution control board authorities. An efficient effluent management technology can prevent and control the disposal of industrial effluent into the environment. The basis of effluent management is sustainable and cleaner production, treatment of effluent and proper handling of these effluents [3].

12.2 Water in textiles

12.2.1 Use of water

The requirement of water in the processing of textile materials is very high, and it varies from one mill to another depending on the following factors:

- Source and availability of water
- Quality and quantity of fabric to be produced
- Type of processing and its sequence[4].

In textile processing, water is mainly used in wet processing, namely desizing, scouring, bleaching, dyeing, printing and finishing[5]. The washing of the fabric in each process requires a large quantity of water (Fig. 12.1).

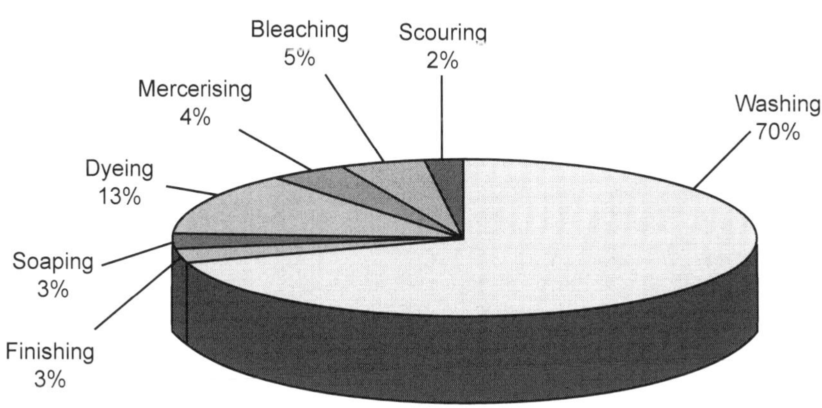

Figure 12.1 Water consumption in cotton and synthetic textiles

The water consumption depends on the M:L (material to liquor) ratio required by the washing machine. Here the material is a textile substrate, and liquor is water-based treatment liquor. The water consumption to produce one meter of fabric is in the range of 12 to 65 litres. The amount of water consumed in the wet processing of fabric depends on the process sequence. In the textile industry, water is mainly used for various purposes, as presented in Table 12.1.

Table 12.1 Water usage in cotton and synthetic textiles[6]

S. No.	Purpose	Water usage in %	
		Cotton textiles	Synthetic textiles
1	Steam generation	5.3	8.2
2	Cooling water	6.4	-
3	De-mineralized water for specific processes	7.8	30.6
4	Process water (raw water)	72.3	28.3
5	Sanitary use	7.6	4.9
6	Miscellaneous and fire fighting	0.6	28.0

12.2.2 Sources of water

The sources of water can be broadly classified into surface water and groundwater.

12.2.2.1 Surface water

The surface sources of water include streams, lakes, rivers and ponds. The quality and yield of water vary from source to source and season to season.

12.2.2.2 Groundwater

The rainwater or melted snow enters the ground (infiltration), a portion of which is percolated through the ground and is stored as groundwater at the hard stratum. The groundwater can be further classified as springs, infiltration galleries, porous pipe galleries and wells (Dug or Draw Wells, Artesian Wells, Infiltration Wells and Tube Wells.)

12.2.3 Characteristics of water

The groundwater is free from coarse suspended particles, but contains a large number of impurities in the form of

- Fine suspended particle
- Colloidal particles
- Dissolved solids

These contaminants are harmful to the consumer and cause problems during various industrial processes. Water quality standards have been prescribed for human consumption and different industrial processes. Hence, the evaluation of water quality is a must to ensure its suitability for use in our daily life as well as in the industry.

12.2.4 Water quality requirement for drinking purposes

The norms for drinking water are summarised in Table 12.2.

Table 12.2 Essential characteristics of drinking water[7]

S. No.	Characteristics	Desirable Limits
1	Colour, HU, max.	5
2	Odour	Unobjectionable
3	Taste	Agreeable
4	Turbidity, NTU, max.	1
5	pH value	6.5 - 8.5
6	Total Dissolved Solids. max	500
7	Total Hardness (as CaCO3), mg/L, max.	200
8	Chloride, (as Cl) mg/L, max.	250
9	Iron (as Fe) mg/L, max.	0.3
10	Residual Chlorine, mg/L, min.	0.2
11	MPN Coliform Organisms/100 ml, max.	<2

For drinking purpose, water should have following qualities:

- It should be cool, colourless, sparkling, tasty, odorless, clear, and free from suspended particles, harmful impurities and living organisms.
- It should contain 5-6 mg/L dissolved oxygen so that it remains fresh.
- Absolutely pure water (e.g., distilled water) is not good for health.
- It must contain a certain amount of essential dissolved mineral salts and oxygen which gives water a pleasant taste and supply a portion of mineral constituents necessary for body growth.
- An excessive amounts of dissolved salts are harmful to public health.

- It should be free from all objectionable matter and bacteria which may cause disease.
- There should not be any active virus. One hour contact period with 0.5 mg/L of free chlorine is enough to inactivate the virus.
- It should not contain pollutants like heavy metals, detergents and mineral oils.

12.2.5 Water quality requirements for textile processing

Water requirement of the textile industry is generally met through municipal supply or tube wells. In general, the water for textile processing should meet the following criteria (Table 12.3).

- It should be non-corrosive.
- It should be free from algae and mould spores.
- It should be free from objectionable odour.
- It should be free from colour and turbidity.
- Its pH should be in a neutral range.
- It should be free from substances such as iron, manganese, calcium and magnesium salts.
- It should be odourless and free from substances that are responsible for foam generation[8].

Table 12.3 Water quality requirements for textile processing[9]

S. No.	Test parameter	General purposes	Bleaching, dyeing and subsequent processing	Wool scouring
1	Colour, HU, Max	20	5	70
2	Turbidity, NTU, Max.	2	2	2
3	pH value	6.0 – 8.5	6.0 – 8.5	6.0 – 8.5
4	Alkalinity (as $CaCO_3$), mg/L, Max.	150	150	150
5	Hardness (as $CaCO_3$), mg/L, Max.	50	2	2
6	Chloride (as Cl), mg/L, Max.	100	-	-
7	Sulfate (as SO_4), mg/L, Max.	100	-	-
8	Aluminium (as Al), mg/L, Max.	0.1	0.1	0.1
9	Iron (as Fe), mg/L, Max.	0.25	0.1	0.1
10	Manganese (as Mn), mg/L, Max.	0.1	0.1	0.1
11	Iron + Manganese, mg/L, Max.	0.25	0.1	0.1

Note: For dyeing dark or medium shades, a relaxed limit of 20 HU is permissible for colour, but for whites and light shades, the colour should not exceed 5 HU.

12.3 Source of steam: Boiler

For steam generation boiler is used. A boiler is a device which generates steam. It consists of two principal parts (Fig. 12.2):

- Furnace to provide heat by burning fuel.
- Boiler proper to change water into steam. This steam is then re-circulated out of the boiler to use in various textile processes in heating applications [10].

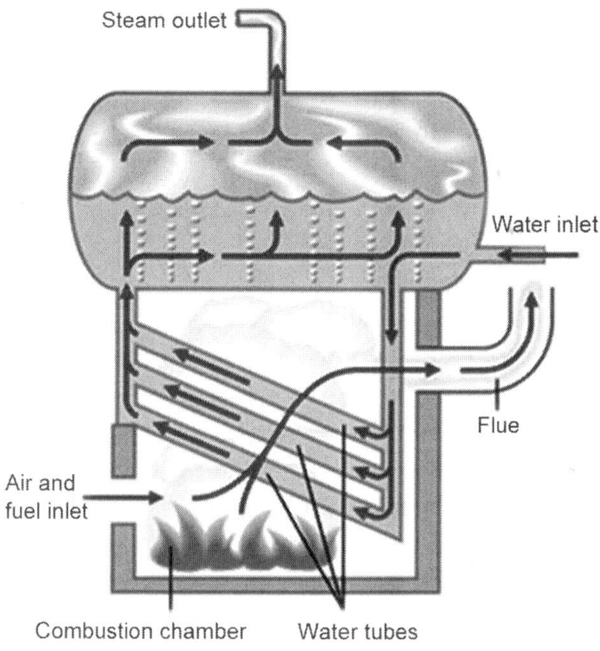

Figure 12.2 Diagram of boiler

12.3.1 Feed water for boilers

The demineralised water supplied to a boiler for steam generation is known as boiler feed water. The boiler feed water is stored, pre-heated and conditioned in a tank to feed into the boiler by a pump. The feed water received in a boiler is a mixture of condensed water (return water) and freshwater (make-up water) in various proportions. The make-up water is either natural raw water or treated water. Therefore, the composition of the feed water depends on the amount of condensed water and the quality of the make-up water [11-12].

12.3.2 Effect of impurities in feed water

The impurities present in feed water reduce the efficiency of boilers. Non-volatile salts get deposited in boilers when the feed water is evaporated. The heat transfer and flow rate become slow due to these deposits and eventually block boiler tubes. These impurities need to be removed because high concentrations of the impurities require a larger quantity of water to avoid their deposition. Even worse are minerals (calcium salts) that form sludge and scales in boilers. Therefore, the make-up water added to replace any losses of feed water has to be demineralised or deionised.

12.3.3 Water quality requirement for boilers

Good quality of water is required in boilers for their efficient and safe operation. The requirements of boiler feed water for low and medium-pressure boilers are given in Table 12.4.

Table 12.4 Requirements of boiler feed water[13]

S. No.	Characteristics	Requirement for boiler pressure		
		Up to 20 kg/cm^2	Up to 21-39 kg/cm^2	Up to 40-59 kg/cm^2
1	Silica (as SiO_2), mg/L, Max	-	5	0.5
2	Dissolved Oxygen, mg/L, Max	0.1	0.02	0.01
3	pH value	8.5 – 9.5	8.5 – 9.5	8.5 – 9.5
4	Hardness (as $CaCO_3$), mg/L, Max	10	1.0	0.5

12.4 Effluent

12.4.1 Sources of effluent generation

The main sources of wastewater in a textile mill are:

- Wet processing of textiles such as desizing, scouring/kiering, bleaching, mercerization, dyeing and printing.
- Washing operations in each step of textile wet processes.
- Various waste streams namely
  - Cooling water
  - Boiler blowdown
  - House-keeping (Cleaning water)
  - Spills and leaks
  - Sanitary waste

12.4.2 Quantity of effluent

The quantity of effluent discharged from each unit process is very small compared to the volume of wastes resulting from the washings and rinsing of cloths. The quantities of wastewater are highly variable from one mill to another, depending upon the water consumption. The wastewater discharged by the mills lies in the range of 57 to 800 m^3 or kL per kg of cloth produced, as shown in Table 12.5.

Table 12.5 Quantity of effluent discharged by some textile Mills[14]

Mill No.	Production kg/day	Water usage m^3/1000 kg of cloth	Waste water discharged	
			m^3/ day(gross)	m^3/ 1000 kg of cloth
1	42000	57 to 65.5	2050 to 2300	49 to 55
2	12000	187.5	1845	155
3	25000	600	6250	245
4	7500	244	1800	240
5	6700	164	900	134

12.4.3 Characteristics of textile processing effluents

The characteristics of the effluent from the various processes are controlled by factors such as

- Quantity and type of chemicals used
- Use of water
- The intensity of the washing process
- Type of wet processing operations

The liquid effluent from the textile industry is a mixture of chemicals present in the dissolved, colloidal and suspended forms in the water. It contains small fibres, finishing agents, inorganic salts, reaction by-products, unspent dyes and chemicals, and domestic sewage. The grey cloth passes through a series of wet processing operations such as desizing, scouring/ kiering, bleaching, dyeing, printing and finishing, and the characteristics of each process effluent is different from each other[15].

12.4.3.1 Desizing wastes

Desizing is the process of removing the size materials from the warp yarns in grey woven fabrics. The size materials used to protect the yarn are various kinds of water-soluble polymers such as modified starch, polyvinyl alcohol (PVA), carboxy methyl cellulose (CMC) and acrylates. Wax, added to reduce the abrasiveness of the warp yarns, is also removed. Desizing wastes consist mainly of washings and contain colloidal and dissolved organic matter resulting from the hydrolysis of starch. It contains much organic matter derived from the natural impurities of the cotton and sizing materials in fabrics. The volume of waste is small and contains high BOD, which contributes to 35 to 50 % of the total BOD. Desizing of synthetic fibres or blends does not produce so concentrated effluent as that of cotton. However, the amount of organic matter from the size and anti-static agents in the synthetic cloth may still be appreciable [16-17].

12.4.3.2 Scouring/Kier wastes

The scouring/ kiering process is a process to remove oils, waxes, fats and other impurities. It consists of boiling the fabric in an alkaline solution containing 1-3 % caustic soda, soda ash, sodium silicate, sodium peroxide and other chemicals for 4-8 hrs in the presence of steam. The liquors from kier and washings are dark brown in colour with strong alkaline pH. This liquor contains caustic soda and suspended solids due to fragments of cotton. The wastes have high pH, TDS, alkalinity and BOD. The BOD of the kier liquor contributes almost 30 to 35 % of the total load[17].

12.4.3.3 Bleaching wastes

Bleaching is the process of whitening textile substrates by destroying the coloured impurities and it is done mainly by using oxidising agents. Bleaching is carried out with hypochlorite or peroxide solution in an alkaline pH. Hydrogen peroxide is an universal bleaching agent and generally preferred for bleaching of delicate fabrics. In bleaching with peroxide other chemicals like sodium peroxide, caustic soda and peroxide stabilisers, sequestering agents, etc., are used. Wastes from bleaching are small in volume, but they are highly alkaline. It contains organic matter and the oxidising agents used, such as sodium hypochlorite, sodium chlorite, chlorine and hydrogen peroxide, as well as alkalis. Residual peroxide is not detrimental, but the other oxidising agents can be reduced or kept out of the treatment plant. Excess acid or alkali always has to be neutralised. BOD load contributed by the bleaching wastes is only ~5 % [18].

12.4.3.4 Mercerizing wastes

Mercerizing is the process of cotton or linen yarn or fabric in which highly concentrated caustic soda (generally 250-300 g/L) is used. This process is followed by washing with profuse quantities of water. The mercerizing wastes are almost negligible both in its volume and waste load, which is less than 1

% of the BOD load. They usually contain considerable amounts of caustic soda, and this has to be neutralised. It can be collected separately, and some recovered or reused in the process house. The weaker wash liquors may not be worth collecting, but they have to be neutralised before sending them to the effluent treatment plant[19].

12.4.3.5 Dyeing wastes

Dyeing is a process of colouration of textile substrates carried out using various dyes, such as direct, basic, vat, sulphur, azoic, etc. The dyeing process contributes to a large volume of effluent. It exerts an oxygen demand due to the organic matter of the dyes, reducing agents like sulphide, hydrosulphite, nitrite, acetic acid and soaps. The BOD contribution from the dye house is about 6 % of the total waste load. Its composition and characteristics are highly variable, but the components common to most are dye residues and the dyeing assistants, usually surfactants. In addition, certain dye liquors contain substances that can be troublesome in treatment plants. For example,

- Sulphur dye liquors containing sodium sulphide
- Chrome dye liquors
- After-treatment solutions containing heavy metals

They can interfere seriously in biochemical processes and has to be removed or diluted before treatment [20-21].

12.4.3.6 Printing wastes

Printing of textile materials is generally done using block printing, roller printing and screen printing. The colour is applied on fabric in the form of viscous paste along with a thickener and chemicals. The wastewater derived from the printing process consists of colours, starch, gum, thickener and oil used in the printing pastes, and a substantial amount of dye, pigments and soaps. The quantity of waste from the printing process is small but has high BOD and COD [22-23]. There are three sources of colour in the effluents:

- Unused colour pastes
- Residual colour on machines, screens and rollers
- Loose colour on fabrics

12.4.3.7 Finishing wastes

Finishing is the process that is performed on fabric to improve its aesthetic and functional properties, including its look, performance, or "hand" (feel).

It is such a wide field that it is difficult to pick out individual substances responsible for the effluent load. The residues of various starch fillings are probably found in the greatest quantity. Most of these come from surplus materials in mixing and machine boxes. Other finishes being more expensive are employed less wastefully, so that little unused material is discarded. The finishing of cotton fabrics contributes a small volume of wastewater containing traces of starch, tallow, and salts, etc. [24-25].

12.4.3.8 Combined effluent

All the process effluents are mixed in an equalization tank and are known as combined effluent. The combined effluent contains chemicals used by different processing sections of the mill. Its colour, pH, alkalinity, TDS, COD and BOD varies sharply depending upon the characteristics of individual process effluents, viz. desizing, scouring, bleaching, dyeing, printing, etc. [26].

Table 12.6 Characteristics of textile processing effluents[27]

S. No.	Unit operation of wet processing	Possible pollutants	Volume of effluent	Nature of effluent
1	Desizing	Starch, glucose, polyvinyl alcohol, carboxymetyl cellulose, resins, fats, waxes	Very small, ~50 L/kg of cloth	High BOD (35-50 % of total) polyvinyl alcohol, carboxymetyl cellulose do not exert high BOD
2	Scouring/Kiering	Caustic soda, soda ash, sodium silicate, waxes, grease and broken fibres	Small, 10 L/kg of cloth	Strongly alkaline, dark colour, high BOD (30 % of total)
3	Bleaching	Hypochlorite, chlorine, caustic soda, hydrogen peroxide, etc.	Mostly washing	Alkaline, waste load ~5 % of the total BOD,
4	Mercerizing	Caustic soda	Small, mostly washing	Strongly alkaline, low BOD (less than 1 % of total)
5	Dyeing	Unexhausted dyes of various classes, mordants, reducing agents like sulphides, hydrosulphites, acetic acid and soap	Large	Intensively coloured, fairly high BOD (6 % of total
6	Printing	Colours, thickeners, oil, china clay, mordants, acids, metallic salts	Very small	Intensively coloured, fairly high BOD, oily appearance
7	Finishing	Traces of starch, tallow, common salt, glauber salt, etc.	Very small	Slightly alkaline, low BOD

12.5 Effect of effluent

12.5.1 Effect of textile effluents on the environment

The cotton textile industry produces effluents that are highly polluting in nature and create harmful effects on the environment.

Their harmful effects may be categorised as

- Effect on surface waters
- Effect on sewers
- Effects on agricultural land

12.5.1.1 Effect of textile effluents on surface waters

The increased pH of effluent is due to the high alkalinity of incoming water in the effluent. If the pH of receiving stream is more than 9, it causes an adverse effect on aquatic life. The soluble dyes and colour present in the effluent persist in the stream and prevent the penetration of sunlight for photosynthesis. The suspended and colloidal matter in the effluent causes an increase in turbidity of the water and produces an unsightly appearance. The textile effluent is responsible depletion of dissolved oxygen in the receiving body of water.

Immediate oxygen demand is exerted due to the presence of organic substances present in the textile effluent, such as starch, dextrin, and inorganic chemicals like sulphide, hydrosulphite and nitrite. The long-term oxygen demand is exerted due to the presence of dyes and chemicals. These changes in the oxygen balance of receiving streams harm aquatic life and also interfere with the self-purification of water [28-29]. Toxic chemicals like sulphide, chlorine, chromium and aniline dyes also affect aquatic life. Oil and grease present in the effluent form an oily scum on the surface of the water if discharged into water bodies. This causes an unsightly appearance and interferes with the mechanism of oxygen transfer at the air/water interface.

12.5.1.2 Effect of textile effluents on sewers

Due to high pH value, alkalinity and TDS, the textile effluents cause incrustation of sewer lines. Incrustation is the deposition of materials on the interior of pipes from chemicals present in the conveyed effluent. Sulphur dyes and sulphur compounds present in the effluent gradually lead to crown corrosion. The crown of the sewer, which contains more bacteria, comes in contact with H_2S gas. This H_2S gas is converted into sulphuric acid by the action of bacteria. The corrosion due to this sulphuric acid in the crown of the pipe is called crown corrosion. Hence, it is desirable that process waste streams containing sulphur dyes and sulphides are excluded from the effluent to be discharged into sewers [30-31].

12.5.1.3 Effect of textile effluents on agricultural land

Textile effluents have adverse effects on land in many ways. The pores of soil may get clogged due to suspended and colloidal matter. High alkalinity is harmful to crops, and high salinity impairs their growth. Sodium hardens the texture of the soil and prevents the penetration of roots. The soil loses its water holding capacity. As a cumulative effect, the soil loses its productivity [32-34].

12.5.1.4 Effect on a sewage treatment plant

The machinery of the municipal sewage treatment plant (STP) is exposed to corrosive action caused by acids and hydrogen sulphide. The pumps and pipelines may be subjected to incrustation. In any case, the cumulative effect of textile effluents on STP is increased cost of maintenance and repair.

12.5.2 Pollution abatement

The general approach to the problem of pollution abatement of textile effluents is based on the following criteria:

- Reduction in waste volume
- Reduction in waste load
- Treatment of effluents by primary and secondary treatment
- Application of advanced treatment methods.

12.6 Reduction in waste volume

The waste volume generated from the textile wet processing industry can be reduced by using various approaches; a few of them are explained below.

12.6.1 Reuse of recovered water from textile effluents

The scarcity of water has made it imperative to recover water from the textile effluents and reuse it in the textile industry. However, the use of a wide variety of auxiliary chemicals in textile wet processing results in unwanted enhancement of TDS to a very high level [35-36]. The removal of dyes from the effluent requires specific treatment for its de-colourisation. Mostly physicochemical treatment is given to the effluent, which is followed by biological treatment. But, the treated effluent still contains residual suspended and dissolved substances, which may cause a problem during its recycling in the process house. Highly concentrated wastewater is segregated from the less contaminated wastewater [37]. The dilute effluent after minor physico-chemical treatment is given advanced or tertiary treatment to remove these impurities. Some common technologies available for advanced treatment are

- Sand filtration for removal of residual suspended solids
- Activated carbon treatment for removal of residual colour and odour
- Softening/ cation exchange process to remove the hardness (Calcium and Magnesium).
- Ion-exchange/ reverse osmosis to remove dissolved inorganic salts (TDS)

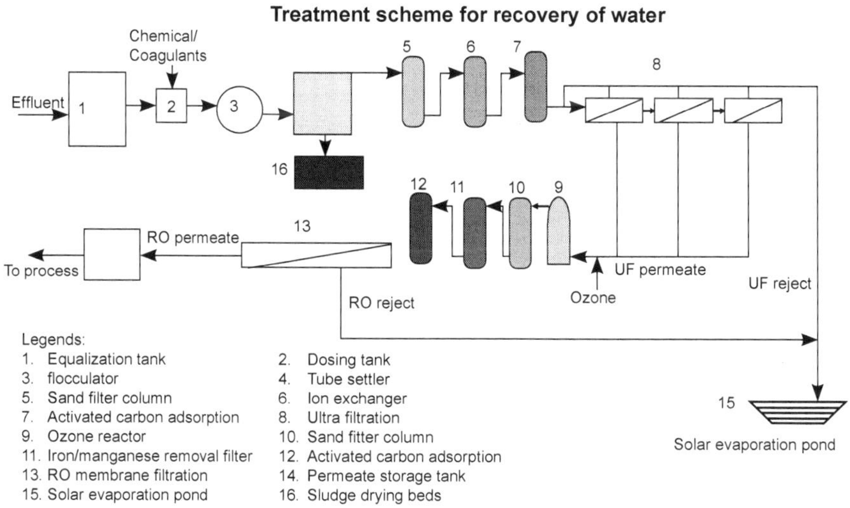

Figure 12.3 Treatment scheme for recovery of waste water

The wastewater volume can be reduced by the reuse of diluted streams which is economical in terms of water usage as well. Reuse of water is possible to the extent of 20 to 40 % in a cotton textile mill. But, the gross waste load in terms of kg/L of BOD and suspended solids remain unchanged [38]. On an average, the recovered water may be reused in:

- Textile processing if it meets the water quality requirement for textile processing (IS 201:1992)
- Flushing of toilets
- Gardening
- Floor washing
- Washing of process machines
- Miscellaneous areas

12.6.2 Process modification

A huge quantity of water is consumed in the washing which is required after each process during the wet processing of textiles.

Fabric moves at 10-20 kg/min in continuous washing which requires high consumption of water, i.e., 90 to 135 L/min. Besides, every chemical process requires a minimum of one rinse with an adequate volume of water. Multiple intermediate washing is required during the bleaching process. Attempts to reduce the use of water can be a step toward the reduction in the duration of each of the processing sequences[39]. The following process modification may be done:

- Use of counter-current method of washing
- Systematic control of bleach baths
- Re-circulation of washings and scouring wastes
- Reuse of dye baths

These modifications may result in the reduction of water consumption. Good housekeeping and prevention of leaks and spills are important steps toward the reduction in waste volume[40].

12.7 Reduction in waste load

12.7.1 By Recovery and Reuse of Process Chemicals

12.7.1.1 Recovery of synthetic sizing agents

For example, carboxymethyl cellulose can be recovered from desizing waste liquor by using precipitation with metallic salts like aluminium sulphate and its conversion into sodium salt by treatment with caustic soda.

12.7.1.2 Caustic recovery

Caustic can be recovered from mercerization drainage liquor using evaporation and dialysis techniques[41].

12.7.1.3 Recovery of dyes

In the vat dyeing process, only 65 to 75 % of the dye is consumed by the fibre and the remaining is drained out. The unspent dye may be recovered and reused in the same dyeing process. Similarly, bright-coloured lakes, lacquers, etc., may be recovered from basic, azo and reactive dye waste liquors by precipitation of their metallic derivatives [42-44].

12.7.1.4 Process modification

Modifications of textile processes in mills, such as the use of machines with low material to liquor (M:L) ratio, reduce chemical consumption which results in reducing the effluent load [45-46].

12.7.1.5 Substitution of process chemicals

Few suggestions for reducing BOD load of textile effluents to the extent of 50 % are given in Table 12.7.

Table 12.7 Substitution of process chemicals [47-49]

Process/class of chemicals	Conventional chemical used	Eco-friendly alternative
Sizing	Starch	Water-soluble PVA
Desizing	Hydrochloric acid	Amylases(Enzymatic desizing)
Scouring of cotton	Sodium hydroxide	Pectinases(Bio-scouring)
Bleaching	Hypochlorites	Hydrogen peroxide
Wetting agents and detergents	Alkyl phenol ethoxylates	Fatty alcohol phenol ethoxylates
Neutralization agent	Acetic acid	Formic acid
Peroxide killer	Sodium thiosulphate	Catalases
Mercerization	Sodium hydroxide	Liquid ammonia
Dyeing	Powder form of sulphur dyes	Pre-reduced dyes
Reducing agents	Sodium sulphide	Glucose, acetyl acetone, thiourea dioxide
Oxidation of vat and sulphur dyes	Potassium dichromate	Hydrogen peroxide, sodium perborate
Thickener in pigment printing	Kerosene	Water-based polyacrylate copolymers
Hydrotropic agent	Urea	Dicyanamide (partially)
Water repellents	C8 fluorocarbons	C6 fluorocarbons
Easy-care finish	Formaldehyde-based resin	Polycarboxylic acid
Flame retardant finish	Bromated diphenyl ethers	Combination of inorganic salts and phosphonates
Shrink proofing	Chlorination	Plasma treatment

12.8 Standards for discharge of textile effluents

Standard limits prescribed by the Central Pollution Control Board for the discharge of effluents from all integrated textile units, units of cotton, woollen, carpets, polyester, units having printing, dyeing, bleaching process, or garment manufacturing units are given in Table 12.8.

Table 12.8 Standards for discharge of effluents from textile industry [50-51]

S. No.	Parameters	Standard (applicable for all modes of disposal*)
1	pH	6.5 to 8.5
2	Suspended Solids, mg/L	100, max
3	Colour, P.C.U (Platinum Cobalt Units)	150, max
4	Bio-Chemical Oxygen Demand [3 days at 27°C] (BOD_3), mg/L	30, max
5	Oil and Grease, mg/L	10, max
6	Chemical Oxygen Demand (COD), mg/L	250, max
7	Total Chromium as (Cr), mg/L	2.0, max
8	Sulphide (as S), mg/L	2.0, max
9	Phenolic Compounds (as C_6H_5OH), mg/L	1.0, max
10	Total Dissolved Solids, Inorganic (TDS), mg/L	2100, max
11	Sodium Absorption Ratio (SAR), mg/L	26, max
12	Ammonical Nitrogen (as N), mg/L	50, max

12.8.1 Purpose of pollution control board standard limits

The purpose of standard limits by the pollution control board is to protect the environment from pollutants present in the effluent. The following pollutants have been considered in the standard due to their harmful effects.

12.8.1.1 Colour

Textile dyeing involves the use of several classes of dyes and pigments. The unspent dyes and pigment come out in their waste, making the effluent coloured. The effluents from sulphur and reactive dyeing are intensively coloured. Acid and vat dyeing effluents are coloured and transparent. Those of naphthol and disperse dyeing are coloured and turbid. Certain dyes, particularly hydrolysed reactive dyes and some acid dyes, are difficult to biodegrade and not absorbed by activated sludge during biological treatment. Thus, they pass through the treatment plant into river streams[52].

The textile effluents impart colour to the water body in which they are disposed off. The soluble dyes and colour present in the effluent persist in the stream and interfere with the penetration of sunlight essential for photosynthesis. By photosynthesis, the self-correction of rivers takes place. Colour in rivers severely affects the aquatic and marine life[53].

12.8.1.2 pH value

The pH of textile processing effluents varies from 3 to 13 depending upon the number of acids and alkalis used during the process. Acids and alkalis used in the process are acetic acid, hydrochloric acid, lactic acid, nitric acid, oxalic acid, sulphuric acid, caustic soda, soda ash, sodium bicarbonate, sodium silicate, trisodium phosphate, etc. Most of the cotton processing effluents coming from the processes like conventional scouring, mercerizing, hypochlorite bleaching, peroxide bleaching, reactive dyeing and sulphur dyeing wastes are alkaline. Desizing, chlorite bleaching, disperse dyeing and acid dyeing effluents are acidic[54].

A pH value of more than 9.0 in receiving stream causes harmful effects on aquatic life. The alkaline effluent also causes incrustation of sewer pipes. The hydroxyl alkalinity precipitates the bicarbonates in sewage as calcium carbonate and thus forms deposits in sewers. If the effluent is highly acidic, then it causes corrosion of sewers and the effluent treatment plants. Alkaline effluents also cause damage to crops by impairing their growth[55].

12.8.1.3 Total dissolved solids

The textile processing effluents have very high TDS due to excessive use of auxiliaries and chemicals during the process, as given below:

- Acids like acetic acid, hydrochloric acid, lactic acid, nitric acid, oxalic acid, sulfuric acid, etc.
- Alkalis like caustic soda, soda ash, sodium bicarbonate, sodium silicate, trisodium phosphate, etc.
- Salts like common salt, diammonium hydrogen phosphate, Glauber salt, tartar-emetic, etc.
- Oxidizing agents like hydrogen peroxide, sodium chlorite, sodium hypochlorite, etc.
- Reducing agents like sodium sulphite, sodium dithionite, sodium metabisulphite, sodium sulphide, etc.

High TDS of the effluent increases the tendency of incrustation of sewer pipes. Most of the dissolved solids are in the form of sodium salts. Such effluent with a high percentage of sodium affects the soil quality adversely, making it barren in the long run. Due to this, calcium and magnesium contents of soil are reduced which hardens the soil and prevents penetration of roots. On prolonged application, sodium also exhibits deflocculating effect, and soil loses its water holding capacity. As a cumulative effect, the soil loses its productivity [56].

12.8.1.4 Total suspended solids

The textile effluents contain a lot of suspended solids (TSS). Some of the sources of suspended matter in the textile effluent are:

- Spent kier liquor and washings containing fragments of cotton as suspended solids.
- Carbonizing of wool, which generates short fibre fragments as suspended matter.
- Insoluble dyes and other foreign matters like dust may also contribute to the suspended matter.

The suspended matter (TSS) present in the effluent, along with colour, increases the turbidity of water and gives it a bad appearance and foul smell. Small suspended particles and colloidal matter clog the pores of soil by forming an impervious bed. The large suspended matter may choke the sewer line [57-60].

12.8.1.5 Chemical oxygen demand

The Chemical Oxygen Demand (COD) is due to organic matters present in the textile effluent. The desizing liquor from a cotton process house contains much organic matter due to the removal of water-soluble natural impurities of the cotton (pectins, proteins, waxes, seed capsules, etc.) and from sizing materials in the fabric (polyvinyl alcohols, starch, modified cellulose, poly-acrylates, polyesters, etc.). It exerts up to 70 % of the total COD load. The COD of scouring effluents is due to not readily bio-degradable surfactants and complexing agents. COD of peroxide bleaching effluent is due to complexing agents, wetting agents and stabilising agents. The high oxygen demand in the textile effluent is exerted due to the organic matter of the dyes, reducing agents (like sulphide, hydrosulphite, nitrites), acetic acid and soaps. Organic matters like starch, gum and oil also come in the effluent from the printing pastes. Mercerizing and some of the dyeing effluents have low COD loads (less than 1500 mg/L)[61].

The most serious effect of textile effluent on the receiving body of water is the depletion of dissolved oxygen due to organic matters present in the effluent. Immediate oxygen demand in textile effluent is exerted due to the presence of organic substances such as starch, dextrin, and inorganic chemicals like sulphide, hydrosulphite and nitrite. The long-term oxygen demand is exerted due to the presence of dyes and colours. Such changes in the oxygen balance of receiving streams become deleterious to aquatic life and also interfere with the self-purification of water.

12.8.1.6 Biochemical oxygen demand

The oxygen required by microorganisms for the oxidation of biodegradable organic matter is known as biochemical oxygen demand (BOD). BOD in the textile effluents is mainly contributed by desizing liquors (35-50% of total BOD) and kier liquors (30-35 % of the total BOD). The high BOD of desizing liquor is due to the removal of water-soluble natural impurities of the cotton (pectins, proteins, waxes, seed capsules, etc.) and sizing materials (starch) from the fabric. Synthetic sizing agents (CMC, PA, PVA) and scouring liquor (surfactants and complexing agents) are not readily bio-degradable, hence exert lesser BOD load. Mercerizing effluents have low BOD values (50-100 mg/L) and some of the dyeing effluents (e.g. disperse/ vat) have 75-200 mg/L BOD values[61].

The harmful effects of BOD are the same as that of COD. The depletion of dissolved oxygen because of biodegradable organic matters present in the effluent is also a severe effect of textile effluent on the receiving body of water[62].

12.8.1.7 Oil and grease

Oil and grease are primarily composed of fatty matters from animal and vegetable sources and hydrocarbons of petroleum origin. The main sources of oil and grease in textile effluents are:

- Desizing and kiering processes that remove grease, waxes, and natural fats from the grey fabric
- Preparation agents used in the pretreatment process of man-made fibres
- Oil present in additives and softeners used in sizing
- Mineral oil used as an additive in thickeners for making printing pastes
- Antifoaming agents based on mineral oil
- Anionic detergents, which are based on sulphated vegetable oil
- Lubricants which are usually refined natural oils (e.g., Fatty acid triglycerides) and ester oils (e.g. Octyl stearate, tridecyl stearate, etc.)

Oil and grease present in the effluent form an oily scum on the surface of the water if discharged into water bodies. This causes an unsightly appearance and interferes with the mechanism of oxygen transfer at the air/ water interface. It may also foul instruments and equipments by accumulating in unwanted places, causing hazards or performance problems. If present in excessive amounts, it may also interfere with treatment processes like gravity settling and biological processes (aerobic and anaerobic) and decrease their efficiency [63-64].

12.8.1.8 Residual chlorine

Residual chlorine in textile effluents may come from the following processes:

- Pretreatment with chlorine-containing substances like dichloroisocyanurate.
- Chlorite bleaching, where a large quantity of bleaching agents like sodium chlorite, sodium hypochlorite and chlorine are used.
- Anti-felting treatment where chlorine compounds (e.g., dichloroisocyanurate) are used.
- Stripping (removal) of dyes and auxiliaries from the fibre where Sodium chlorite is used as an oxidising agent.
- Chlorination process, which is the most common method for disinfection of water.

The presence of an excess amount of chlorine in the effluent may produce an adverse effect on aquatic and marine life, if discharged into water bodies. The taste and odor of phenols and other organic compounds present in the water may be enhanced. Chlorinated effluents normally contain only combined chlorine. After chlorination of ammonia or amine, chlorine is formed which is dangerous for some aquatic life [65-66].

12.8.1.9 Total chromium

Chromium in textile effluents may come from the following sources:

- Dyeing and printing processes where chromium compounds ($KCrO_4$, $K_2Cr_2O_7$) have been used as mordant.
- Chromium complex dyes and after-chrome dyes used in the dyeing of wool.
- Oxidation of reduced forms of vat dyes, leuco-sulphur dyes and sulphur dyes using sodium chromate or potassium dichromate.
- After-treatment processes using chromium salts for improving the wash fastness properties.
- Cooling water where chromium compounds are frequently added for corrosion control.

If effluent containing chromium is discharged into water bodies, it makes the water toxic and destroys the aquatic and marine life. Chromium may exist in the effluent in both hexavalent (Cr VI) and trivalent states (Cr III). The toxicity of Cr (III) is less than Cr (VI), probably due to its low solubility. Cr (VI) is carcinogenic by inhalation and is corrosive to body tissues. Chromium toxicity results in poor growth and damage to the liver and kidney. It may also seriously interfere with biochemical processes and has to be removed or diluted before treatment [67-69].

12.8.1.10 Sulphides

The main processes responsible for sulphides in the effluent are:

- Sulphur dyeing where the dye contains sulphur compounds and is applied along with a reducing agent like sodium sulphide, sodium hydrosulphite (sodium dithionite), etc.
- Vat dyeing in which strong reducing agents like sodium sulphide, sodium dithionite (sodium hydrosulphite) are used to make them water-soluble.
- Reduction clearance after treatment using sulphur compounds during disperse dyeing.
- Reductive bleaching with sodium dithionite in case of protein fibre, polyamide, wool and elastane.

Sulphide is present in higher concentrations when the organic load of the effluent is high. The microorganisms use sulphate as an electron acceptor to decompose organic matter and convert sulphate into sulphide. Its common presence in the effluent is partially due to the decomposition of organic matter, but it is mostly due to the bacterial reduction of sulphate[70].

Sulphide has an objectionable odour resembling that of a rotten egg which decreases the aesthetic value and renders the water unfit for consumption. Odour nuisance is caused due to the escaped H_2S into the air from sulphide-containing effluent. H_2S is very toxic to fish and aquatic organisms if dissolved in water. In the gaseous form, it is harmful to sewer many workers and has claimed worker's lives. H_2S gets oxidised to form H_2SO_4 and causes serious corrosion on the pipe walls and in the concrete of sewers[71].

12.8.1.11 Phenols

The presence of phenols in the textile effluents is due to the use of the following auxiliaries and chemicals in the process:

- Non-ionic surfactants (wetting agents) having alkylphenol as a hydrophobic part.
- Non-ionic detergents based on nonylphenol-ethoxylate/Nonyl phenols used in printing paste.
- Carriers based on o-Phenyl phenol and alkylphenol ethoxylate.

Alkylphenol ethoxylate is used as a dispersing agent in coating pastes and carriers in exhaust dyeing for the acceleration of dye adsorption and dye diffusion.

Chlorination of water containing phenols may produce odour and objectionable taste due to the formation of chloro-phenols. The objectionable taste and odour are sometimes rejected on the grounds of aesthetic value. The taste and odour in water are also not suitable for food, beverages and pharmaceutical industries.

12.9 Effluent treatments

The conventional treatment methods used to treat the industrial effluents are:

- Preliminary treatments
- Primary (physico-chemical) treatments
- Secondary (biological) treatments
- Advanced (tertiary) treatments

12.9.1 Preliminary treatments

The physical separation of impurities is termed as preliminary treatment (Fig. 12.4). This involves the following steps:

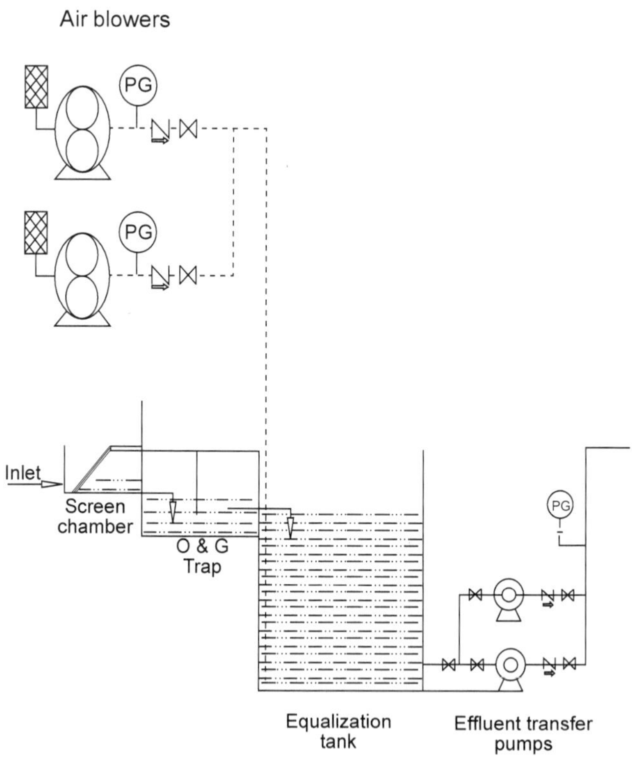

Figure 12.4 Preliminary treatment

12.9.1.1 Screening and grit removal

It removes coarse floating and suspended matter like lint and fibres in addition to heavy and readily settleable grit and dirt.

12.9.1.2 Oil and grease removal tap

It removes oil and greasy matter which gets collected as scum on the top surface of the water by gravity separation or air floatation. The scum is removed by oil skimmer or manually.

12.9.1.3 Mixing and equalization

Mixing of effluents from various process streams and their equalization for 6-8 hr is essential to avoid wide variations in characteristics of the effluent and to prevent shock loads.

12.9.1.4 pH correction

After equalization, pH correction is necessary to increase the efficiency of subsequent treatments.

12.9.2 Primary or chemical treatments

The chemical coagulation of effluent is termed as primary treatment (Figs. 12.5 and 12.6). This is very effective in the removal of colour, suspended particles and colloidal organic matter like starch and gums. The common coagulants are alum, ferrous sulphate, ferric chloride and chlorinated copperas. Coagulation aids like poly-electrolytes are used to improve the efficiency of coagulation. Primary treatment using lime and ferrous sulphate or alum removes the colour very effectively and reduces the COD load up to 70% and BOD load up to 40%. Chemical treatment results in sludge production, which needs to be separated out and dried on sludge drying beds and disposed of as a landfill [72-73]. UV or ozone are used to inactivate the indicator organism's *Fecal coliforms, Enterococci, MS-2 coliphage, or Clostridium perfringens* in the effluent from this plant. PAA doses to reach the target fecal coliform level of 9000CFU/100mL exceeded 6 mg/L; similar results were obtained for *enterococci* , and no inactivation of *Clostridium perfringens* was observed. However, a 1-log reduction of MS-2 occurred at PAA doses of 1.5mg/L and higher. It was expected that this effluent would have a high ozone demand and require relatively high UV fluences, because of relatively high effluent COD, iron and suspended solids concentrations, and low UV transmittance. For UV, the inactivation curve for fecal coliforms showed the typical two-stage shape, with the target of 1000CFU/100mL (to account for photoreactivation).

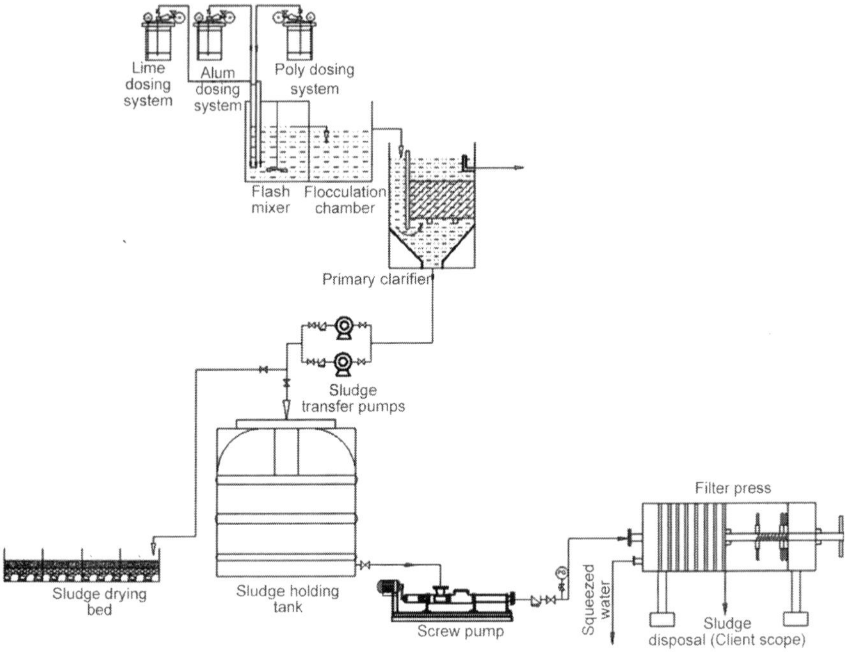

Figure 12.5 Primary treatments with sludge treatment

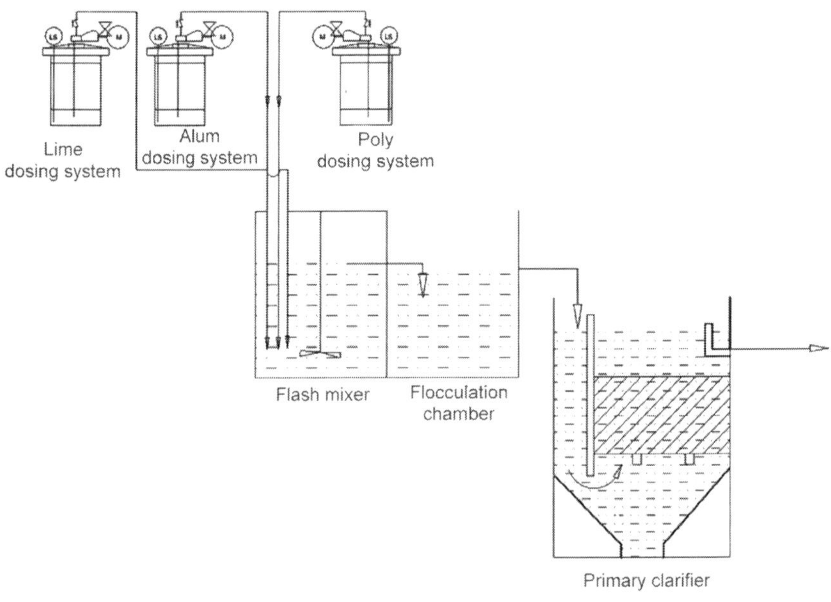

Figure 12.6 Primary treatments without sludge treatment

12.9.3 Secondary or biological treatments

Biological treatment of effluents is termed as secondary treatment (Fig. 12.7). After primary treatment, the organic matter present in the effluent is largely in colloidal or dissolved form. Biological treatment is a process of precipitating organic matter on biological slime. The colloidal and dissolved matters are converted into various gases and cell tissues by the action of microorganisms. These converted cell tissues are heavier than water, hence can be removed by settling. The biological treatment can reduce the BOD by more than 90% [74-76]. A bisulphite-catalysed sodium borohydride reduction followed by an activated sludge technique was used in order to remove the colour at ambient temperature and pressure. This experimental study consisted of two major parts: reduction treatment and biological oxidation. Both synthetic and actual wastewater was used in this research. Synthetic wastewater was made by several groups of dyes such as direct, basic and reactive colours. Actual wastewater was collected from two different textile industries in the city of Isfahan, Iran. The characterisation of raw and treated wastewater was carried out by infrared and ultraviolet spectrometers. The results of this study demonstrated that the newly developed treatment technique decreased colour, and BOD. Commonly used biological treatment methods are presented in further sub-sections.

12.9.3.1 Aerobic treatment process

It is carried out in the presence of oxygen.

For example:

- Percolating or trickling filters
- Activated sludge process

12.9.3.2 Percolating or trickling filters

These filters consist of rocks such as coke sludge or granite on which active bacteria or protozoa form colonies. The settled effluent is sprayed or percolated over these rocks at regular intervals. During the course of flow of the effluent, the organic matter is precipitated or oxidised over the surface of biological slime attached to the rocks[77]. In recent years, high-rate anaerobic-aerobic bioreactors have been increasingly employed for wastewaters with high COD.

12.9.3.3 Activated sludge process

In this process, the biological slimes are formed as flocs and are maintained in suspension in tanks by blowing air for the oxidation of organic matter by the microbes on the surface of the flocs [78-80].

12.9.3.4 Anaerobic digestion

It is carried out in the absence of oxygen, e.g. bioreactors. In this process, the microorganisms operate in the absence of oxygen and decompose the organic matter into simple products such as carbon dioxide or methane gas. This method is very slow and has low efficiency [81-83].

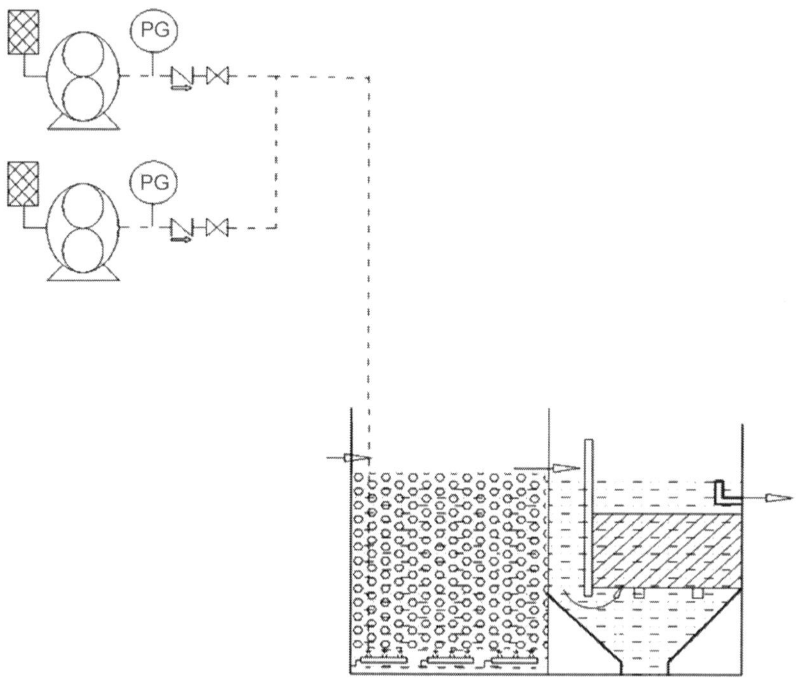

Figure 12.7 Secondary treatments

12.9.4 Advanced or tertiary treatments

This is an additional treatment recommended to remove suspended and dissolved substances residual after conventional secondary treatment or physico-chemical treatment. These substances may contain organic matter, suspended solids, and inorganic ions such as calcium, sodium, potassium, chloride, sulphate, or highly complex synthetic organic compounds.

The need for advanced effluent treatment (Fig. 12.8) arises for the recovery of water from waste water and its reuse for various purposes[84]. The various available technologies for advanced treatment are:

- Granular medium sand filtration for removal of residual suspended solids[85].
- Granular activated carbon filtration for removal of residual colour, toxic compounds and refractory (poorly biodegradable) organics[86].
- Softening by ion exchange, sequestration or soda lime treatment for removal of hardness[87] and, in particular, magnetic ion exchange (MIEX).
- Demineralisation/ Reverse osmosis for removal of total dissolved solids and metal ions[88].

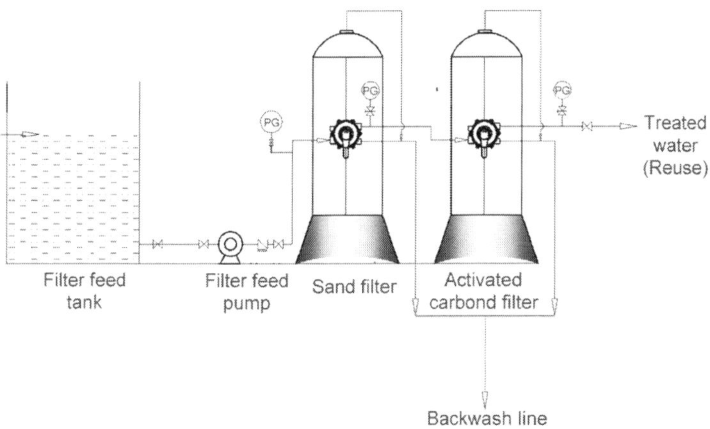

Figure 12.8 Tertiary treatments

12.9.5 Nanotechnology in effluent treatment

Nanomaterials are typically defined as materials smaller than 100 nm in at least one dimension. The application of nanomaterials in effluent treatment is an emerging field for scientists. Nanomaterials can be used in various ways in the effluent treatment: a few of them are as follows:

12.9.5.1 Nanofiltration

Filtration is the most common process for water purification. Nanofiltration (NF) is a recent advancement in the water purification technique which is pressure driven. The properties of NF membrane are between the non-porous RO membranes and porous ultrafiltration membranes. It works mainly based on solution diffusion mechanics. By using nanofiltration, it is possible to recover 90-97% of water from the effluent stream; this water can be recycled, and permeate can be used for dyeing [89-92].

12.9.5.2 Photocatalysis

Photocatalysis is an advanced technique for water purification using nano-structured catalysis which degrades various pollutants present in the water. Metallic nanoparticles have promising performance due to their high and broad activities for various pollutants present in the textile effluent. Various nanoparticles such as TiO_2, and ZnO have demonstrated very good photocatalytic properties for decolourising the effluent water from dye houses[93-94].

12.9.5.3 Nanomaterials for adsorption of pollutants

Carbon nanotubes (CNT) have high adsorption capacity due to their larger surface area and diverse interaction with various contaminants. If oxidised CNT is used, it can absorb the heavy metal ions. Metal-based nanomaterials such as gold, silver, TiO_2 and metal oxides are also used as adsorbents. Dendrimers and composite nanoparticles with improved properties can be used as an adsorbent in the treatment of effluent from the textile process house [95-98].

12.10 Disposal of textile effluents

The disposal methods of textile effluents in the country are as follows.

- Disposal into public sewers in cities like Ahmadabad, Mumbai, Bangaluru, Kolkata, Delhi, Kanpur, Chennai and Madurai.
- Directly on land in unsewered towns like Kalol and other textile centres in Gujarat and other states.
- Ultimate disposal along with sewage into the sea in Mumbai and Chennai
- On land for irrigation in cities like Ahmadabad, etc

Mostly textile wet processing effluents are disposed off into sewers and ultimately into rivers or streams. The effluents from textile processing units have very high TDS in the form of sodium salt; therefore, they are not preferred to dispose off on land. Nowadays, efforts are being made to reuse the treated effluent in the process house to reduce water pollution[99]. The government is planning to make it mandatory for processing industries to implement "Zero Liquid Discharge (ZLD)".

12.10.1 Treatment methods recommended

Recommended treatment methods for effluents from textile mills in India are based on the ultimate disposal[100].

12.10.1 1 Disposal into public sewers

The suggested minimum treatment is preliminary and primary treatments, including screening, grit removal, chemical coagulation, flocculation and sedimentation. The treatment process flow diagram for disposal into sewers is shown in Fig. 12.9.

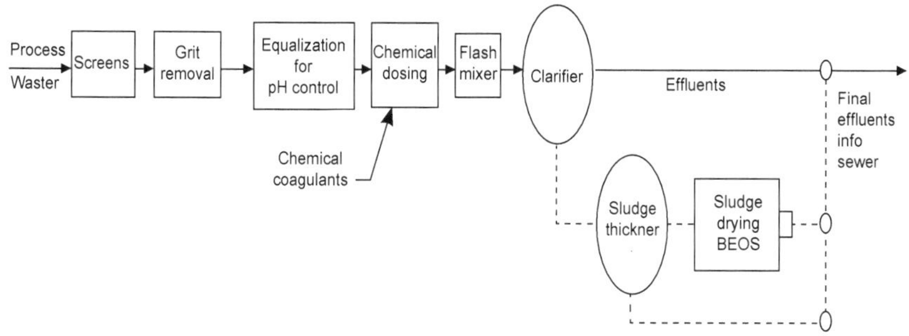

Figure 12.9 Treatment process flow diagram for disposal into sewers

12.10.1.2 Disposal into streams

The suggested treatment is preliminary and primary treatments, including screening, grit removal, chemical coagulation, flocculation and sedimentation. Primary treatment is followed by secondary treatment for BOD reduction. The treatment process flow sheet for discharge into inland surface water is shown in Fig. 12.10.

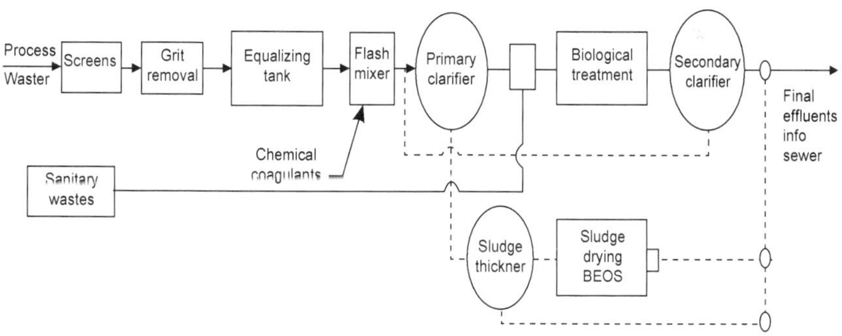

Figure 12.10 Treatment process flow sheet for discharge into inland surface water

12.10.1.3 Land disposal

Preliminary and primary treatments or secondary treatments, or both followed by treatment with gypsum, are suggested. Gypsum treatment is given to maintain a sodium absorption ratio of less than 26 by increasing the amount of calcium in the treated effluent. The treatment process flow sheet for land disposal is shown in Fig. 12.11.

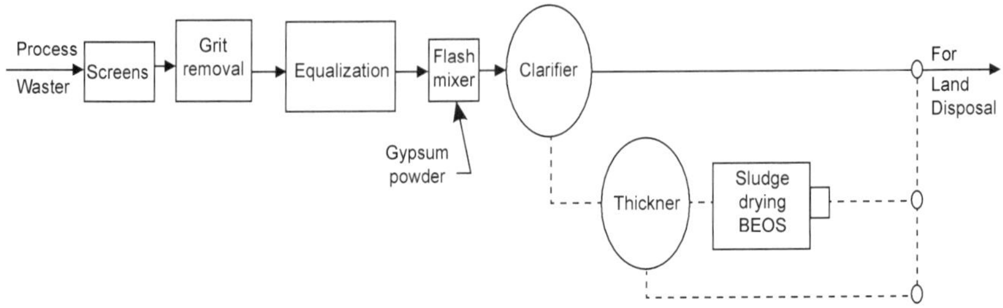

Figure 12.11 Treatment process flow sheet for land disposal

12.11 Problems faced by the industry

12.11.1 Non-availability of good quality water

It is mainly due to the following reasons:

- Scarcity of low TDS water in some areas, particularly in arid zones
- Dwindling natural water resource
- Uneven rainfall and drought in some parts of the country
- Contamination of surface and sub-soil water due to pollution
- Falling underground water table

12.11.2 Disposal of liquid effluents

It is mainly due to the following reasons-

- Stringent effluent disposal norms
- The pressure to reuse the treated effluent to
  - Reduce fresh make-up water requirement
  - Prevent water pollution.
- Costly recovery of water from high TDS effluent using RO system
- Declaration of some areas as "Zero Effluent Discharge Zone"

12.11.3 Disposal of ETP sludge

It is mainly due to the following reasons:

- Generation of a huge quantity of sludge in conventional effluent treatment plants (ETP)
- Space constraint for drying the sludge.
- Toxic nature of chemical (primary) sludge.

Concluding remarks and future perspectives

The textile chemical processing requires a huge amount of good quality water and generates an equally large volume of effluent. The effluent is mostly intensively coloured and alkaline with high TDS contents. Desizing and scouring wastes have the highest BOD load. They pollute the environment if untreated textile effluents are discharged into surface water, sewer or on agricultural land. The pollution level can be reduced by taking suitable measures such as:

- Reduction in effluent volume by recovery and reuse of water
- Reduction in waste load by recovery and reuse of chemicals and auxiliaries
- Process modification
- Substitution of process chemicals
- End of the pipe treatment

In order to check the pollution, standards for disposal of textile effluents have been prescribed vide Textile Gazette Notification dated October 10, 2016. The common standard has been prescribed for all types of textile industry, which is applicable for all modes of disposal, whether it is surface water sewer or land. The parameters which have been considered in the standard are colour, pH, TDS, TSS, COD, BOD, O&G, chromium, sulphide, phenols, SAR and NH_3-N. It is mandatory to meet the standard norms for the disposal of textile effluents into the environment. In some sensitive areas, “Zero Liquid Discharge” has been imposed.

The conventional treatment technologies for the treatment of textile effluents are preliminary treatment, physico-chemical treatment and biological treatment. Advanced treatment is also given to the treated effluent for recycling and reuse of the treated effluent. The disposal method varies from city to city, depending upon the availability of natural resources and norms prescribed by the pollution control board. The treatment technologies have also been prescribed for various modes of disposal methods. The main problems being faced by the textile industry are non-availability of good quality water, disposal of liquid effluents and disposal of ETP sludge. All these problems can be solved by an efficient effluent management in the textile industry.

References

1. Meena CR, Girendra, Adivarekar RV,Sekar N (2013), ‘Synthesis and Application of Vinylsulphone Disperse Reactive Dyes for Polyester’, *Int. J. ChemTech Res.* 5 2, 585–591.

2. Kumar PS, Saravanan A, (2017), 'Sustainable wastewater treatments in textile sector', *Sustainable Fibres and Textiles,* 323-346, Elsevier Ltd, Amsterdam, Netherlands, https://doi.org/10.1016/B978-0-08-102041-8.00011-1

3. Chaudhari DH, Dhoble RM (2010), 'Performance evaluation of effluent treatment plant of dairy industry',*Cur. W.Environ.* 5,2,373–378. http://dx.doi.org/10.12944/CWE.5.2.26

4. Ashfaq A,Khatoon A(2014), 'Waste Management of Textiles : A Solution to The Environmental Pollution', *Int. J. Curr.Microbiol. App. Sc.,* 3,7, 780–787. https://www.ijcmas.com/vol-37/Ahmad%20Ashfaq%20and%20Amna%20Khatoon2.pdf

5. Kiron MI (2014), 'Water in Consumption Textile Processing Industry"[Online]. https://textilelearner.blogspot.com/2014/04/water-consumption-in-textile-industry.html. (Accessed May 2020).

6. Shaikh MA (2009), 'Water conservation in textile industry', *Pak. Tex. J.*, 48–51. https://sswm.info/sites/default/files/reference_attachments/SHAKIH%202009%20Water%20conservation%20in%20the%20textile%20industry.pdf

7. House IE (1991), 'Bureau of Indian Standards Specification for Drinking Water BIS 10500 : 1991',1,1–5. https://law.resource.org/pub/in/bis/S06/is.10500.1991.pdf

8. Begum M, Hossen M, Khatton A(2009),'Water Characteristics Suitability For Textile Wet Processing and Its Remedy Measures', *Cellulose Chem. Tech.*, 43 (4-6),189-192. https://www.cellulosechemtechnol.ro/pdf/CCT4-6-2009/P.189-192.pdf

9. BIS (2012), *Indian Standards Drinking Water Specifications IS 10500*,2. http://cgwb.gov.in/Documents/WQ-standards.pdf

10. Kispotta N (2014),'Common Boiler Feed Water Treatment in the Industry', *Int. Jour. Inn. Res. Sci. Tech.* 1(6), 59–62.

11. Muhaisen NM (2014), 'Potential Saving For Steam Boilers Based On The Impact Of Most Affecting Factors : A Case Study' *Proceedings of the 2014 Int Conference on Ind. Eng. Ops Man. Bali, Indonesia,* 481–487. http://ieomsociety.org/ieom2014/pdfs/105.pdf

12. Kukic D,Sciban M, Mitrovic BB, Prodanovic JM (2012), 'Treatment Possibility of improvement of boiler water treatment process — ion exchange vs . reverse osmosis' *Desalination and Water Treatment*, 51(1),1-7. http://dx.doi.org/10.1080/19443994.2012.714860

13. Hall SM (2018),'Chapter 21 – Boilers' *Rules Thumb Chem. Eng.*,371–386. https://dl.icdst.org/pdfs/files1/6a51bf52b79d86bdf1913fc25a1700ba.pdf

14. Bureau of indian standards,(1980),'S 9508: Guide for Treatment and Disposal of Effluents of Cotton and Synthetic Textile Industry', https://archive.org/details/gov.in.is.9508.1980

15. Oras K, Gruner E (2005),'The Estimation of the Wastewater Generation and Pollution Load by the Branches of Industry', IWG-Env,International Work Session on Water Statistics,Vienna,1–10. https://mdgs.un.org/unsd/environment/envpdf/pap_wasess4b2estonia.pdf

16. Doble A, Mukesh (2005), 'Textile Effluent', *Bio. Ind. Eff.,* 123–132. https://doi.org/10.1016/B978-075067838-4/50012-9.

17. Mondal P, Baksi S, Bose D (2017), 'Study of environmental issues in textile industries and recent wastewater treatment technology', *World Sci. News* 61(2), 98–109. http://www.worldscientificnews.com/wp-content/uploads/2016/11/WSN-612-2017-98-109.pdf

18. Karmakar SR (1999),'Bleaching of textiles',*Tex.Sci.Tech.,12,*160–216. https://doi.org/10.1016/S0920-4083(99)80007-3

19. Souther ARH et al (2014), 'Textile Wastes : Recovery and Treatment', S*ewage and Industrial Wastes* 29(8), 918–935. https://www.jstor.org/stable/25033407.

20. Lin SH, Lin CM (1993), 'Treatment of Textile Waste Effluents by Ozonation and Chemical Coagulation', *Water Res.*, 1–5. https://doi.org/10.1016/0043-1354(93)90112-U

21. Robinson T, Mcmullan G, Marchant R, Nigam P (2001), 'Remediation of dyes in textile effluent : a critical review on current treatment technologies with a proposed alternative', *Biores. Tech.* 77, 247–255. https://doi.org/10.1016/s0960-8524(00)00080-8

22. Sostar S, Simonic M, Petrinic I (2005), 'Wastewater treatment after reactive printing', *Dyes Pigm.*, 64(2),147-152. https://doi.org/10.1016/j.dyepig.2004.04.001

23. Kabdasli I, Gürel M,Tünay O (2010),'Characterization and Treatment of Textile Printing Wastewaters', *Env. Tech.*, 21,10,37 41.https://doi.org/10.1080/09593330.2000.9619001

24. Bes-Piá A, Mendoza-Roca JA, Alcaina-Miranda MI, Iborra-Clar A, Iborra-Clar MI(2003).'Combination of physico-chemical treatment and nanofiltration to reuse wastewater of a printing, dyeing and finishing textile industry' *Desalination* 157(1-3) ,73–80. https://doi.org/10.1016/S0011-9164(03)00385-0

25. Paul SA, Chavan SK, Khambe SD(2012),'Studies on characterization of textile industrial waste water in solapur city' *Int. J. of Chem. Sci.*10(2),635–642. https://www.tsijournals.com/articles/studies-on characterization-of-textile-industrial-waste-water-in-solapur-city.pdf

26. Chethan K(2017), 'The Effect of Industrial Textile Effluent Interaction with both Laterite Soil and Expansive Soil' *Int.J. of Advance Research and Development*,2(3),19–25.

27. Yaseen DA, Scholz M(2019),'Textile dye wastewater characteristics and constituents of synthetic effluents: a critical review' *Int. J. of Env. Sci & Tech.*,16(2), 1193-1226. http://dx.doi.org/10.1007/s13762-018-2130-z

28. Berryman D, Deblois C, Shea MO(2004),'Nonylphenolic compounds in drinking and surface waters downstream of treated textile and pulp and paper effluents : a survey and preliminary assessment of their potential effects on public health and aquatic life' c*hemosphere,*56(3), 247–255. https://doi.org/10.1016/j.chemosphere.2004.02.030

29. Rathore J(2011),'Assessment of water quality of River Bandi affected by textile dyeing and printing effluents , Pali , Western Rajasthan , India' *International Journal of Enviromental Sciences,* 2(2), 560–568.

30. Jason Kane (2014), 'Hydrogen sulphide gas in sewers – the challenges of odour and corrosion' 39th Annual WIOA Queensland Water Industry Operations Conference and Exhibition Logan Metro Indoor Sports Centre, Logan 3 to 5 June, 2014,75, 75–81. https://www.wioa.org.au/conference_papers/2014_qld/documents/Jason_Kane.pdf

31. Moran C, Hall ME and Howell R(1997),'Effects of sewage treatment on textile effluent', *Journal of the Society of Dyers and Colourists,* 113, 272–274. https://doi.org/10.1111/j.1478-4408.1997.tb01847.x

32. Ravi D, Parthasarathy R, Vijayabharathi V, Suresh S (2014), 'Effect of Textile Dye Effluent on Soybean Crop' *J. Pharm., Chem. Bio. Sci.* 2,111–117. https://www.jpcbs.info/2014_2_2_7_Ravi.pdf

33. Length F(2014),'Textile effluents changes physiochemical parameters of water and soil : Threat for agriculture', *Afri. J. Agro.*,2(10), 219–223. https://citeseerx.ist.psu.edu/viewdoc/download?doi=10.1.1.682.1740&rep=rep1&type=pdf

34. Roohi M et al (2016),'Varied effects of untreated textile wastewater onto soil carbon mineralization and associated biochemical properties of a dryland agricultural soil', *J. Environ. Management,* 183, 530-540. https://doi.org/10.1016/j.jenvman.2016.09.005

35. Harane R Adivarekar R(2017),'A frugal way of reusing wastewater in textile pre-treatment process' *J. Water Process Eng.* 16, 163–169. https://doi.org/10.1016/j.jwpe.2017.01.002

36. Harane RS Adivarekar RV(2016),'Sustainable processes for pre-treatment of cotton fabric' *Text. Cloth. Sustain.*,1-9. https://doi.org/10.1186/s40689-016-0012-7

37. Tian Y, Hu H, Zhang J(2017),'Solution to water resource scarcity : water reclamation and reuse' *Env. Sci. Poll. Res*, 24, 5095–5097. https://doi.org/10.1007/s11356-016-8331-4

38. Kumar P, Omprakash P(2017),'Quality and management of wastewater in sugar industry' *Appl. Water Sci.*,7,461–468. https://doi.org/10.1007/s13201-015-0264-4

39. Alkaya E, Demirer GN (2013), 'Sustainable textile production : a case study from a woven fabric manufacturing mill in Turkey' *J. Clean. Prod.*,65,595-603. https://doi.org/10.1016/j.jclepro.2013.07.008

40. Babu BR, Parande AK, Raghu S, Kumar TP (2007), 'Cotton Textile Processing: Waste Generation and Effluent Treatment' *J. Cott. Sci.*153, 141–153.

41. Costa JM, Química DP, De Tecnologia F,Antoine A (2013) 'Reactive and Vat Dyestuff in the Dyeing of Cotton : A Review of Energy and Water Consumption , Ecological Analysis and Effluent Treatment' *Int. Clean. Prod. Into Sustain. Strat.* http://www.advancesincleanerproduction.net/fourth/files/sessoes/5A/6/rosa_jm_et_al_work.pdf

42. Mukherjee BA. Gupta KB, Chowdhury S.M.S.(1999), 'Separation Of Dyes From Cotton Dyeing Effluent Using Cationic Polyelectrolytes' *American Dyestuff Report,*25-28. https://citcseerx.ist.psu.edu/viewdoc/download?doi=10.1.1.404.4464&rep=rep1&type=pdf

43. Li M, Wang X,Porter CJ, Cheng W, Zhang X, Wang L, Elimelech M (2019), 'Concentration and Recovery of Dyes from Textile Wastewater Using a Self-Standing,

Support-Free Forward Osmosis Membrane, *Environ. Sci. Technol.*, 53(6), 3078–3086. https://doi.org/10.1021/acs.est.9b00446

44. Taylor P, Uzal N, Yilmaz L, and Yetis U (2013),'Separation Science and Technology Nanofiltration and Reverse Osmosis for Reuse of Indigo Dye Rinsing Waters' *Saperation Science andy Technolog*, 45(3) 37–41. https://doi.org/10.1080/01496390903484818

45. Adivarekar RV,Harane RS(2014),'Standing Bath Method for Economical Pre-treatment of Cotton' *Asian Dyer*,11,31–34.

46. Harane RS, Mehra NR, Tayade PB, Adivarekar RV(2014), 'A facile energy and water-conserving process for cotton dyeing' *Int. J. Energy Environ. Eng.*, 5, 96. https://doi.org/10.1007/s40095-014-0096-2

47. Ozturk E, Yetis U, Dilek FB, Demirer GN (2009),'A chemical substitution study for a wet processing textile mill in Turkey' *J. Clean. Prod.*17(2), 239–247. https://doi.org/10.1016/j.jclepro.2008.05.001

48. Ferrero F, Periolatto M, Rovero G, Giansetti M (2011), 'Alcohol-assisted dyeing processes : a chemical substitution study' *J. Clean. Prod.*19(12),1377–1384.https://doi.org/10.1016/j.jclepro.2011.04.008

49. Rott U, Minke R (1999), 'Overview of wastewater treatment and recycling in the Textile Processing Industry', *Water Sci. Technol.*, 40(1), 137–144. https://doi.org/10.1016/S0273-1223(99)00381-9

50. Neill CO, Hawkes FR, Hawkes DL, Lourenc ND (1999), 'Review Colour in textile effluents – sources , measurement , discharge consents and simulation : a review' *J. Chem. Technol. Biotechnol.* 74, 1009–1018. https://doi.org/10.1002/(SICI)1097-4660(199911)74:11%3C1009::AID-JCTB153%3E3.0.CO;2-N

51. Sivri N, Toroz I (2007), 'Pollutants of Textile Industry Wastewater and Assessment of its Discharge Limits by Water Quality Standards' *Tur. Journ. Fish. aqua. Sci.*,7, 97–103. https://citeseerx.ist.psu.edu/viewdoc/download?doi=10.1.1.468.3596&rep=rep1&type=pdf

52. Korucu M E, Gurses M, Dogar C,Sharma SK,Acıkyıldız M (2015), 'Removal of Organic Dyes from Industrial Effluents: An Overview of Physical and Biotechnological Applications', *Green Chem. for Dy. Rem. from Waste Wat. Res. Tredns and Appl.*, Scrivener Publishing LLC,1-34. https://doi.org/10.1002/9781118721001.ch1

53. Singh R, Singh P (2019),'Treatment and Recycling of Wastewater from Textile Industry', *Ad. Bio. Treat. Ind. Waste Wat. their Recycl. Sustain. Future,* Springer, Singapore,225–266. DOI: 10.1007/978-981-13-1468-1_8

54. Prabha S, Gogoi A, Mazumder P, Ramanathan AL(2017), 'Assessment of the impact of textile effluents on microbial diversity in Tirupur district, Tamil Nadu' *Appl. Water Sci.*7(5), 2267–2277.

55. Panthi SR (2003), 'Carbonate chemistry and calcium carbonate saturation state of rural water supply projects in nepal', *Seventh Int. Wat. Tech. Conference Egypt,*545–560. https://citeseerx.ist.psu.edu/viewdoc/download?doi=10.1.1.302.2356&rep=rep1&type=pdf

56. Kaur B, Chanchal (2016),'Textile wet processing : environmental and health impacts' *Int. J. Civ. Struct. Env. Infra. Eng. Res. Dev.*, 6(5), 19–22. http://www.tjprc.org/publishpapers/2-11-1482313598-2.%20Civil%20-%20IJCSEIERD ENVIRONMENTAL%20AND%20HEALTH%20CONCERNS%20OF%20%20 THE%20TEXTILE%20INDUSTRY..pdf

57. Islam I, Mehedy E, Chowdhury S, Sen P, Jannatul H, Mrityunjoy S (2015), 'Physicochemical Analysis of Textile Dye Effluent and Screening the Textile Dye Degrading Microbial Species' *Journ. Env. Sci. Toxic. Food Tech.*, 9(3), 51–55. https://www.iosrjournals.org/iosr-jestft/papers/vol9-issue3/Version-2/H09325155.pdf

58. Zielina M (2010), 'Measurement of suspended particles in water filtration 1 . introduction and literature review',Fourteenth International Water Technology Conference, IWTC 14 2010, Cairo, Egypt, 105–112. https://citeseerx.ist.psu.edu/viewdoc/download?doi=10.1.1.302.3861&rep=rep1&type=pdf

59. Islam MG, Mostafa MR (2019), 'Textile Dyeing Effluents and Environment Concerns A Review', *J. Environ. Sci. Nat. Resource* 11, 131-134. https://doi.org/10.3329/jesnr.v11i1-2.43380

60. Chockalingam S, Banerjee N, Muruhan (2019), 'Characterization of Physicochemical Parameters of Textile Effluents and Its Impacts on Environment'. *Environment and Natural Resources Journal*,17(2),41-53.

DOI: 10.32526/Ennrj.17.2.2019.11

61. Akhtaruzzaman M, Clemett A, Knapp J, Mahmood MA, Ahmed S (2004), 'Choosing an effluent treatment plant',1-34. https://assets.publishing.service.gov.uk/media/57a08cc940f0b652dd001568/R8161-Treatmentplants.pdf

62. Aslam MM, Athar M, Baig A, Qazi IA (2010), 'Textile Wastewater Characterization and Reduction of its COD & BOD by Oxidation' *Bangladesh J. Bot.*, 42, 279-285.

63. Yusuff RO, Sonibare JA (2005), 'Characterization of textile industries effluents in Kaduna , Nigeria and pollution implications', *Global Nest Int. J.* 6(3), 212–221. https://doi.org/10.30955/gnj.000284

64. Taylor P, Correia VM, Stephenson T, Judd SJ (2008),'Characterisation of textile wastewaters □ a review characterisation of textile', *Env. Tech.,* 15,(10),37–41. https://doi.org/10.1080/09593339409385500

65. Ramachandra RE, Chandran TV, Joshi MD, Rajagopal N, Kumar N(2012), 'Water, Soil and Sediment Characterization: Sharavathi River Basin, Western Ghats' *ENVIS - The Ministry of Env. For. and Clim. Change, Technical report* 21. http://wgbis.ces.iisc.ernet.in/biodiversity/pubs/ETR/ETR21/index.htm

66. Gutie M, Vilaseca M,Gutiérrez MC,Victor López-Grimau VL,Mesas ML,Crespi M (2010), 'Biological Treatment of a Textile Effluent After Electrochemical Oxidation of Reactive Dyes' *Wat. Env. Res.* 82(2),176-182, https://doi.org/10.2175/106143009X447902

67. H. Summary, "Chromium Compounds," no. Cr III. (https://www.epa.gov/sites/default/files/2016-09/documents/chromium-compounds.pdf),09/2016

68. Fibbi D, Doumett S, Colzi I, Coppini E, Pucci S, Gonnelli C, Lepri L, Del Bubba M (2011),'Total and hexavalent chromium removal in a subsurface horizontal flow (h-SSF) constructed wetland operating as post-treatment of textile wastewater for water reuse', *Water Science and Technology* 64(4),826–832. https://doi.org/10.2166/wst.2011.548

69. Porter RS (2000), 'Biological treatment of effluent containing textile dyes', *Coloration Tech.*,116(12),393–397. https://doi.org/10.1111/j.1478-4408.2000.tb00016.x

70. Dey S, Islam A (2015), 'A Review on Textile Wastewater Characterization in Bangladesh' *Resources and Environment*,5,115-44. doi:10.5923/j.re.20150501.03

71. Carliell-marquet CM, Buckley CA (1996),'Treatment of exhausted reactive dyebath effluent using anaerobic digestion : Laboratory and full-scale trials', *Water SA*,22, 225-233.

72. Gehr R, Wagner M, Veerasubramanian P, Payment P (2003), 'Disinfection efficiency of peracetic acid, UV and ozone after enhanced primary treatment of municipal wastewater' *Water Research*, 37,4573–4586. doi:10.1016/S0043-1354(03)00394-4

73. Sonune A, Ghate R (2004), 'Developments in wastewater treatment methods', *Desalination,*167, 55-63. https://doi.org/10.1016/j.desal.2004.06.113

74. Ghoreishi SM, Haghighi R (2003), 'Chemical catalytic reaction and biological oxidation for treatment of non-biodegradable textile effluent', *Chemical Engineering Journal,*95(1-3),163-169. https://doi.org/10.1016/S1385-8947(03)00100-1

75. Andleeb S,Atiq N,Ishtiaq Ali M, Razi-ul-hassan R†,Shafique M, Ahmad B,Ghumro PB,Husain M, Hameed A, Ahmad A.(2010), 'Biological treatment of textile effluent in stirred tank bioreactor', *International Journal of Agriculture and Biology,* 12, 256-260

76. Sarayu K, Sandhya S (2012), 'Current technologies for biological treatment of textile wastewater-A review', *Applied Biochemestry and Biotechnology,* 167, 645–661. https://doi.org/10.1007/s12010-012-9716-6

77. Chan YJ, Chong MF, Law CL, Hassell DG (2009), 'A review on anaerobic-aerobic treatment of industrial and municipal wastewater', *Chemical Engineering Journal* 155,1-18. https://doi.org/10.1016/j.cej.2009.06.041

78. Gujer W., Henze M, Mino T, Van Loosdrecht M (1999), 'Activated Sludge Model No. 3', *Water Science and Technology,* 39(1),183-193. https://doi.org/10.1016/S0273-1223(98)00785-9

79. Li B, Zhang T (2010), 'Biodegradation and adsorption of antibiotics in the activated sludge process', *Environ. Sci. Technol.* 44(9), 3468–3473. https://doi.org/10.1021/es903490h

80. Liu Y (2003), 'Chemically reduced excess sludge production in the activated sludge process',*Chemosphere,*50, 1-7. doi: 10.1016/s0045-6535(02)00551-9

81. Georgiou D, Metallinou C, Aivasidis A, Voudrias E, Gimouhopoulos K (2004), 'Decolorization of azo-reactive dyes and cotton-textile wastewater using anaerobic digestion and acetate-consuming bacteria', *Biochemical Engineering Journal,*19,75-79.

82. Bhatia D, Sharma NR, Singh J, Kanwar RS (2017), 'Biological methods for textile dye removal from wastewater: A review' *Crit. Rev. Environ. Sci. Technol.* 47(19),1836-1876. DOI: 10.1080/10643389.2017.1393263

83. Supaka N, Juntongjin K, Damronglerd S, Delia ML, Strehaiano P (2004), 'Microbial decolorization of reactive azo dyes in a sequential anaerobic-aerobic system', *Chemical Engineering Journal,* 99(2),169-176. DO - 10.1016/j.cej.2003.09.010

84. Amar N B, Kechaou N, Palmeri J, Deratani A, Sghaier A(2009), 'Comparison of tertiary treatment by nanofiltration and reverse osmosis for water reuse in denim textile industry', *Journal of Hazardous Matereals.* 170(1), 111-117. https://doi.org/10.1016/j.jhazmat.2009.04.130

85. Matsumoto M. R., Galeziewski T. M., Tchobanoglous G, Ross DS (1982) 'Filtration of primary effluent' *Journal of Water Pollution Control Federation,* 54(12), 1581-1591. http://www.jstor.org/stable/25041764

86. Nawaz MS, Ahsan M (2014), 'Comparison of physico-chemical, advanced oxidation and biological techniques for the textile wastewater treatment', *Alexandria Eng. J.* 53(3), 717-722. https://doi.org/10.1016/j.aej.2014.06.007

87. Apell JN, Boyer TH (2010), 'Combined ion exchange treatment for removal of dissolved organic matter and hardness', *Water Res.* 44(8), 2419-2430. https://doi.org/10.1016/j.watres.2010.01.004

88. Abid MF, Zablouk MA, Abid-Alameer AM,(2012), 'Experimental study of dye removal from industrial wastewater by membrane technologies of reverse osmosis and nanofiltration' *Iranian J. Env. Health Sci. Eng.,* 9-17. https://doi.org/10.1186/1735-2746-9-17

89. Xu Y, Lebrun RE, Gallo PJ, Blond P, (1999), 'Treatment of textile dye plant effluent by nanofiltration membrane', *Sep. Sci. Technol.* 34(13), 2501–2519.

90. Bes-Piá A, Mendoza-Roca JA., Roig-Alcover L, Iborra-Clar A(2003),'Comparison between nanofiltration and ozonation of biologically treated textile wastewater for its reuse in the industry', *Desalination,* 157, 81–86. https://doi.org/10.1016/S0011-9164(03)00386-2

91. Suthanthararajan R., Ravindranath E., Chitra K, Umamaheswari B, Ramesh T,Rajamani S(2004),'Membrane application for recovery and reuse of water from treated tannery wastewater', *Desalination,* 164(2),151–156. https://doi.org/10.1016/S0011-9164(04)00174-2

92. Tang C, Chen V(2002),'Nanofiltration of textile wastewater for water reuse', *Desal.*143(1),11–20. https://doi.org/10.1016/S0011-9164(02)00216-3

93. Anjum M, Miandad R, Waqas M, Gehany F, Barakat MA(2016),'Remediation of wastewater using various nano-materials', *Arabian Journal of Chemistry.* 12(8), 4897-4919. https://doi.org/10.1016/j.arabjc.2016.10.004

94. Bora T, Dutta J(2014),'Applications of nanotechnology in wastewater treatment-A review', *J. Nanosci. Nanotechnol.* 14(1), 613–626. https://doi.org/10.1166/jnn.2014.8898

95. Gangadhar G, Maheshwari U and Gupta S (2013),'Application of Nanomaterials for the Removal of Pollutants from Effluent Streams', *Nanoscience & Nanotechnology-Asia*, 2(2), 140–150. http://dx.doi.org/10.2174/2210681211202020140

96. Kale RD, Kane PB (2019), 'Colour removal of phthalocyanine based reactive dye by nanoparticles', *Groundwater for Sustainable Developement.*, 8, 309–318. https://doi.org/10.1016/j.gsd.2018.11.007

97. Kale RD, Kane PB (2018), 'Synthesis of PVP stabilized bimetallic nanoparticles for removal of azo based reactive dye from aqueous solution', *Sustain. Chem. Pharm.* 10,153–162.

98. Kale RD, Kane PB (2017), 'Colour removal using nanoparticles', *Text. Cloth. Sustain.*,2,1-14. https://doi.org/10.1186/s40689-016-0015-4

99. Ademoroti CMA, Ukponmwan DO, Omode AA (1992), 'Studies of textile effluent discharges in nigeria', *Int. J. Environ. Stud.*,39(4),291-296. *https://doi.org/10.1080/00207239208710704*

100. Indian Standard,(1980), Guide For Treatment And Disposal Of Effluents Of Cotton And Synthetic Textile Industry, *Water Sectional Committee*, CDC 26, https://law.resource.org/pub/in/bis/S02/is.9508.1980.html

13

Right First Time (RFT) for process sustainability

Ashok Athalye
Department of Fibres and Textile Processing Technology, Institute of Chemical Technology,
Nathalal Parekh Marg, Matunga, Mumbai -400019
E-mail : ar.athalye@ictmumbai.edu.in

Abstract: Textile is considered as the basic need of mankind, and its consumption is increasing day by day with the growing world population. Textile manufacturing is considered to be the mother of industrialisation, and it is one of the largest global industrial sectors. Apart from the basic requirements of product quality, supply consistency and eco compliance, the major differentiating factor is its cost-effectiveness. The manufacturing cost economy can be achieved by ensuring strict conformance to the Product-Process-Performance parameters and avoiding any alterations, repetitions, and resource wastage.

13.1 Introduction

The global human population is over 7.9 bn and growing steadily; the present per capita consumption of textile material is about 11 kg per person per year and increasing steadily by about 4.6 % CAGR. The textile and clothing trade has already crossed the USD 1000 bn mark, and the increasing young population, demographics, rapid urbanisation, and emerging fashion trends are considered to boost future growth.

The textile market is broadly divided into two segments – commodities and specialities. About 80% of the trade is in regular daily use apparel clothing, home furnishing as well as commonly required technical textiles, while the remaining 20 % of trade is considered to be in the speciality value-added niche segment which mainly covers *haute couture* as well as all textile material produced with intended performance effect [1-4].

The major competition is in the large volume commodities segment, where the challenges faced by the textile manufacturer are in terms of meeting the end-user expectations in terms of product quality, supply consistency, on-time delivery and adherence to regulatory eco compliances. Sustainability parameters, ecological impact in terms of carbon footprint and water footprint generated during the entire textile supply chain, cradle-to-grave as well as cradle-to-cradle product life cycle analysis, circular economy, etc., are used as common buzzwords [5-7].

At the same time, to survive the cut-throat competition, it is essential to maintain cost-effectiveness and provide value for money to the demanding consumers. Essentially it amounts to the "survival of the fittest" to operate and sustain in this market segment [8-10].

The commonly adopted methods to maintain such competitiveness are

- Provide appropriate product, at the required place, at the desired price
- Derive benefits arising from economies of scale through increased productivity
- Operate closer to the market to reduce logistics and transport cost
- Adopt the latest technology to enhance product output and process yield
- Minimise manufacturing bottlenecks and ensure uninterrupted production
- Avoid operational inconsistency, production loss, work repetition and utility wastage

Thus, the major area which is well within the control of a textile manufacturer is to avoid production losses and supply delays due to the internal factors affecting the non-conformance of the product quality and the resultant poor product reproducibility.

Within the entire value chain of textile manufacturing, from fibre to fashion, wet processing is considered to be the weakest link owing to highly complicated technologies involving an interplay between various chemical, mechanical, thermal, electrical and environmental aspects. The complexities involved during textile processing include -the type of fibre and its construction, the extent of impurities present, its physico-chemical characteristics, quality of input water, the machinery involved, temperature gradients, type of basic as well as auxiliary chemicals used, specific application needs of different type of colouring components, processing sequence, machinery used, thermo-mechanical finishing treatment, etc.

The textile industry comprises a diverse and fragmented group of establishments that converts fibre into fabric and then follows a complex set of processing steps to appropriately finish the garments as per the fashion trends. During this process, the textile material is subjected to various chemical, mechanical and thermal processing treatments that involve the application of colours at various stages of garment manufacturing [11-12]. Thus, the important factors demanding quality consistency are

- Modernisation and automation
- Fast-changing consumer tastes
- Stringent demand as per international brands and retailers
- Conservation of water and energy

While the challenges faced by the textile processor are

- Increased productivity
- Ability to process 'Short Run'
- Quick lab-dip for customer approval
- Robustness of process parameters
- Higher fresh packing realisation

In order to achieve such consistency in product quality under the optimum consumption of necessary utilities, the textile processor needs to ensure the Right First Time (RFT) as well as Right Every Time (RET) approach to be able to deliver On-Time-In-Full (OTIF) supply of goods to meet the end-use competitive price point.

13.2 Right first time approach

This is a concept that helps ensure adequate and consistent product quality under optimum resource consumption and provides a timely response for achieving maximum productivity. Thus, it reduces the avoidable wastages, unnecessary re-processing as well as input utility costs and helps increase process efficiency, thereby conferring desired profitability. Therefore, the RFT approach makes the process economical, eco-friendly, and energy-efficient by producing the goods with desired quality that meet the target quality norms at the "Right First Time".

13.3 RFT for water and energy conservation

Textile manufacturing consumes a considerable amount of water and energy during various stages of processing. In case of any deviations in quality requiring partial or full re-processing, again, the large volume of water and energy gets re-consumed which not only increases manufacturing cost but impacts the critical environmental sustainability norms. Owing to all these concerns, the best and wise way for the textile manufacturer is to strictly adhere to the RFT concept.

When the target product quality parameters are met right at the first time of processing, it automatically eliminates the unnecessary consumption of water and energy for the re-processing part, thereby saving time, manpower, and machinery utilisation. Generally, the adherence to obsolete or outdated technology is also considered as one reason for quality inconsistency resulting in product re-processing which exorbitantly increases the inputs of dyes and chemicals, shooting up the cost of production. Thus, the percentage of RFT is directly proportional to the savings in the extra cost of the process.

The advantages and benefits arising from the Right First Time approach are

Financial

- Increased productivity
- Reduced costs or manufacturing
- Improved profit margin

Non-financial

- Consistent quality
- Enhanced customer satisfaction
- Reduced environmental impact

The poor RFT performance is directly linked with a poor economy and the resultant impact is in terms of

- Increase in utility cost
- Increase in chemical cost
- Decrease in the selling price of the product

RFT approach requires proper selection of raw material, machinery, technology, process sequence, quality control and skilled manpower to manage the critical process parameters.

Raw Material (RM) – this is the basic component of the textile manufacturing process, and it plays a vital role in determining the quality of the final product.

Process Sequence (PS) – the process yielding non-reproducible results would directly impact the cost of manufacturing. Thus, the practical viability, as well as adoption based on prevalent machinery in use, should be considered.

Process control (PC) - the parameters need to be properly controlled and may be classified as specifically assignable and other random variables. The assignable components include the pH of the fabric, the volume of water used, the gradient of temperature involved, the extent of electrolyte used, and the dosing/addition sequence of the dyes and chemicals. These can be measured by employing various process control checkpoints and by using a microprocessor controller. The random variables, on the other hand, are the impurities in the textile substrate, quality of water, colourants and chemicals. Control of these random variables requires the intervention of the management [13,14].

13.4 Reproducibility for increased productivity

In the coloured textile material, inadequate shade reproducibility is the main reason for the production loss and increasing processing cost. Generally, when a dyed/printed colour does not get reproduced and requires a corrective action such as the addition of an extra quantity of colourant or chemical, extra machine run time, additional dye bath heating or cooling, stripping of faulty dyeing, re-dyeing or over-drying. Such an additional processing step for corrective or repair procedures requires extra time and utilities which further increases the risk of physical damage to the textile material and resultant increase in effluent load.

Such re-processing and addition/alterations are not acceptable options, and therefore the RFT processing is essential. The financial significance of such shade corrections and re-processing has a direct and major impact on the overall cost of dyeing. This assumes the approximate ballpark cost of RFT as a baseline figure for simplicity of calculation. Table 13.1 indicates a tentative increase in cost due to loss in productivity and impact on the profitability of the manufacturing unit due to such nonconformity to RFT dyeing (exhaust method).

Table 13.1 The cost of non-reproducibility

Process	Cost/kg	Productivity %	Profit/(Loss) %
Right first time (RFT)	100	100	100
One Recipe correction	110	80	50
Multiple additions	135	40	(45)
Stripping and redyeing	206	10	(375)

From the above representative data, it is evident that even a single corrective dye addition impacts up to a 10% increase in the total cost and a resultant 20% decrease in productivity. It thereby reduces the profits by almost 50% as compared to the right-first-time dyeing. This clearly explains the reason behind the reduction in productivity and loss in revenue earning, which occurs when machines that should be regularly processing the production lots are occupied with corrective treatments. These factors also have impact on the 'designed capacity' of the plant and the capital investment required.

To achieve RFT processing using blind dyeing techniques, a high level of reproducibility must be obtained in the dyeing and related processes. For this, a number of critical factors need to be considered, thoroughly observed and carefully controlled.

The major factors which control RFT are summarised as given below:

1. Materials and inputs

(i) Substrate
- Quality / characteristics
- Absorbency
- Core pH
- Residual chemical components
- Whiteness/colourimetric co-ordinates

(ii) Dyestuff
- Tinctorial strength and moisture content of dyes
- Adulteration / impurities
- Sensitivity to process variables
- Compatible dye selection
- Number of dye components in the dyeing recipe
- Metameric index of the dye combination

(iii) Water and Steam

Table 13.2 Quality and impurities

Characteristics	Permissible limit
Colour	Colourless
Odour	Odourless
pH (1% Solution)	7 – 8
Hardness	< 5° dH
Organic component	< 20 ppm ($KMnO_4$ consumption)
Inorganic component (salt)	< 500 ppm
Iron (Fe)	< 0.1 ppm

(iv) Chemicals
- Types of basic and auxiliary
- Strength and purity
- Amount/weight dosing

2. Machinery and equipment
- Avoid leakage – from valves
- Monitor location of temperature sensor
- Optimise the performance of pump/heat exchanger at operating temperature
- Ensure accuracy and timely calibration of

- Level gauge
- Weighing balances
- Temperature and pressure indicator
- Liquor flow controller
- Spectrophotometer
- Glassware

3. Process conditions

- Material to liquor ratio
- Temperature of input water
- Temperature during dye fixation
- Increase in the temperature gradient
- Concentrations of dye, electrolyte, alkali and auxiliaries
- Addition profile of dyestuff and auxiliary chemicals
- Conductivity of the dye bath
- Addition profile of alkali
- Time–temperature profile at each processing stage
- Comparative material load size and liquor level
- Flow of dye liquor and its reversal sequence
- Post dyeing wash-off sequence/conditions
- Method/condition of material drying

4. Methods and practices

- Frequent changes in the supply of dyes/chemical
- Inadequate storage of dyes/chemicals
- Improper marking/labelling of dyes/chemicals
- Inaccurate weighing/ faulty balances
- Improper location of weighing balance, airflow or machine vibration causing turbulence
- Spillage of dyes /chemicals outside the machine
- Inadequate dye dissolution
- Contamination of dyes
- Improper dye strength evaluation and colour preparation
- Calculation errors leading to inaccurate dye recipe formulation
- Faulty dye/chemical dispensing
- Inadequate conditioning of the sample before colour assessment
- Improper drying of sample for colour assessment
- Soiled fabric sample for shade comparison
- Dots/fluff around the sample area

- Inaccurate database for computerised recipe prediction
- Poor housekeeping and post dyeing operations
- Improper training/understanding of machine operations
- Negligence, the lethargy of the shop floor team
- Unsafe practices and conditions

13.5 Achieving cost-effectiveness through RFT

The awareness is growing in terms of the need for Right First Time (RFT) for the cost optimised processing and increasing production capacities. The RFT production allows the above concepts to be met by shortening the manufacturing cycle from concept to delivery, but failure to obtain RFT production also incurs serious financial penalties. The RFT concept is introduced mainly to reduce the unintended and unnecessary wastage of dyes, chemicals, energy, machinery utility and production time. Such wastage increases the total cost of processing which unnecessarily gets added to the substrate on and above the minimum cost at which the fabric would have been processed. The impact of non-conformance to RFT in terms of an overall increase in the cost of dyeing is shown in Fig. 13.1. RFT, in short, proves to be a cost-effective method to achieve the target quality with less waste. The indicative financial implications and the benefits in terms of productivity enhancement are summarised in Table 13.3.

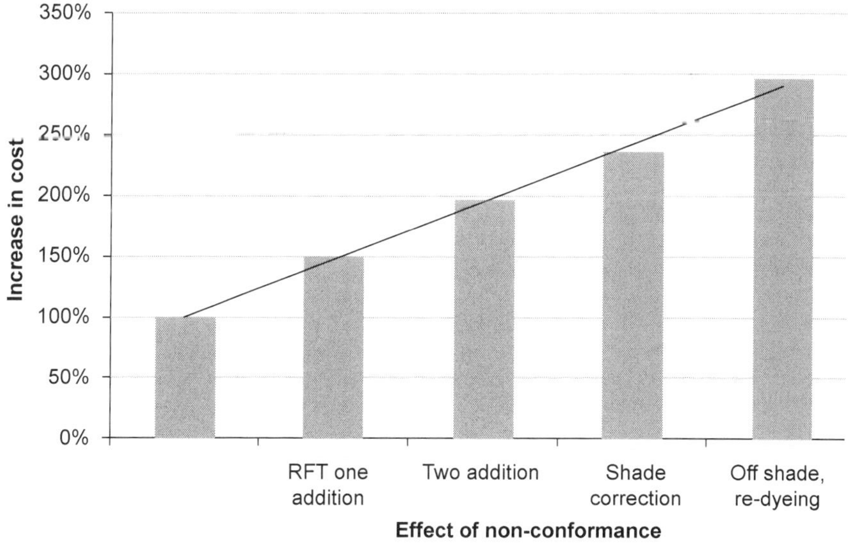

Figure 13.1 Effect of non-conformance to RFT on the cost of dyeing

Table 13.3 Impact on increase in productivity and reduction in manufacturing cost

Increase in Productivity	10%	20%	30%	40%
Reduction in total cost %	9	17	25	33
Additional batches/week	1.4	2.8	4.2	5.6

The factors affecting RFT processing of dyeing operation are summarised in Table 13.4. Similar parameters, would be considered for the textile preparation and finishing treatment and the necessary limits of accuracy are determined. Such relevant parameters should be incorporated in the standard operating procedures (SOP) and work instructions for laboratory as well as bulk production processes and must be followed-up rigorously. Most of these factors are referred to as 'assignable variables', which should be monitored through the SOP, and there would be relatively few factors known as 'random variables', which would require regular evaluation. The random variables include the quality of the textile substrate, water, colours and chemicals. It is often the failure to monitor these random variables that prevent textile processors from achieving a high level of RFT production through blind dyeing techniques [15].

Table 13.4 Important factors for RFT processing

Factor	Controlled by laboratory testing	Monitored by standard operating procedures
Input components		
Water purity	Yes	Yes
Pretreatment of textile material	Yes	
Dyeability of a textile substrate	Yes	
Standardisation of dyes /chemicals	Yes	
Dyeing process control		
Compatible dye selection	Yes	
Dye recipe optimisation	Yes	Yes
Weighing of material		Yes
Weighing/dispensing of dyes/chemicals		Yes
Control of pH, liquor ratio		Yes
Time/temperature profile		Yes
Liquor flow/ substrate circulation		Yes
Method of application		
Colour control (dyeing/printing)		Yes
Dye combination		Yes
Accuracy of laboratory dyeing recipe		Yes
Scaling of the lab to bulk recipe transfer		Yes
Batch to batch reproducibility	Yes	Yes
Method of colour assessment	Yes	
Determination of metameric index	Yes	

Water is an essential input component of textile wet processing, and the quality of textiles produced is profoundly affected by the quality of water. The amount and the type of water-born impurities vary depending on their source. Usually, the cationic components - alkali earth metal ions (calcium and magnesium carbonates), heavy metals (iron, copper, manganese) and the anionic components like sulphide/halides form a major part of impurities. The other unwanted components in water are the sediments, clay, suspended matter, oil and grease [16].

Table 13.5 Effect of water quality in textile processing

Concern	Probable reason
Inadequate removal of starch-based size	Alkali earth metal ions Heavy metals
Harsh hand-feel of the substrate	High solid content in rinsing bath
Inconsistent absorbency after scouring	Water hardness Alkali and alkaline earth metals
Soil deposition on the substrate	Water hardness Alkali and alkaline earth metals Greasy contaminants
Reduced dye solubility	Water hardness Electrolyte content of water
Ineffective detergency of surfactants	Alkali and alkaline earth metals Residual electrolyte content
Fibre degradation, strength loss and reduction in whiteness	Alkaline earth metals (other than magnesium) Heavy metal ions
Inconsistent shade reproducibility	Heavy metals Uncontrolled pH fluctuation Presence of optical brighteners
Blotches and uneven washing off	Precipitation due to the presence of coagulants, flocking agents and organic contaminants
Dye filtering in package dyeing	High salt content in the washing bath
Dye stains and/or spots	High salt content in the washing bath Metallic impurities
Loss in colour depth	Calcium ions
Inadequate removal of unfixed dye	Calcium ions, electrolytes
Decrease in the colour wash fastness	Calcium ions
Foaming tendency	Residual surfactants
Corrosion of machine	Dissolved carbon dioxide, dissolved oxygen

13.6 Laboratory support

The laboratory is called the 'nerve centre of a dye-house'. It plays a pivotal role in establishing the limits of accuracy required to obtain a high degree of reproducibility. It also helps in establishing and maintaining the SOP for bulk production and to ensure that a high success rate of RFT production is achieved. The dye house laboratory is expected to provide adequate support through close coordination with the production departments. The laboratory has the responsibility of developing standard procedures, monitoring checkpoints and ensuring satisfactory performance. It also needs to ensure that all processes and related data are adequately documented and properly stored [17].

The major functions of the support laboratory are as follows:

- Prepare an adequate database
- Test and select suitable/compatible products
- Develop desired colour matching
- Analyse input dyes, chemicals and other raw materials
- Monitor - work in progress and finished products
- Evaluate finished products against a specification
- Develop new testing methods as per emerging market demand
- Develop new processes and products
- Upgrade standard operating procedures (SOP) as per the need
- Investigate faults and complaints
- Provide technical support to sales and marketing

To obtain RFT production, the following three major objectives must be met:

- High degree of accuracy and reproducibility
- Accurate transfer and scale-up for bulk processing
- Reproducibility between the production lots

Most of the important factors for achieving reproducible dyeing and RFT production can be controlled by means of SOP, whilst relatively few require regular monitoring in the laboratory. Routine evaluation of water quality, substrate dyeability against a standard and the quality of incoming dyes against a standard, particularly for strength and hue, must be monitored for RFT to be achieved. It is often the lack of this monitoring that is considered to prevent the textile processing mills from achieving the desired level of RFT production.

The technical staff must establish SOP for bulk production, and these must be translated into SOP for laboratory dyeing. Inadequate preparation of SOP may result in variations between lab and bulk procedures, the factors most commonly at variance being listed below:

- Incoming water supply
- Dyeability of textile substrates
- Quality of dye supply
- Method of dye application
- Consistent material to liquor ratio
- Specific dyebath pH
- Time/Temperature profile
- Adequate weighing of substrate
- Method of colour evaluation

The global textile apparel and clothing brands aim to ensure that the products they promote and deliver meet the consumer expectations pertaining to the aspects of quality, performance reliability, sustainability, and conformance to ecological norms. This is achieved by meeting the standard specifications devised by the supply chain regulators and the various stakeholders like textile processors, colour/chemical suppliers, garment makers, fashion designers and the local governmental regulatory bodies, by following desired norms and ensuring strict control over the product standardisation.

13.7 Product and process standardisation

Dyestuffs and pigments form a major part of textile colouration for enhancing the aesthetic appeal, and different ranges of dyes are used for dyeing different types of fibres in varied forms and on different machines and by different application methods. Thus, the natural, regenerated, and synthetic textile materials in woven, knitted and nonwoven forms are processed by a batch, semi-continuous or continuous method on a suitable application machine.

Generally, it is observed that different colourant manufacturers provide a dye or pigment of the same colour index, but it does not provide the same application performance in terms of expected colour shade and strength. This is probably dependent on the extent of impurities and isomeric components present in the dye, which is based on the varying raw materials, route of manufacturing, machine material of construction and the method of synthesis. Thus, the textile processor needs to check and evaluate dyestuff before bulk usage so as to meet its application needs.

The textile processor in tern demands strict quality consistency and supply of standardised dyestuff from the colourant manufacturer. The importance of product and process standardisation is believed to be in terms of receiving a uniformly consistent quality of supply of input components, getting uniform and reproducible results of the manufactured goods, increasing the productivity of textile material by avoiding undesired shade rejections and achieving an overall reduction in the consumption of water and energy at the same time reducing overall processing time without affecting the intended results. Furthermore, the ecological factors related to the generation as well as reduction of solid waste, decreased liquid effluent load and minimal gaseous emission are important parameters for the overall consideration. The standardisation tests involve

- Analysis of incoming basic chemicals
- Conformance specification of intermediate components
- In-process testing to ensure the consistency during dye/chemical production
- Performance evaluation for optimising colour yield and achieving desired fastness properties

The standardisation of colourants and chemicals involves the use of sophisticated instruments and testing methods which include physical aspects, chemical analysis, fastness properties, biodegradability, toxicological hazards, effluent treatment requirements, and then a comparative analysis is made against an industry-accepted internal or the consumer specified external standards.

Physical characteristics – generally, synthetic dyes are manufactured and provided in various physical forms. Therefore, maintaining consistency in the physical aspects of the product and ensuring consistent quality supply is critical for its effective usage and consumption.

(i) Appearance – the dyes and chemicals are supplied in different physical forms like flakes, powder, grains, beads, thick paste, free-flowing liquid, etc., and the consistency in supply from batch to batch is maintained as per the approved standard.

(ii) Aspect – this relates to the measurement and maintenance of specified forms of the product, like bulkiness or the fluffiness of the powder, shape and size of the granules, viscosity of the paste and density of the liquids.

(iii) Colour – the appearance of the product is determined in terms of whether it is lighter or darker, brighter or duller. This is necessary as it impacts the impression of the unskilled shop floor worker.

(iv) Odour – during manufacturing, if any unintended or un-reacted component of the raw material or the dye intermediate remains in the final product or subsequently if any other ingredients are added for dilution, dispersion, emulsification, pH adjustment, reduction/ oxidation prevention, sequestration, etc. or for the prevention of any bacterial/fungal growth when the product is supplied and stored in a paste or liquid form then depending on the conditions and duration of such storage there is a possibility of formation of a gaseous by-product and a resultant offensive or obnoxious smell.

(v) Moisture content – usually, the colourants are supplied and sold on a weight basis; when it is in the solid form, either as powder or flakes, it tends to absorb moisture from the surrounding environment depending on the storage facility and the climatic conditions which may vary depending on the prevailing atmospheric humidity, this tends to affect the weighing accuracy of such products during their actual usage.

Chemical analysis – for ensuring fault-free manufacturing efficiency and productivity, characterisation and standardisation is carried out using various chemical analysis and instrumental evaluations with the help of internationally approved and specified standard methods. Some of these instrumental methods involve

(i) Chromatographic techniques – for the identification, differentiation and separation of the chemical components in a product, different methods are employed depending on the type of colourant. Such evaluation involves the use of various chromatographic instruments like gas chromatography (GC), thin layer chromatography (TLC), high-performance liquid chromatography (HPLC), etc.

(ii) Spectroscopic techniques – this method of analysis involves the use of various sophisticated instruments, such as ultraviolet-visible (UV-Vis) spectroscopy, infrared (IR) spectroscopy, Fourier transform infrared (FT-IR) spectroscopy mass spectroscopy and transmission spectroscopy to evaluate the product being standardised.

(iii) Analytical techniques – apart from the analysis of the chemical elements in the product, the qualitative and quantitative assessment of alkali and heavy metal ions are needed. This is performed by sophisticated and high-precision instruments like atomic absorption spectroscopy (AAS). Further, for the determination, estimation and quantification of the particle size distribution of dye/pigment, particle size analyser (PSA) is used, while the use of a dust analyser is needed for the evaluation of the dusting behavior of dye powder.

Gravimetric analysis – is recommended for the quantitative determination of the main colouring chemical component as well as estimation of the type and extent of other ingredients or impurities in the product. These methods involve estimation and evaluation of ash content, determination of the extent of electrolyte content, amount of the water-insoluble material, etc. [18-20].

Tinctorial specification – this is the main criteria for the determination of the colouring components during its standardisation. For this, the commonly used methods are

(i) Visual assessment –this is the first and foremost parameter of colourant estimation, where the shade and strength of the dye under evaluation are visually compared against the standard under varying illuminant conditions.

(ii) Optical density – in the case of water-soluble dyes, the dissolved dye is evaluated by spectrophotometer and compared against the standard for its intrinsic colourimetric estimation.

(iii) Instrumental measurement –the quantitative assessment of the actual amount of colourant on the textile material is best performed by the computer colour matching (CCM) measurement. This instrumental analysis measures the reflectance data and monitors the colour electronically. The spectrophotometers are for assessment and standardisation within predetermined limits for approval or rejection of the manufactured dye. Generally, the internationally acceptable tolerance limits for the colour strength variance are within ± 2.5 %, and the shade deviation by dE value is expected to be less than 1.0 by the CIE Lab measurement system.

Apart from the usual characterisation tests like the determination of the ionic nature of the product and the pH of the aqueous dye solution, many other application performance tests are carried out depending on the type of colourant, intended substrate, machinery used and the method of processing. Given below are a few test methods with their brief description:

(i) Solubility parameter - the water-soluble dyes have an inherent tendency and extent of aqueous solubility which varies depending on the dye characteristics. This is a critical parameter from the applicability and textile processing point of view depending on the type of substrate, machinery used, material to liquor ratio employed and the intended method of application. This is measured by dissolving the specified quantity of dye in distilled water and then spotting it on a filter paper of a specified type to check whether the dye precipitates and deposits at the centre or spreads uniformly in a wide circle. The results are evaluated and compared against the standard product [21-23].

(ii) Temperature - the effect of kinetic energy in terms of the extent of improvement in the aqueous solubility of the dye is measured and estimated under varying temperature conditions starting from ambient to atmospheric boiling conditions.

(iii) Electrolyte – the effect of the addition of different types (common salt, Glauber's salt) and concentrations of salt is determined considering the exhaust application method.

(iv) Alkali – the dyes intended for the batch-wise, semi-continuous and continuous application of reactive dyes, the stability of the colourant in a different type of alkali (sodium carbonate, caustic soda, sodium silicate) is evaluated, assessed, and compared.

(v) Strength and shade evaluation – the colour (K/S) value and the tone of the dye are compared against the standard product by the actual method of application on the specified textile substrate. Generally, the comparison is made at different concentrations covering low, medium and dark (LMD) depths. Generally, the lower depth is checked for the colour tone or hue, the middle depth is assessed for the strength evaluation and the higher depth for the dye build-up properties.

(vi) Exhaust (batch-wise) dyeing - the specified textile material is dyed as per the recommended and intended bulk dyeing method, maintaining the various dye bath parameters like material to liquor (MLR) ratio, the temperature of application, time of dyeing, concentration, and the dosing sequence of basic chemicals (salt, alkali, reducing agent) and the dye bath additive auxiliary chemicals (dispersing, levelling, sequestering, pH controlling). The dyeing is followed by an after-treatment process involving neutralisation of pH, soaping, oxidation, reduction clearing, etc., depending on the type and concentration of the dyestuffs. Finally, the substrate is dried, conditioned to regain atmospheric moisture and assessed for its comparative shade and strength [24-27].

(vii) Semi-continuous and continuous dyeing- depending on the type of textile substrate, class and sub-class of dyestuff, machinery used and the method of application, different methods like cold-pad-batch (CPB), pad-steam (PS), pad-dry-pad-steam (PDPS), pad-dry-thermosol-pad-steam (PDTPS), etc., are employed.

(viii) Fastness assessment – from the consumer point of view, uniformity, levelness, durability and colourfastness are the major expectations. The bleeding and cross-staining of colour from one part of the dyed/printed portion to the adjacent differently dyed or undyed part affects the aesthetic appeal of the garment and spoils the fabric's appearance. For this, various comparative colourfastness tests are performed which include washing under different temperature/time conditions, exposure to different light sources for varying time duration, dry and wet rubbing, acidic and alkaline perspiration, saliva, etc. These tests are also performed as per the method of testing specified by the brands or retailers, and the methods include parameters specified as per the International Organization for Standardization (ISO), American Association of Textile Chemists and Colorists (AATCC), Marks and Spencer (M&S), etc. Depending on the requirement and the expectation of the end-user, these tests can be carried out internally by the colour manufacturer as well as from the externally specified certified and accredited facility.

(ix) Storage stability – from the time the colourant is manufactured till it is actually consumed by the textile processor, there could be a considerable time lag; this could be due to the storage and transportation while logistical distribution and shipping delays during overseas dispatches. Further, depending on the seasonality and the location of the stock points, there could be an effect of varying temperature and humidity conditions. Therefore, it is important to evaluate and estimate the extent of storage stability and expected shelf life of the product.

(x) Effluent related – after completion of dyeing/printing and post washing off process, the amount and the concentration of coloured effluent generation depends on the extent of leftover dyes which are unfixed or unreacted at the end of the textile application. The superficially deposited dyes get removed during the washing-off process to achieve the intended fastness properties, while the removed colourant gets transferred to the effluent. This thus necessitates an adequate effluent treatment process that depends not only on the decolourisation of the water but also on the removal and degradation of the chromophoric chemical components of the dye. The tests involve evaluation and estimation of total suspended solids (TSS) during the primary treatment, chemical oxygen demand (COD), biochemical oxygen demand (BOD), total organic content (TOC), absorbable organohalogen (AOX) during the secondary or biological treatment while the total dissolved solids (TDS)and the extent of sludge formation during the tertiary effluent treatment stage [28-31].

(xi) Sustainability – here, various issues related to the ecological and toxicological parameters and the associated evaluation tests for the chemical ingredients present in the colourant and chemicals are carried out to assess their impact on the safety, security and health of the human being, other animals and aquatic life along with the effect on the environment. The determination of the type and extent of heavy metal ions present in dyes and the acceptable tolerance limits, presence of harmful banned amines, phenolic compounds, non-biodegradable dye components, etc., needs to be evaluated and their conformance ensured. From the end users' point of view, the toxicological and dermatological effects like skin irritation, carcinogenicity and impact on biological immune systems, are the major concerns. This is governed by the norms, test methods and tolerance limits specified by various authorities. The parameters and specification limits vary and depend on the end-use requirements related to health and hygiene.

Therefore, the regular and critical analysis, testing, evaluation and standardisation of input colour/chemical components during textile processing needs to be ascertained for achieving desired RFT, RET and OTIF performance.

13.8 The role of chemicals and process parameters in RFT

The final textile article, be it a ready-to-wear apparel garment, home textile or a technical performance material, undergoes many stages of processing [32-35]. The steps involved (Fig. 13.2) are dependant on various factors such as

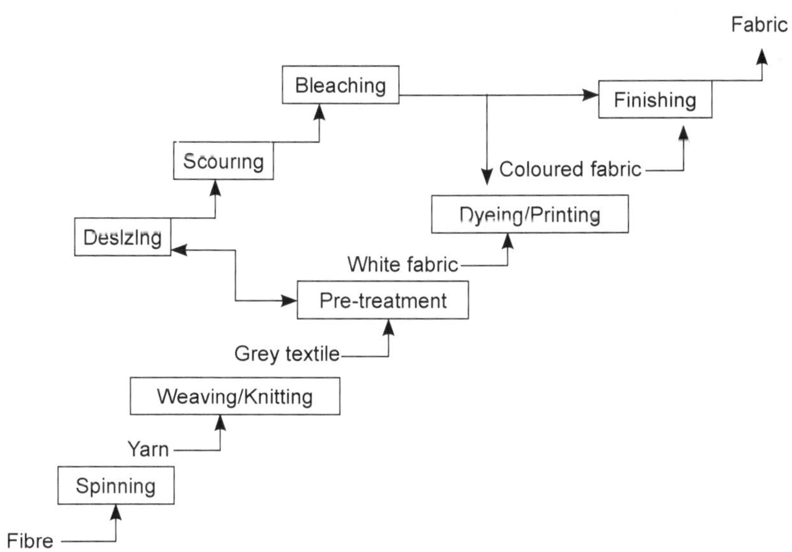

Figure 13.2 Various processes involved during the conversion

of fibre to finished fabric

- Origin of fibre – plant, animal, synthetic
- Form of material – yarn, woven, knit, towel, garment, carpet
- Method of processing – batch, semi-continuous, continuous
- Machinery used – liquor ratio, extent of turbulence, temperature and pressure

The objective of these processing steps is aimed at

- Controlling the residual moisture content in fibre
- Achieving absorbency desired for the subsequent process
- Obtaining and maintaining uniform whiteness and brightness
- Optimising extraction and removal of undesired component/ impurities
- Minimising fibre damage and loss in the tensile strength
- Maintaining uniform and consistent pH from core to the surface of the fibre

The textile preparatory process is highly water, energy and chemical-intensive. The desired and expected outcome is possible only when adequate process parameters are set. This involves the use of basic chemicals -alkali (caustic soda, soda ash), acid (acetic, formic), Redox chemicals (hydrogen peroxide, etc.) and auxiliary chemicals (sequesterant, stabilisers, dispersants, enzymes) categorized based on the intended purpose or processing step for which they are being used. A broad list of chemicals generally used during cotton processing is given below.

- Enzymatic desizers
- Polymeric dispersant
- Surfactants and detergents
- Demineraliser and sequesterants
- Peroxide stabiliser and bleach cleaners
- Deaerators and defoamers
- Buffered pH controllers
- Levelling and retarding agents
- Dye bath controllers and migration inhibitor
- Diffusion accelerator and penetrants
- Lubricants and crease controllers
- Washing-off agents and dyefixers
- Reducing and oxidizing agents

- Anti-back staining agents
- Colourfastness enhancers

At the end of the textile processing, after the intended application performance, the basic as well as auxiliary chemicals need to be removed from the textile and drained into the effluent stream. This necessitates an understanding of their disposal as well as their effect on solid waste generation, liquid effluent formation and gaseous emission.

Concluding remarks and future perspectives

The application of effective process control is essential in ensuring the right product quality and consistency of textile processing. It helps achieve a high degree of reproducibility, both in the laboratory as well as in bulk production. This results in RFT processing and blind dyeing techniques eventually attained. RFT production makes major savings for the textile processor, contributing significantly to its quality aspects and consistent production. The resultant benefit is in terms of reduced cost and improved profitability.

References

1. https://www.worldometers.info/population/ (accessed Dec 2020).
2. www.thefiberyear.com/fileadmin/produkte/Part3. (accessed Dec 2020).
3. https://www.technopak.com/textile-apparel-compendium-2021/(accessed May 2021).
4. www.oica.net/category/production statistics. (accessed Jan 2020).
5. www.chemicalland21.com. (accessed Jan 2020).
6. www.technical-textiles.net/new (accessed Jan 2021).
7. Chow CK (2009), 'Estimation of carbon emission during textile manufacturing and its use', *Textile Asia* 3 (1), 23-27.
8. Athalye AR (2012), 'Carbon footprint in textile processing', *Colourage* 59(12), 45-47.
9. Ruparelia A (2010), 'Home textiles-a promising future', fibre2fashion.com. (accessed Dec 2020).
10. www.marketsandmarkets.com/reports. (accessed May 2021).
11. www.ibef.org/industry/textiles. (accessed May 2021).
12. Lüscher M (1973), *Color-the mother tongue of the unconscious*, Parke-Davis gelatin capsules: monographie, volume 5, Capsugel NV.
13. Collishaw PS, Phillips D, Bradbury MJ (2008), 'Controlled coloration: a success strategy for the dyeing of cellulosic fibres with reactive dyes', *J. Soc. Dyers Colour.* 109, 284-292.

14. Bradbury MJ, Collishaw PS, Moorhouse S (1995), 'Dynamic response – process optimisation in the exhaust dyeing of cellulose', *J. Soc. Dyers Colour.* 5, 130-134.

15. Park J, Shore J (2009), 'Evolution of right-first-time dyeing production', *Color. Technol.* 125(3), 133-140.

16. www.textiletoday.com.bd. (accessed May 2021).

17. Shamey R, Hussein T (2005), 'Critical solutions in the dyeing of cotton textile materials', *Text. Prog.* 37(1-2), 1-84.

18. Allen NS (1987), 'Photofading mechanisms of dyes in solution and polymer media', *Rev. Prog. Color.* 17, 61.

19. Stead CV (1990), Halogenated Heterocycles in Reactive Dyes, In *Colourants and Auxiliaries*, Shore J (ed.), volume 1, SDC, Bradford.

20. Oakes J (2001), 'Photofading of textile dyes', *Rev. Prog. Color. Relat. Top.* 31, 21-28.

21. Warren SP (1991), A review of textile dyeing processes. *Textile Colouration and Finishing*, Carolina Academic Press, pp.23-27.

22. Athalye AR (2012), 'Desizers are deciders', fibre2fashion.com. (accessed Nov 2012).

23. El-Molla MM, Haggag K, Mahamoud ZM (2015), 'Dyeing of polyester fabrics using microwave irradiation technique', *Int. J. Sci. Research* 4(4), 80-81.

24. Park J (2013), 'Exploiting right-first-time processing in textile coloration', *Textile today* 6(2), 38-42.

25. www.eia.doe.gov/emeu/aer/txt/ptb1204.html. (accessed Dec 2020).

26. Athalye AR (2017), 'Protective textile', *J. Text. Assoc.* 78(1), 31-33.

27. www.naturalfibres2009.org/en/iynf/sustainable.html. (accessed Dec 2020).

28. Rupp J (2008), 'Ecology and economy in textile finishing', *Textile World* 4, 73-75.

29. www.domain-b.com/environment/20090403_carbon_footprint.html. (accessed May 2021).

30. Shamey R, Hussein T (2005), 'Critical solutions in the dyeing of cotton textile materials', *Text. Prog.* 37(1-2), 1-84.

31. Park J, Park K (2005), 'Concept to consumer (C2C): latest technology in colour matching', *Colourage* 52, 71-76.

32. Park J (2008), 'Right-first-time production: the control strategy', *Colourage* 55(2), 34-37.

33. Athalye AR (2015), 'Standardisation of dyes', *Colourage* 62(10), 50-53.

34. Athalye AR (2017), 'Colourants- end use perspective', fibre2fashion.com. (accessed Dec 2020).

35. Athalye AR (2018), 'Performance evaluation of textile chemicals', fibre2fashion.com. (accessed Dec 2020).

14

Dry processes for chemical processing of textiles

A K Patra*, Supriyo Chakraborty
U P Textile Technology Institute (Formerly GCTI), Kanpur 128001, India
*Corresponding Author, Email: arunkpatra1@gmail.com

Abstract: Textile processing conceptually is undergoing a paradigm shift with changing environmental, economic and social needs. The global population almost reaching 7.8 billion and with growing affluence in the developing world, the amount of textile materials required and consequently dyes and chemicals used is enormous. The carbon footprints and water footprints of the industrial practices add to the concern. In fact, water is soon to be a scarce resource and hence an urgent need to save it. This necessitates to have completely dry chemical processing of textiles, thus eliminating the use of water, problems of water pollution and their subsequent treatment. The use of plasma in textile finishing is one such step for textile finishing, which has been discussed in this chapter. Similarly, supercritical Carbon dioxide, which has been developed for quite some time now, is being worked out as a livable technology for dyeing of polyester. This doesn't require water at all and hence holds promise primarily for disperse dyeing. As regards textile printing, digital printing has come as a game changer. The transfer printing technology, when used in digital printing, can make the complete operation a dry process. The installations of such printers are growing exponentially in process houses, thus minimising water usage, downtime, inventory requirement, response and development time. This is likely to change the age-old practice of chemical wet processing, although rotary screen printing is still in vogue.

14.1 Introduction

Conventionally, textile chemical processing is considered to be a highly water-intensive operation and many attempts have been made to reduce the water usage in this step of textile manufacturing. Often referred as textile wet processing, this step has essentially been presumed to be a process involving use of substantial quantity of water. Unfortunately, water as a natural resource is getting scarce due to intensified demand. As per one of the surveys, 47% of the global population will face severe water shortage in a decade's time [1,2]. Textile processing not only uses a large quantity of fresh water and chemicals but also discharges heavily contaminated effluent. To overcome this problem, best practices in industrial processing such as counter-current washing, low liquor-ratio dyeing and automated water flow dosing have been adopted over a period of time, while many dye and concept developments have been done by dye manufacturers in the form of high fixation dyes requiring less washing, innovative process cycle etc. Techniques for reusing water by ultrafiltration and reverse osmosis have met with reasonable success in areas where water availability is limited. Although these have paid dividends, in the present context, they appear to be too little and too late. So, the lone alternative is to have a chemical process where water is not used at all. The concept is therefore moving from 'Reduce – Reuse – Recycle' to total 'elimination'.

14.2 Water footprint

The term water footprint owes its genesis to Arjen Hoekstra, a professor of water management at the University of Twente in the Netherlands. Coining it in 2002, he defined the water footprint of an individual, community or business as the total volume of freshwater used to produce the goods and services consumed by an individual or a community or produced by the business. It can be quantified for a particular product, for any well-defined group of consumers (like individual, family, village, city, province, state or nation) or producers such as a public organisation, private enterprise or economic sector. Water footprint can be categorized to the following types.

- Green water footprint: It is the volume of rainwater consumed during the production process.
- Blue water footprint: It is the amount of surface and groundwater used for the production of goods or services.
- Grey water footprint: This refers to pollution of water resources, defined as the quantity of freshwater required for assimilating or diluting the pollutant load got generated by a given process, resulting and maintaining concentrations naturally present and agreed to the water quality standards [2, 3].

As regards textile industry, it is the third largest consumer of water in the world. A large quantity of water is consumed during the various wet treatment operations like desizing, scouring, bleaching, mercerization, dyeing and finishing. About hundred litres of clean water is used to process one kilogram of cotton textiles material, starting from the preparatory to finishing operations. As it is finally discharged it contains effluent contaminated with unused dyes, pigment, acids, alkalies, size and other chemical auxiliaries. In one of the old estimates, it was found that dyeing of textiles consumes more than 9 trillion litres of water every year, which might have increased by now due to higher production. Besides a large amount of water being used, the impact of textile wastewater on the quality of water is also very significant. On completion of the processes, water is returned to the ecosystem if untreated, which may contain harmful chemicals like formaldehyde, chlorine and heavy metals. It has been found that about 72 toxic chemicals that are present in our water are solely from textile colouration processes; 30 of these toxic chemicals cannot be removed. ZDHC (Zero Discharge of Hazardous Chemicals) has been one of the concepts for responsible chemical management [4]. A further improvisation on the process, with stress on the reuse of water, has now led to ZLD (Zero Liquid Discharge), with new state-of-the-art plants coming up with no water discharge from the process house. Interestingly, some of the innovations done earlier are now fast catching up for completely waterless chemical treatment of textiles, thus reversing the conventional idea of wet processing.

14.3 Dry process types

The idea of non-aqueous chemical process is mainly related to plasma treatment, water-less dyeing and digital printing by transfer technique. The first one is aimed at the surface finishing of fabrics by use of plasma. This treatment can be given even to non-textile materials like plastics, paper and board, metals, ceramics etc. One of the major applications has been in micro-electronics. Plasma, often referred as the fourth state of matter, is essentially an ionized gas with an approximately equal number of positive and negative charges. Some common examples of plasma include a neon sign, a lightning bolt and a flame. In waterless dyeing, the use of supercritical CO_2 for disperse dye application is an established technical possibility while digital printing of polyester by transfer of colour from paper to fabric is gradually becoming commercially popular, although the concept of transfer printing was developed decades back.

14.4 Plasma treatment

The concept of plasma as a state of matter was first suggested by Sir William Crookes in 1879. He identified it while working with an electrical discharge tube where the air inside it is ionized by the application of a high voltage through a voltage coil. However, the term plasma was first coined by the American scientist and Nobel laureate Irving Langmuir in 1928. Plasma contains a combination of reactive species such as heavy particles, free radicals, and electrons. Hence it is hailed to be a unique and diverse media for surface modification of various substrates. As regards their use, the main industrial applications of plasma started in the 1960s in the micro-electronics industry. In the 80s, their usefulness was spread to include many other kinds of surface treatments, particularly in the fields of polymers and metals. The subsequent experimental and industrial applications have been in plastics, paper and boards, metals, ceramics and biomaterials. The prospects of good results in the micro-electronics industry, in fact triggered efforts worldwide to processing of textiles and nonwovens using plasma.

The characteristics and properties of textile materials are quite different from those of metal and plastic surfaces in general. Despite the fact that the surface properties of textile material have a negligible contribution to the overall mass of the material, it often influences the performance of textile products and thus creates diverse end applications. The surface properties have an important role in influencing the performance of textile materials. Properties like wettability, water resistance, adhesion, friction, printability, static charge generation, pill propensity and soil resistance are all governed by the surface characteristics of textile materials. The potential application of plasma on textiles is mainly on the surface. The kind and extent of the effect of plasma on the textile substrate depends mostly on the type of interaction between plasma and the textile surfaces [5].

Plasma is produced by coupling of electromagnetic power into a process gas volume and basically is an ionized gas comprising of a dynamic mix of electrons, neutrons, photons, ions, free radicals, meta-stable excited species and molecular and polymeric fragments. Although ionized, it has an equal density of positive and negative charges. When textile substrate is treated with plasma in batch or continuous manner, a variety of modifications like etching, grafting of various functional groups, deposition of conformal coatings, cleaning via dissolution of surface contaminants, and surface activation through breakage of bonds and generation of reactive sites takes place [6]. The advantages of plasma treatment against classical wet processing are as follows:

- No wet chemistry involved, as water is not used. The treatment is based on the excited gas phase only. Hence only negligible water consumption is involved.
- Applicable to all substrates with almost a free choice.
- Optimisation and changing of surface characteristics of the textile materials without real alteration or any negative effect on bulk characteristics. Hence, the reaction takes place locally. In conventional wet processing, bulk of the material is generally affected.
- Works on polymers that are unable or very difficult to modify with wet chemicals.
- Very low chemical consumption due to the physical process.
- Energy consumption is very low compared to the high energy requirement in wet processing.
- Great potential for new processes against literally static technology of conventional water-intensive chemical processing.

- The process being performed in a dry, closed system, is environment-friendly.

14.4.1 Plasma generation and classification

In simple terms, plasma can be generated on application of an electric field over two electrodes with a gaseous medium in between. It is possible to achieve plasma at atmospheric pressure; otherwise, it can be created in a closed vessel at some reduced pressure. In either cases, the characteristics of the plasma are determined by the gaseous medium used, the applied electric power and the type of electrodes [7].

14.4.1.1 Low-pressure plasma

It is a very well-established technology in the micro-electronics industry. People explored its possibilities and many researchers have used plasma techniques for textile and polymeric surfaces for diverse applications [8,9]. In this method, the plasma condition is achieved when a gas is under sufficiently low pressure and at the same time, electromagnetic energy is provided to the gas. In such a condition, the process gas gets partially decomposed into radicals and atoms and also becomes partially ionised. Often, the working gas pressure is kept as low as 0.1 mbar for the radiofrequency range, while it is maintained between 0.5 to 1 bar for microwave sources. The low-pressure cold plasma technology (vacuum plasma technology) involves equipment set up consisting mainly of a vacuum chamber and a pumping unit to evacuate the vessel. There are some valves that operate between the pumping group and the vacuum chamber isolates these two sections. Moreover, a set of mass flow controllers and valves allows the process gas to admit in a precise way. An electrode system powered by electromagnetic generators is placed in the chamber. The whole equipment may be controlled by a microcontroller or a personal computer. The overall cost of the equipment is usually high. The plasma so generated is characterised by its uniformity over a large volume. Nobel gases can be used for plasma generation. Other than these, nitrogen and oxygen gas can also be used, thus widening the scope of use for such plasma. A roll-to-roll system of application on textiles is a possibility, although there are practical difficulties. The roll has to be placed in a vacuum chamber and then the chamber may be pumped to the desired pressure, after which the process can begin [7].

Advantages and disadvantages:

Low-pressure plasma has the following merits:

- Glow obtained is uniform
- Low breakdown voltage
- High concentration of reaction species
- Non-thermal plasma generated

A major disadvantage with this is, it cannot be employed or implemented with the necessities of continuous processing of textile materials. Also, creating and sustaining a vacuum or low-pressure conditions for plasma affects machine productivity [5]. The maintenance cost of vacuum system is also high, while the dimensions of textile materials/samples that can be processed are limited compared to the size of vacuum chamber.

14.4.1.2 Atmospheric pressure plasma

Atmospheric pressure plasma (APP) has come up as a rather attractive alternative to low-pressure plasma for textile applications. The APP technology offers many advantages over low-pressure systems, such as working at atmospheric pressure, the possibility of continuous processing and integration with existing textile processing as well as cost-effectiveness. Some major types of APPs used are – corona discharge, dielectric barrier discharge and plasma jet.

14.4.1.3 Corona discharge

It is a process in which a current develops between two high potential electrodes in a neutral fluid, usually air from which plasma is generated. Corona discharge (Fig. 14.1) essentially involves two electrodes whose geometry is highly asymmetric, one with high curvature (like the tip of a needle or a narrow wire), while the other of low curvature (such as a plate). Due to high curvature, a high potential gradient is created around one electrode for the generation of plasma. Following is a schematic figure of a point-to-plane corona.

The existence of plasma is usually there in a region of the gas extending out only by about 0.5 mm from the metal point. The discharge being quite narrow, the time available for the plasma treatment of fabric would be too short to be used on a commercial scale. Moreover, the power level that can be applied is quite limited. Hence it may not be quite suitable for textile surface application. However, coronas are used for activation of certain polymer surfaces, manufacture of ozone, scrubbing particles from the air-in-air conditioning systems and to remove undesirable volatile organics [10,11].

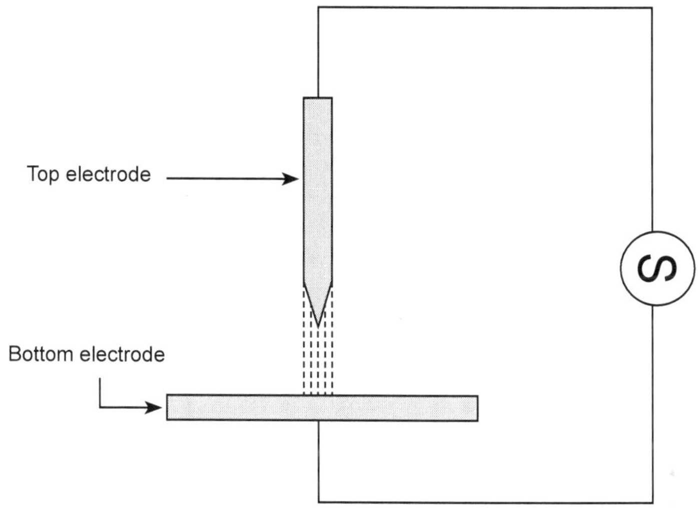

Figure 14.1 Arrangement for a corona discharge

14.4.1.4 Dielectric barrier discharge

This is a low-temperature plasma generation with some similarity to corona discharge. However, in case of Dielectric barrier discharge (DBD), out of the two electrodes it has, at least one is coated with an insulator, thus causing a dielectric barrier for electric currents (Fig. 14.2). The gap between the two is of the order of several mm and the applied voltage is about 20 kV. The dielectric barriers do not allow the operation of this electrode configuration with DC-fields. Thus, they are usually operated with sinusoidal wave current, square wave current or pulse wave forms.

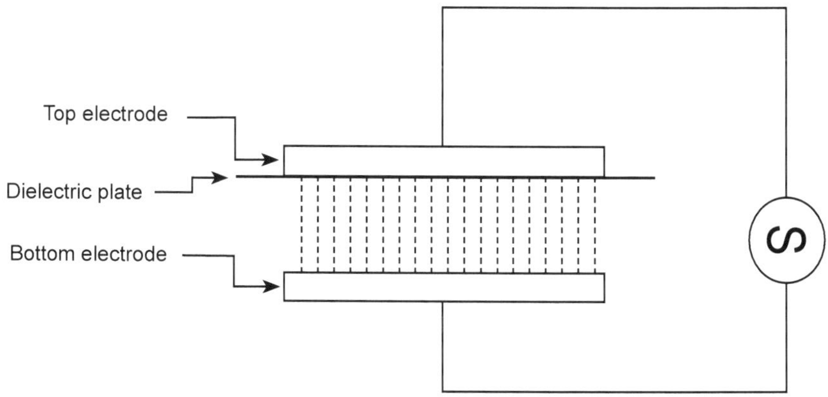

Figure 14.2 Schematic of DBD

Dielectric barrier discharge is an effective and applicable technology for the treatment of fabrics. Ambient air is often used as the process gas in a DBD due to its easy availability. The gas temperature can be moderate, ranging from room temperature to 100 °C. This enables the treatment of textiles without causing thermal damage [6]. However, the plasma generated is not quite uniform which limits its application in etching and deposition.

14.4.1.5 Plasma Jet [11]

This is a relatively new source consisting of two concentric electrodes through which a mixture of helium, oxygen and other gases flows. The gas discharge is ignited by applying 13.56 MHz RF power to the inner electrode at a voltage between 100 V and 250 V. The ionized gas from the plasma jet comes out through a nozzle, where it is directed onto a substrate a few millimetres below. In a usual working condition, the gas velocity is about 12 m/s. However, as of now it has been used for non-textile applications such as etching polyimide, tungsten and silicon dioxide as well as to deposit silicon dioxide films. A typical arrangement for film deposition is shown in Fig.14.3.

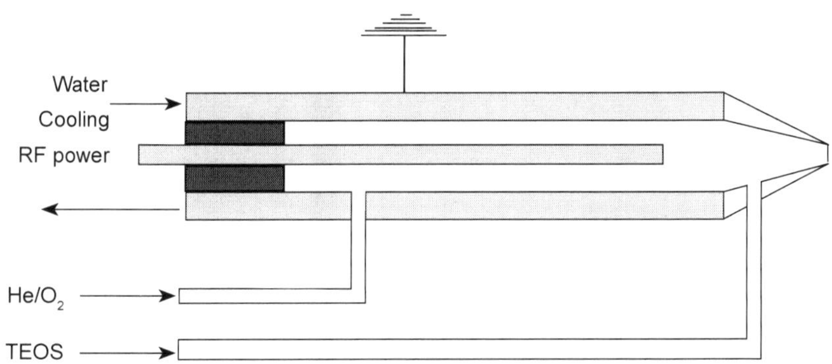

Figure 14.3 Plasma jet operating at atmospheric pressure for the deposition of silica films

14.4.2 Processes for textile applications

Broadly, plasma brings out two kinds of interactions with the textile surface. The first type of interaction involves the chain scission on the surface which results in etching of the surface, cleaning and activation of the surface. This is obtained by the use of non-polymerising gases like helium, argon, oxygen, air and nitrogen. Plasma-induced polymerisation or grafting is another type of interaction, which is possible by using different polymerising gases and precursors such as hydrocarbons, fluorocarbons, and silicon-containing monomers. Figs. 14.4a and 14.4b show the two types of interactions with textile substrates during plasma application, which are self-explanatory [5].

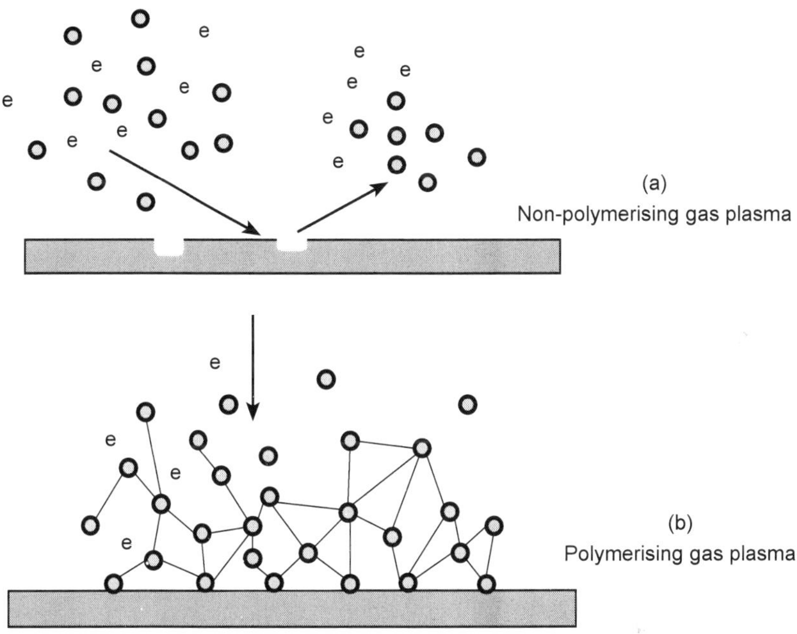

Figure 14.4 (a) Surface cleaning, etching and ablation with plasma and (b) Polymerisation and grafting with plasma

In both types of above interactions, carrier gas plays an important role. Inert gases like helium or argon can be used as a carrier gas for both types of interactions, either involving etching or polymerising plasmas. Helium has high energy metastable state and shows excellent heat conductivity which makes it a preferred gas over others. It has a special place in the technology of atmospheric pressure plasma. The desired activation by inert gases involves the generation of free radicals on the textile surface by means of chain scission. On the other hand, reactive gases such as oxygen and ammonia can result in the incorporation of oxygen and nitrogen-containing groups. The changes on the substrate's surface because of this, lead to various effects such as enhanced dyeability, better adhesion, improved biocompatibility, etc. [11].

The action of plasma can be further classified into four major categories, i.e., cleaning, activation, grafting and deposition. The cleaning process involves the use of inert (Ar, He) and oxygen plasmas, where the removal of organic contaminants like oils and other production releases from most industrial materials takes place. During the process, the abstraction of hydrogen takes place from the surface contaminants resulting in the formation of free radicals and repetitive chain scissions due to the influence of ions, free radicals and electrons of the plasma (Fig. 14.5). Coming to activation plasma processes, it happens when a surface is treated with a gas, such as oxygen, ammonia or nitrous oxide, etc., that does not contain carbon. Here, the various moieties of the process gas get incorporated on the surface of the material being treated. For example, the surface of polyethylene which consists of carbon and hydrogen only, may get activated on plasma treatment, by anchoring of functional groups like hydroxyl, carboxylic, carbonyl, peroxyl, amino and amines as shown in Fig. 14.6. This is a technique useful in producing technical fabrics with higher adhesion and permanency. As regards the third category, i.e., grafting, many free radicals are created on the material surface when an inert gas such as argon is used as process gas. In such case, if a monomer capable of reacting with the free radicals is introduced into the chamber, the monomer can get grafted. This process is applicable for low-pressure plasma treatment as well as atmospheric pressure plasma. The monomers are acrylic acid, allyl amine and allyl alcohol (Fig. 14.7). Finally, for surface deposition, more complex molecules are employed as process gas like methane or carbon tetrafluoride. The fragmentation of gas takes place in the plasma, resulting in unique film deposition and the process is referred as plasma-enhanced chemical vapour deposition (PECVD) [12].

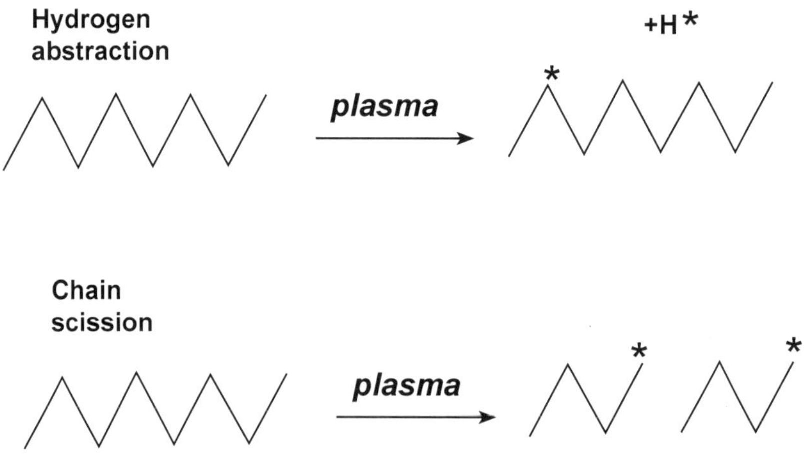

Figure 14.5 Free radicals formation by action of plasma

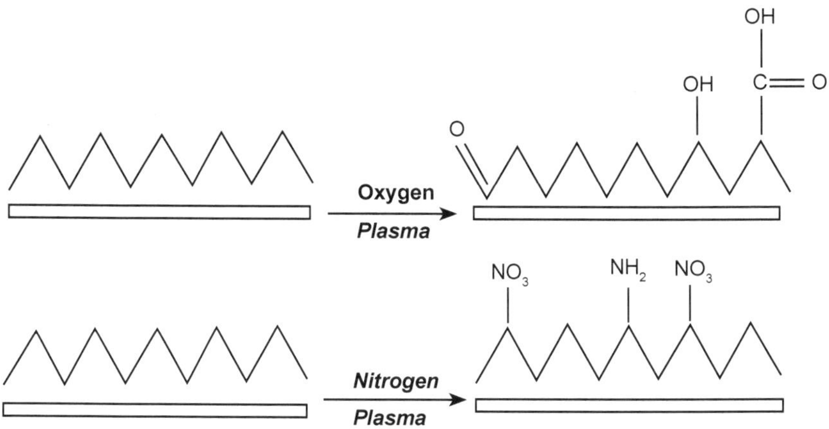

Figure 14.6 Surface activation by the substitution of hydrogen in a polymeric chain with other groups such as O, OH, COOH, NO_3 and NH_2

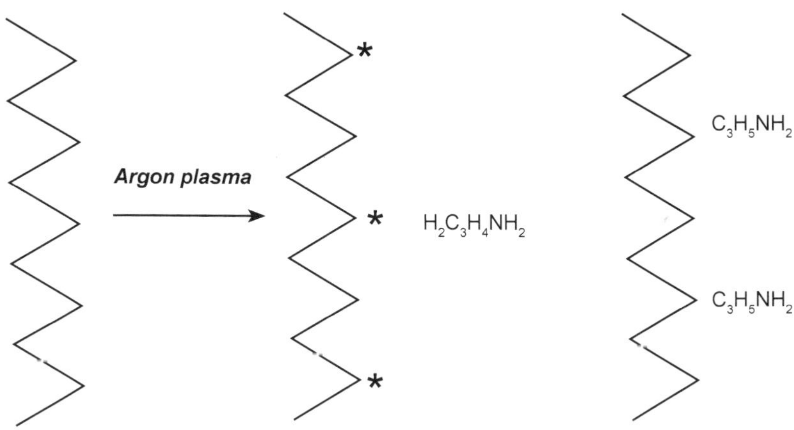

Figure 14.7 Grafting of a monomer on the surface by producing radicals with argon plasma

14.4.3 Property enhancement in textiles

The various improvements and functionalities that can be offered to textiles by plasma application have been worked out by various researchers. They are as follows:

(i) Improved mechanical properties: Softening of cotton and other cellulosic polymers can be achieved by the treatment with oxygen plasma. Anti-felting of wool by can also be achieved by the same treatment. Dipping in DMSO and subsequent treatment with nitrogen plasma has offered surface resistance in cotton, wool and silk fabrics.

(ii) Electrical properties: Antistatic finish of viscose rayon with chloromethyl dimethylsilane in plasma.

(iii) Wetting properties: Argon as vacuum gas and acrylic acid as a monomer on polyester increase wettability. This hydrophilicity also enhances soil-release and dirt resistance. Improvement of surface wetting in synthetic polymers like polyethylene, polypropylene, polyester etc., by treatment with oxygen, air and NH_3 plasma was also reported.

(iv) Dyeability and printability: These properties are improved in wool and cotton by treatment with oxygen plasma. Treatment of polyester with $SiCl_4$ plasma and polyamide with Argon plasma improves their dyeability.

(v) Fire retardancy: Fire retardant coating on nylon is possible by using monomers like fluoroacrylates in the presence of vinyl crosslinking agents.

(vi) Shrink proofing: The level of shrink resistance in wool can be increased by suitable plasma application, like treatment with oxygen plasma.

(vii) Others: Application of metal coated organic polymers, adhesion in composites and laminates, UV-protection of polyester-cotton blends, use in membrane and environment technology are a few other proven possibilities of plasma action on textile materials.

14.5 Supercritical carbon dioxide dyeing

Numerous advancements have taken place place over the last three decades in the technology of supercritical fluids. The supercritical fluid extraction and drying technologies have applications in various engineering and production processes such as in the petrochemical industry, pharmaceuticals, extracting fragrance and flavours from plants and natural resources, nutrients and pesticides, nicotine extraction from tobacco; in biological studies, coffee and tea decaffeination, making of aerogel and never the less for dyeing of textile material and dry cleaning of textiles [13-16]. Supercritical fluid technology is increasingly being used in various branches of engineering and technology, such as environmental studies, cosmetics, material science, electronics and polymer processing, etc. [17,18].

Carbon dioxide is typically a gas in the ambient condition or can be a solid, in the frozen form, called dry ice. In a confined environment with the simultaneous increase in the pressure and temperature from the ambient condition above the critical point for CO_2, the liquid/gas boundary tends to vanish and it assumes the characteristics in between gas and liquid. The classic *p-T* curve for CO_2 is shown in Fig. 14.8. The so-called critical point was first coined way back in 1869 [19] and the critical point of carbon dioxide was determined a few years later. Supercritical fluid exhibits properties of a fluid (solvent) and transport properties of a gas. Gas-like property causes effusion through the solids, whereas the liquid has the solvating property and dissolves solids. Due to this unique combination of liquid and gas-like qualities, supercritical fluids (e.g., $scCO_2$) become very appropriate for several technical applications, as mentioned earlier. It is noted that carbon dioxide assumes the supercritical fluid state as it is heated more than 31.2 ºC and compressed at high pressure at the same time (>74 bars).

Table 14.1 Critical temperature and pressure of common chemicals and solvents [20]

	Critical temperature (°C)	Critical pressure (Bar)
Ammonia	132.5	112.8
Benzene	289.0	48.9
Carbon dioxide	32.1	73.8
Cyclohexane	280.3	40.7
Ethane	32.2	48.8
Ethylene	9.4	50.7
Isopropanol	235.2	53.7
Propane	96.7	42.5
Propyline	92.4	46.3
Toluene	318.6	41.1
Water	374.2	220.5

Table 14.1 shows the critical temperature and pressure of some typical chemicals. Of these compounds generally used as a supercritical fluid media (e.g., ammonia, water, methanol, ethane, etc.), carbon dioxide is one of the most common choices due to its convenience, i.e., its critical temperature, T_c being close to ambient temperature, truly non-toxic and non-flammable as compared to others. Supercritical CO_2 is generally considered to offer environmental advantages as compared to many other traditional dissolving agents. A few other gases, such as nitrous oxide or lower alkanes like ethane, can also be used which have comparable boiling points and critical temperatures like supercritical carbon dioxide (e.g., N_2O, 72.45 atm at 36.4 °C; C_2H_4 48.3 atm 32.4 °C). Being non-toxic, non-flammable, and from a practical point of view (lower operating temperature), $scCO_2$ becomes the evident choice.

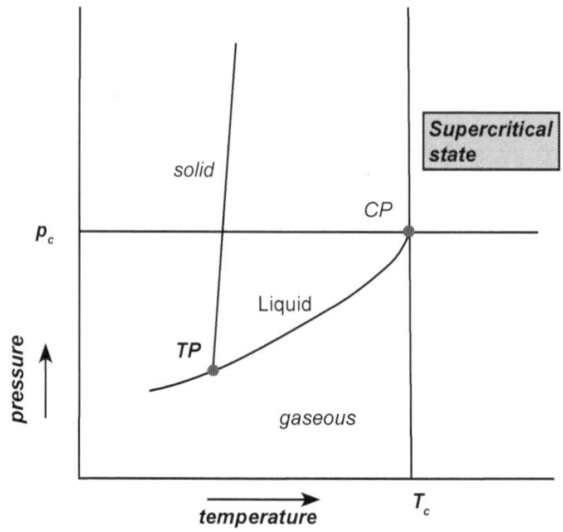

Figure 14.8 – p-T diagram of carbon dioxide: Critical values of pressure p_c and temperature T_c has been shown, along with triple point TP, and critical point CP [21]

Fundamentally, the behaviour of supercritical fluid is best understood from the phase diagram (Fig. 14.8). In a closed system, liquid carbon dioxide (e.g., CO_2 is liquid at lower temperatures and very high pressure (say at 10 °C and 80 bars) as can be observed from the phase diagram) will get converted to supercritical fluid as it is taken beyond both the critical pressure (p_c) and temperature (T_c). After the supercritical state is achieved, additional pressure will result in an increase in the density and the dielectric constant. Following this technique (when CO_2 is in use) capacity to dissolve hydrophobic molecules becomes appreciable. Carbon dioxide is chemically a very stable molecule, environmentally benign and does not provide high polarity. The addition of co-solvents (also known as modifiers or entrainers) often helps in increasing the polarity and enhancing its interaction with the solutes.

As already mentioned, carbon dioxide gas can dissolve many hydrophobic substances in the liquid and supercritical states [22]. Supercritical carbon dioxide can be used in place of water, for example, for dyeing of polyester fibres with many advantages. As $scCO_2$ exhibits gas-like properties and also has liquid-like densities, that can be of great advantage. It is capable of dissolving hydrophobic dyes, simultaneously with a gas-like character having negligible viscosities and a high diffusibility. This results in shorter dyeing times compared to that done in aqueous media. The advantage of the $scCO_2$ dyeing as compared to dyeing processes using aqueous media is that the removal of oil marks created during spinning, colouring and the simultaneous removal of excess dyestuff can be done in a single plant. The parameters that need to be controlled in supercritical dyeing are only temperature and pressure at the supercritical region for CO_2, as shown in Fig. 14.8. The dyeing process does not involve the use of water; thus, no drying is involved (only gaseous CO_2 is discharged), making the process economical, and it ensures saving of energy. The carbon dioxide that is released from the above process can be reprocessed without much difficulty. About ninety percent of the collected gaseous CO_2 is reused after precipitation or segregation of the excess dyestuff by using a suitable separator.

14.5.1 Interaction with polymers/fibres

Carbon dioxide gas has a considerable affinity for many polymeric materials, including textile fibres such as polyester, at high pressure and temperature near critical point. Solubility of CO_2 in the polymer increases with the number of polar groups present in the polymer. As mentioned by Berens and Huvard [23], the near-critical point CO_2 works like a highly volatile, organic polar solvent as it interacts with polymeric materials. In rubbery poly (ethylene-vinyl acetate) PEVA polymer with different vinyl acetate (VA) proportions, the presence of carbonyl group and polymer crystallinity have a significant impact on CO_2 solubility at equilibrium conditions, as reported by Shieh and Lin [24]. At or below the critical pressure, the sorption process was noticed to be markedly affected by the carbonyl groups present, whereas above the critical pressure, it is affected mainly by the polymer's degree of crystallinity. The interaction of CO_2 with the carbonyl groups present in the polymer is suggested to be of Lewis acid-base type [25]. There are various methods of quantitative determination of the solubility of CO_2 in the polymer, such as barometric (pressure decay) method, gravimetric method, frequency modulation methods and chromatographic techniques.

Supercritical carbon dioxide ($scCO_2$) can reduce the glass transition temperature (T_g) of the crystalline polymers, swell and plasticise it which is important and related to the saturation and modification of polymers for useful applications. In dyeing of the polyester fibres using supercritical CO_2, it may reduce the glass transition temperature (T_g) of polymers by 20-30°C, and at more than 70°C unoriented and amorphous yarn will result. Carbon dioxide can more easily spear into the polymer systems with the increase in the amorphous regions resulting greater drop in T_g.

The super-molecular polymeric structure can be observed to suffer marked changes after treatment in $scCO_2$. Significant morphological changes in polyester fibre and swelling of the PET films have been noticed, with a drop in T_g. The swelling of PET polymeric material has been found to increase in the presence of modifiers (methanol, ethanol, etc.). It has been noted that the changes in the polymer structure due to SCF CO_2 processes are similar to that of annealing at higher temperatures. Interaction of the CO_2 molecules with the basic sites present in the polymeric material is the main cause that creates a plasticising effect resulting in decreased inter-chain attractions and enhanced movements of the polymeric chains [26].

Depression of melting temperature(T_m), glass transition temperature (T_g) and polymer plasticisation due to supercritical carbon dioxide has been studied using techniques such as dielectric relaxation, high-pressure partition chromatography, Nuclear Magnetic Resonance (NMR), and high-pressure calorimetric measurements, and never the less by use of DSC [27]. Natural and synthetic textile fibres are not as such damaged by the action of supercritical CO_2 in optimum dyeing conditions except when the working temperature during dyeing is close to or more than the heat setting temperature in case of the thermoplastic fibres. Significant changes in amorphous polymers occur, whereas textile fibres with a high degree of crystallinity are less affected. As a rule, polyester must be heat set if it is to be dyed by scCO_2 technique to avoid any substantial change in fibre structure. Supercritical CO_2 treatment of heat set PET increases total crystallinity with the decrease in crystal size and development of new smaller, imperfectly formed ones. Shrinkage on non-heat set polyester yarn is much higher compared to the heat set yarns in supercritical dyeing conditions.

14.5.2 Dye solubility in scCO_2

Solubility of dye is a fundamental criterion that must be considered for textile colouration, and it is a function of solvent density in case of supercritical CO_2 processing. In general, scCO_2 shows the features of the hydrocarbon solvents such as toluene. Beyond the critical point, scCO_2 shows density and solvating capacity similar to liquid solvents and at the same time, shows very rapid diffusion characteristics and viscosity comparable to the gases. A fair amount of information on the solubility of disperse dyes are available in papers and books. Most of the data can be found on the solubility of pure disperse dyestuffs and some data can also be found in case where a co-solvent is applied. Carbon dioxide being non-polar in nature, disperse dyes get dissolved without any dispersing agents; at the same time, presence of polar groups, such as hydroxyl and carbonyl groups, decreases solubility. Solubility information related to the chemically modified dyes suitable for the colouration of the natural textile fibres and nylons is also available [13, 28]. Solubility of dyes in scCO_2 basically depends on the operating temperature and pressure of the $scCO_2$ and a huge amount of solubility data is available for azo, anthraquinone, mordant and benzothiazoleazo disperse dyes. Temperature-dependent solubility of the dyes occurs through vaporisation and due to the gas-like character of supercritical fluids, whereas pressure-related solubility and solute/solvent exchanges are because of the liquid-like character of the $scCO_2$ fluid. Dye solubility is usually low (around 10^{-4} to 10^{-7} mols dye/ per mols of CO_2) in the temperature range 120~140 ºC and pressure around 300 bar. Some dyestuffs with larger molecular weight seem to form aggregates in the SCF. Static apparatus without CO_2 circulation agglomeration, melting and crystallisation results in lower solubility in CO_2. Finely grounded dyes result in increased solubility. The presence of polar groups, as stated earlier, causes a decrease in solubility (most prominent when –COOH groups are present), whereas halogen and nitro groups have positive effects. It has been found that in the pressure range up to 400 bar, dyestuff solubility increases continuously.

In case of the temperature controllable disperse dyes, solubility drops with the decrease in temperature and hence can be managed by controlling the temperature. In the other case of density controllable dyes, solubility can be manipulated by varying operating pressure at some constant temperature [29]. Both approaches can be practised to control the dye partition in favour of the fibres. In cases where more than one dye is used in combination for colouration, dyestuffs of comparable or similar pressure and temperature dependence in $scCO_2$ should be preferred.

14.5.3 Setup for $scCO_2$ dyeing

The supercritical CO_2 dyeing unit essentially contains a CO_2 gas source supplied from a cylinder, a cooling unit, co-solvent inlet, pressurizing pump, dye vessel, stirrer, a dye pot with an oven, a pressurization pump, temperature and pressure control devices, a recirculation system and separating units [14]. A schematic diagram of the supercritical dyeing setup is shown in Fig. 14.9. The plant feed line consists of a tank with a heat exchanger and a compression pump. The tank is used to store CO_2 at a pressure of about 70 bars. The temperature of the input line is kept to about 10°C. A compression pump (CP) is also cooled with the circulating liquid in a cryogenic bath. Compressed CO_2 then passes through another heat exchanger (*H2*), where it is brought to the working temperature and finally reaches the dyeing chamber (*DV*). At first, the CO_2 fluid from the tank is allowed to fill the vessel and recycled until it reaches desired working pressure and temperature. Then, the dye tank is brought into the cycle and dissolved dye in supercritical CO_2 charges in the dye pot and dyeing is carried out by controlling temperature and pressure. During this period of dyeing, the inlet for fluid and the outlet valves are kept closed. The dyed materials are washed by repeating the first step while the dye vessel remains cut off to remove deposition of dyes on the yarns. Finally, the system is completely discharged letting all the gas to pass through, or carbon dioxide is recovered after the separation of excess dyes. It was found that of among the dyeing factors, the flow rate of CO_2 gas has the greatest influence on the dyeing levelness, dye partition and dye exhaustion; otherwise, it may take even a few days to achieve levelness [30, 31]. A typical example of process parameters and procedural steps for bulk dyeing are demonstrated in Table 14.2.

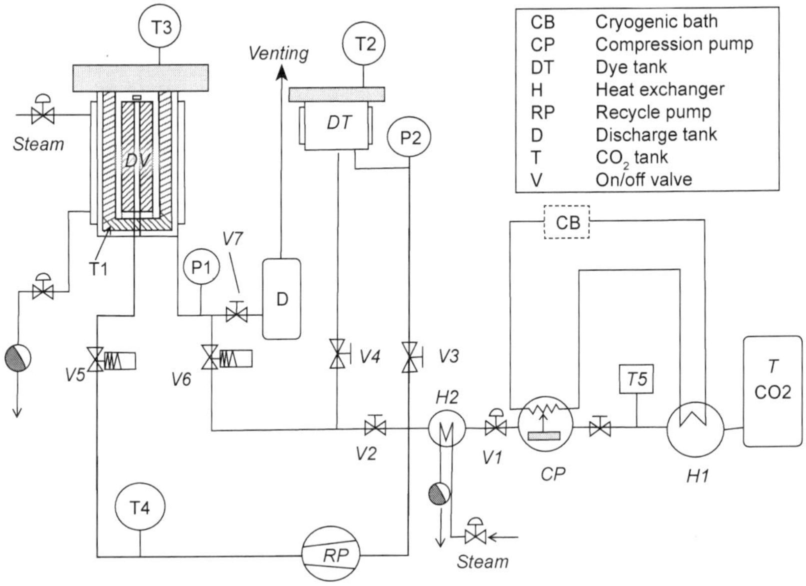

Figure 14.9 Schematic representation of the $scCO_2$ dyeing [30]

Table: 14.2 Super-critical CO_2 Polyester Yarn Dyeing Process [14]

Conditions	Procedure
Temp. - 100-130 °C	Loading of yarn and the dye
Pressure : 200-300 bars	Filling with CO_2 to required density
density of CO_2: 0.3-0.6 g/cc	Starting of the circulation with flow reversal
Flow Rate : 50-80 l $min^{-1}kg^{-1}$	Heating to the dyeing temp.
Holding Time : 10-30 min	Holding at that temperature
	Depressurizing
	Unloading

Except for the high pressure that is required to dissolve the dye, $scCO_2$ dyeing method is not very different from the conventional aqueous dyeing of polyester fibre. The processing temperatures and volumetric flow rates of the liquid are similar to that is used in case of polyester dyeing. The $scCO_2$ dyeing process shows very good development of shade, great levelness and reduced crocking (wet or dry). Additionally, after the $scCO_2$ dyeing process, the physical properties of yarn remain unaffected and are comparable with that of conventional aqueous dyeing.

14.5.4 Diffusion of dye in the polymer matrix

The mechanism of disperse dyeing of polyester in supercritical conditions can be considered to be the same as that of the HTHP (High Temperature High Pressure) dyeing of PETs. Dye molecule transport to the fibre from the $scCO_2$ media is governed by some process parameters. Factors such as duration of dyeing, temperature, pressure and gas density influence the solubility of the dye, its affinity towards the fibre, and the diffusion coefficient. The dye diffusion coefficient for disperse dyes in polyester films may vary from 10^{-11} to 10^{-14} m^2s^{-1} using supercritical CO_2 techniques. These values are one to three orders of magnitude greater than the corresponding values for dyeing by conventional methods. The dye structure and capability to form chemical bonds with the polymer can significantly affect the uptake or diffusion. Equilibrium concentrations of dyes are often experimentally determined for the development of optimum colour intensity. With increasing pressure, dye molecules comfortably diffuse in the amorphous regions of the polymeric structure. It is to be noted that the activation energy of diffusion of the dyes in supercritical CO_2 is much lower as compared to that in usual aqueous processes, and the diffusion can take place much more easily [32, 33].

Both the factors, e.g., temperature and pressure, cause the dye to move preferentially towards the fibre, and when the dye is exhausted from the gas, it gets absorbed by the fibre [29]. In certain parametric conditions, dye molecules can exhaust from the solution at a greater rate towards the fibre, which may result in uneven dyeing and poor fastness. A typical sequence of controlling temperature and pressure in $scCO_2$ dyeing is shown in Fig. 14.10. Optimum dyeing time can be in the range of 45-60 min.

It can be noted that though non-polar (disperse) dyes get conveniently dissolved in supercritical carbon dioxide but acid and direct dyes, and some others being polar, they have poor solubility in it. In many cases, polar co-solvents or modifiers or entrainers are used in small amount (0-20%, ethanol or methanol or hexane) has been shown to improve dyeing performance in case of polyester dyeing. The relatively poor solubility of acid or direct dyes gets improved as the polarity and thus the solubilising power of $scCO_2$ increases with the addition of polar co-solvents [28]. The way the modifiers/entrainers increases solubility and affect the fixation of disperse dyestuff on polyester fibres in the $scCO_2$ conditions can be compared with the effects brought about by the carriers in PET dyeing. Also, various other mechanisms have been proposed to explain the action of modifiers, such as increased solubility of dyestuffs, swelling of the fibre, changes in fibre structure, plasticisation of fibre, and so on. The presence of co-solvent results in increased swelling, greater interaction between the solid and fluid phases, and increased plasticisation. It reduces the minimum pressure by a large extent at which the dye will be soluble. Increased solubility of disperse dyes (azobenzene) using many alcohols probably stem from the interaction of the non-polar part of alcohol with azobenzene [34]. The effects of entrainers in $scCO_2$ dyeing are similar to what happens as water is being added during the colouration of polyester fibres using some solvents like perchloroethylene(C_2Cl_4), where the presence of polar water molecules reduces the solubility of the dye in the perchloroethylene bath.

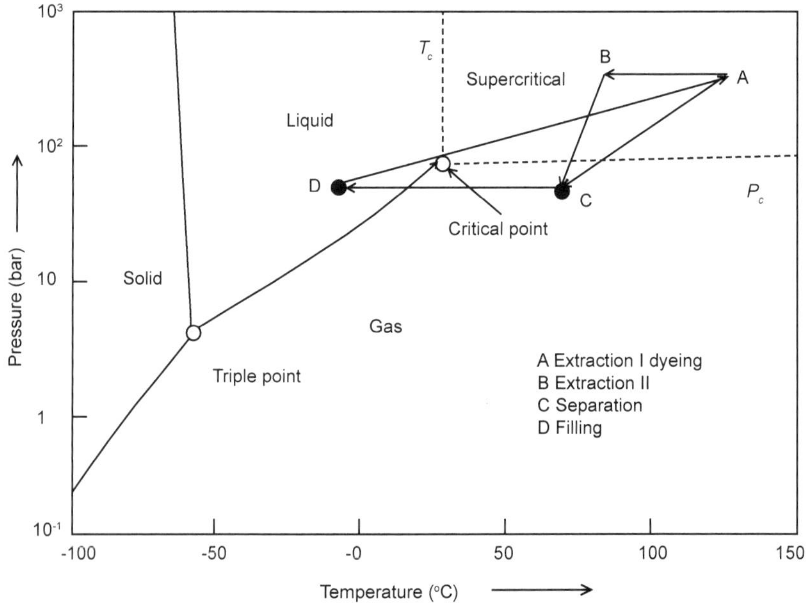

Figure 14.10 Typical dyeing sequence shown with the pressure-temperature phase diagram of CO_2 sorption [22]

14.5.5 Supercritical CO_2 dyeing of natural fibres

It is well known that polyester can be dyed using $scCO_2$ process, whereas the natural fibres namely cellulosic and protein fibres, cannot be dyed easily. As natural fibres are polar in nature, $scCO_2$ cannot swell the fibres effectively, and diffusion of dyestuff remains a major issue. In general, various polar dyes such as direct, basic, acid and vat dyes those are used to colour natural fibres are sparingly soluble in supercritical CO_2 [28]. There are some strategic ways following which natural fibres can also be dyed using the $scCO_2$ process. In one method, natural fibre textile material can be pretreated with, say, polyethylene oxide or PEG by soaking the fibrous material, which helps diffusion of dye molecules or introducing hydrophobic groups in cotton. In a different approach, disperse dyestuff can be synthesised which is CO_2 soluble, as well as possessing functional groups that can react with the fibres by forming chemical bonds. There are other practices also; amongst them, two main approaches are of significance. One of them is the solubilisation of the regular polar dyestuffs in $scCO_2$ by means of microemulsions of water/CO_2 (the reverse micellar systems); the other is the synthesis of some novel reactive dyes. The second method is coupled with a physical pretreatment of the textile materials and/or the adding of suitable co-solvents. In some instances, the cotton fibre has been modified with small amounts of trichlorotriazine by attaching a reactive group that is able to chemically react with the disperse dyes containing at least one hydroxyl or an amino group in their structure [35]. In another approach, Liu [36] and co-workers changed the mercerized ramie fibre (a natural cellulosic fibre with high crystallinity) using benzoyl chloride to increase chemical affinity of the fibre towards the disperse dye. The lack of polarity of the disperse dyes permits easy solubilisation in the supercritical media without using any additives or co-solvents, while the presence of a suitable functional group permits the formation of a chemical linkage with the reactive sites on the fibre ensuring the fixation of the dye molecules [28].

Reversed micelles are in fact self-organised nanometer-size water droplets (<10 nm) created by the action of surfactant molecules containing an inner core of water molecules dispersed in a continuous apolar media. The polar part of the surfactant molecule is able to hold and solubilise water in the inner core while the apolar part remains in the organic media. This micellar system provides a stable aqueous micro-environment that can be so-called 'water pools'. A small amount of water dispersed in $scCO_2$ can produce a similar effect of the water in oil micro-emulsion. Such a micro-emulsion system is suitable for dyeing polar natural fibre-based textile materials as the polar dyes can be solubilised in micelles and spread in super-critical CO_2. Some organic dyes that are rather insoluble in pure $scCO_2$, namely thymol blue, methyl orange and others, can be suitably dissolved in water-CO_2 reverse micro-emulsions system with proper co-surfactants [37]. Solubility of dyes in $scCO_2$ can be determined using a SCF set up by entrapping the solubilised dyes using UV-Vis spectroscopy [30, 38]. The amphiphilic surfactants (fluorinated/non-fluorinated) contain both hydrophobic and hydrophilic groups at the ends that are essential to form water reverse micellar systems in $scCO_2$. Fluorinated surfactants do not require co-solvents, whereas non-fluorinated ones have limitations, including lower solubility in $scCO_2$, requiring a co-solvents such as alcohols or some fluro-derivatives and associated with the complexity of recycling the chemicals [39]. The process parameters such as temperature or pressure do not have an appreciable effect on polar dye's solubility in the micellar system during $scCO_2$ dyeing. For increasing solubility of the dyes, formation of micelles is essential [40].

Polyamide fibres such as Nylons can be dyed with disperse dye in $scCO_2$, but it results in poor fastness and lower dye absorption. A high degree of crystallinity can be correlated with the inferior fastness. Specifically produced disperse reactive dyes have found some success [41]. It was noted that adding a small quantity of water may act like a swelling agent, and it increases the dye-fibre association in a better way. To dye polypropylene fibres which are characteristically non-polar and crystalline, dyes with more hydrophobic nature and aliphaticity than it is commonly used for polyester fibres can be taken [42].

14.6 Digital printing in vapour phase

The conventional analog printing of textiles has been there for a long time and is still largely in practice. The digital printing technology is relatively new in that sense, although it is more than three decades now since it was first commercialised. The introduction of a variety of inkjet printers dates from the mid to late 1980s [43]. Now we find a lot of installations of such machines in production houses across the globe. With matching lengths often getting shorter and the requirement of niche products increasing, digital printing is picking up pretty well. Screen engraving for sampling is done away with reducing development time to market [44]. In addition to other printings like reactive, pigment etc., polyester printing with disperse dyes is quite common among digital printing machines. Disperse printing of polyester by digital printing is again of two types- one being direct printing and the other is sublimation transfer printing. The second one is a completely dry process, complying with our topic of discussion. The vapour phase transfer printing with disperse colours has the simplest processing route: printing the disperse dye ink onto a specific paper, followed by thermal transfer (at 210 °C for

30 s) onto the fabric. Because of transfer process efficiency, no wash-off is required [45]. Moreover, for vapour phase transfer printing with disperse dyes, no pretreatment of the polyester is needed, unlike other digital printings, including that of direct disperse dye printing. Elimination of the pre- and post-treatments make the transfer digital printing a water-free process.

A wide range of sublimation transfer printers is available now, starting with a basic model of just two print heads. The number of heads can go up to eight and accordingly printing speed increases. These machines can print at 60 m^2/hr to around 200 m^2/hr. In this arrangement, the paper used for transferring the design onto fabric is available in different qualities like 50, 70 and 90 gsm. Both, a paper printing unit and heat transfer fusing machine (for design transfer from paper to fabric) are required in this system. The heat transfer machines are available in the market with a wide range of cost, quality and speed.

The sublimation ink used for this printing is different from the disperse dye used for direct digital printing. The former is of low energy type so that it can easily sublime off. The cost of the inks is, however, comparable for both. Four colour printing of C (Cyan), M (Magenta), Y (Yellow), K (Black) is used for all possible colour range on the paper. Normally, printing, drying and roll winding functions all happen in the printer itself. As regards the heat transfer machine for paper to fabric transfer, the speed usually depends on the drum size. For a typical drum size of 600 mm diameter, the speed is 150 linear metres per hr. The advantages of sublimation transfer printing on polyester over direct disperse printing (both digital) are as follows:

(i) Shorter process as no prior treatment or post treatment required for the polyester fabric
(ii) No ager or washing unit required
(iii) Space requirement is less
(iv) Machine cost is less, as in direct digital printing for disperse dyes, the belt used for fabric also significantly adds to cost other than remaining components
(v) The print head cost is about one-third as that of the direct digital printer and hence the replacement of header and consequently maintenance cost is less
(vi) Labour cost is reduced, as no ageing and washing are required.

Concluding remarks and future perspectives

The three water-free processes covering dyeing, printing and finishing are practical alternatives to the existing methods. Out of these, plasma technology, although widely experimented and effects proven, needs further improvisation to bring in commercial line of production. Super critical carbon dioxide dyeing, although still in pilot plant stage, is gradually being scaled up to higher volumes of production. Digital printing in general and dry transfer printing in the context of dry processes is very much in practice. Although, digital printing has a present share of less than 5% in textile printed fabrics, it is likely to grow in times to come. The water-free operations in textile chemical processing are here to stay, as there is a scarcity of water globally which is aggravating day by day. Hence, the processes discussed in this chapter appear to be the technology options for the present and future processing industry.

References

1. Hoekstra AY, Mekonnen MM (2011), Value of water Research report Series No.53,UNESCO-IHE Institute of Water Education, Netherlands.
2. Thapa P, ThakurPS (2014), 'Reducing water footprint in Textile Industry', *Ind. Text. J.*,124, 88-91.
3. http://www.waterfootprint.org (accessed August, 2019)
4. Pant P (2018), 'Responsible textile manufacturing through sustainable chemicals management', *Abstracts - International Conference on Advances in Textile Materials and Processes,* 19-20 Nov 2018, UPTTI Kanpur, pp. 7-8.
5. Kale KH, Desai AN (2011), 'Atmospheric pressure plasma treatment of textiles using non-polymerising gases', *Ind. J. Fibre Text. Res.*, 36 (3), 289-299.
6. Shishoo R (2007), *Plasma technologies for textiles*. Elsevier, England.
7. Buyle G (2009), 'Nanoscale finishing of textiles via plasma treatment', *Mater. Technol.*, 24(1), 46-51.
8. Bahrami N, NouriKhorasani S, Mahdavi, H, Ghiaci M,Mokhtari R (2019), 'Low-pressure plasma surface modification of polyurethane films with chitosan and collagen biomolecules', *J. Appl. Polym. Sci.*, 136(21), 47567.
9. Bhat NV, Upadhyay DJ (2002), 'Plasma-induced surface modification and adhesion enhancement of polypropylene surface', *J. Appl. Polym. Sci.*, 86(4), 925-936.
10. Schutze A, Jeong JY, Babayan SE, Park J, Selwyn GS, Hick RF (1998), 'The atmospheric-pressure plasma jet: a review and comparison to other plasma sources'. *IEEE Trans. Plasma. Sci.*, 26(6), 1685-1694.
11. Samanta K, Jassal M, Agrawal AK (2006), 'Atmospheric pressure glow discharge plasma and its applications in textile', *Ind. J. Fibre Text. Res.*, 31(1), 83-98.
12. https://www.researchgate.net/publication/1908142_Plasma treatment advantages for textiles (accessed October, 2019)
13. Gupta RB, Shim JJ (2006). *Solubility in supercritical carbon dioxide*. CRC press, Boca Raton.

14. Hendrix WA (2001), 'Progress in Supercritical CO2 Dyeing', *J. Indu. Text.*, 31(1), 43-56.

15. Williams JR, Clifford T (2000), *Supercritical fluid methods and protocols*. Humana Press. Totowa, New Jersey: pp. 1-16

16. Brunner G (2013), *Gas extraction: an introduction to fundamentals of supercritical fluids and the application to separation processes* (Vol. 4). Springer Science & Business Media. Berlin: pp. 1-58.

17. DeSimone JM, Guan Z, Elsbernd CS (1992), 'Synthesis of fluoropolymers in supercritical carbon dioxide', *Science*, 257(5072), 945-947.

18. Kemmere MF, Meyer T (ed) (2006), *Supercritical carbon dioxide: in polymer reaction engineering,* John Wiley & Sons, Weinheim, Germany: pp. 1-14.

19. Andrews T (1869),'The Bakerian Lecture.—On the continuity of the gaseous and liquid states of matter', *Philos. Trans. R. Soc.*, 159, 575-590.

20. https://www.chemengonline.com/supercritical-co2-a-green-solvent/(accessed November 30, 2019)

21. Gebert B, Saus W, Knittel D, Buschmann HJ, Schollmeyer E (1994), 'Dyeing natural fibers with disperse dyes in supercritical carbon dioxide', *Text. Res. J.*, 64(7), 371-374.

22. Bach E, Cleve E, Schollmeyer E (2002), 'Past, present and future of supercritical fluid dyeing technology–an overview',*Review of Progress in Coloration and Related Topics*, 32(1),88-102.

23. Berens AR, Huvard GS, Korsmeyer RW, Kunig FW (1992), 'Application of compressed carbon dioxide in the incorporation of additives into polymers', *J. Appl. Polym. Sci.*, 46(2), 231-242.

24. Shieh YT, Lin YG (2002), 'Equilibrium solubility of CO2 in rubbery EVA over a wide pressure range: effects of carbonyl group content and crystallinity', *Polymer*, 43(6), 1849-1856.

25. Tomasko DL, Li H, Liu D, Han X, Wingert MJ, Lee LJ, Koelling KW (2003), 'A review of CO2 applications in the processing of polymers', *Ind. Eng. Chem. Res.*, 42(25), 6431-6456.

26. Kazarian SG, Brantley NH,Eckert CA (1999), 'Applications of vibrational spectroscopy to characterize poly (ethylene terephthalate) processed with supercritical CO2', *Vib. Spectrosc.*, 19(2), 277-283.

27. Shieh YT, Li YH, Huang CC, Wang, TL (2010), 'Effects of CO2 treatments on the dual glassy relaxations and dual melting peaks in poly (ethylene terephthalate)', *J. Supercrit. Fluids.,* 55(1), 373-380.

28. Banchero M (2013), 'Supercritical fluid dyeing of synthetic and natural textiles–a review',*Coloration Technology*, 129(1), 2-17.

29. Montero GA, Smith CB, Hendrix WA, Butcher DL (2000), 'Supercritical fluid technology in textile processing: an overview', *Ind. Eng. Chem. Res.*, 39(12), 4806-4812.

30. Banchero M, Sicardi S, Ferri A, Manna L (2008), 'Supercritical dyeing of textiles—from the laboratory apparatus to the pilot plant', *Text. Res. J.*, 78(3), 217-223.

31. Tabata I, Lyu J, Cho S, Tominaga T, Hori T (2001),'Relationship between the solubility of disperse dyes and the equilibrium dye adsorption in supercritical fluid dyeing',*Coloration Technology*, 117(6), 346-351.

32. Von Schnitzler J, Eggers R (1999), 'Mass transfer in polymers in a supercritical CO2-atmosphere', *J. Supercrit. Fluids.*, 16(1), 81-92.

33. Özcan, A.S., Özcan, A., (2005), Adsorption behaviour of a disperse dye on polyester in supercritical carbon dioxide. *J. Supercrit. Fluids.*, 35(2), 133-139.

34. Takagishi T, Katayama A, Matsuoha M, Konishi K, Kuroki N (1969), 'Effects of alcohols on the aqueous solubility of disperse dyes', *Kolloid-Zeitschrift und ZeitschriftfürPolymere*, 232(1),699-703.

35. Schmidt A, Bach E, Schollmeyer E (2003), 'Supercritical fluid dyeing of cotton modified with 2, 4, 6-trichloro-1, 3, 5-triazine', *Coloration Technology*, 119(1), 31-36.

36. Liu ZT, Zhang L, Liu Z, Gao Z, Dong W, Xiong H, Peng Y, Tang S (2006), 'Supercritical CO2 dyeing of ramie fiber with disperse dye', *Ind. Eng. Chem. Res.*, 45(26), 8932-8938.

37. Liu J, Ikushima Y, Shervani Z (2004), 'Investigation on the solubilization of organic dyes and micro-polarity in AOT water-in-CO2microemulsions with fluorinated co-surfactant by using UV-Vis spectroscopy', *J. Supercrit. Fluids.*, 32(1-3), 97-103.

38. Carrott MJ, Wai CM (1998), 'UV− visible spectroscopic measurement of solubilities in supercritical CO2 using high-pressure fiber-optic cells', *Anal. Chem.*, 70(11), 2421-2425.

39. Sawada K, Jun JH, Ueda M (2003), 'Dyeing of natural fibres from perfluoro-polyether reverse micelles in supercritical carbon dioxide', *Coloration Technology*, 119(6), 336-340.

40. Sawada K, Takagi T, Jun JH, Ueda M and Lewis DM (2002), 'Dyeing natural fibres in supercritical carbon dioxide using a nonionic surfactant reverse micellar system', *Coloration Technology*, 118(5), 233-237.

41. Van der Kraan M, Cid MF, Woerlee GF, Veugelers WJT, Witkamp GJ (2007), 'Dyeing of natural and synthetic textiles in supercritical carbon dioxide with disperse reactive dyes', *J. Supercrit. Fluids.*, 40(3), 470-476.

42. Miyazaki K, Tabata I, Hori T (2012), 'Effects of molecular structure on dyeing performance and colour fastness of yellow dyestuffs applied to polypropylene fibres in supercritical carbon dioxide', *Coloration Technology*, 128(1), 51-59.

43. Dawson TL, Glover B (2004) (ed), *Textile Ink Jet Printing – A review of ink jet printing of textiles, including ITMA 2003*, SDC Publication, Bradford, UK.

44. Chapman K (2002), 'Printing 2002: A digital odyssey', *AATCC Review*, 2(5), 12-15.

45. Siemensmeyer K, Provost JR, Raulfs FW, Weiser J (2001), 'Thermodynamics of the Digital transfer printing process', *Image Science and Technology NIP 17*, Digital Printing Conference, 426-431, Ft Lauderdale, USA.

15

Sustainability standards for textile processing

Prasad Pant[1]*, Gururaj Kanchi[2]
[1]The ZDHC Foundation, Amsterdam
[2]Independent Consultant and Textile Professional
*Corresponding author, Email: pant@zdhc.org

Abstract: Textile processing has always been perceived as a hugely polluting industry and a heavy resource-consuming activity in terms of manpower, water, chemicals and energy. Since the past decade, there has been a recognition of the need to incorporate sustainable practices in textile processing to reduce its impact on the environment as well as to ensure worker and consumer safety. A plethora of eco-labels, legislation, industry standards, NGO campaigns and Brand/Retailer requirements have been gaining ground and driving the apparel and textile industry towards more sustainable practices. This chapter gives an overview of the important statutory requirements and voluntary standards and programmes for sustainability in textile production and products.

15.1 Introduction

Since the discovery of synthetic dyes in 1856 (thanks to the efforts of Sir William Henry Perkin in discovering Mauveine dye), the synthetic dyes and chemicals industry has grown phenomenally as has textile and apparel production. It is estimated that more than 700,000 tons per annum of colourants [1] are being manufactured and used in textile production. Since approximately 80% of these dyes get fixed on the textile substrate, huge amounts of colourants (approx. 300,000 tons per annum) are released into the environment through the discharged effluent, if the effluent is not appropriately treated in a wastewater treatment plant. Thus, textile processing is widely perceived as a hugely polluting industry as well as resource intensive as it uses vast quantities of water, energy and workforce.

In a Cradle-to-Grave Study conducted by the global denim brand Levis Strauss [2] on its 501^R Jeans in 2006, it was found that to make one single pair of denim jeans:

- 3480.5 litres of water is required
- 32.3 kg of CO_2 equivalent is released
- 400.1 MJ of energy is used up

The focus of this study is the need to incorporate sustainable practices in textile processing to optimise the use of resources as well as reduce its impact on human health and the environment. Various NGO campaigns in the recent past, notably the Greenpeace Detox Campaign [3], have highlighted the use and discharge of hazardous chemicals in textile production that has led to unmitigated pollution of rivers and lakes in producing countries, causing widespread diseases and environmental disasters. The focus of this campaign has been to drive global apparel brands and retailers towards sustainable chemical management practices in their supply chain and the elimination of these hazardous substances in manufacturing and release across all pathways.

"Eco-friendly", "sustainability" and "green" are fast becoming the buzzwords in the textile and apparel industry today. This has led to a plethora of eco-labels and sustainability standards that the textile and chemical companies are latching onto to demonstrate their commitment to these issues. Eco-labels help companies to increase the confidence of their Buyers and Customers in their products and/or manufacturing practices. They also create awareness about sustainability issues in the marketplace.

Such kind of standards can be classified into 4 categories:

1. *Obligatory standards:* These are legal or mandatory requirements that need to be complied with for the manufacture or marketing of a product. Some examples are USDA National Organic Program (NOP), GB Standards in China, California Proposition 65, Japanese Agricultural Standards (JAS), REACH Directive 834/2007, India's Environment Protection Act, Washington Children's Safe Products Act (WCSPA), Korean Chemicals Control Act, Consumer Product Safety Improvement Act (CPSIA), The Toxic Substances Control Act (TSCA), etc.
2. *Quality standards:* These are established by international standards organisations to provide a structured framework for facilities to implement. Examples are: ISO 14001, ISO 18001, ISO/TS 14067 (Carbon Footprint), Energy Star, EMAS
3. *Voluntary standards or Eco-Labelling Programs:* These are run by private set-ups or by government patronage and can be global or regional in their outlook. They are usually demanded by Brands/ Retailers to ensure that their products meet specific sustainability criteria. These standards can be focused only on end-product or cover the entire production chain. Examples are Global Organic Textile Standard (GOTS), OE 100 Organic Content Standard, OekoTex 100/ Made in Green/ Oekotex STeP, EU Flower, Nordic Blue Angel, bluesign, Better Cotton Initiative (BCI), etc.
4. *Ethical Sourcing or Production standards:* These are concerned with fair trade practices (producers are paid fair wages and minimum

price), social responsibility (safe working conditions, equality for women, no child labour, maternity protection for women, reasonable working hours, freedom from discrimination, etc.) and corporate responsibility (impact of company on environment and community). Examples: SA 800 social accountability standard, Business Social Compliance Initiative (BSCI), Fair Trade, Fair Wear, etc.

15.1.1 History and origins of legislation on chemical restrictions

It is estimated that approximately 4 kgs of chemicals are used in the production of 300 gms Cotton T-Shirt, right from cotton growing to garment production. A lot of these chemicals have been found to have a negative impact on the environment and aquatic ecosystem. Restrictions on harmful chemicals started in the 1960s when the US and Nordic countries introduced legislation on DDT, Mercury and Polychlorinated Biphenyls (PCBs).

From the 1970s, several global treaties, protocols, agreements, declarations and conventions were initiated to regulate the use and production of different chemicals. Important among these were the Stockholm Conventions of 1972 and 1990. The Stockholm Convention of 1972 focused on chemical pollution and environmental issues, and the Convention of 1990 led to a ban on 12 Persistent Organic Pollutants, that is, POPs.

The Montreal Protocol of 1985 focused on ozone-depleting substances and POPs. The Rio Declaration of 1992 established principles on the safe use of chemicals. The Kyoto Protocol of 1997 focused on the reduction of emission of six greenhouse gases, and the Rotterdam Convention of 1998 focused on the usage of pesticides.

In 2002, the World Summit on Sustainable Development (WSSD) which was held in Johannesburg, has adopted a historic objective of using and producing chemicals using methods and processes which will minimise the negative effects of chemicals on the ecosystem via risk management procedures, by the year 2020.

In 2006, significant legislation by the EU parliament enacted the EU REACH. As per this regulation, all chemicals made, used, sold and imported into the EU must be registered with the European Chemical Agency (ECHA). These chemicals will be evaluated for ecological and toxicological data based on the Registration Dossier submitted by the chemical companies registering their substance with ECHA. After evaluation, these chemicals may be either authorised or restricted for usage in mixtures, preparations and articles.

The first regulated chemical in textiles was formaldehyde as it was a suspected carcinogen and respiratory sensitiser. It was regulated in 1975 by the Act on Control of Household Products containing Harmful Substances, popularly known as Japanese Law 112. As per this law, formaldehyde should not be detected in baby products. In addition, the textile products which come in direct contact with the skin should not have more than 75 ppm concentration of formaldehyde. Formaldehyde was also regulated in Germany in 1993 under the Commodity Goods Ordinance. As per this law, a product containing more than 1500 ppm of formaldehyde should be labelled as 'contains formaldehyde'.

In addition, certain amines released from azo dyes were banned under the Commodity Goods Ordinance in Germany as they were found to cause bladder cancer. 20 amines were listed under this ordinance and the requirement was that each of these amines should not exceed 30 milligrams per kilo in an article. The Indian government issued Gazette Notifications in 1993 and 1997, prohibiting the "handling" of 112 azo dyes. The term "handling" includes manufacturing, use, storage and transportation.

Pentachlorophenol or PCP was regulated in Sweden in 1985 under the Act on Chemical Products, in Germany in 1993 under the Ordinance on Prohibition of Chemicals, and under the PCP law in Holland in 1994, as it was found to be toxic to humans through ingestion or inhalation. As per these laws, the PCP in an article should not exceed 5 milligrams per kilogram.

In the past decade, many countries across the globe have enacted legislation on hazardous chemicals in textile products. These include Canada Consumer Product Safety Act, California Proposition 65, Washington Children's Safe Product Act, Japan Chemical Substances Control Law, Consumer Product Safety Improvement Act, China GB Standards, Taiwan CNS 15290 Safety of Textiles, EU REACH and Korea Safety Quality Mark Act.

15.1.2 The need for sustainable practices in textile processing

The wet processing segment of textile and apparel production is one of the most resource-intensive and complex segments. Different substrates require a specific type of processes, colourants and processing aids. All these chemicals must be added to the substrate through the water as the medium. Many processes require high temperatures to be effective. Additionally, after the processing is over, the wastewater containing residues of these chemicals must be treated to ensure that the discharged water quality does not harm the environment. Thus, energy, water and chemicals are used in high quantities in a wet processing facility.

You will be wonderstruck to know that to make a 250 gms cotton knitted T-Shirt:

- 1.7 kgs of fossil fuel are used up
- 4 kgs equivalent CO_2 is liberated into the atmosphere
- 2600 litres of water is used (right from cotton fibre growing)
- 142 gms of chemicals are added
- 0.4 kgs of solid waste go into landfills

It is estimated that the annual global consumption of water used in textile processing would be about 3/4th year's (219 days) drinking water requirement of the world's population of 6.2 billion [4]. The production countries such as India, China and Bangladesh are fast turning from water surplus areas to water-scarce areas [5] and the day will soon come when water for industrial production may not be available. This has pushed the industry to implement water conservation techniques (such as ultra-low-liquor dyeing machines or supercritical CO_2 dyeing), processes (pad-dry-pad-steam, pad-humidity-fix, cold-pad-batch, continuous bleaching) and water recoveries technologies such as ultrafiltration and reverse osmosis.

Safe chemical usage in textile production has also garnered importance over the past decade. Product Safety and restriction of hazardous substances in textiles were started by Levi Strauss & Co. in the year 2000 through the publication of their 'Restricted Substances List' or RSL. Subsequently, many brands and retailers have published their RSLs which are used to communicate Brands' requirements to their supply chain partners on products being free of such harmful substances, thereby ensuring consumer safety. The conformance to these RSLs is through random testing of finished articles before shipment from the garment manufacturer. The RSLs became the responsibility of the Garment Cut and Sewed units, who usually got declarations from their upstream (fabric/raw material) suppliers to establish transparency and conformance to these RSLs.

There are more than 600 chemical substances that are currently used in textile and leather production that are known to cause harmful effect to the environment and/or human health. Many of these substances impact us in many ways such as:

- Toxic (oral/ dermal/ inhalation)
- Skin or eye irritation
- Respiratory or skin sensitisation
- Carcinogenicity
- Mutagenicity and genotoxicity

- Reproductive toxicity
- Endocrine disruptors
- Specific target organ toxicity (STOT)
- Neurotoxicity
- Teratogenicity
- Developmental toxicity/ CNS Inhibitor
- Persistent, bio-accumulative, aquatic toxic (PBT)

These 'hazard end-points' have been defined and categorised in the classification of Globally Harmonized System (GHS), Labelling and Packaging of chemicals, which was published under the aegis of the United Nations. The GHS harmonizes the hazard categories, hazard and precautionary statements and pictograms/labelling across the global so that there is a most associated approach to classification and communication of hazards in chemicals. Other sources of hazard endpoints are the OECD Screening Information Data Sets (SIDS) and the Alternatives Assessment criteria for hazard evaluation followed by USEPA Design for Environment Program (DfE). Based on these hazard endpoints, several chemical groups have been identified as hazardous and which are the focus of elimination by brands as well as regulatory bodies. Some of these are described in the table below:

Table 15.1 List of harmful chemical groups, their application and harmful effect

Chemical Group	Use/ Application	Harmful Effect
AP/APEOs	Wetting, Detergency, Emulsification	Endocrine Disruptors, Aquatic toxicity
Chlorophenols and Chlorotoluenes	Preservatives, Pesticides	CMR
Banned amines from azo dyes	Direct, Basic, Disperse, Acid dyes	Bladder cancer
Carcinogenic Dyes	Direct, Basic, Disperse dyes	Cancer
Allergenic Disperse Dyes	Disperse dyes	Skin allergy
Glycols/ Solvents/ VOC	Spot cleaners, adhesives, softeners	CMR/ Ozone depletion
Heavy Metals	Dyes, Pigments, caustic soda	STOT, Toxic
Organotin Compounds	PU Coatings, PVC	Mutagenic
Polycyclic Aromatic Hydrocarbons (PAHs)	Pigment black, Dispersing agents	CMR
Per and Polyfluorinated Compounds (PFCs)	Oil and Water repellants	vPvB, Thyroid problems
Phthalates	Plastisol prints, softeners	CMR, Endocrine Disruption

Waste from textile production is also a matter of grave concern. Various processes in wet processing result in waste generation in terms of spent chemicals, used chemical drums, wastewater and sludge from effluent treatment operations. Safe management and disposal of this waste is a formidable problem. Waste minimisation efforts can be implemented in a textile facility through:

- *Chemical recovery and re-use*: For example, recovery of caustic soda from mercerization process or Glauber's Salt through reverse osmosis process in a ZLD plant.
- *Recipe optimisation:* Improving Right-First-Time (RFT) Dyeing rate and Lab-to- bulk reproducibility percentage to reduce re-dyeing and additions in batch processes, leading to a reduction in chemical and water usage
- *Improve scheduling:* Assessing the planning of uninterrupted processes like sizing, desizing, padding and so on; to confirm that the identical chemical bath is employed as repeatedly as possible, thus decreasing the number of wastes to the drain per day.
- *Process changes:* Incorporate low- liquor dyeing or cold -pad-batch or continuous dyeing processes, resulting in less water usage and thus fewer chemicals
- *Correct handling and storage:* Implement safe storage and handling practices to reduce spillages, breakages and wastages during internal transportation of chemical containers
- *Automated dosing systems:* Processes can be optimised and wastage of chemicals drastically reduced through the mixing of the chemicals centrally and then directly pumping them onto machines through calibrated automatic dispensers, without any manual handling.
- *Packaging material:* Ordering raw materials in bulk quantity or returnable intermediate bulk containers (IBCs) can help reduce wastages and spillages, besides reducing the amount of storage space required. Plastic carboys that can be returned to chemical vendors without washing can also reduce wastage.

15.2 The need for standards

Having understood the need for sustainable practices in textile processing, it is imperative to measure or demonstrate these practices in finished products or production stages. This requires sustainability standards that outline a framework or specific criteria, which can be followed by textile producers in their production processes and end-products. Some of these standards can lead to a certification or labelling on end-product so that producers can demonstrate their conformance to these standards to their buyers and consumers.

The drivers for such sustainability standards are:

- Growing consumer awareness
- Requirements of brands and retailers as well as legal obligations
- Impact of NGO Campaigns

15.2.1 Growing consumer awareness

A survey conducted by Eurobarometer [6] for the European Commission conducted in October 2012 among a representative sample of EU citizens in 27 member states found that 85% of the respondents were worried about chemicals present in clothes and shoes. More than 50% of these respondents felt that it is the responsibility of national authorities to ensure the safety of chemicals used in clothes and shoes, rather than it is only the responsibility of manufacturers.

There is no doubt that there is increasing consumer awareness and concern with respect to the use of safe chemicals in products. They want to know which chemicals have been used and whether the manufacturing process was sustainable. The consumer concern on information available on chemicals presents in daily routine products was highlighted by UN Chemicals in Products (CiP) Programme [7]. The CiP is actively involved in the improvement of information access and availability throughout the life cycle of the products – in order that they can properly manage those products and the chemicals present in them. The CiP programme aims to bring in transparency about chemicals used in products by letting consumers know what is in the products they buy and seek safer alternatives.

Eco-labels such as OekoTex 100, EU Flower, Global Organic Textile Standard (GOTS) and bluesign offer "tags" on the consumer articles that assure consumers that the products have been "either tested for harmful substances" or have been manufactured by a supply chain which has implemented sustainable practices as per the standards and norms of these eco-labels. Such eco-labels help to improve consumer confidence about sustainable products as well as increase the demand for eco-friendly products.

15.2.2 Requirements of brands and retailers as well as legal obligations

Brands and retailers need to communicate to consumers that their products have been made ethically and sustainably as well as comply with all legal requirements. For this, they have to communicate to their supply chain partners specific norms for product safety as well as worker and environmental safety.

Apparel and footwear brands have relied on individual Restricted Substances Lists (RSLs) as their standard to ensure consumer safety. The RSL document is a complex list of chemical substances, CAS Numbers, limit values and test methods that are handed down to their CMT (Cutting/Making/Trimming) suppliers, who are then expected to communicate the RSL to their upstream supply chain partners, including fabric mills and laundries. This paper trail often led to RSL failures and product recalls as the understanding and implementation of the RSL document by a fragmented supply chain was a challenge. Individual brands published their RSLs with varying substances lists and limit values that led to confusion and misunderstanding with their supply chain partners.

The chemical industry also needs standards to communicate their conformance to safer chemistry. Third-party certification agencies such as GOTS, EcoPassport by Oeko-Tex and bluesign have well-defined standards that apply to chemical formulations that are manufactured and marketed to the textile industry. Often, these standards are part of the overarching standard or label for the finished product. Conformance to these standards helps the chemical industry to improve their sustainable chemical practices as well as increase the confidence levels in their products. With the ZDHC MRSL (Manufacturing Restricted Substances List) now becoming the industry standard for input chemical formulations, a unified approach to sustainable chemistry is fast becoming a reality and helps reduce duplication of efforts and costs for the chemical industry.

Global regulations on chemical restrictions also drive brands and their supply chain partners to adopt sustainable practices. Laws such as REACH, Cal Prop 65, WCSPA, CPSIA, GB Standards as well as pollution control norms for the discharge of effluent and air emissions need to be complied with by the supply chain partners. This makes the processes and finished products sustainable and safe. Governments enact regulations due to their obligation to protect their citizens and surrounding communities from air and water pollution caused by the manufacturing industry. The regulatory standards are developed through committees consisting of experts in their respective fields and through a study of standards followed across other countries

15.2.3 Impact of NGO Campaigns

Non-Government Organisations (NGOs) such as NRDC, Greenpeace, Friends of the Earth, WWF and Friends of Nature are working actively to highlight the issue of pollution caused by the textile industry, especially in producing countries like China, Bangladesh, Mexico, Indonesia and India. These NGOs challenge brands and governments to ensure supply of safe products to consumers and the adoption of cleaner chemicals and production methods.

The most impactful NGO campaign has been the *Greenpeace Detox Challenge* [8]. This campaign has revealed that although many brands have policies and commitments that ban the use of toxic chemicals in their supply chain, such chemicals are still being widely used and discharged in the manufacturing supplier facilities of these brands.

In China, Greenpeace has initiated a program to outline the issue of water pollution resulting from the discharge of harmful chemicals by the textile and apparel industry. The investigations focused on two facilities in China on the Pearl river that were suspected to be releasing a variety of persistent and harmful chemicals with hormone-disrupting properties. Greenpeace has published their report "*Dirty laundry Report 1- Unraveling the corporate connections to toxic water in China: testing of effluent in China*" which showed the results of these investigations. The obtained results are indicative of a bigger problem that might be of serious concern for our environment.

After this investigation, Greenpeace undertook another campaign with a focused approach on Nonylphenol – a proven endocrine disruptor. They collected 78 articles from various brands' flagship stores and tested them for nonylphenol ethoxylates (NPEs). The results of these findings were published in their '*Dirty Laundry Report 2 – Hung Out to Dry: Unraveling the toxic trail from pipes to products*". The results showed that out of 78 analysed articles, two-thirds of articles displayed the presence of NPEs above its detection limit, which is 1mg NPEs/kg of material. This indicated that alkylphenol ethoxylates (APEOs) are being extensively utilised in the textile processing unit, even for leading brands – irrespective of their claims of implementing RSLs at the supplier level.

In March 2011, Greenpeace undertook a further study ("*Dirty Laundry Reloaded: How big brands are making consumers unwitting accomplices in the toxic water cycle*") on these garments to check the release of these NPEs on domestic laundering. 14 samples out of the 54 positive tested garments were subjected to domestic laundering, and the release of NPEs at each wash was measured. The results from this testing showed that after first wash almost 17% to 94% of NPE residues are leached from the fabric, with > 80% NPEs are being washed out from 50% of the samples.

Greenpeace questioned the brands to obligate their whole inventory network to move to the utilisation of zero-toxic synthetic substances by launching the "Detox Now!" programme. The activists of Greenpeace asked potential consumers to "re-think" their choice to buy the contaminated apparel and clothing. In the wake of raging stores, the activists distributed program pamphlets to consumers and gave store staff yellow admonition cards that advised the brand line of apparel to "play clean". Greenpeace urged purchasers to confront significant brands to "Detox" their inventory network and items and to support a pollution-free future.

The impact of such NGO campaigns is a defining moment for an industry. With brands being at the top of the supply chain pyramid, they are able to generate change in the industry through their purchasing power. Using this philosophy, NGO campaigns target brands and push them to set standards for the industry to achieve new sustainability paradigms.

The Greenpeace Detox campaign resulted in a collaboration of global apparel and footwear brands that made a promise to take out all harmful discharge from its entire product life cycle and its supply chain by 2020. This collaboration – called as the Zero Discharge of Hazardous Chemicals (ZDHC) – has now grown to 29 Signatory brands and has culminated in the formation of a legal entity in 2015 called the ZDHC Foundation [9].

The ZDHC Roadmap to Zero Programme has published standards for the textile industry with respect to sustainable chemical management. The backbone of the Programme is the MRSL or Manufacturing Restricted Substances List, which lists the substances that should not be utilised deliberately in commercial chemical preparations while allowing for threshold limits (at ppm level) for impurities or unintentional residues. Another standard is the ZDHC Wastewater Guidelines that define wastewater quality discharged from a textile manufacturing unit in terms of conventional parameters as well as the MRSL Chemical Groups. These Guidelines establish sampling points, limit values and test methods for wastewater that needs to be monitored by textile processing facilities.

15.3 Types of standards

Numerous certifications and standards are available, which can be helpful to manage sustainability in product supply chains. To maintain the sustainable supply chains, the standard should focus on the pillars of sustainability such as environmental, social, products and the process.

A standard is a collection of requirements, guidelines or criteria that determine how something should be performed or how procedures should take place, whereas certification is a contractual specification culminating in a document or label that conveys that the certification standard or criteria have been met.

Consequently, a substance or procedure may be approved against a specific standard or a number of different standards based on environmental and social conditions. Although certification's goal is to ensure that certain conditions have been met, joining specific schemes and projects is a way to gain support, benefit, learn from others, and cooperate in the supply chain with others for sustainability.

Many of the initiatives and standards were established in the supply chains of major global fashion companies as a response to poor practices [10].

Some of the critical Standards that one is likely to come across in the textile industry are:

- **Statutory standards:** REACH, Cal Prop 65, CPSIA, WCSPA, TSCA, GB Standards, Act on Control of Household Products containing Harmful Substances (Act no 112)- Japanese Law 112, India's Environment Protection Act
- **International standards**: ISO 14001, ISO 18001, ISO/TS 14067 (Carbon footprint standard)
- **Voluntary standards:** Standard 100 by Oeko-Tex, STeP by Oeko-Tex, GOTS, bluesign, Greenscreen, Global Recycled Standard, Cradle to Cradle, EU Flower, Nordic Swan, Higg Index, ZDHC

Given below is a quick overview of some of the Standards mentioned above.

15.3.1 Statutory standards

Statutory standards are laws enacted by state and central governments to ensure the safety and health of its citizens and minimise degradation of air, water and soil quality to the extent that it becomes harmful to humans, flora and fauna.

15.3.1.1 REACH (Registration, Evaluation, Authorisation and Restriction of Chemicals)[11]

This is a European Regulation enacted on 18th December 2006 as a Directive No. EC 1907/2006. It came into force on 1st June 2007. It harmonises about 40 legislations and directives that existed in the European Union on restrictions in chemicals. REACH addresses the production and use of chemicals in the industry as well as for domestic use (such as paints, cleaning agents, etc.) and their potential effects and impact on human health and the environment. REACH is applicable to manufacturers and importers of chemicals and articles in the Member States of the European Union. The scope of REACH covers pure substances, substances in preparations or mixtures and substances in articles.

ECHA periodically releases a Candidate list of "Substances of Very High Concern (SVHC)". SVHCs are substances that are harmful to human and environmental health such as carcinogenic, mutagenic, reprotoxic, endocrine disruption, persistence, aquatic toxic, etc. Any SVHC used above 0.1% concentration by weight in an article and which is > 1 tonne per annum has to be notified to European Chemical Agency (ECHA). In case the tonnage of the SVHC is < 1 tonne per annum, then the use of the SVHC has to be communicated to all recipients of the article across the supply chain, along with safe use practices

Under REACH, every chemical manufacturer or importer must register his substance or substances in preparations or articles with ECHA or stop the manufacture, import or use. All required information about the substance should be communicated to ECHA through a Registration Dossier, composed of a Technical Dossier and a Chemical Safety Report. ECHA will then evaluate these Registration Dossiers for risks of these substances to human health and environment for the propose of:

- Identifying the substance as an SVHC
- Authorisation of its use under Annex XIV, based on a 'Sunset Date' for progressive phase-out
- Restriction of use under Annex XVII

15.3.1.2 CPSIA (Consumer Protection Safety Improvement Act) [12]

This is an important product safety federal law in the USA that was amended in 2008 and provides CPSC (Consumer Safety Protection Council) with a new regulatory and enforcement tool. The CPSIA includes requirements for lead, among other things, addressing, toy safety, phthalates, toddler products, and durable infant, certification and third-party testing, imports, tracking labels, criminal and civil penalties. Currently, under section 101 of the CPSIA, the main content in children's goods should not surpass 100 ppm. Similarly, the main content in any paint or surface coating used in children's goods should not surpass 90 ppm.

As per section 108 of the CPSIA, usage of six phthalates, namely, DBP, DEHP, DIDP, BBP, DNOP and DINP, is restricted in products such as toys and childcare articles. The concentration of these phthalates (as a sum) should not exceed 0.1% or 1000 ppm in the articles.

CPSIA defines the term "children's product" and generally requires that children's products complying with all applicable children's product safety rules. They should be tested for agreement by a CPSC-accepted qualified laboratory unless subject to an exception. The product should have perpetual tracking information attached to the product and its packaging where practicable. And the written Children's Product Certificate must be there that delivers evidence of the product's compliance.

CPSIA also needs domestic importers or manufacturers of non-children's products to provide a General Certificate of Conformity (GCC). These GCC's are applicable to products that are subjected to a consumer product safety rule or any similar CPSC standard, ban, rule, or regulation imposed by the Commission. Finally, for some toddler products, durable infant products including cribs the CPSIA lists special necessities in Section 104.

CPSIA also mandates having tracking labels for children's products. These labels must provide identifying information and to be attached to the products and its packaging. All tracking labels must comprise certain necessary information, such as:

- Manufacturer's name
- Manufacturing location code
- Manufacturing date
- Comprehensive information on the manufacturing process, such as a run or batch number, or other identifying features

15.3.1.3 California Proposition 65 (Cal Prop 65) [13]

This is a State Law and applicable only to the State of California. It was passed in 1986 and originally called "Safe Drinking Water and Toxic Enforcement Act 1986 – Proposition 65." The purpose of this is to protect the natural water resources from harmful chemicals that cause deadly diseases and to decrease the exposure to such chemicals in products and give warning in advance to such exposures. Proposition 65 prohibits companies carrying out manufacturing activity within California from significantly releasing these listed toxic chemicals into drinking water bodies. Enforcement of the law is carried out by lawsuits for injunctive relief and civil penalties, which can be more than 2500 dollars per violation per day.

The California Environmental Protection Agency has published a list containing more than 820 chemical substances until now, which are identified to cause deadly diseases. The chemicals used in textiles which are listed in the Cal Prop 65 List are carcinogenic dyes, carcinogenic amines released from azo dyes, phthalates and other chemicals such as lead, cadmium, PCBs, toluene, vinyl chloride, asbestos and PAH. It is the responsibility of the importer to prove that the imported article is free from listed chemicals to the Office of Environmental Health Hazard Assessment.

If an article has any of the listed chemicals, then a warning label must be put on the product affirming that, "This product contains a chemical known to the State of California to cause congenital disabilities, cancer, or other reproductive harm". Once a chemical is listed under the Cal Prop 65 List, companies have 12 months to comply with the cautioning provisions.

15.3.1.4 TSCA - Toxic Substance Control Act [14]

This is a federal law of the United States passed by the United States Congress in 1976 and directed by the U.S. Environmental Protection Agency (EPA). The law controls existing as well as the newly introduced chemicals. The prime aim of this law is to determine and regulate new chemicals prior to their entry in the market, to impose a regulation on already existing chemicals before 1976 such as PCBs, lead, mercury, etc.) which posed an unacceptable hazard to health and the environment, and to control the supply and utilisation of these chemicals.

15.3.1.5 WCSPA: Washington Children's Safe Product Act [15]

WCSPA was enforced in April 2008 and is only applicable in the State of Washington. The main aim of this law is to reduce the risk posed by harmful chemicals present in children's goods.

WCSPA has been implemented in two parts:

The first part puts restrictions on children's products comprising cadmium, lead, and phthalates.

The second part relates to a list of "Chemicals of High Concern to Children" or CHCC.

Currently, as per this law, lead content in an article should be less than or equal to 90 ppm, cadmium content should be less than or equal to 40 ppm, and concentration of 8 restricted phthalates should be less than or equal to 1000 ppm individually or in combination.

If a CHCC is deliberately added as a toxin exceeding 100 ppm in children's products, such as garment, toys and jewellery, then the manufacturers or distributors must provide a notice to the Washington Department of Ecology. Reporting a CHCC is not required if a product contains a CHCC above 100 ppm as a contaminant, provided the manufacturers or the distributors can demonstrate that they have a systematic manufacturing control program to minimise the presence of the contaminant.

15.3.1.6 Act on Control of Household Products containing Harmful Substances (Act no 112) [16]

This Act, popularly known as Japanese Law 112, is applicable for textile articles manufactured or imported in Japan. It was enacted by the Japan Ministry of Health, Labour and Welfare in 1973 and amended in April 2015.

The law prohibits different substances for use in textile, including:

- Organo- mercury compounds
- Formaldehyde (banned in baby clothes, age 0-24 months)
- Tris (1-Azinydyl) Phosphine Oxide
- Tris (2,3-dibromo propyl) phosphate
- Dieldrin
- Triphenyl tin
- Bis(2,3-dibromopropyl) phosphate
- 4,6-Dichloride-7-(2,4,5-trichlophenoxy)
- Trifluoromethylbenzimizole
- Azo dyes*

*The Act designates azo compounds as hazardous substances and restricts the existence of any of the twenty-four (24) specified aromatic amines to less than 30 mg/kg. The testing is to be performed using gas chromatograph mass spectrometer (GC-MS).

Textile products covered under the Act are underwear, caps, diapers, sleepwear, diaper covers, gloves, outer garments, socks, hats, intermediate garments, floor coverings, bedding, handkerchiefs, tablecloths, towels, collar ornaments, and bath mats.

Fur and/or leather products covered under the Act are intermediate garments, and floor coverings underwear, caps, gloves, hats and outer garments.

15.3.1.7 Indian Environment Protection Act 1986

This is an Act enacted by the Indian government according to Article 253 of the Constitution. It entered into force on 19th Nov 1986 and contained 26 sections, and amunded in 1991.

The purpose of the Act is to:

- Protect and improve the quality of the environment
- Control, abate and prevent environmental pollution

The Act is an "umbrella" legislation intended to offer an outline for the coordination of the activities of many state and central authorities established under preceding laws, such as the Air Act and the Water Act by the central government.

The scope of this standard is:

- Laying down standards for quality of the environment
- Setting standards and maximum allowable limits for release of pollutants in different media (air, water, soil)
- Co-ordination of actions by central and state authorities
- Establishing protocols and protections for the prevention of accidents which may cause pollution, with corrective actions for such accidents
- Laying down safeguards and procedures for the handling of harmful substances
- Restriction of areas for industrial operations
- Establish or recognise environmental laboratories

The Environment Protection Act includes major rules related to the management of chemicals. These are:

1. The Manufacture, Storage and Import of Hazardous Chemical Rules, 1989
2. The Hazardous Waste (Management and Handling) Rules, 1989
3. The Chemical Accidents (Emergency Planning, Preparedness and Response) Rules, 1986.

15.3.1.8 G B Standards [17]

G.B standards are issued by the Chinese National Committee for ISO, Standardization Administration of China (SAC) and IEC. GB stands for *Guobiao* or "Standard Code". There are 4 levels of Chinese standards:

- National Standards referred to as GB standards: They are consistent across all of China and maybe mandatory or voluntary standards. About 15% of standards are mandatory and 85% voluntary. The Mandatory National Standards are prefixed 'GB', Voluntary National Standards are prefixed 'GB/T' (T means "*Tuijian*" or "recommended" in Chinese) and National Guiding Technical Documents are prefixed as 'GB/Z'. Many GB National Standards are based on ISO and IEC or other International Standards. For textile and apparels, the GB 18401-2010 standard is the mandatory standard for general safety. It is divided into three types: products for infant, products for skin contact and products for non-skin contact. Currently, this standard cover product safety in terms of pH of the aqueous extract; formaldehyde content; colour fastness to rubbing, to water, to odour, to perspiration, to saliva; and dyestuffs cleaved into aryl amines from azoic colourants.

- Professional Standards or industry standards: These are applied when no National GB Standards are available. These are pre-fixed as 'FZ". A necessary voluntary standard for infant clothing is FZ/T 81014-2008. This standard applies to:
  - All textile woven fabrics as the primary raw material
  - Apparel products for infants aged 24 months or younger
  - As per this standard, products for the infant should be labelled as 'Infant Product' in Chinese.
  - In addition, the textile articles for infants must comply with limits currently specified for formaldehyde, pH value, carcinogenic amines from azo dyes and heavy metals.
- Local Standards or 'Provincial Standards' are developed when both Professional Standards and National Standards are not available, but combined requirements for hygiene and safety industrial products are desirable within a local area. Local Standards are delineated with either 'DB+*/T' (voluntary) or "DB+* (mandatory).
- Enterprise Standards may be developed by an enterprise when none of the above standards is available. They are designated by 'Q+*'

It is mandatory for the materials exported to China to comply with the relevant GB Standards to be accepted. Similarly, internal stakeholders also have to comply with relevant GB Standards to be marketed inside China.

15.3.2 International standards

ISO mainly prepares these standards. They are available worldwide for use and consideration. These standards can be used either directly or through a process of amending the standards to encounter local needs.

These are one way to overcome the barriers in international trade caused by differences between standards developed and technical regulations separately by each company, national standard organisation or a country. Establishing international standards helps in having a level playing field and a common platform to conduct trade.

International standards certification is not a legal requirement but is widely accepted and trusted in trade and commerce worldwide that specific standard processes are being followed, which is audited and certified by an independent third party.

Below is a brief description of some of the international standards related to environment and sustainability.

15.3.2.1 ISO 14001 - ISO 14000

This is a collection of environmental standards prepared and published on behalf of organisations by the International Organization for Standardization (ISO). These standards deliver a guideline or outline for organisations needing to systematise and improve their efforts in environmental management and establish an environmental management policy. The standard also supports the organisation to comply with the regulations, applicable laws and other ecologically related necessities and to continuous improvement in the above. It also aids in having a common framework for comparison of the environmental management systems of different organisations.

The ISO 14001 structure is split into 10 sections. The first 3 are introductory and the other 7 contain requirements for EMS. The EMS sections (4-10) deal with:

1. Context of the organisation: define scope, processes, stakeholders and issues
2. Leadership: Top management commitment, environmental policy and responsibilities throughout the organisation
3. Planning: Risks and opportunities for EMS, improvement needs, legal and other commitments
4. Support: Competence, awareness, documentation, communication requirements
5. Operation: Controls required, potential emergencies and responses
6. Performance evaluation: monitoring of EMS, internal audits, management review
7. Improvement: Continuous developments, corrective and preventive actions for non-conformities

15.3.2.2 OHSAS 18001

Occupational Health and Safety Assessment Series 18001 is a British Standard developed by British Standards Institute (BSI) for the management of the work-related safety and health. It was called as BS OHSAS 18001. BS OHSAS compliance requires companies to show that a company has a system in place for occupational safety and health.

When ISO prepared Occupational health and safety management system ISO 45001, BSI decided to cancel BS OHSAS 18001 and adopt ISO 45001 as BS OHSAS 18001.

ISO 45001:2018/BS OHSAS18001 is a safety and occupational management system which lays down guidelines for documentation, processes for implementing and continual improvement in health and safety systems in an organisation and it's certification by a recognised third-party organisation.

15.3.2.3 ISO 14067

2018 Greenhouse gases – Carbon Footprint of Products set out the requirements, values, and standards for quantifying and recording a product's carbon footprint in a manner consistent with Life Cycle Assessment (LCA) (ISO 14040 and ISO14044).

CFP is defined as the sum of and the removals of greenhouse gas emissions in a system expressed as CO_2 equivalent and based on the LCA (3.5.3)

This international standard reports the single category of climate change impact and does not evaluate other potential environmental, economic and social impacts arising from product delivery. The CFPs calculated in compliance with this international standard does not give a measure of the impacts of the overall products on the environment.

15.3.3 Voluntary Standards

Voluntary standards are standards generally set by private sector organisations that are available for use by any government or private, organisation or individual. The popularity of voluntary standards has grown in recent years and many have almost become a necessity to carry out global international trade.

Voluntary standards are many times more stringent than statutory standards. Sometimes, this means deploying additional resources within an organisation to comply with these standards and further expenditure to the organisation.

Below is a brief description of some of the voluntary standards which are popularly adopted by the textile and apparel supply chain facilities and pushed by brands:

15.3.3.1 Standard 100 by Oeko-Tex® [18]

This is one of the most widely accepted standards worldwide by international brands. This standard, initiated by Oeko-Tex Association in 1992, is now a collaboration of 18 independent textile testing and research institutes based in Japan, Europe and their representative offices worldwide.

The Standard assures consumers that textile products at various stages of production, viz. raw materials, semi-finished goods, finished fabrics, all accessories like stitching threads, buttons, linings, zippers, labels, etc., in garments have harmful chemicals within strictly defined limits.

This standard is mainly for the end use of products. The human-ecological needs to be fulfilled by a product get stricter as the skin contact of a product gets closer. Accordingly, the standard's limits and certification of a product are divided into four classes:

Product Class I, products for infants and babies have the strictest limits and also need to pass saliva fastness test while Product Class IV articles which do not come in contact with skin (curtains, tablecloths, upholstery, etc.) have more relaxed threshold limits. The other two classes, namely Product Class II and Product Class III have threshold limits between these two classes.

Oeko-Tex Association has another standard – STeP by Oeko-Tex® - that defines technical conditions for the certification of leather and textile manufacturing units for eco-friendly process implementation, safety, health and socially accountable working environments. It provides tools for assessment and transparent disclosure of improvements in these areas to the industry and consumers.

15.3.3.2 GOTS (Global Organic Textile Standard) [19]

This is a standard, started in 2002, applicable to the processing of garments and articles from organic cotton. Organic cotton is characterised as organically grown cotton from non-genetically altered seeds and without the usage of harmful chemicals such as fertilizers or pesticides. It prohibits the cross-contamination of genetically modified or conventional cotton during farming, harvesting, ginning, baling and packing of raw cotton fibre.

The standard defines criteria for:

- certification of organic cotton through the OE100 standard
- classification of organic content into two categories: "Organic" and "Made with x% organic"
- each stage of production from ginning to fabric manufacturing, pre-treatment to printing, finishing to packing, labelling and transportation of finished products.
- input dyes and chemicals used during wet processing with respect to toxicity and environmental fate
- quality parameters for finished articles with respect to fastness
- residues of restricted substances in certified finished articles
- quality of materials and accessories that may be used during garment production to qualify to label it as an organic cotton garment.
- environment management, health, safety and social parameters

The standard ensures the transparency and credibility through the issue of *Transaction Certificates (TC)* at each stage of production namely raw fibre, yarn, fabric and garment stage. TC at each stage gives the material balance of the organic cotton issued and final product at each stage. This ensures that the quantity of final output of organic cotton product does not exceed the input quantity. Dyes and chemicals have to be certified for the GOTS criteria through approved Certification Bodies(CB) based on Safety Data Sheets and/ or testing and only GOTS-certified dyes and chemicals must be used in the wet processing of GOTS finished goods.

15.3.3.3 bluesign® [20]

This is a chemical, environmental and health management system formulated by Bluesign Technologies Ag, Switzerland in the year 2000. The aim of the standard was to develop textile products with the lowest environmental footprint and to motivate suppliers, manufacturers and top brands to reduce the overall footprint of textiles, with emphasis on chemicals.

Bluesign® system uses the Input Stream Management approach that eliminates harmful substances in the beginning itself. This is done by analysing and then selecting the chemicals, dyes and auxiliaries that are free of toxic and harmful effects before starting production so that the final product is safe for human use and minimises adverse effects on the environment. bluesign® system encourages the system partners to continually improve the safety during use of chemicals, optimise processes and use of resources to reduce environmental impact.

bluesign® system aids chemical suppliers in homologation (grading) of chemicals in three categories using the bluesign® bluetool:

- Blue: All criteria for final product, worker and environmental release are acceptable
- Grey: Can be used in manufacturing under certain pre-conditions
- Black: Fails criteria; control would be difficult or unlikely – No use is acceptable

It also helps in compliance of Restricted Substance List (RSL) and with SVHC candidate list from REACH. The system mandates suppliers to use only those chemicals listed in bluesign® bluefinder, an internet platform for bluesign® approved list of chemical products. Finished products are labelled as bluesign® approved through a hang-tag.

15.3.3.4 Greenscreen® Certified [21]

It was launched by an independent NGO (Non-Governmental Organization) named Clean Production Action in April 2017. It is a certification standard which endorses the practice of chemicals inherently safer in articles. The certification meets growing demand for harmless chemicals in production of products and a simple platform that allow the communication of harmless chemicals across supply chains of products and their formulations.

GreenScreen® Certified for Safer Chemicals is based on the GreenScreen®, a globally recognised tool that recognises harmful and safer chemicals by a laborious bench-marking scoring system.

The Greenscreen tool builds on the USEPA DfE approach and is a method for Comparative Chemical Hazard Assessment. It uses 18 Hazard End-points (based on GHS, OECD SIDS and USEPA DfE criteria) related to environmental toxicity and fate, physical hazards and human health to assess all the ingredients in a chemical formulation against standard public databases on hazards classifications and then benchmark the formulation as:

- Benchmark 1: Avoid/phase-out (contains chemicals of serious concern)
- Benchmark 2: Use but search for safer alternatives
- Benchmark 3: Use, but still an opportunity for improvement (getting there!)
- Benchmark 4: Preferred (Safer chemical-inherently low hazard)

This benchmark score can be used, without toxicological training, to adopt preferred materials, guide new product development and drive innovations of new materials. They can also be used for alternatives assessment.

15.3.3.5 Global Recycled Standard® (GRS)

Control Union originally developed this standard in 2008 and passed ownership onto Textile Exchange in 2011.

This is a full product standard which establishes third-party certification requirements for custody chain, chemical restrictions, recycled content and environmental and social policies. The GRS aims at increasing the utilisation of recycled materials and reducing/eliminating the damage instigated by its manufacturing.

The latest iteration - GRS 4.0 -was released on 1st July 2017. The standard applicable for products that hold $\geq$ 20% Recycled Content. However, only products that have at least 50% Recycled Content succeed for GRS labelling.

Recycled material is characterised as a material that, through a manufacturing process, has been reprocessed from Reclaimed Material and rendered into a final article or a part for incorporation into an article. Recycled material may be recycled cotton, wool, polyester, etc. For the material to be labelled as "Made from recycled material", all the entities engaged in production have to be certified for GRS viz. material recycling units, spinning, knitting and weaving, dyeing, printing and garment production units.

15.3.3.6 Cradle to Cradle Certified® (C2C) Product Standard [22]

This standard is rooted in the principles of Cradle to Cradle® design developed by Dr Michael Braungart and William McDonough of "The Cradle to Cradle Products Innovation Institute" located in Oakland, California. The standard is now owned and administered by the Institute which is governed by an independent board of directors.

Currently, in its third iteration, C2C is a continuous improvement mechanism that looks at a commodity across five categories of quality—social fairness, material health, water stewardship, renewable energy and carbon management and material reutilisation. A product obtains a level of achievement in 5 categories — Basic, Bronze, Silver, Gold, or Platinum.

Material health: Through the identification of material as technical and biological nutrients, material health gives an understanding of the chemical ingredients of every material in the product which helps to optimise them towards the safer materials. In addition, it helps to understand how chemical risks combine with probable exposures to assess potential negative impacts on the environment and human health.

Material reutilisation is to (1) develop the product with naturally occurring materials that can be safely returned to the industry or nature (2) maximise the percentage of renewable materials that can be reused, recycled or composted after disposing of a product.

Renewable Energy and Carbon Management is a (1) planning of the use of renewable energy for product manufacturing (2) offset carbon emissions and renewable energy source for the final stage of product manufacturing.

Water Stewardship is about managing clean water as a valuable resource and an important human right. (1) Addressing local geographical and industrial water impacts of each production unit (2) Recognise, evaluate, and optimise any industrial chemicals in the effluent of the production unit.

Social Fairness requires designing of operations to honour all people and natural systems affected by a product being created, used, disposed of or reused. (1) Utilise worldwide recognised resources to conduct self-assessments to identify supply chain problems and third-party inspections to ensure optimum conditions (2) Make an encouraging change in the local community and the lives of employees.

15.3.3.7 EU Flower / EU Ecolabel [23]

This is a voluntary standard that producers, importers and retailers can choose to apply for labelling their products. The EU Ecolabel was established in

1992 and is the label of environmental excellence awarded to services or products that meet high standards from raw material extraction to production, distribution and disposal throughout their life cycle. This encourages a circular economy by encouraging producers in the manufacturing process to produce less waste and CO_2. The focus of the EU label measures for products is on those stages that have a maximum environmental impact and this differs from product to product.

In textiles, the life cycle begins with the cultivation of cotton, and it continues to ginning, spinning, weaving or knitting, bleaching, dyeing, printing, finishing, stitching garments, packing, etc. It is well known that in fabric production, maximum environmental impacts is during bleaching, dyeing, printing and finishing processes. So, great emphasis is placed on minimising the harmful environmental impacts of dyes, chemicals and auxiliaries in this standard. An EU Ecolabel® labelled products give consumers confidence that the product has been made with minimum adverse impacts on humans, flora, fauna and aquatic organisms.

The EU Ecolabel guarantees:

- Limited use of harmful chemicals to the environment
- Reduced air and water pollution
- Textile shrink resistance during washing and drying
- Colour resistance to perspiration, light exposure, washing, dry and wet rubbing and perspiration

It enforces strict criteria for each life-cycle step in textile production that includes:

1. Fibres: Types of fibres allowed, restriction in toxic residues of fibres, reduction in air and water pollution during the fibre manufacturing process
2. Manufacturing (Process and chemicals): restriction in the use of chemicals harmful to health and environment, reduction in energy and water consumption
3. End-use: Performance, durability and re-cycling

15.3.3.8 Nordic Swan® Ecolabel

It is an eco-label that is quite popular in Nordic countries and Germany. It applies to textile, hides/skin and leather products across 60 different product groups covering numerous product types.

In 1989 the Nordic Council of Ministers established Nordic Swan. It is managed by a non-profit organisation, the Nordic Ecolabelling Board. Nordic Swan® Ecolabel is administered in Sweden, Denmark, Norway, Finland and Iceland.

Nordic Swan® Ecolabel labelling of textile products was started in 1994, and Version 4.0 is under discussion for approval to be adopted. Nordic Swan® Ecolabel labelled cotton products should have a minimum of 10% organic cotton having traceability documents. Also, the certified organisations have to comply with the use of chemicals that conform to REACH regulation on SVHCs and have a system in place towards a responsible social commitment to improve health, safety and provide a safe working environment. It should also have a goal of continually minimising environmental impacts due to its activities.

15.3.3.9 ZDHC (Zero Discharge of Hazardous Chemicals) [24]

ZDHC is a partnership between global apparel and footwear brands and retailers, as well as other industry stakeholders, with the vision for the broad application of environmental best practices and sustainable chemistry in the leather and textile supply chain. It is a holistic approach to chemical management at a manufacturing facility that uses and discharge chemicals in the production process. It has shifted the focus on input chemical management rather than end-of-pipe treatments. ZDHC goes beyond only product safety to supply chain management and innovation leadership.

The "ZDHC Roadmap to Zero Programme" establishes standards for the industry to achieve its goals of environmental, worker and consumer safety. For input chemical management, the Manufacturing Restricted Substances List or MRSL is the industry standard that lists out substances in 16 Priority Chemicals Groups that are not to be used intentionally in commercial chemical products. It also established threshold values for unintentional contaminations or impurities in these products. The MRSL is aspirational but achievable, as only those substances where safer alternatives are available are listed. The MRSL Groups listed are (1) APEOs (2) Phthalates (3) Banned amines from azo dyes (4) Brominated and chlorinated fire retardants (5) Chlorobenzenes and chlorotoluenes (6) Halogenated solvents (7) Chlorophenols (8) Short-chained chlorinated paraffins (SCCPs) (9) Heavy metals (Chromium, Lead, Cadmium, Mercury) (10) Organotin compounds (11) Perfluorinated chemicals (12) Glycols (13) Carcinogenic dyes (14) Allergenic disperse dyes (15) Navy Blue colourant (16). ZDHC has also established a platform called the ZDHC Gateway that is a database of MRSL conformant chemical formulations, verified through ZDHC-approved third- party certifications. Supplier facilities can register on the Gateway and use this database to make informed purchasing decisions on their input chemical inventory.

Another standard established is the ZDHC Wastewater Guidelines that cover the quality of the wastewater discharged from a manufacturing facility

– both in terms of conventional parameters as well as the MRSL parameters. The Guidelines establishes sampling points, limit values and test methods for all these parameters. Facilities supplying to ZDHC Signatory brands need to test their wastewater to the ZDHC Guidelines and upload their test report on the ZDHC Gateway.

15.3.3.10 Higg Index FEM 3.0 [25]

The Sustainable Apparel Coalition (SAC) in July 2012 introduced a self-assessment tool for assessing the environment and social sustainability at brands as well as manufacturing facilities. It was called the 'Higg Index' and a Facility Environmental Module (FEM Version 3.0) and was published in 2016. The tool aids a textile processing facility to assess its sustainability performance in 7 areas: (1) Environment Management System (EMS) (2) Energy and Greenhouse Gas emissions (3) Water usage (4) Wastewater management (5) Waste management (6) Air emissions and (7) Chemical management. Based on this preliminary analysis, one can decide on the areas of improvement and the necessity of third-party experts to get suggestions on ways to improve.

The section on Higg FEM Chemicals Management is a joint effort between the Outdoor Industry Association (OIA), the Sustainable Apparel Coalition (SAC), and the Zero Discharge of Hazardous Chemicals (ZDHC) Programme to converge their respective chemical tools into a single evaluation questionnaire. The questions are graded into three Levels: Level 1 (basic) to Level 3 (advanced). Each question needs to be answered as "yes", "partial yes" or "no", depending on the level of completion or action being done at the facility. Documents have to be uploaded on the SAC portal to provide proof of actions being taken for the answer. Once the self- assessment is completed and all the documents are uploaded, a verification exercise is done through a SAC- approved Verifier. The verification not only provides a reality check on the self- assessment score but also an improvement plan for the facility to implement.

15.4 Overview of five important standards and their requirements

Of all the sustainability standards and certifications, five popular and important sustainability standards are described in detail here. These are REACH, Oekotex, GOTS, Higg Index FEM 3.0 and the ZDHC Roadmap to Zero Programme.

15.4.1 REACH (Registration, Evaluation, Authorisation and Restriction of Chemicals) [26]

15.4.1.1 What is REACH?

REACH is a regulation that is applicable in the European Union (EC no 1907/2006), enacted on 8th December 2006 and brought into force on 1st June 2007. REACH is concerned with:

- Registration, Evaluation, Authorisation and Restriction of Chemicals
- Establishing a European Chemicals Agency (ECHA)
- Amending Directive 1999/45/EC
- Repealing Council Regulation (EEC) No 793/93, Commission regulation (EC) No 1498/94, Council Directives 76/769/EEC, 91/155/EEC, 93/67/EEC. 93/105/EEC and 2000/21/EC

The purpose of the REACH Directive is:

- To ensure a high level of protection of human health and the environment.
- To ensure free movement of substances while enhancing competitiveness and innovation.
- To promote the development of alternative methods for the assessment of hazards.
- To achieve sustainable development.
- To contribute to the fulfilment of the Strategic Approach to International Chemical Management (SAICAM).
- To encourage replacement of SVHC (Substances of Very High Concern) *

**An SVHC is defined in Article 57 of the REACH regulation as any substance that is Category 1 and 2 Carcinogen, Category 1 and 2 Mutagen, Category 1 and 2 Reprotoxic, Bioaccumulative, Persistent and Toxic or PBT, very bioaccumulative, Very persistent and or vPvB and any substance of comparable concern such as Endocrine Disruptors. ECHA lists SVHCs under a 'Candidate List', which is then circulated to all EU member states for review. Based on their inputs, an SVHC is put under Annexure XIV (Authorisation) or Annexure XVII (Restriction) of the regulation. As on 1st August 2019, there are 201 chemical substances listed as SVHCs.*

The scope of REACH covers pure substances, materials in preparations or mixtures and materials in articles. The regulation has different requirements for each of these categories.

15.4.1.2 How does REACH apply to a chemical manufacturer?

Under REACH, every manufacturer or importer of chemicals into the EU above 1 tonne per annum must register his substances with the ECHA or stop production, import or use. The hazards and risks of the substances must be assessed by the manufacturer and communicated to ECHA through a 'Registration Dossier'. Registration is based on the principle of "one substance, one registration". For chemicals produced or imported into the EU with a tonnage more than 10 tonnes a year, a supplementary 'Chemical Safety Report' must be sent to ECHA consisting the substance's evaluation data for persistent, human health and physico-chemical hazard, bio-accumulative and toxic (PBT), environmental hazard, and very bio-accumulative (vPvB) assessment and very persistent.

If from CSA, the registrant concludes that the substance is classified as dangerous, following must be performed:

- an exposure assessment including the generation of exposure scenarios or the identification of relevant categories of use and exposure and estimation of exposure
- a risk characterisation

The information requirements to be submitted in the Registration Dossier are listed in Table 15.2.

Table 15.2 Information required in the registration dossier for REACH

Physicochemical endpoints	Invertebrate animal endpoints	Vertebrate animal endpoints
Description of the state of the substances at 20 °C/103.3 kPa	*In-vitro* skin irritation/corrosion	Acute toxicity (Oral)
Melting/ Freezing Point	*In-vitro* Eye irritation	
Boiling Point (if applicable)	Skin sensitisation	
Relative Density	*In-vitro* gene mutation in bacteria	
Vapor Pressure (if applicable)	Short- term toxicity on invertebrates	
Surface Tension (if applicable)	Inhibition of growth study on aquatic plants	
Water Solubility	Ready biodegradability	
Partition Co-efficient		
Flammability		
Granulometry (if applicable)		
Flash Point		
Self-ignition temperature		
Oxidative/Explosive properties		

After the Registration phase for substances (the deadline for substances between 1–10 tonnes per annum was May 2018), the next step is of Evaluation of the Registration Dossiers by ECHA.

If ECHA or the assessing Member State considers that the use of the substance poses a risk after review of the available and new data submitted in the Technical Dossier, it may then proceed with follow-up actions such as:

1. A proposal for harmonized classification and labelling for reproductive, respiratory sensitizer or other effects that are carcinogenic, mutagenic or toxic.
2. A proposal to identify the substance as a 'Substance of very great concern' (SVHC).
3. A proposal to authorise usage (list in Annex XIV) or restrict the substance (Annex XVII).
4. Actions outside REACH's scope such as a proposal for EU-wide occupational exposure limits, national measures or voluntary actions by industry.

Authorisation aims to ensure that risks are controlled, and that appropriate safer alternatives progressively substitute them. These are listed under Annex XIV of REACH. Such substances can be used under Authorisation from ECHA until the "Sunset Date" after which its use is prohibited.

Restrictions may limit or ban a product from production, putting it on the market or use. These are classified under REACH Annex XVII.

15.4.1.3 How does REACH apply to a textile article?

Any manufacturer or importer of articles shall submit to ECHA a registration for any substances contained in those articles if both conditions are met:

1. In those products, the substance is present within amount greater than 1 tonne per manufacturer or importer per year.
2. The material contained in the article shall be published under usual or practically foreseeable conditions of use.

Manufacturer or importer of textile articles also needs to comply with the SVHCs listed under REACH regulation. There are three types of obligations required for SVHCs:

1. If an SVHC on the Candidate List is used in an article produced or imported into the EU above 0.1% concentration, it is an obligation for EU producers and importers of articles to communicate the name of the SVHC and safe use of the article to the recipients of the article across the supply chain. Also, if the SVHC use in the article works out to > 1 Tonne per annum, then the producer or importer should notify ECHA about the SVHC in the article

2. If an SVHC or substance listed in Annexure XIV of REACH must be used in an article produced in the EU, but not imported into the EU, then proper Authorisation from ECHA needs to be taken by the producer of the article. The objective of the authorisation process is to ensure that usage of SVHCs is under control and that the alternative substances are progressively replacing SVHCs. ECHA grants authorisation for using an SVHC only if the risks from the SVHC are controlled or the socioeconomic benefits of using the SVHC outweigh the risks. An SVHC listed in Annexure XIV can also be used if it has not reached the Sunset Date.
3. Any SVHC or substance listed in Annexure XVII of REACH should not be used in an article produced or imported into the EU. Article 67(1) of the REACH regulation states that a substance is not to be manufactured, placed on the market or used on its own, in the preparation of an article set out in Annexure XVII.

It is also prudent for exporters of textile articles to the EU to check if any SVHCs are present or used in their articles or if there is any intended release from the article.

15.4.2 Oeko-Tex

In 1989, the ÖTI (Institute for Ecology, Technology and Innovation GmbH - OETI, Vienna) developed a testing and certification system for pollutants according to ÖTN 100 standard in order to respond to the increasing public interest in textile ecology and health. ÖTN is the abbreviation of Österreichische Textilforschungsinstituts-Norm. A total of 16 ÖTNs (standards) under the ÖTN 100 series were developed for pollutant testing of textiles. Similarly, in 1991, Germany›s Hohenstein Research Institute established and performed the pollution analysis according to 'Hohenstein Pollution Check'.

In 1992, these two research institutes founded "International Association for Research and Testing in the Field of Textile Ecology", also called Öko-Tex or Oeko-Tex.

Over the years, the association has introduced different standards and certifications that assures the consumer of the safety and sustainability of the products at different stages of the textile value chain - raw materials, semi-finished goods, finished fabrics, garments, accessories used in garments – and covers safe chemical usage, energy conservation, water and air pollution plus training, welfare of employees and humane working conditions.

The various standards and certifications offered by Oeko-Tex Association members are

- Standard 100 by OEKO-TEX® (1992)
- STeP® by OEKO-TEX® (2013) earlier named OEKO-TEX Standard 1000 (1995)
- Made in Green by OEKO-TEX® (2015) earlier called as OEKO-TEX Standard 100Plus
- Eco Passport by OEKO-TEX® (2016)
- Detox to Zero by OEKO-TEX® (2016)
- Leather Standard by OEKO-TEX® (2017)

The above standards by OEKO-TEX® Association are described in further sub-sections giving the salient points in each of the above standards and certification

15.4.2.1 Standard 100 by OEKO-TEX®

Standard 100 by OEKO-TEX® (previously called OEKO-TEX 100® Standard) standard and certification is to ensure that textile products at various stages of production viz. raw materials, semi-finished goods, finished fabrics, all the accessories like stitching threads, buttons, linings, zippers, labels, etc., in garments have harmful chemicals within strictly defined limits and is safe for human use.

Since its introduction in 1992, the central focus of by OEKO-TEX® STANDARD 100 has been on the development of scientifically based test criteria, test methods and limit values. The STANDARD 100 by OEKO-TEX® takes into account based on its extensive and stringent list of steps, with various hundred controlled individual substances:

- Important legal regulations, such as banned nickel, azo colourants, cadmium, pentachlorophenol, formaldehyde.
- Various hazardous chemicals, even if they are not yet legally regulated.
- Requirements of Annexes XVII and XIV of the EU REACH as well as of the ECHA SVHC Candidate List insofar as they are assessed by expert groups of the OEKO-TEX® Association to be relevant for fabrics, textiles, garments or accessories.
- Requirements from the US Consumer Product Safety Improvement Act (CPSIA)
- Various also environmentally relevant substance classes

The standard OEKO-TEX® limitations and test methods are based on the end-use of the textile materials and products. The human-ecological requirements to be complied by a product get stricter with as the skin contact of a product gets more intensive and more sensitivity of the skin. Accordingly, the standard's limits and certification of a product is divided into four classes:

Product Class I: These are products for babies and toddlers up to the age of 3 years such as terry products underwear, bed linen, rompers clothing, etc. The standard limits for this class are the strictest and also include additional criteria such as fastness to saliva test.

Product Class II: These are products that are worn close to the skin such as socks, underwear, T-shirts, bed linen, etc. The standard limits for this class are slightly relaxed compared to that of Class I, but stricter than Class III and Class IV products

Product Class III: These are products used away from the skin such as coats, jackets, etc. The standard limits for a product are further relaxed compared to that of Class I and Class II.

Product Class IV: These are furnishing materials and decoration articles. The limits for a product in this class is higher than the other three classes.

As an example, the limit value for free formaldehyde in Product Class I is not detected (< 16 ppm), for Class II is 75 ppm, For Class III is 150 ppm and Class IV is 300 ppm.

If a product submitted meets the OEKO-TEX® Standard 100 parameters than one can use the OEKO-TEX label that has a unique Test Number to identify the user, the certificate is valid for 1 year and has to be renewed by submitting fresh samples and certificate fee. The Hohenstein Institute ensures the maintenance of standards by random checking of products from retail outlets and visiting the factory to collect the samples.

Advantages of standard 100 by OEKO-TEX® certificate are as follows:

1. It ensures that the articles have been screened for toxic chemicals and parameters are maintained within certain specified limits i.e. the textile is safe to wear/use.
2. The certificate is a sort of passport to export textiles to Europe. It also ensures that you become a preferred supplier to many well- known branded garment manufacturers
3. In case of export of made-ups, it guarantees the buyer that the fabric and all the accessories used viz. buttons, labels, zippers are free of harmful substances.
4. In having 4 Product Groups, it prevents one from any misunderstandings regarding the parameters to be maintained.
5. The Standard also ensures that the same testing procedures are followed at all centres and uniformity is maintained.

15.4.2.2 STeP by OEKO-TEX®

OEKO-TEX®'s Sustainable Textile Production or STeP [27] was introduced in 1995 by the Oeko-Tex Association to define technical conditions for the certification of production facilities to implement environmentally friendly processes, socially responsible working conditions and optimal health and safety of workers.

STeP by OEKO-TEX® evaluates manufacturing facilities for the entire production chain including dry spinning, twisting, special yarn production, winding, etc., wet spinning, pretreatment, dyeing, printing and finishing, garment and accessories manufacturing, foam and mattresses.

The standard evaluates, tests and certifies the following modules in textile production:

- Chemical Management - This module evaluates chemicals (dyes, auxiliaries, accessories) used at various stages of production chain for toxicity, carcinogenic, mutagenic effects, etc. It also evaluates and tests these for the safety of humans, flora and fauna and the environment.
- Environmental Performance - This section addresses the facility's environmental impacts in the processing and storage of chemicals, consumption and use of resources such as water, energy, water, carbon emissions, generation of waste, transport and prevention of unintended incidents such as chemical and oil spills, effluent discharge, sludge.
- Environment Management Systems - The organization should have a clearly defined EMS that includes Environment Policy, well-defined processes and procedures, continual improvement plans based on initial benchmarking of environment, gap analysis, etc. The management shall also ensure adequate financial and personnel support to make EMS a day-to-day implementation and also make provisions for regular audits, maintain the facility's environmental performance and fulfil external and internal legal obligations. It should also have a system of regular review of environmental performance and improvement actions based on these reviews.
- Social Responsibility Management System - This module covers the management policy towards social responsibility, compliance with national, legal and other requirements that apply to the organisation. It also deals with prevention of child labour, communication with stakeholders, work contracts, safe working conditions for juvenile labour, prevention of harassment and abuse, wages and benefits, freedom of association/ right to collective bargaining, compulsory and prison labour and exploitation, prevention of discrimination, Prevention of forced, sanitary facilities, health benefits, provision of dormitories, canteen/eating areas, creche, etc.

- Quality Management System - This module is based on the implementation of the Quality Management System accredited to ISO 9001. Though STeP certification does not need formal certification of ISO 9001 standard, it helps to confirm that the organisation has a QMS in place that formalises Quality Policy, Process description, Documentation, Delegation of responsibility known throughout the facility, Product traceability, Quality control system, Employee training and Continual improvements. It also demonstrates that risk management, corporate governance, and business excellence are essential parts of strategic business decisions.
- Health and Safety - This covers safety and health at work of the manufacturing unit. It evaluates conditions in the workplace like dust, noise, lighting, chemical risks, heat stress, care for employee safety and health via the facility of protective clothing (e.g. PPE) and injury prevention through safe equipment/machinery. The module also includes the protection of buildings in relation to emergency procedures, fire prevention, building safety and the safety of workers' health in the event of such incidents. Furthermore, it assesses the safety of production and its installations.

15.4.2.3 Made in Green by OEKO-TEX®

Made in Green by OEKO-TEX® is a label [28] for products produced using sustainable processes and tested for the existence of harmful materials. The label offers consumers transparency with unique Product ID or a QR code which traces to production facilities and the countries.

Made in Green by OEKO-TEX® certification is eligible for production facilities that are already certified for STeP by OEKO-TEX® and Standard 100 by OEKO-TEX®.

Benefits of Made in Green by OEKO-TEX®:

- The Made in Green gives a transparent communication tool to educate customers and other stakeholder groups about responsible behavior credibly.
- By certifying a product with Oeko-Tex standard 100 legal compliance and effective consumers against harmful chemicals are ensured.
- At the same time, sustainable production and working conditions of the supply chain are being ensured through the STeP certification of production facilities.
- More effective manufacturing processes offered by STeP certification provide additional economic benefits, savings and an optimum position in global competition.

15.4.2.4 Eco Passport Standard by OEKO-TEX®

Most of the standards discussed so far concentrate more on the presence of toxic substances in the fabrics/made-up articles and the impact of the same on humans and the environment. It would a logical step if the toxic substances were controlled and limits observed in the ingredients used in the manufacture of dyes, chemicals and auxiliaries.

EcoPassport [29] by OEKO-TEX® introduced in the year 2016, by Oeko-Tex Association, bridges this gap and defines the technical conditions to be complied with for certification of textile and leather chemicals, auxiliaries, colourants and licensing of the use of its trademark. The certificate excludes colourants and auxiliaries containing genetically modified organisms, flame retardants, biocides, pesticides and other bio-active substances. The certification is a 3-stage process, of which the third stage is optional. These are:

Stage I: Products are screened at the level of the ingredient through a screening of the CAS number and compared to the EcoPassport list of restricted substances (RSL).

Stage II: Analytical test in an OEKO-TEX® Institute laboratory presence of restricted substances as per the EcoPassport RSL. As long as all of the requirement of this standard are met (and the optional STAGE III has not been selected), the OEKO-TEX® Institute research issues the EcoPassport certificate for that chemical product.

Stage III: This is to evaluate if product stewardship systems exist at the chemical manufacturer. This is done through a document assessment of the systems (as per OEKO-TEX® requirements) and an on-site visit of the chemical manufacturer manufacturing facility to verify ecological and product stewardship actions by the factory.

15.4.3 Global Organic Textile Standard (GOTS) [30]

The starting point of GOTS was a workshop launched in 2002, Intercot conference, in Dusseldorf (Germany). It was made up of representatives of organic cotton producers, the textile industry, standards organisations, consumers and certifiers to discuss the need for harmonisation and worldwide recognised single organic textile standard to replace a plethora of different organic textile standards prevalent at that time. The different standards created confusion in producers of raw cotton, processors, traders and consumers who were interested in organic textiles and was a barrier to a smooth exchange between interested domestic and international stakeholders.

After a number of meetings, in October 2006 the International Working Group (IWG) on Global Organic Textile Standards launched the first version of Global Organic Textile Standard. They also released the GOTS label in 2008 along with Version 2.0 of GOTS standard. The latest is GOTS Version 6.0 released on 1st March 2020 to be complied by 1st March 2021.

The GOTS standard is comprehensive in the sense that it is applicable right from ginning stage through spinning, weaving, wet processing, producing of made-up articles, packaging and labelling, trading and distribution. In addition, dyes, auxiliaries and accessories for made-ups have to fulfil stringent criteria and must be approved by GOTS for usage in manufacturing of GOTS certified organic textiles.

Cotton should be certified by recognised national or international standards (IFOAM family of standards, EEC 834/2007, USDA NOP, APEDA in India) for organic status. During cotton cultivation, no chemical fertilizers are to be used and cross-contamination with conventional fibres is prohibited. An important aspect is that Genetically Modified (GM) cotton seeds should not be used to grow the cotton i.e. cotton should be GM-free. All social norms such as minimum wage, 8-hour working, overtime payment, etc., should be followed at the cotton cultivation farm.

The standard provides for a sub-division into two label-grades:

- "Organic", which means the fabric or yarn is containing ≤ 5 % non-organic natural or synthetic fibres and ≥ 95% certified organic fibres,
- "Made with x % organic", which means the fabric has ≤ 30 % non-organic fibres and ≥ 70% certified organic fibres, but a maximum of 10% synthetic fibres (respectively 25% for sportswear, socks and leggings)

It is not permitted to blend organic and conventional fibres of the same type in the same article.

The criteria for other stages of production are as follows (described in Section 2.4 and summarised in Table 15.3):

Table 15.3 GOTS criteria for each production stage

Production Stage	GOTS criteria/ requirement
Spinning	Paraffin products: residual oil should be 0.5%
Sizing/Weaving/ Knitting	1. PVA/PAC < 25% of total-sizing combination
	2. Machine oils must be free from heavy metals
Pre-Treatment	1. Ammonia treatment prohibited (except for wool)
	2. Only oxygen bleaches permitted (no chlorine bleaches)
	3. Washing detergents must not contain phosphates
	4. Chlorination of wool prohibited
	5. Caustic in mercerization must be recycled.

Contd...

Contd...

Production Stage	GOTS criteria/ requirement
Dyeing	1. Azo dyes, allergenic disperse dyes and dyes which contain heavy metal in molecule prohibited: specific exemption Cu – permitted up to 5% per weight in T. Blue 2. Natural dyes based on Red List of UN prohibited.
Printing	1. Discharge print using aromatic solvents prohibited 2. Plastisol printing using phthalates and PVA prohibited 3. Pigments that release banned amines prohibited
Finishing	1. Synthetic inputs for antimicrobial finish prohibited 2. Biocides, coating, lustring, stiffening prohibited 3. Sandblasting of denim prohibited 4. Flame retardants are prohibited
Accessories, Appliques, Buttons, Linings, Zips	1. Natural materials allowed 2. Chrome, nickel (as a component of metal) and PVC prohibited.
Packaging and transport	1. Packaging material must not contain PVC. 2. Compliance of Pesticides/ Biocides used in storehouses/ transport 3. Recycled paper or cardboard in packaging, hangtags, etc. are certified by according to FSC or PEFC.
Wastewater Treatment	1. Must meet all local legal requirements of discharge 2. COD from wet process to be< 20 mg/kg of textile output

Other general criteria

- During processing and manufacturing stages, organic cotton must be separated from conventional /GM cotton at all stages and identified.
- The production facility shall have environmental policy, wastewater treatment plant, sludge removal and meet the legal emissions limits. The facility shall keep records of water and energy consumption and continual improvements made to reduce the usage.
- All the dyes, chemicals must be assessed and fulfil the basic needs for toxicity, biodegradability and eliminability outlined under Section 2.3.
- Technical quality parameters (washing fastness, rubbing fastness, perspiration fastness, etc.) as described in the Standard and Manual shall be met.
- The final product does not have any harmful and toxic residues above the prescribed limits as given in the GOTS standard for finished products, as given in Section 2.4.15
- The production facility shall also comply with all the social requirements such as no child labour, safe working conditions, etc., as per ILO guidelines.

Criteria for addition of non-organic fibre materials.

- Additional fibre materials maybe mixed with organic fibres to the fabric or used in some parts of the fabric
- Blending organic cotton with conventional cotton, conventional angora hair fibre, virgin polyester, acrylic fibre, asbestos fibre, carbon fibre, silver, mulesed wool fibres is prohibited.

Allowed are individually or in combination up to 30% (≤ 30%), the following

1. non-GMO natural vegetable and natural animal fibres.
2. Lyocell or protein based fibres derived from non-GMO sources and certified organic raw material from pre or post-consumer waste.
3. recycled synthetic fibre derived from pre- or post-consumer waste. Synthetic recycled polymer includes polyamide, polypropylene, elastomer, elastomultiester, (DuPont's elastrell-p)
4. Polylactic acid (PLA) derived from organic biomass.

Allowed are Individually or in combination as a sum total up to 10% (≤ 10%)

1. regenerated fibres like lyocell, viscose, modal where raw material used in non-GMO.
2. virgin synthetic polymer fibres such as polyamide, polypropylene, elastomultiester (DuPont's elastrell-p), elastase.
3. stainless steel fibre and mineral fibre

Criteria for input chemicals

All dyes and auxiliaries that are used for the production of GOTS-certified finished goods must fulfil the criteria laid out in the standard for toxicity and biodegradability parameters. The list of substances that are prohibited for usage in chemical or colourant formulations are:

- Genetically Modified Organisms (GMO) such as enzymes or starch/ oils made from GMO or GM plants
- Aromatic solvents
- Chlorophenols
- APEOs, LABS and a-MES in active detergents
- Complexing agents such as EDTA, DPTA and similar persistent agents
- Formaldehyde and glyoxal
- Short chained chlorinated paraffins (SCCPs)
- Organotin compounds

- Plasticizers (phthalates, bisphenol A PAH, and other plasticizers with endocrine-disrupting potential)
- Functional nanoparticles (with a size of < 100 nm)
- Fungicides and biocides
- Heavy metals (with exception of Fe and Cu in Turquoise blue dyes)
- Fluorocarbons
- Quaternary ammonium compounds
- Halogenated solvents
- All SVHCs listed under REACH Directive
- Permanent AOX (restricted for inputs that contribute > 1% per weight of effluent)

In addition, the GOTS 5.0 standard prohibits all substances and preparations that are classified with the following Hazard Statements (as per GHS):

- H300, 310, 330, 340, 341, 350, 351, 360, 361, 370, 371, 371
- H400, 410, 411, 413
- EU H 059

GOTS standard also includes environmental toxicity and fate for chemical products. All products with an Acute Oral Toxicity value LD_{50} > 2000 mg/kg only are allowed to be used in the production of GOTS goods.

Also, substances or preparations with Aquatic Toxicity value LC_{50} < 1 mg/L are not allowed. Those with LC_{50} > 100 mg/L are allowed, irrespective of their biodegradability. Preparations with LC_{50} values between 1 -100 mg/L are allowed with the below relation (Table 15.4) with biodegradability of the preparation:

Table 15.4 Relation between aquatic toxicity and biodegradability in GOTS

LC_{50} value	Biodegradability*
1-10 mg/L	>95%
10-100 mg/L	>70%

**Accepted test methods for biodegradability are OECD 301 (A-F), ISO 7827, OECD 302A, OECD 302B*

The GOTS 5.0 criteria for input chemicals are now aligned with the ZDHC MRSL V 1.1, and the same is detailed in the GOTS Implementation Manual that is published with the Standard. Thus, a chemical formulation that is certified for GOTS 5.0 version is accepted as Level 1 MRSL conformant chemical in the ZDHC Gateway- Chemical Module.

15.4.4 Higg Index FEM 3.0 [31]

Developed by the Sustainable Apparel Coalition (SAC) – an industry alliance of apparel, footwear and textile industry stakeholders for sustainable production - the Higg Index is a set of tools allowing retailers, brands and services of all sizes, at every point of their sustainability journey, to precisely measure and rank the sustainability success of a business or product. The Higg Index provides a holistic summary that empowers businesses to make meaningful enhancements that defend the life of the employees, the environment and local communities. The Higg tool are of 3 types: Facility Tools, Product Tools, Retail and Brand Tools.

Here, an overview of the Higg Facility Environmental Module (FEM) Version 3.0 that is applicable to individual apparel, footwear and textile production facilities is given. The Higg FEM 3.0 is designed to:

- Quantify and measure the sustainability impacts of a facility
- Reduce duplication in sustainability assessment and reporting
- Foster business value by reducing risk and revealing efficiencies
- Build a common language and means for communicating sustainability to investors

This is an online self-assessment tool consisting of questions on 7 sections:

1. Environmental Management System (EMS)
2. Energy and GHG emissions
3. Water use
4. Wastewater
5. Waste management
6. Air emissions (if applicable)
7. Chemicals use and management

The answer to each question is either "yes", "partial yes" or "no", depending on how much the facility is ready with the implementation status for the topic covered by that question. Facilities should be honest and transparent when the assessments are completed. *The Higg FEM is NOT a pass/fail evaluation, but rather an instrument that identifies opportunities for improvement.* If in question as to whether the response counts as "Yes," it is recommended that you take a more conservative approach and response answer "Partial Yes" or "No or Unknown," where possible. For each question in each section, specific document uploads or supporting information are requested which the facility selects to choose "Yes" or "Partially Yes" to a question.

Based on the answers, a score is generated for each section. The questions in each section are arranged in 3 Levels, with the Level 3 going towards aspirational expectations from a facility for the sustainability parameters. Users conduct the evaluations at least once a year, and SAC-approved, on-site evaluators then confirm these evaluations. Benchmarking by type of facility enables managers to equate their results to that of their peers.

Section 1: EMS

This section evaluates a facility's holistic strategy to track and manage the environmental impacts of their manufacturing processes over time. The Higg Environmental Management System (EMS) section requires facility to:

- Identify personnel for managing environmental management activities and assure technical expertise
- Identify important environmental impacts related to current operations
- Set a long-term strategy for handling the climate
- Create an enforcement framework for all rules, legislation, guidelines, codes and other statutory and regulatory requirements
- All factory equipment is continuously maintained
- Engage environment strategy and performance leadership facilities, workers
- Engage environmental performance with subcontractors and upstream suppliers via the Higg Index
- Engage local stakeholders on improvements in environmental performance

Section 2: Energy use and GHG emissions

This section evaluates how a facility is driving energy conservation, implementation of renewable energy, reducing GHG emissions and mitigating risks of fossil fuel and energy costs increases. The Higg Energy and GHG section require facility to:

- Track all sources of fuel and energy and report quantities used during the last calendar year
- Identify which factors contribute most to energy use on-site (e.g., processes, machines, or operations using the most energy)
- Set a standardised energy usage benchmark, such as "80 MJ per unit of output in 2016"
- Set standardised energy reduction targets, such as "Reduce energy consumed per production unit by 70% by 2020."

- Create an action plan with specific measures and objectives for meeting energy reduction goals
- Establish energy saving against the baseline, such as "Last year we used 60 MJ per unit of output, an annual decrease of 25%.

The tool provides a GHG calculation for both direct and indirect emissions based on emissions factors taken from the best publicly-available, free sources.

Section 3: Water Use

At the beginning of this section, a facility is asked to evaluate its water risk using either the "WRI Aqueduct Tool" or the "WWF Water Risk Filter". Facilities with high water use and those located in areas of high/very high water risk are asked to complete the full Water section to ensure appropriate water management. Facilities with low water use that are located in areas of low water risk need to answer only Level 1 questions.

The Higg Water section requires facility to:

- Track all report quantities and sources of water used in the last calendar year
- Identify the factors that require the most water use on-site
- Set a standardised water-use baseline, e.g "20 cubic meters per production unit in 2016"
- Set standardised targets for water reduction, such as "Reduce water used per production unit by 70% in 2020"
- Develop an action plan with specific measures and strategies for achieving water reduction targets
- Determine baseline water reductions as "Last year we used 15 cubic meters per unit of output which represents an annual reduction of 25%".

Level 3 in this section deals with how the facility maintains a Water balance to determine how water is used and managed in the entire facility.

Section 4: Wastewater

Before answering the assessment questions in this section, the facility is first asked to define its approach to wastewater treatment and discharge (off- site or on-site or Zero Liquid Discharge)

The Higg Wastewater section requires facility to:

- Monitor the amount of wastewater discharge from industrial and/or household operations

- Report all parameters of wastewater quality found to be non-compliant with permits or industry standard(s), such as the ZDHC Wastewater Guidelines, in the most recent quality test Report name and quality of offsite wastewater treatment plant (if applicable)
- Define backup process if the treatment fails regularly (if applicable)
- Ensure proper disposal of the sludge (if applicable)
- Describe whether your site reuses wastewater and/or recycles it as process water (if applicable)

Section 5: Waste Management

The United States Environmental Protection Agency describes toxic waste as "waste that is unsafe or possibly harmful to our health or the atmosphere. Hazardous wastes can be solids, liquids, sludge or gases.

The Higg Waste section requires a facility to:

- Monitor all flows of harmful and non-harmful waste
- Describe the amount produce and the method of disposal of all hazardous and non-hazardous wastes
- Separate, properly store, and train workers to deal with all hazardous and non-hazardous waste streams
- Prohibit open dumping and burning waste on location and adequately monitor any incineration on site
- Establish standardised baselines for waste generated (e.g., 20 kgs of domestic waste per production unit generated in 2016) and disposal methods (e.g., 80% of landfilled occupied by domestic waste in 2016)
- Set up standardised waste reduction targets and improve the disposal methods
- Establish an action plan with specific measures and approaches to meet the targets for waste reduction
- Demonstrate baseline of waste reductions such as "we generated 16 kgs of domestic waste per unit of production last year, which represents a 20% annual reduction since 2016."
- Develop best industry practices

Section 6: Air emissions

Air emissions are commonly generated in textile production facilities from:

- Facility operations: cooling systems, generators and boilers (typically emit dust/particulates (PM10, PM2.5), various oxides of Sulphur ("SOx"), ozone depleting substances ("ODS"), various oxides of nitrogen ("NOx"), and other toxic air pollutants).

- Production processes: production line equipment and manufacturing processes (typically emit volatile organic compounds ("VOCs"), ozone depleting substances ("ODS"), dust/particulates (PM10, PM2.5), and other toxic air pollutants).

The Higg Air Emissions section requires facility to:

1. Monitor the amount of emissions from operations at facilities and from refrigeration, if applicable.
2. Monitor the amount of emissions from manufacturing processes, if applicable.
3. List abatement processes/control devices and monitor frequency for emissions from operation and refrigeration
4. Specify advanced performance achievements in particulate matter (PM), nitrogen oxides (NOx) and sulphur oxides (SOx)
5. Specify if your facility has a process for modernising the air emissions improvement equipment.

Before starting this section, facility need to pick which processes or operations are present in the factory which emit air pollution. These selections will guide the facility towards the Higg questions most applicable for your facility

1. If you have an air-emitting process (e.g., boiler), you will be answering questions about all forms of operational pollution.
2. If you have an air-emitting processes (e.g., solvents or adhesives), you will be able to answer questions about Level 1 production emissions.
3. You will not need to complete this section if you do not have any operation or air emissions from production.

Section 7: Chemical Use and Management

The objective of this section is to promote responsible chemicals management programs at production facilities to eliminate the discharge and use of hazardous and toxic to the human health and the environment. Chemicals management touches all parts of the business unlike other sections in Higg - from inventory and purchasing, to the production floor, to storage and waste sites.

The section on Higg FEM Chemicals Management is a joint effort between among the Sustainable Apparel Coalition (SAC), the Outdoor Industry Association (OIA), and the Zero Discharge of Hazardous Chemicals (ZDHC) Program to converge their respective chemicals tools into single evaluation questionnaire. It assesses the following areas:

- Employee training and communication, chemical management policies, compliance procedures.

- Purchasing practices and chemical procurement
- Chemical use, handling and storage practices
- Management of chemical inventory
- Accidents and spills remediation plan, Emergency Response Plan (ERP), product integrity and quality
- Process and chemicals innovation

The three levels in this section guide a facility from basic to advanced practices.

15.4.5 ZDHC Roadmap to Zero Programme [32]

The ZDHC Roadmap to Zero Programme is a collaboration between major fashion brands, textile, chemical, service sector, and associations affiliated with the value chain. The vision of ZDHC is widespread implementation of safer chemical management practices in the apparel, textile, footwear and leather value chains through creation of industry standards, developing tools for implementation of these standards, capacity-building for chemical management understanding and support to facilities to implement frameworks for chemical management.

Restrictions on use of dangerous chemicals in the leather and textile supply chain has been enforced traditionally by brands and retailers through their Restricted Substances List or RSL to ensure consumer safety. However, the practices at manufacturing facilities on their use and discharge of chemicals for worker and environmental safety was out of scope for brands. The Greenpeace Detox campaign led to a response from brands to form a collaborative initiative (called Zero Discharge of Hazardous Chemicals or ZDHC) to tackle this challenge and gain more transparency on chemical management practices in the supply chains of these brands.

The ZDHC Roadmap to Zero Programme began in 2011 as a collaboration of six companies participating in a Joint Roadmap, which was updated in 2015 after which the ZDHC transitioned to an Amsterdam registered not-for-profit legal entity – the ZDHC Foundation. This has now become a multi stakeholder collaboration of brands, textile manufacturers, chemical companies, service providers, academia, NGOs and associations.

The ZDHC Programme is a holistic approach to chemical management that focuses on three areas: Input, Process and Output.

Input: elimination of hazardous substances in input chemical formulations used in a textile, apparel, leather and footwear manufacturing plant such as wet processing, garment washing, printing, footwear assembling units and tanneries.

Process : implementation of best practices in chemical management for purchase, use, storage and disposal of chemicals in the daily operations of Suppliers (production facilities) to ensure worker and environmental safety

Output : monitoring and controlling the discharge of hazardous substances into wastewater, sludge, air and the product (all pathways) to ensure consumer and environmental safety.

In each Focus Area, ZDHC has developed standards and tools to achieve its goals. Some of them are :

ZDHC Manufacturing Restricted Substances List (MRSL): it is a list of substances that are hazardous to human health and environment, and must not be used intentionally in commercial chemical and colourant formulations. For unintentional contaminations or impurities, the MRSL outlines threshold limits at ppm level to which chemical manufacturers must conform through testing or ZDHC-approved third-party certifications. The MRSL standard contains 16 Chemical Groups, each group having individual analytes with CAS number details and maximum allowable concentrations in ppm. General test methods for testing these analytes are also given in the MRSL Table.

The MRSL, although aspirational, is achievable since all the analytes listed therein have existing suitable safer alternatives. It is divided into two parts: one for Textile formulations and the other for Leather formulations.

ZDHC Wastewater Guidelines: it is a unified standard for wastewater and sludge quality that is discharged from a textile manufacturing unit. It goes beyond regulatory compliance and includes not just conventional parameters but also the MRSL substances. It outlines limit values, sampling points and test methods.

The conventional parameters are at 3 Levels: Foundational (which must be met by all supplier facilities to ZDHC Brands), Progressive and Aspirational. The limit values under these 3 levels get increasingly stricter. The Foundational level limits may also be stricter in some cases than local pollution control board norms.

The Conventional parameters that need to be tested are: Temperature [°C], TSS, COD , total-N, pH, colour [m-1] (436nm; 525 nm; 620nm), BOD_5, ammonium-N, total-P, AOX, oil and grease, phenol, coliform [bacteria/100 ml], persistent foam, metals, sulphide, sulphite, cyanide.

The MRSL Groups to be tested in wastewater and sludge are: chlorobenzenes and chlorotoluenes, chlorophenols, dyes – azo (forming restricted amines), alkylphenol (AP) and alkylphenol ethoxylates (APEOs), Dyes-disperse (sensitising), Dyes-carcinogenic or equivalent concern, flame retardants, glycols, organotin compounds, halogenated solvents, perfluorinated chemicals (PFCs), ortho-phthalates-including all ortho esters of polycyclic aromatic hydrocarbons (PAHs), phthalic acid, volatile organic compounds (VOC). The limit values for the MRSL analytes are in parts per billion (ppb).

ZDHC Gateway : it is an online platform for manufacturing facilities to search MRSL conformant products and make informed decisions on their input chemical inventory. The platform has a database of MRSL conformant products uploaded by chemical companies and verified through third-party certifications, which are graded at 3 "Confidence Levels" for conformance to the MRSL (since Certification systems differ in their approach and depth of their reviews of chemical formulations). The higher the conformance level, the more confidence there is that the chemical formulation will consistently meet and conform to the ZDHC MRSL.

Level 1 Conformance: it is at the product level and the certification system evaluates the product conformance either through a review of the Safety Data Sheet (SDS) or the ingredients used or a test report. For examples, GOTS, EcoPassport and Greenscreen certifications are approved by ZDHC for MRSL Level 1 conformance

Level 2 Conformance: it requires not just a Level 1 conformance criteria for the product, but a review of the product stewardship practices (health, safety and environment) by the third- party certifier. This may include (but not limited to):

- Analytical test data
- Evidence that manufacturing is conducted as per ISO or other quality management systems
- Commitment to global safe manufacturing standards, such as Responsible Care
- Demonstrate proper wastewater management systems and worker safety

Level 3 Conformance: it requires all the elements of MRSL Conformance Level 1 and 2 and an on-site visit by the third-party certifier to verify the product stewardship systems first-hand

The Gateway is also used by supplier facilities to publish their wastewater test data and upload Corrective Action Plans (CAP) for non-conformities to the ZDHC Wastewater Guidelines.

Concluding remarks and future perspectives

The textile sector is of vital importance to humanity but is also an industry that uses a huge amount of chemicals, water and energy and causes an impact on human health and environment through the discharge of wastewater, sludge and air emissions. There is a need to incorporate sustainable practices in textile manufacturing to control the use of natural resources as well as reduce its impact on the environment. The recent increase in legislative enforcement, such as REACH, CPSIA, Cal Prop 65, GB Standards and Japan Chemical Substances Control Law, has ensured compliance with sustainable production practices in the textile industry. However, the industry is also going beyond regulatory frameworks by adopting voluntary standards and programmes for sustainable production and products. Growing consumer awareness, NGO campaigns and brand and retailer requirements are pushing industry players towards implementation of eco-labels such as GOTS, Oeko-Tex 100, bluesign, Cradle-to-Cradle, Greenscreen, EU Flower, or adopting industry-driven initiatives such as Higg Index and ZDHC.

The industry must understand these different standards for proper adoption and conformance.

References

1. Carmen Z, Daniela S (2012). Textile Organic Dyes – Characteristics, Polluting Effects and Separation/Elimination Procedures from Industrial Effluents – A Critical Overview. In Organic Pollutants Ten Years After the Stockholm Convention - Environmental and Analytical Update (Puzyn T & Mostrag-Szlichtyng A. (Eds.). IntechOpen. https://doi.org/10.5772/32373
2. LS&Co's Life Cycle Assessment of Levis 501 Jeans for the US market, 2006 production year https://www.levistrauss.com/wp-content (Accessed Sept 2019)
3. https://www.greenpeace.org/international/publication/7168/dirty-laundry/ (Accessed Oct 2019)
4. https://www.colorzen.com/ (Accessed Sept 2019)
5. Coca Cola Study of water scare areas, Water Stewardship Programme (2007) https://www.coca-colacompany.com/sustainability/water-stewardship (Accessed Sept 2019)
6. Eurobarometer Flash Survey Report 361 on "Chemicals" (February 2013) https://www.ab.gov.tr/files/ardb/evt/Chemicals_2013.pdf (Accessed Oct 2019)
7. https://www.unenvironment.org/explore-topics/chemicals-waste/what-we-do/emerging-issues/chemicals-products (Accessed Sept 2019)
8. https://www.greenpeace.org/archive-international/en/campaigns/detox/water/detox/intro/ (Accessed Sept 2019)
9. https://www.roadmaptozero.com/(Accessed Oct 2019)
10. https://www.futurelearn.com (Accessed Sept 2019)
11. https://www.echa.europa.eu/(Accessed Sept 2019)

12. https://cpsc.gov/Regulations-Laws--Standards/Statutes/The-Consumer-Product-Safety-Improvement-Act (Accessed Oct 2019)
13. https://oehha.ca.gov/proposition-65/proposition-65-list/ (Accessed Sept 2019)
14. https://www.epa.gov/laws-regulations/summary-toxic-substances-control-act (Accessed Sept 2019)
15. https://ecology.wa.gov/Waste-Toxics/Reducing-toxic-chemicals/Childrens-Safe-Products-Act (Accessed Sept 2019)
16. https://www.otexa.trade.gov/overseas_mkts/japan.pdf (Accessed Sept 2019)
17. https://www.standardsportal.org/usa_en/prc_standards_system/standards_used_in_china.aspx (Accessed Sept 2019)
18. https://www.oeko-tex.com/en/business/certifications_and_services/ots_100/ots_100_start.xhtml (Accessed Sept 2019)
19. https://global-standard.org/ (Accessed Sept 2019)
20. https://www.bluesign.com/en (Accessed Sept 2019)
21. https://www.greenscreenchemicals.org/learn (Accessed Sept 2019)
22. https://www.c2ccertified.org/get-certified/product-certification (Accessed Sept 2019)
23. https://ec.europa.eu/environment/ecolabel/index_en.htm (Accessed Sept 2019)
24. https://www.roadmaptozero.com/programme/ (Accessed Sept 2019)
25. https://apparelcoalition.org/the-higg-index/ (Accessed Sept 2019)
26. https://www.echa.europa.eu/ (Accessed Sept 2019)
27. Brochure on Standard STeP by Oeko-Tex® (Edition 02.2019); Oeko-Tex Association https://www.oeko-tex.com/en/(Accessed Oct 2019)
28. Fact Sheet on Made in Green by Oeko-Tex® (Edition 01.2019); Oeko-Tex Association https://www.oeko-tex.com/en/(Accessed Oct 2019)
29. Brochure on Standard Eco Passport by Oeko-Tex (Edition 02.2019); Oeko-Tex Association https://www.oeko-tex.com/en/(Accessed Oct 2019)
30. https://www.global-standard.org/about-us/history.html (Accessed Sept 2019)
31. https://apparelcoalition.org/the-higg-index/ (Accessed Oct 2019)
32. https://www.roadmaptozero.com/about/ (Accessed Oct 2019)

16

Ethical issues in achieving sustainable textile processing

Mangesh D. Teli
Institute of Chemical Technology, Matunga, Mumbai, 400019; India
Email: mangesh95teli@yahoo.com

Abstract: The necessity of adapting to sustainable processing has become inevitable given the alarming situation of environmental degradation, GHG emissions, depletion of nonrenewable resources, etc. It is the responsibility of our generation to arrest this environment degradation as well as to take care of the interests of all the stake holders through social compliance. Time and again however, it has been shown that the *business as usual* practices are continued simply for making profits, totally disregarding people and planet. Textile processing operations consumes enormous amount of water, energy as well dyes and chemicals, many a times they are hazardous. Using latest technologies the inputs of utilities including water are to be optimised and least amount of dyes and chemicals should be let into the effluent. In fact everywhere more and more zero liquid discharge operations are demanded, which itself poses challenges to the developing countries where such manufacture and processing is done in large quantities. Sustainability has basically three components, profitability, people and planet (environment) and it is very important that there has to be ethical commitment of the management of the companies to see that social and environmental compliance is indeed followed in the letter and spirit. How honestly the legislated laws are implemented and to what extent individual units consider their social responsibility, coupled with corruption free governance can indeed boost sustainable textile processing. The consumers should also be ready to pay the premium for such materials, as it is in the interest of all. Ethics thus becomes at the core of sustainability issues, the companies who are ethically committed, do not find something challenging to walk this path. But majority of the Units, do find it challenging, and their mind set has to change and thus introspection and commitment to ethics have no substitute. It is here, belief systems of the individuals also matters in shaping their behavior as conscious global citizen.

16.1 Introduction

We all know that Textile industry plays a paramount role in the life of human being. The civilised human irrespective of which level of societal pyramid he is living, has always been found to be in the need of textile material to cover his body,-initially for protection from severe climatic conditions; but successively for a number of functional aspects right from expression of his personality, comfort properties or functionality in terms of gym and sports wears, night wears, protective wears, home furnishing, wellness clothing and so on. Accordingly various textile and apparel manufacturing companies are offering the textile and apparels to suit the requirements of the customers with their ever advancing life style demands. While it is inevitable to acquire that competitive edge in terms of the functionality of such textiles and garments, it is equally important to understand that such consumption of textiles and apparel is growing rapidly. Thus, as the global population is growing and so also the standard of living, the per-capita consumption of such materials is also increasing.

The demands of the modern consumers include apparels with best of designs, strength, comfort, drapability, and with very high order of quality in terms of fastness towards light ,wash, perspiration, rubbing, etc. To achieve this standard is indeed though challenging, but textile manufacturing is being pretty matured technology, with the use of best of technologies, dyes, application methods and specialty chemicals, these expectations can be met with customers' delight. In other words, what can be expected from branded clothing is available in the market for the consumer in term of the properties he/she expects. With increasing advent of products having met the standard degree of these parameters, one will not need to be further convinced that most of the branded clothings which are available in the market are indeed the ones which do not lose their service life due to the loss of functional properties, colours, strength, comfort properties etc., but in fact due to shear loss of interest on the part of the wearer, in wearing the same apparel for quite some time.

We normally say minimum 20 washing cycles should be enough for specialty finish to indicate its antibacterial property or UV protective or mosquito repellence or fire retardant properties. In these cases what is envisaged is the stability of such performance properties to around twenty times of washing as that many times a given clothing may be used, say in the period of six months to one year. Similarly for regular wears, wherein such functional properties are not specifically demanded, performance consistency up to about 40 washing cycles is more than welcomed and that means a good solid 40 weeks of life for that garment. However, what happens after that period? The first owner gives it out as a waste material once that has lost value for him. Should it go for landfill? Actually most of the clothings after such useful service life is extinguished, indeed are buried in the landfill.

16.2 Burden on earth during apparel manufacture, service- life and there after

The worldwide consumption of the fibres is 62% synthetic fibre, 26% cotton, 6% wood based, 1 % wool and 5 % other natural fibres [1]. It also goes without saying that the countries having relatively low GNI(Gross net income) more categorised as lower middle income group(eg.,Bolivia, Egypt, India, Indonesia, Philippine),consumed annually least fibre per capita and it ranged from 4 to 8 kg.Upper middle income group nations' (Argentina, Brazil, PRChina, Columbia, Mexico, Peru, Russia, and Thailand), per capita annual consumption is raised to 18kg. Whereas advanced and developed countries more known as high income countries (Australia, Canada, France, Germany, ltaly, Japan, Korea, Spain, Switzerland, Taiwan, UK and USA)consumed around 35 to 40 kg., way above the world average fibre consumption [2]. All the above and similar innumerable write-ups and reports indicate that the earth, our only planet which we are inhabitants of, is being exploited by us to meet our demands not only in terms of food, clothing and shelter, but also for bearing the humongous amount of waste which we are producing year after year. Everyone knows that the synthetic fibres are non-biodegradable and thus if sent in the landfill, will never be assimilated by the earth and will continue to deteriorate the quality of the land. How much will this planet can take up such a nasty load which is indeed not the part of the sustainable textile production system is indeed a question before all of us. Hence there arise a number of ethical issues relating to sustainable textile processing, the use of such goods and disposal of such material.

Under the title of a safe operating space for humanity, Johan Rockström et.al [3] indicate that out of these nine boundaries in which human development is expected to operate, namely, climate change; rate of biodiversity loss (terrestrial and marine); interference with the nitrogen and phosphorus cycles; stratospheric ozone depletion; ocean acidification; global freshwater use; change in land use; chemical pollution; and atmospheric aerosol loading, we have already crossed the three of them namely climate change, rate of human interference in Nitrogen cycle and loss of biodiversity. The recent fire in Australian jungles has been eye splitting example of fastest rate of loss of biodiversity.

Brundtland Report released by World Commission on Environment and Development, UNO in 1987 defined sustainability as "Sustainable development is development that meets the needs of the present without compromising the ability of future generations to meet their own needs [4]."

This globally accepted definition of sustainable development talks about "meeting the needs" of the present and that of the future. The needs are different than what greedy person's wants are. Mahatma Gandhi says, "there is enough on this planet for every ones' needs, but not for greed". This infcat calls for introspection of our ways in which we are living and consuming the goods in this era. *Take-make-waste* [5] or *business as usual* practices will no longer prove to be beneficial, neither to us nor to the mother earth.

Three bottom lines of measurement of business performance in todays's world are the commitment of the industry to profitability, planet(environment) and people (societal). Profit, Planet and People are to be given balanced attention in the development of any enterprise. Last two to three decades you could see innumerable examples in which business enterprises hardly gave any attention to environment as well as social issues and hence although they made money and profit for shareholders, it was all, at the cost of environment and society.

And those who appeared to have complied with these requirements, were most of the time found to give more focus on their social responsibility (CSR) and environment responsibility was ignored. Now the term corporate sustainability is thus introduced which includes environmental aspects too. In fact many of the organisations are categorised as the ESG ones, i.e. they are supposed to be conscious of their environment, social and governance responsibilities.

Whichever stage we work- our operation will become sustainable only when we allow the spirit of sustainability to penetrate through downstream and upstream processes and also on the either sides.

16.3 Textile & apparel industry and environment

Let us now apply this parameter in case of Textile and apparel industry. In this case we needed to look into kind of water consumed which has been way above the generally consumed in continuous processing. One kg of cotton textiles needed water more than 150 litres in the batch wise production. Of course with latest minimum liquor application technology, the ratio of textile material to water has tremendously come down right from 1:50 to 1:5 ,which itself is quite attractive. Naturally, the volume of water used for pretreatment, dyeing, printing, finishing and washing has been tremendously cut down. That also resulted in decrease in the amount of dyes and chemicals used and also in the effluent volume to be treated leading to cost reduction. The latest technological developments have indeed helped in reducing utility costs, energy, chemicals as well as dyes since the low liquor volume processing is practiced. Now these new technologies and dyes of special category (say for eg. high exhaustion HE and ME dyes) would certainly reduce effluent load; although effluent treatment cost will be reduced, some increase in cost for technology up-gradation and special category dyes and chemicals would be eventual.

There could be cited a number of incidents where in pollution control board officials were found hand in gloves with the processors and thus instead of actually taking steps to treat the effluent properly and then discharge once the basic criteria of the treated water are met, many a times untreated raw effluent would be let into such common places of discharge. In India in number of textile and apparel processing regions, such malpractices used to be taking place, quite often a couple of decades ago. Surprisingly now also, in some regions it is still going on. The after effect of this has been, the land becoming totally infertile in that region and ground water becoming totally polluted completely unsuitable for drinking, and water level itself also has depleted to an alarming level. Surat, Jetpur and Tirupur clusters in Gujarat, Rajasthan and Tamilnadu states, respectively became infamous because of such exploitation of the environment during the textile processing. In fact Supreme court of India had to issue closure order to these polluting industries a few years back which sent a strong signal that every processor has to take care of the effluent they generate so that environment is not degraded. Of course when this happened in Tirupur, again there were a number of issues involved. While environment degradation was anticipated to be stopped by this order, many industries' closure left thousands of workers jobless and thus it became a social issue. However, just as more important thing has to give in, for most important thing, here environment degradation got preference to job losses and economic losses, and come what may, the decision was implemented with iron fist. It is this move taken in 2011, today we see that Tirupur cluster is slowly returning to its previous glory with environmental compliance.

16.4 Textile industry and exploitation of labour including children

During the turn of this millennium, world was also witnessing issues like child labour or exploitation of labour in general. Indeed, no one would like the children to go to industry to earn their livelihood when they are expected to be in the school. Here , also, from advanced and developed world point of view, this was not at all acceptable; however the poverty stricken developing countries still continued with this practice if not formally, but informally. Many a times, they defended such practice saying and justifying that rather than leaving the young ones to die hungry, they are at least providing them the means to live. Slowly NGOs and Civil societies have taken up their cause very seriously and because of this on global level ,the awareness of children's rights and this weaker section is so intense that in the schools incentives such as free education and nutritional meals are provided to the children to report to the school. There has also been equal pressure from Brands on the manufacturers and processors, that they will discontinue doing the business with them, if their supplier is found employing child labour or not giving proper wages to the workers or not providing them right kind of atmosphere at the work place and medical help. These are some of the positive pressure tactics which Brands are employing in order to bring in changes required in the system.

Seeing that this works if not voluntarily, but under the muscle power of the money, there has been additional need on the part of the brands to see that their textile and apparel supply chain gets increasingly geared up and integrated to the sustainability requirements. In other words which ever level they process these products, their processing should be environment friendly as well as they should be socially accountable.

16.5 One problem and two mindsets

There has been clear divide between those countries which are developed and advanced on one hand and the other under developed and developing world. While there is a clear need that any processing activities carried out are free from Carbon foot prints, the fact remains that whatever said and done, still they contribute toward greenhouse emissions and impact environment. The developed countries meet their demands of textiles and apparels by importing them from third world countries. The manufacturing base is shifted to the these countries, which although equally are in need of protection of environment, due to their shear economic needs and social needs to provide jobs for the people, get involved in manufacturing the textiles and apparels. Naturally , due to these reasons, there are less stringent ecological considerations in this area and manufacturing goes unabated. Although at present more incidents of disregarding environment laws are seen in developing countries, if this problem is analyzed more minutely, a lot of blame would go to those in the past who tested the fruits of unbridled industrialisation which contributed significantly into this problem, much before the major part of the world.

There has been a global debate that due to the early industrial revolution, contribution to environmental damage done by the developed countries is so enormous that it is there bounden duty to assist the developing nations to manufacture the textiles and garments for the world at large using environment friendly technology. Of course such technology comes with a premium and it is the duty of these advanced nations to assist the developing nations to balance greenhouse gas(GHG) emissions. The Kyoto Protocol [6] was born in 1997, out of this understanding and it is an international agreement linked to the United Nations Framework Convention on Climate Change, which commits

its parties by setting internationally binding emission reduction targets. There are more than 140 countries signatories to this agreement who have vouched to reduce emission of greenhouse gases. So in a sense we all have agreed to be agent of this change without any more wasting time in blaming to the individual nations. The challenge is how do we implement the elements of this understanding into the practice so that on ground, the situation changes for better.

This needed some kind of self-volition on the part of the participating nations. For being self-disciplined United Nations Global Compact [7] encouraged businesses worldwide to adopt sustainable and socially responsible policies, and to report on their implementation. It was first announced in July 2000.There are ten principles concerning human rights, labour, environment, anticorruption. It is world's largest corporate sustainability initiative with 13000 corporate participants and other stakeholders in over 170 countries. These principles around which businesses of the participants are expected to run, include two of them relating to human rights: respect internationally accepted human rights and not to be complacent about human rights abuses. Four of them relate to labour their: freedom of association and collective bargaining; elimination of forced labour; abomination of child labour and elimination of discrimination in any form of employment and occupation. Next three relate to environment: precautionary approach to environmental challenges; initiate promotion of environmental responsibility and promote development and diffusion of ecofriendly technology. Tenth one is on anticorruption: work against any forms of corruption including extortion and bribery.

When we look at these principles, most of them can only be implemented, if companies are ethically committed. The need to make corporate working "corruption free" in this competitive world may sound idealistic or ridiculous. But, there are many successful companies whose core strength is good governance and ethical commitment. The recent example of removal of person heading a reputed retail company just because he was suspected to be involved in bribing government officials indicates that there is zero tolerance for unethical practices in such companies. Sustainability of the multinational businesses can only be attained when such stringent measures are in place. We all must thus accept, that sustainability is a serious issue worthy of total commitment and it is not a wishful thinking.

16.6 Response to challenges of sustainable textile production

Many of the fast developing countries such as those in south east Asia, initially thought that such requirements with respect to environment protection and social accountability while running the textile and apparel processing , are nothing but tariff barriers for them. Hence in 1994, when first Germany banned azo dyes which were supposed to release amines which are carcinogenic, India and similar other counties thought it is another barrier for their economic development. This is how it took almost a decade to be serious about this ban and when WTO came in full force in 2005, almost every single export assignment to Germany and European countries required to be free from banned azo dyes. Similarly came into existence Oeko-Tex Standard 100, which sets up criteria for classification of products on the basis of their ecological properties and their impact on end users. Many other compounds in textile processing were thus substituted due to this prohibition. Formaldehyde free finishing agents thus came in with the advent of polycarboxyl based finishing agents. Pentachlorophenol based preservatives of gums were substituted. APEO-free wetting and leveling agents were sought after. Now a days CFC-based and phthalate-based finishing agents are being also banned and getting substituted. Enzyme based desizing, scouring, etc. is being welcomed. Synthetic thickeners in place of kerosene/water emulsion thickeners in pigment printing are being used, as kerosene is banned. Sand blasting technology in denim finishing is replaced. Most of these changes in the textile and apparels processing came into being not because of the realisation of good effects of the same on environment and society, but those importing countries refused to accept such goods processed following *business as usual* practices. When our business came in danger, all of us had to toe the line whether we liked it or not. Now it is becoming minimum expectation if we need to be in the serious game of processing of textiles and apparels. Now we see more and more efforts of reducing water consumption in processing are being tried out. Supercritical CO_2, Digital printing, Plasma technology, Biotechnology and Nanotechnology, etc. in textile processing are being increasingly explored for looking into their application potential. All of them are aimed at decreasing the Carbon footprints and thus to take a step towards sustainability.

Most of my work on natural fibre modification towards ecofriendly oil sorbents, or biopolymer modification to obtain super absorbents or natural dyeing of natural fibres, and functionalisation of these fibres clearly traced the path of sustainability. It has shown a ray of hope for the small entrepreneurs in rural area to enhance their income further. It is here what I mean that, ethics is something which cannot remain on the periphery of the Textile processing. Spirit of ethics should be running as a spinal cord of the very enterprise, so that it supports every single activity with inherent strength and contributes to the vitality of the business.

Todays' flourishing and growing businesses have certain things common. They employ modern technology and their products are of high quality as well as cost competitive; their growth is also due to their possessing legitimacy of operation due to environment friendly processing and due to social accountability; due to their brand image and reputation of their Board of Directors. Their growth in near future and ability to withstand competition, is certain due to their continued care and nurturing of their core competence and investment in disruptive technology. They can expect to get entry into newer markets and mainly in the area where majority of the market lies at the bottom of pyramid. In other sense, the millions of masses having aspiration to grow and use branded goods are the ones hoping for similar products at most affordable price. Whosoever does it, that enterprise becomes sustainable. Hence, the issue is whether I keep trying to feed rich ones and create new products for them or take care of millions of aspiring individuals who are struggling to raise their living standard. Again, it has ethical dilemma –whether to be in volume manufacturing (commodity) or in value manufacturing (niche products).

Jenny Grönwall and Anna C. Jonsson [8], in their paper on 'Regulating Effluents from India's Textile Sector: New Commands and Compliance Monitoring for Zero Liquid Discharge', throw light on how important it is to be conscious of creating regulations which are sensitive to the needs of the industry as well as environment and their applicability in the presence of best available technology (BAT) in the given setting. The compliance issues are definitely there for ample of reasons and the controlling mechanism needs to be equally transparent and can't be arbitrary in application of their regulations which has tremendous potential for inbreeding bribery and bias. As stated above that simply command and control mechanisms are not going to work with respect to Zero Liquid Discharge policy in many of the Textile Processing regions. For example implementation of ZLD requires a multistage treatment and filtration of the effluent, the use of expensive RO system, the management of the rejects ,etc and although individual units may not be able to bear such heavy costs, common effluent treatment plants (CETP)can be worked out in a given cluster with a spirit of cooperation. These measures as well as the regulations were already there and Tirupur's 2011 example where in all the polluting units were ordered to be closed down is a fresh testimony that we cannot take environment for granted. Why did such situation arise in the first place? All what was required was an honest application of the laws which were in existence. Sincere intention to modify the legislation so that while environment is better taken care of, the industry is also shown the ways to handle such issues could have been much more useful. Thus society which is stake holder for the jobs and dependent on the growth of such Industry would not have been ignored and had to suffer. As we now have the company law [9] asking the industry to utilise 2 % of their profit for CSR activities, which include maintenance of water quality, environment sustainability and preservation of natural resources, I hope such a provision is better utilised in letter and spirit.

16.7 Ethical standards take center stage in sustainable processing

When we analyse such a situation at different levels, the need to maintain ethical standards become much more evident. It has to be all encompassing if we really wish the sustainable solution to it rather than doing trouble shooting. Assessment of the polluting industry and level of pollutants have to be pragmatic. With informed data, the tolerance limits for the finally treated effluent water should be realistic and achievable as well as meaningful which can make positive contribution to our environment and natural resources. The legislation should be free from being influenced by lobbyists and made sensitive to tackle the issues at the level it requires and at a time it requires. Extent of deterrent to the erring should be accordingly balanced. Monitoring should be highly objective and transparent and the compliance should always be celebrated while non-compliance need to punished. Everyone at various levels should be held accountable if found deviating from certain level of expected ethical commitment which is required.

All these steps involve the human element to manage the show and it is here, these individuals are got to have social responsibility and commitment. If they do not deliver their duties well and become prey to the bribes offered to shield the culprits, no law or system how well intentioned it may be, will be able to deliver the fruits of such system. Hence, now a days Governance issues have also become equally important ones.

It is very much funny, but true, that while we have attained the capacity to measure 1 ppb level traces of impurities in the compounds, we have no guarantee, that despite these capacities available in the system, we may not let lose someone having the traces of impurities on their products may be 1000 folds more than permitted. What preventive mechanism can work here except the inner desire to be ethical? We know, if one is caught, he/she will be punished; but thousand times before he/she is caught, he/she goes scot free doing the same. So the damage to the system, to the environment and especially rotting of the governance system which continues, may go unnoticed. How far thus, then this external moral policing will work? This is a question every citizen should ask herself or himself. Can there be then more effective system of self-discipline? The one which does not need external monitoring. The only answer to this is sincere commitment to ethics.

Many of my colleagues would say science and technology can solve all these problems and we should not be overly concerned. If today's textile processing industry is causing so much pollution and endangering the sustainability, is it because there is lack of technology available? Because we have not realised that this alarming level of environment degradation will not only bring the textile activity at halt, but very life of the society, despite the fact that enough data is backed by the scientific studies?. Is it not proved that the greenhouse gases are causing global warming? No, we all agree to these facts. But simple knowledge of this common danger, is not sufficient enough for us to mend our ways of doing business. Knowledge of an issue and capacity to solve the same are not the only factors sufficient enough to act. Unless there is greater will to do good for all-something which is in the interest of all, no action will be taken. The paralysis of will to change for better and commit for ethically inspired sustainable development is indeed what we need to practice.

16.8 Commitment to sustainability - a moral responsibility of Brands and every one in supply chain

Moral responsibility of business and corporate sustainability are closely intertwined [10]. In a globalised world, where in fragmented business environment is existing, all the units in the supply chain are equally responsible to make the corporations truly commit for sustainability. Of course however such corporations cannot be truly successful unless they also provide opportunities for their growth in the supply chain as the significant contributor. Moral responsibility of these units or firms irrespective at what stage they are working should be so great that they would not deviate from the agreed terms and conditions which are required to be maintained for sustainable textile production. Here needs commitment on the part of the corporations and their employees. The employees themselves are expected to understand the benefits of such functioning. If they don't find the business of their unit sustainable, if they do not find prospects for their growth, why would they show loyalty and commitment for the principles of moral responsibility?

The issue of who should drive this agenda of sustainability still remains unanswered. What is required is the total paradigm shift in the way we do business. We need to take all the members of our supply chain as true business partners and need to see that for our own sustainability, the sustainability of the supply chain units is equally important. In a supply chain if on left side you have someone as supplier, on the right side you are a supplier to someone else and this goes up to the retails Brands, who finally supply to the final consumers in retail market. In any stage if we try to squeeze our suppliers beyond bearable limit, he will leave the principles of moral responsibility when it is the question of his survival and growth. I have been debating this fact that, ofcourse Brands ought to want to optimise cost of the products they sell. However, the major slice of profit comes to them, as they put on significant amount of profit on the final purchase price at which they buy from garment manufacturers. And at the retail outlets, they sell these branded products with handsome margin of profit. Their selling price is at least 5 to 10 times what they pay to the garment manufacturer. Where as in between the supply chain members are squeezed in such a fashion that they hardly get 15 to 20% of profit. The price at which these brands purchase garments from manufacturers is so arm-twisting and stringently negotiated up to final cent's level, that the margin of the manufacturer is further shrunk at all the different stages people in the supply chain. This does not give sustainability to the whole supply chain and thus it is important that the moral responsibility of the brands who actually provide business to all the supply chain people, to see that each level in supply chain there is respectable margin existing for the people to remain motivated in the business. It is the brand which shoulders the social and environmental responsibility as well as good that of good governance. Brands are playing their role to dictate the agenda of sustainability. However, they themselves have to be morally and ethically committed for their endorsement to the sustainability principles.

They should in fact sacrifice a good chunk of their hefty profit and distribute it across the whole supply chain partners and dictate that profitability is offered to all these firms in the supply chain so that everyone remains committed to sustainability requirements while progressing in their business.

Say for example if the Brand is talking about organic cotton garments, they should see the farmers of such organic cotton get right incentive, the spinners maintain separate lines for spinning and take care of what is required during the manufacture of such yarns so that they get higher price for such conversion or the yarn spun with meticulous standard. The fabric manufacturer as well as textile fabric processors also have to carry their operations in such a way that organic cotton processing and maintenance of eco-friendly processing route is adhered to. This needs special care, special technology and machinery so that it is not mixed in day to day casual operations. But this also needs to be incentivised and they should find premium in processing such goods. The garment manufacturer will pay relatively higher price taking into consideration all the input costs and he too should be well assured that he will get good amount of repeat business at premium price. When it is win -win situation for all these players, everyone will have future to grow and stability and security of their income to which each and every individual's life is attached. Two examples come to my mind which are practical. In India there has been direct transfer of credit of subsidies of different kind, to the people whom this subsidy should go(mainly people below poverty line or close to it) thanks to digitisation and linking of Aadhar card to bank account. This was not possible earlier as the middle men used to grab it. The Brands should identify supply chain Units, their employees and suppliers and that whatever premium they are offering should directly be going to the final employee of these units. They should not restrict their liability up to the people who are working with them in their corporate affairs. All those workforce working in supply chain are morally their extended partners and thus they need to be cared for and shared with the certain level of their profitability.

16.9 Transparency and Traceability

Sustainable textile processing needs committed participation of all the supply chain players. When some dispute arises over the product or process or a raw material, it is important that one can fall back on the information which is already documented and supplied. One such company [11] is having such documentation for their fibre product. According to them all this information with respect to pulp coming from which forest, batch code number and date, where it was processed, by which method and recipe, what quality parameters were maintained, etc. are put into one QR code so that they can trace back and see the origin of the problem. This kind of traceability and transparency is useful for building long term relationship and loyalty and settle the disputes and claims more efficiently and in rational ways. It also allows people working in different department at different level, to share this information in real time to convince them as to what stage of processing is their product at that time and where; This way it is also possible for them to assess whether their delivery schedule of their product is going to be maintained or not.

Another simple example I would like to state here is one of my friends (I am concealing his identity just to respect his privacy) who has been working with India's private sector Infrastructure giant (again name of the company is withheld just to respect privacy). This man used to be doing their Interior decoration work and renovation work of their offices. He was working as a vendor and doing this job uninterruptedly for more than 35 years. I knew his ethical commitment; but that was not sufficient for his success. He needed conscious suppliers, honest and sincere workers having expertise with commitment so that he could do quality work which was beyond expectations of the officers who offered him such contract year after year. He told me revealing the secret of his success, a simple rule which he followed.... Transparency. He was totally transparent to the company officers. He would tell them how he would do quality work and thus would also need quality material which would cost more. He would reveal to them at what costs he would buy these materials and in fact would show them the actual bills of materials bought for their work. Labour costs would also be shared with them and then charge them exactly 20% profit level for all his efforts over and above all the expenses. He had built up such an atmosphere of trust among his suppliers of materials as well as with the company officers. Since he followed this ethical ground rule of business, things became quite simple for everyone. Sit across the table, look at the volume of work, do costing and add 20% above your actual costs ,add taxes and that leads one to come up to the tender value. The best part and the luckiest part of him was, while he remained "honest and ethical, the other side people at the company office also did not expect any personal favours and they too remained committed to ethics due to the culture of their company. Normally this does not happen. Just a few days back when he wanted to wind up his operations because of his advanced age, he gave his business to senior most employee and requested the same company to cooperate with him. I see these are simple but very strong examples of sustainability. Once the culture of ethical commitment is nurtured, trustworthiness thrives in the system and sustainability becomes attainable dream.

16.10 Assessment of sustainability

Nannan Yang and Jung E. Ha-Brookshire [12] studied the sustainability reports of 86 top textile and apparel units in China with respect to what they perceived about sustainability and implications of their sustainability endeavors. Three important criteria were put: Whether they believe sustainability as their perfect duty or not? Whether they have put clear goals to attain higher level of sustainability? Whether they have put in required structures to achieve the same or not? While 66 of the 86 reports described sustainability as their perfect duty, 11 denied it whereas 9 had no opinion on it. Out of 66, 19 showed their clear goals as well as consequent structures to achieve them, thus qualifying clearly committed to sustainable textile processing; whereas 43 of them lacked such clear goals, though had some structures and thus called as occasionally sustainable as referred by JUNG Ha-Brookshire [10]. However, remaining could be some where on the spectrum of MRCS scale showing moral responsibility towards sustainability. These findings as authors conclude provided Chinese government as well as textile and apparel associations the poor sustainability capability existing among these top companies. This also envisages the need to increase consciousness of these corporations of moral responsibility towards sustainability and see that they have clear cut goals towards it and structures to achieve them. The importance of moral responsibility towards sustainability also has to percolate in Textile education system . The moment moral responsibility becomes driving factor towards sustainability, the issue comes before us is to how do we become increasingly conscious of ill effects of immoral behavior towards sustainability? It is here one should look in to the origin of such "need for ethical behavior". Most of the time when we see no perfect answers in outward world, we are drawn to introspect and look within. It is here right from ancient times, we have been asked to explore the answers to complicated issues by peeping into the conscience deep within.

Textile and apparel processing is being minutely studied all over the globe and various ways and means of reduction of the ill effects of the effluents on environment are actively implemented by using different systems of effluent treatment. For example when we see the series of stages of processing we know that in sizing effluents constitute yarn waste and unused starch-based sizes; in desizing effluent we find enzymes, starch, waxes, ammonia; in scouring effluents Sodium hydroxide, Oily fats, BOD , high pH, temp (70-80°C), dark colour, etc. are the challenges; Since in bleaching H_2O_2, AOX, NaOCl, organics are used, in effluent we do get their residuals along with high pH,and high TDS; dyeing and printing effluents give high toxicity, BOD (6% of total), high dissolved solids, high pH; the finishing effluents give toxic chemicals like formaldehyde and some solvents etc. In other words, textile processing effluents are complex in nature. There are many companies offering sustainable solutions for water, air and waste treatment, as well as for energy recovery at the global level. This way the large foot prints of textile and apparels on water and energy can be drastically reduced .

The ethics of manufacture is many a times ignored by many companies which are merely engaged in 'green-washing' to try to profit from the growing eco-conscious market. Underpaid workers, forced labour of children and disregarding of environment during manufacture of such garments are common practices and it is here the role of Fair-trade, Global Organic Textile Standard, SA 8000, and a whole hosts of international labeling certifications guaranteeing the ethics of a garment's manufacturer becomes highly crucial and responsible. However it is not enough; Many a times it so happens that once these certifications are obtained, if the commitment of these supply chain players is not serious, it could be also used as marketing gimmick, without actually taking care of the sustainability principles. One should also thus find out about the exact situation by asking the right questions to the labels you are interested in rather than simply relying on their publicly displayed certificates [13].

Sustainable fashion doesn't just happen in the stores, and it is critical that we commit to mending and repurposing our old garments if we genuinely want to reduce the impact of our ward robes. My personal experience is worth mentioning here. Two of my students for Master degree, undertook the subject for research on sustainable clothing. They refurbished some of the old garments purchased as waste material. They first bleached and disinfected them and then segregated them with respect to deficiency in them. After examining as to why did these garment lose their service life, they added value in them by various ways of refurbishing eg. little topping with bright colours; putting a small patch of printing or imparting attractive coloured design , or dyeing it in totally different colours, etc. After ironing and packing them nicely, they put them on sale and to our surprise, they found takers for such garments at a price which could fetch them profitability to the tune of 500 to 700%.Mind well high profitability was mainly due to initial clothing was bought at waste value. Here it goes to prove, that such an enterprise for refurbishing old garments would bring in positive prosperity, if not positive Luxury. Every penny earned as a profit will increasingly relieve the earth from immediately bearing a lot of these waste garments at least for a few more years. In addition they will serve the immediate needs of the poorer class. Positive Luxury(certified with blue butter fly mark) indicates responsible shoppers, that such an item has been made with sustainable processing [14]. It represents philanthropy, environmental framework, innovation, social frame work and good governance. There are a number of Brands of these types which have to undergo stringent scrutiny for such certification. Elvis and Kresse [15] is yet another sustainable luxury brand which I accidently came across when my daughter was doing her MBA project on sustainability in UK University. It represents positive luxury and was started in 2005 by putting in use decommissioned fire-hose pipes from landfill; After employing highly skilled, traditional craftsmanship, on this fire- hose pipes seemingly waste material, they prepared Luxury products like belts, purses, wallets, etc and these items were being sold in Luxury stores like Harrods in UK. Half of the profit went into the charity fund set up for welfare of widows of the Fire fighters' who laid their lives while saving the victims of fire in London. On the other hand the issue of getting rid of such a huge waste before London Fire Department was addressed in such a unique way. This is a kind of positive luxury. I call it also source of positive prosperity as it is not at the cost of society or environment; in fact it enables one to be socially committed, ecologically sensitive, philanthropic and also income generating for sustainable living.

16.11 Conscious urge to be ethical-a driving force behind sustainability

In short ethics provides purpose of our deeds, whereas the technology provides the "know how" and how should we do them efficiently and in a cost effective way. As we repeatedly mentioned, textile processing to be sustainable requires this commitment to ethics or so called moral responsibility.

We have been governed under the environment law of the land which is making us to comply with these laws. However, these laws may not see the full picture, in fact it is always concentrating on fragments of development spectrum. Human happiness, end of the day will depend upon one's holistic feeling of serving various stake holders. That is what true prosperity is and in that sense promoting moral and social values, such as social responsibility, human rights, animal welfare, compliance with the law, public health and safety etc., we shall be more happy looking for most of the stake holders. becoming part of our progress and it is a win-win situation for every stake holder.

However, this consciousness to be ethical which is the strongest force in the world of today stems from our duel reality of man. While man is a body, he is also a spirit (soul). The physical body is fed on all the foods and material as well as intellectual knowledge for its growth. This enables man to be able to develop the tools of technology to achieve his goals. But goals are decided with the clarity of "purpose" and that originates from his spirit (soul). And an intense desire to attain true happiness stems from the spirit and true happiness can only be obtained in giving selfless service to the society and environment.

The spiritual reality or soul of human being is also considered to be immortal by all the Holy Scriptures in the world. It is this spiritual reality which is fed by the individual's belief system. Whatever religion or belief system he possesses, requires him to follow that Golden rule to "treat your fellow being as you would like to be treated". Hence love, compassion, kindness, generosity, honesty, integrity, trustworthiness, justice, fair-mindedness, conviction, commitment, courage, courtesy, love for unity, all of these spiritual qualities originate from these belief systems of ours. It is our religion or the Faith which thus governs our agenda or purpose of living.

As a believer in latest universal Faith [16], "The betterment of the world can be accomplished through pure and goodly deeds and through commendable and seemly conduct." It makes me believe that the religion is the way of life of the people and it is their choice what they wish to practice. All religions have one single foundation. Thus to believe in the unity of religions and human kind becomes inevitable part of my belief system. And I am sure millions of people on this planet, practicing different Faiths with enlightened mind, do value and accept this oneness of humanity and our interconnectedness.

If we try to understand as to when we become truly happy, conscious search of this will lead us to realise that "only when we serve others-with no expectations" such happiness is experienced. Our Moral commitment has thus to be reflected through our " deeds" which are pure—not motivated with vested interests. Our conduct to be commendable and it is here professionals doing philanthropic work or service to society considering it their moral obligation is itself praise worthy. And they are the ones who are positive contributors to the welfare of society and human civilisation. These are the ones who are driven by the purpose-which emanates from their spiritual reality the soul. And then their physical body with acquired knowledge, skills and insights, brings into realisation, their goals by carrying out a "purposeful activity".

The moral responsibility of maintaining sustainable development in the world thus cannot be divorced from Faith as we see the inner strength and drive to be ethically committed only comes from our spiritual reality. Understanding thus the strength of the belief system, in addressing the need of sustainability, a paper on Conservation and Sustainable Development was presented [16]. It states that only a comprehensive vision of a global society, supported by universal values and principles, can inspire individuals to take responsibility for the long-term care and protection of the natural environment. Humanity must seek to protect the "heritage [of] future generations;" see in nature a reflection of the divine; approach the earth, the source of material bounties, with humility; temper its actions with moderation; and be guided by the fundamental spiritual truth of our age, the oneness of humanity. The speed and facility with which we establish a sustainable pattern of life will depend, in the final analysis, on the extent to which we are willing to be transformed, through the love of God and obedience to His Laws, into constructive forces in the process of creating an ever-advancing civilisation [17].

Irrespective of our religious beliefs, we need to accept the fact that we are interconnected, whole mankind is one and thus we need to address these global issues with the spirit of cooperation and commitment. The other requirement is to do away with prejudices of all kind-racial/colour/language/class and so on. Unless whole world is seen as one family (*Vasudhaiva Kuttumbakkam*) and interconnected we shall not change our thinking. Unless we are transformed, we cannot really bring in true sustainability.

16.12 Harmony between technology and ethics

One more principle of my belief system has been "harmony between science and religion". Science is mother of technology and hence technological tools which better our daily lives are born out of that science. However, the religion which is mother of morality or ethics is the one inspires us in our daily life from deep within and offers us the purpose of fixing a goal. Hence scientific capability required for material progress should be geared towards meeting the goals set by an enlightened soul. This implies harmony between technology and ethics can only enable us to attain the goal of sustainability. That is why we should address the issue of "environment degradation and social exploitation"- the one which originates from deep within our hearts, with total commitment. It is the soul which feels intensely about this issue of *business as usual* practices which is in variance with the law of nature- the sustainability and thus it cannot go un-noticed forever. The voice of conscience can't get unheard forever. And thus when I heard first time that it advocates harmony between science and religion-it is well understood that I must make use of technology, for achieving the goals inspired by my true human nature and sustainability is one such goal before humanity which deserves our commitment.

How is then possible for us to get transformed in order to align ourselves to address this issue most effectively. Firstly, we need to do away with certain myths which we carry with us. For example years together we have been believing in certain myths and assumptions which are no more valid. They are to be done away with, and a new understanding should be accepted?

Over the past seven decades, great investment of time, energy and resources has been made as the part of our development efforts. Best and most dedicated individuals and organisations have contributed in such efforts. Many approaches have been tried. No stone was left unturned. Over the years many lessons were learned. Indeed tremendous progress in many areas has been achieved. However, the fruits of such development did not remain sustained. Industrial development, scientific development, economic development and many facets of development saw new breakthroughs, but the solutions offered had limited influence and rich continued to become richer and poor becoming poorer. More than 84% of wealth got accumulated with less than 15 % of the global population throwing millions under poverty line.

Why has this been the fruit of such a splendid efforts? This was mainly because our very decision-making process has been based on a fundamentally materialistic view of existence of human being and has essentially ignored the spiritual nature of the human being. This materialistic development paradigm has proven not only insufficient but fundamentally flawed. And thus answers obtained did not holistically addressed the issue. If only part of human reality became centre point of development, the fruits are always going to be having limited influence. A new paradigm is needed. The disparity between the rich and the poor and violence are not the only symptoms of this failure. Another important element is the worldwide erosion of standards and evidences of moral degradation. The corruption is so rampant in all the strata of society, that unfortunately we tend to believe that human being is inherently selfish. Somewhere, the human reality is misunderstood. Problems of the humanity are seen in the mirror of their physical needs. Meeting the physical needs…it is a temporary solution and true development is empowerment of those whose needs are to be addressed.

As we discussed the knowledge flows from two streams-one science and other is religion. While former feeds body the later enlightens the soul. While former equips body to "how to achieve the goal", the latter decides the purpose of "setting a particular goal". Real driving power is thus soul- individuals true and dominating reality. And unless this is enlightened to think his/her welfare as the part of the welfare of the whole, we shall always have "greedy "people finding loop holes in the system and causing system's failure by bribing the others for their narrow benefits. Thus at various levels, we need to be on guard as to where and how we are falling prey to this corrupt system.

Development does not mean give food to those who are hungry, shelter to those who are shelter less; that will neither last too long, nor it is sustainable. Enhancing the capacity of the people to solve their own problems is the real definition of development. It implies empowering them to solve their own problems. But real catch lies in what should they call as a problem? Today under materialistic view point–problem is considered as the difference between what 'one wants to have' and what 'one already has'. For example if someone wants to have two umbrellas and has one, he has a problem. Someone wants to have one umbrella and has one, he has no problem. This is what Mahatma Gandhi pointed out on gift of nature and its self-sufficiency. He said, "there is sufficient to meet our needs, but not sufficient to meet our greed". Hence, it is important that we be cognizant of the damage which we are causing to the environment due to blindly following our materialistic instincts. We have to become responsible consumers, sustainable manufacturers and green crusaders.

16.13 Need to replace invalid premises of development by a new mindset

It is every one's responsibility to bring about change in our attitudes to initiate the transformation in us. Some invalid premises of development thinking and practice are as follows:

- Everything can be understood and dealt within material terms. But it is not correct. We must understand that life has both material and spiritual aspects.
- Man is a machine designed to maximise utility. This too is a myth, which requires to be replaced as "man is a spiritual being with higher aims and purposes".
- Science and technology can solve all problems; this too is no more valid. It is needed to be replaced by understanding that science cannot deal with issues of motivation and purpose.
- Religion is a set of irrational, dogmatic and divisive beliefs. This too is wrong. The reality is, it is a religion that has civilised the character of man.
- The poor are a bundle of needs waiting to be met. It is also untruth. India story is known to many as the country had, about two decades ago millions of young children living at the periphery of the society. Today they are the ones, having purchasing potential and pushing the market, and thus developed countries too are looking at India as the fastest growing market. Hence, new understanding has to be accepted by all that "the poor have innate potential and capacities which must be developed."

I always feel that when we address this issue of sustainable living, spiritual dimension to the development equation needs to be provided. Attention must now be focused upon that which lies at the heart of human purpose and motivation- the human spirit. Nothing short of an awakening of the human spirit can create a desire for true social change and instill in people the confidence that such change is indeed possible. When spiritual principles are fully integrated into community development activities, the ideas, values, and practical measures that emerge are likely to be those that promote self-reliance and safeguard human dignity, thus avoiding patterns of dependency and progressively eliminating conditions of gross inequality [18]. So when we talk about addressing need of ethics in sustainable textile processing, it becomes crystal clear, that ethical commitment would only come when we start listening to the voice of conscience. The very core of our human spirit must drive our activities with the will to instill the necessary change in the society to tackle the challenges of sustainability. Without ethics sustainable textile processing or anything to be truly sustainable with its positive effects cannot be possible. And where does the ethics come from? The source is our religious belief systems. Those who may not believe in such a system but are scientific and rational, can respect the power of this ethical commitment which indeed is the only source to inspire people to march on the path of sustainability. It has to encompass not only the stake holder of textile processing, but the mankind as a whole society and whole universe and its environment right from the depth of the oceans to unlimited height of the sky. And this to happen, we have to commit to universal human values. Once we accept and commit to this conviction with all inclusive approach, this force of ethics will become true driving force and it will need no more external moral policing to implement the laws of sustainability to achieve its goals. Sustainable living principles will then become our culture of living.

16.14 Sustainability as an integral part of human life

Now in this context, when we look at moral responsibility towards sustainability, it makes immense sense that only when we realise, deep down in our heart, that not because we are actively responsible for unsustainable textile processing that we need to take immediate measures because law demands, we need to accept the fact that as a human being, a part of the whole global system, I need to consider this as my moral responsibility. Then it does not matter whether I am a CEO of the company, a member of middle management or simple worker. My obligations are such to remain committed to sustainability goals, come what may. And thus whatever it demands, I must do it not just for compliance point of view, but going beyond the compliance, and that is where path of morality starts. So we should not simply criticise the civil society organisations or NGOs when they oppose certain development projects, they deserve to be heard and their voice deserves patient hearing as they too are stake holders of this planet. In Textile and apparel processing, we must accept the fact that the movements and campaigns run by NGOs such as Green peace and many more, had tremendous effect on building a public pressure on brands in accelerating sustainability agenda. That happened, because generality of public and consumers questioned legitimacy of operation of the brands and brands themselves were experiencing erosion of their value.

Many of the provisions in the laws for implementation of sustainable textile processing have come into play, because of the NGOs and public pressure to stop such exploitation of nature and society. We simultaneously are also fortunate to see in our history many outstanding examples of

Industrialists having such spirit. In 1868, Jamshetji Tata started Tata Steel, the first company from Tata group at Jamshetpur, Jharkhand. When no laws on corporate social responsibility existed at that time, Jamshetji defined business enterprise "as a mean to serve the people". This definition encompasses all those elements of corporate moral responsibility to sustainability, which are now being accepted globally. Thus environmental and social compliance have become part of the culture of the company. Sadly today, meaning of business with materialistic approach towards development is taken as "an enterprise made to earn profit". When such narrow definition has become guiding light of our lives, what would you expect but the disastrous effects of exploitation of environment and weaker sections of society for so called "better economic performance" measured in terms of shear profitability.

Hence, if all the strata of society have to consider elements of sustainability as an integral part of their life, they must undergo lessons of moral science, value education and ethics. Right from preprimary to post graduate level, education systems should be embedded with such educational curriculum. The real catch is, how do we sustain such education in human values for life long and apply the same irrespective of the level or hierarchy of profession we are working. These lessons should be imparted from lowest level to the highest level of society. Students, professionals, people in law enforcing bodies, judiciary and banking systems, management, governance and everywhere, this moral responsibility to the ethical behavior and universal human values got to be there. In fact it should appear to be unwritten precondition to civilised living. How do we put on dress to express our personality? So also we need to adorn ourselves with ornaments of universal human values. Of course, many good institutions are working on education and training of various strata of society in such values. I am fortunate to be associated with one such organisation which imparts training at different levels and works with various enterprises and work force including students and faculty of many Universities. Also I am associated with the Institutions which are imparting value education to the children, junior youths and youths with a hope that world tomorrow would be much better place to live in than what we have today.

When we know that present era is about information, knowledge, data and those who possess that capacity to process the same and use it, they form the part of our creamy layer of society. Educational degrees do play significant role in climbing such ladder of position in society much faster. However, if such people do not have ethical orientation to live the life, they are likely to be prey to their lower greedy nature and society at their hands is likely to pay cost for their doing. Hence while knowledge gives power, the powerful people must be strongly committed to ethics; otherwise we all run in a danger if they abuse the power.

Sustainability demands a new mindset to look at the world too-that whole world is interconnected; that we all coexist; our welfare as well as woes are interconnected, we accept gender equality and abolishing of all kinds of prejudices, we respect all religions and their adherents; we try our hard to display tolerance, integrity, trustworthiness as elements of this modern way of thinking and living in order that we all become positive agents of change towards more sustainable life. Our planet can then breathe a sigh of relief that it would safely be passed over to our next generation. In fact as a trustee of this planet, that is what we are least expected to do. Let us thus accept moral responsibility and ethical commitment towards sustainable living. Of course, sustainable textile processing is subset of such harmonious life and we must keep our efforts on with awakened conscience to be prepared to do what it requires to attain this goal. But we must also know, this is a long journey of 1000 miles, and not a destination and hence let this long journey start from us individually by that first step of our personal moral commitment to the elements of sustainability.

Concluding remarks and future perspectives

Sustainable textile processing involves commitment to prosperity, people and environment. The textile and apparels supply chain is quite long and complex with multiple players in it. However, since every individual born needs to be clothed, with 7.3 bn population of the world, the dimension of this business and the issues originating during the manufacture of such garments, during their service life and there after till they are disposed-off are huge and complex, and all out efforts have to be made at various stages of the manufacture of the garments to reduce their Carbon foot prints on environment while remaining socially accountable. Mere certifications of companies to this effect cannot be reliable source to vouch for sustainability. The deeper issue is to what extent right from CEO to the people at various managerial level are committed to the true spirit of sustainable textile processing? The brands can surely play an important role in driving the sustainability agenda; however, it is not enough, as over whelming majority of the non -branded garment manufacturers require self-discipline and moral responsibility to follow elements of sustainable textile processing. Government legislations backed by impartial and sincere implementation will do their bit; however, the voice of individual's conscience and urge to be ethical are the key elements which will play decisive role in driving down this agenda. It needs new mindset while discarding invalid premises of purely materialistic development and if all the stake holders play their respective role on their own without any external policing, surely we can achieve the goals of sustainable textile processing.

References

1. Garside M., www.statista.com (Accessed April 2019).
2. Fiber year consulting 2017/18, www.Textile media.com (Accessed April 2019).
3. Rockström J, Steffen W, Foley JA et al (2009). A safe operating space for humanity, *Nature* 461, 472–475.
4. Annon (1987). https://sustainabledevelopment.un.org/content/documents/5987our-common-future.pdf (Accessed April 2019).
5. https://www.researchgate.net/figure/The-linear-economy-The-take-make-and-waste-approach-of-production_fig2_323809440, Vide: Thibaut Wautelet, DOI: 10.13140/RG.2.2.17085.15847 (Accessed May 2019).
6. https://europa.eu/capacity4dev/public-environment-climate/document/kyoto-protocol-united-nations-framework-kyoto-1997 (Accessed May 2019).
7. https://en.wikipedia.org/wiki/United_Nations_Global_Compact#The_Ten_Principles (Accessed April 2019).
8. Grönwall J, Jonsson AC (2017). 'Regulating Effluents from India's Textile Sector: New Commands and Compliance Monitoring for Zero Liquid Discharge', *Law, Environment and Development Journal*,13. http://www.lead-journal.org/content/17013.pdf (Accessed May 2019).
9. Schedule VII of the Companies Act, 2013 (18 of 2013) as amended, and the Companies (Corporate Social Responsibility Policy) Rules, 2014. https://www.mca.gov.in (Accessed May 2019).
10. Ha-Brookshire J (2017). Toward Moral Responsibility Theories of Corporate Sustainability and Sustainable Supply Chain. *J Bus Ethics* 145, 227–237. https://doi.org/10.1007/s10551-015-2847-2.
11. https://www.birlacellulose.com/news_pdf/news_11_1577447753.pdf (Accessed May 2019).
12. Yang N., Ha-Brookshire JE (2019). Truly sustainable or not? An exploratory assessment of sustainability capability of textile and apparel corporations in China from the moral responsibility perspective. *Fash Text* 6, 15. https://doi.org/10.1186/s40691-019-0172-64.
13. https://www.tortoiseandladygrey.com/2019/04/23/check-ethics-of-manufacture-before-purchase/ (Accessed May 2019).
14. https://www.positiveluxury.com/ The Butter Fly Mark-Certification from Positive Luxury (Accessed April 2019).
15. https://www.elvisandkresse.com/ (Accessed June 2019).
16. https://www.bahai.org (Accessed May 2019).
17. https://www.bic.org/statements/conservation-and-sustainable-development-bah%C3%A1%C3%AD-faith (Accessed April 2019).
18. https://www.bic.org/statements/religious-values-and-measurement-poverty-and-prosperity (Accessed June 2019).

Index

E

F

G

H

I

J

L

M

N

O

P

Q

R

S

T

U

V

W

X

Z